CAMBRIDGE LIBRARY COLLECTION

Books of enduring scholarly value

Mathematical Sciences

From its pre-historic roots in simple counting to the algorithms powering modern
desktop computers, from the genius of Archimedes to the genius of Einstein, advances
in mathematical understanding and numerical techniques have been directly responsible
for creating the modern world as we know it. This series will provide a library of the most
influential publications and writers on mathematics in its broadest sense. As such, it will show
not only the deep roots from which modern science and technology have grown, but also the
astonishing breadth of application of mathematical techniques in the humanities and social
sciences, and in everyday life.

Scientific Papers

This volume of Lord Rayleigh's collected papers begins with a brief 1892 piece in which the
author addresses the troubling discrepancies between the apparent density of nitrogen derived
from different sources. Intrigued by this anomaly and by earlier observations by Cavendish,
Rayleigh investigated whether it might be due to a previously undiscovered atmospheric
constituent. This led to Rayleigh's discovery of the chemically inert element, argon, to his 1904
Nobel Prize in physics, and to the discovery of all the 'rare' gases. Debate over the nature of
Roentgen rays, is reflected in a short 1898 paper, written in the wake of their discovery. 1900
saw a key contribution, the elegant description of the distribution of longer wavelengths in
blackbody radiation. Now known as the Rayleigh–Jeans' Law, this complemented Wien's
equation describing the shorter wavelengths. Planck's law combined these, in a crucial step
toward the eventual development of quantum mechanics.

Cambridge University Press has long been a pioneer in the reissuing of out-of-print titles from its own backlist, producing digital reprints of books that are still sought after by scholars and students but could not be reprinted economically using traditional technology. The Cambridge Library Collection extends this activity to a wider range of books which are still of importance to researchers and professionals, either for the source material they contain, or as landmarks in the history of their academic discipline.

Drawing from the world-renowned collections in the Cambridge University Library, and guided by the advice of experts in each subject area, Cambridge University Press is using state-of-the-art scanning machines in its own Printing House to capture the content of each book selected for inclusion. The files are processed to give a consistently clear, crisp image, and the books finished to the high quality standard for which the Press is recognised around the world. The latest print-on-demand technology ensures that the books will remain available indefinitely, and that orders for single or multiple copies can quickly be supplied.

The Cambridge Library Collection will bring back to life books of enduring scholarly value across a wide range of disciplines in the humanities and social sciences and in science and technology.

Scientific Papers

VOLUME 4: 1892-1901

BARON JOHN WILLIAM STRUTT RAYLEIGH

CAMBRIDGE
UNIVERSITY PRESS

CAMBRIDGE UNIVERSITY PRESS

Cambridge New York Melbourne Madrid Cape Town Singapore São Paolo Delhi

Published in the United States of America by Cambridge University Press, New York

www.cambridge.org
Information on this title: www.cambridge.org/9781108005456

This edition first published 1903
This digitally printed version 2009

ISBN 978-1-108-00545-6

SCIENTIFIC PAPERS

𝔏ondon: C. J. CLAY AND SONS,

CAMBRIDGE UNIVERSITY PRESS WAREHOUSE,

AVE MARIA LANE.

𝔊lasgow: 50, WELLINGTON STREET.

𝔏eipzig: F. A. BROCKHAUS.

𝔑ew 𝔜ork: THE MACMILLAN COMPANY.

𝔅ombay and 𝔆alcutta: MACMILLAN AND CO., Ltd.

SCIENTIFIC PAPERS

BY

JOHN WILLIAM STRUTT,

BARON RAYLEIGH,

O.M., D.Sc., F.R.S.,

HONORARY FELLOW OF TRINITY COLLEGE, CAMBRIDGE,
PROFESSOR OF NATURAL PHILOSOPHY IN THE ROYAL INSTITUTION.

VOL. IV.

1892—1901.

CAMBRIDGE:
AT THE UNIVERSITY PRESS.
1903

Cambridge:
PRINTED BY J. AND C. F. CLAY,
AT THE UNIVERSITY PRESS.

PREFACE.

BY the present volume the Collection of Papers is brought down to the end of 1901. The diversity of subjects—many of them, it is to be feared, treated in a rather fragmentary manner—is as apparent as ever, and is perhaps intensified by the occurrence of papers recording experimental work on gases. The memoir on Argon (Art. 214) by Sir W. Ramsay and myself is included by special permission of my colleague.

A Classified Table of Contents and an Index of Names are appended. The large number of references to the works of Sir George Stokes, Lord Kelvin and Maxwell, as well as of Helmholtz and some other investigators abroad, will shew to whom I have been most indebted for inspiration.

I desire also to record my obligations to the Syndics and Staff of the University Press for the efficient and ever courteous manner in which they have carried out my wishes in the republication of this long series of memoirs.

TERLING PLACE, WITHAM,
December 1902.

The works of the Lord are great,
Sought out of all them that have pleasure therein.

CONTENTS.

ILLUSTRATIONS.

197.

DENSITY OF NITROGEN.

[*Nature*, XLVI. pp. 512, 513, 1892.]

I AM much puzzled by some recent results as to the density of nitrogen, and shall be obliged if any of your chemical readers can offer suggestions as to the cause. According to two methods of preparation I obtain quite distinct values. The relative difference, amounting to about 1/1000 part, is small in itself; but it lies entirely outside the errors of experiment, and can only be attributed to a variation in the character of the gas.

In the first method the oxygen of atmospheric air is removed in the ordinary way by metallic copper, itself reduced by hydrogen from the oxide. The air, freed from CO_2 by potash, gives up its oxygen to copper heated in hard glass over a large Bunsen, and *then* passes over about a foot of red-hot copper in a furnace. This tube was used merely as an indicator, and the copper in it remained bright throughout. The gas then passed through a wash-bottle containing sulphuric acid, thence again through the furnace over *copper oxide*, and finally over sulphuric acid, potash and phosphoric anhydride.

In the second method of preparation, suggested to me by Prof. Ramsay, everything remained unchanged, except that the *first* tube of hot copper was replaced by a wash-bottle containing liquid *ammonia*, through which air was allowed to bubble. The ammonia method is very convenient, but the nitrogen obtained by means of it was 1/1000 part lighter than the nitrogen of the first method. The question is, to what is the discrepancy due?

The first nitrogen would be too heavy, if it contained residual oxygen. But on this hypothesis, something like 1 per cent. would be required. I could detect none whatever by means of alkaline pyrogallate. It may be remarked that the density of the nitrogen agrees closely with that recently obtained by Leduc using the same method of preparation.

On the other hand, can the ammonia-made nitrogen be too light from the presence of impurity ? There are not many gases lighter than nitrogen, and the absence of hydrogen, ammonia, and water seems to be fully secured. On the whole it seemed the more probable supposition that the impurity was hydrogen, which in this degree of dilution escaped the action of the copper oxide. But a special experiment seems to exclude this explanation.

Into nitrogen prepared by the first method, but before its passage into the furnace tubes, one or two thousandths by volumes of hydrogen were introduced. To effect this in a uniform manner the gas was made to bubble through a small hydrogen generator, which would be set in action under its own electro-motive force by closing an external contact. The rate of hydrogen production was determined by a suitable galvanometer enclosed in the circuit. But the introduction of hydrogen had not the smallest effect upon the density, showing that the copper oxide was capable of performing the part desired of it.

Is it possible that the difference is independent of impurity, the nitrogen itself being to some extent in a different (dissociated) state ?

I ought to have mentioned that during the fillings of the globe, the rate of passage of gas was very uniform, and about 2/3 litre per hour.

198.

ON THE INTENSITY OF LIGHT REFLECTED FROM WATER AND MERCURY AT NEARLY PERPENDICULAR INCIDENCE.

[*Philosophical Magazine*, XXXIV. pp. 309—320, 1892.]

IN a former paper* I gave an account of some experiments upon the reflexion from glass surfaces tending to show that "recently polished glass surfaces have a reflecting-power differing not more than 1 or 2 per cent. from that given by Fresnel's formula; but that after some months or years the reflexion may fall off from 10 to 30 per cent., and that without any apparent tarnish." Results in the main confirmatory have been published by Sir John Conroy†.

The accurate comparison of Fresnel's formula with observation is a matter of great interest from the point of view of optical theory, but it seems scarcely possible to advance the matter much further in the case of solids. Apart from contamination with foreign bodies of a greasy nature, and disintegration under atmospheric influences, we can never be sure that the results are unaffected by the polishing-powder which it is necessary to employ. For these reasons I have long thought it desirable to institute experiments with liquids, of which the surfaces are easily renewed; and the more since I succeeded in proving that (in the case of water at any rate) the deviation from Fresnel's formula found by Jamin in the neighbourhood of the polarizing angle is due to greasy contamination. The very close verification of the theoretical formula in this critical case seemed to render its applicability to perpendicular incidence in a high degree probable. I was thus induced to attack the somewhat troublesome problem of designing a photometric method capable of dealing with the reflexion from a horizontal surface. The details of the apparatus and of the measures will be given presently; but in the meantime it may be well to consider rather closely what is to be expected upon the supposition that Fresnel's formulae are really applicable. Fresnel's formulae are spoken of, because although at strictly perpendicular incidence we should have to do only with Young's expression $(\mu - 1)^2/(\mu + 1)^2$, in

---

* *Proc. Roy. Soc.* November, 1886. [Vol. II. p. 522.]

† *Phil. Trans.* 1889 A, p. 245.

practice we are forced to work at finite angles of incidence. It is thus important to examine the march of Fresnel's expressions, when the angle of incidence (θ) is small.

Writing

$$S = \frac{\sin(\theta - \theta_1)}{\sin(\theta + \theta_1)}, \qquad T = \frac{\tan(\theta - \theta_1)}{\tan(\theta + \theta_1)},$$

where

$$\sin\theta_1 = \sin\theta / \mu,$$

we find

$$S^2 = \left(\frac{\mu-1}{\mu+1}\right)^2 \left\{1 + \frac{2\theta^2}{\mu} + \frac{\theta^4}{6\mu^3}(3 + 12\mu - \mu^2)\right\}, \quad \dots\dots\dots(1)$$

$$T^2 = \left(\frac{\mu-1}{\mu+1}\right)^2 \left\{1 - \frac{2\theta^2}{\mu} - \frac{\theta^4}{6\mu^3}(9 - 12\mu + 5\mu^2)\right\}. \quad \dots\dots\dots(2)$$

Thus S^2 and T^2 differ from the value appropriate to $\theta = 0$ in opposite directions and by quantities of the order θ^2. But on addition we get

$$S^2 + T^2 = 2\left(\frac{\mu-1}{\mu+1}\right)^2 \left\{1 - \frac{\theta^4}{2\mu^3}(1 - 4\mu + \mu^2)\right\}, \quad \dots\dots\dots(3)$$

differing from the value appropriate to $\theta = 0$ by a quantity of the *fourth* order only in θ. When therefore the circumstances are such that it is unnecessary to distinguish the two polarized components, the intensity of reflexion at small incidences is in a high degree independent of the precise angle. If μ is nearly equal to unity, we have

$$S^2 + T^2 = 2\left(\frac{\mu-1}{\mu+1}\right)^2 \{1 + \theta^4\} \quad \dots\dots\dots\dots\dots(4)$$

simply. Again, if $\mu = \frac{4}{3}$,

$$S^2 + T^2 = 2 \times \frac{1}{49}\left\{1 + \frac{69}{128}\theta^4\right\}. \quad \dots\dots\dots\dots(5)$$

A few calculations from the original expressions will serve to indicate the field of these approximations.

$$\mu = \tfrac{4}{3}, \qquad \theta = 10°, \qquad \theta_1 = 7° 29',$$

$$S^2 = \frac{1}{49} \times 1\cdot0467, \qquad T^2 = \frac{1}{49} \times \cdot9541,$$

$$S^2 + T^2 = 2 \times \frac{1}{49} \times 1\cdot0004.$$

From (5) we get as the last factor $1\cdot00050$.

$$\mu = \tfrac{4}{3}, \qquad \theta = 20°, \qquad \theta_1 = 14° 51'\cdot8,$$

$$S^2 = \frac{1}{49} \times 1\cdot2021, \qquad T^2 = \frac{1}{49} \times \cdot8158,$$

$$S^2 + T^2 = 2 \times \frac{1}{49} \times 1\cdot0090.$$

By (5) the last factor is $1\cdot0080$.

Again,

$$\mu = \tfrac{4}{3}, \qquad \theta = 30°, \qquad \theta_1 = 22° \ 1'\cdot 4,$$

$$S^2 = \frac{1}{49} \times 1\cdot5189, \qquad T^2 = \frac{1}{49} \times \cdot5866,$$

$$S^2 + T^2 = 2 \times \frac{1}{49} \times 1\cdot0527.$$

According to (5) the last factor is here $1\cdot0405$.

Fig. 1.

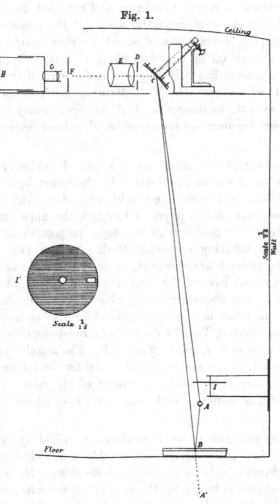

It appears that in the case of water the aggregate reflexion scarcely begins to vary sensibly from its value for $\theta = 0$ until $\theta = 20°$, a property of some importance for our present purpose, as it absolves us from the necessity of striving after very small angles of incidence.

I will now describe the actual arrangement adopted for the experiments. The source of light at A (Fig. 1) is a small incandescent lamp, the current through which is controlled with the aid of a galvanometer. It is so mounted that its equatorial plane coincides with the (vertical) plane of the diagram. Underneath, upon the floor, is placed the liquid (B) whose reflecting power is to be examined. At C, just under the roof, the direct ray AC and the reflected ray BC are turned into the same horizontal direction by two mirrors silvered in front and meeting one another at C under a small angle. The eye situated opposite to the edge C and looking into the double mirror thus sees the direct and reflected images superposed, so far as the different apparent magnitudes allow. D represents a diaphragm and E a photographic portrait-lens of about 3 inches aperture which forms an image of A and A' on or near the plane F. At F is placed a screen perforated with a hole sufficiently large to make sure of including all the rays from A, A' which pass D. To determine this point an eye-piece is focused upon F, so that the images of A, A' are seen nearly in focus. Some margin is necessary because the images of A, A' cannot (both) be accurately in focus at F.

These adjustments being made, an eye placed behind F and focused upon C sees the upper mirror illuminated by the direct light (from A), and the lower illuminated by the reflected light (from A'). And if the aperture at F is less than that of the pupil of the eye, the apparent brightnesses of the two parts of the field are in the same proportion as would be the illuminations on a diffusing screen at C due to the two sources. The advantage of the present arrangement, as compared for example with the double-shadow method, lies in the immense saving of light. In the case of water there is a great disproportion (of about 50 to 1) in the illuminations as seen from F. In order to reduce the direct light to at least approximate equality with the reflected, Talbot's device* of a revolving disk was employed. This is shown in section at I, and in plan at I'. The angular opening may be chosen so as to allow for the loss in reflexion, and for the further disadvantage under which the reflected light acts in respect of distance. The disk finally employed was of zinc, stiffened with wood, and covered on both faces with black velvet.

It was at first proposed to work as above described by eye estimations; but the necessity for a ready adjustment capable of introducing small relative changes of brightness leads to further complications. Moreover, the large disk which it is advisable to use for the sake of accurate measurement of the angular opening, cannot well be rotated at the necessary speed of 20 or 25 revolutions per second. For this reason, and also for the sake of obtaining a record capable of being examined at leisure, it was decided to work by photography. This involves no change of principle. The photographic plate H simply takes the place of the retina of the eye. But now the

---

integration of the effect over a somewhat prolonged exposure (of several minutes) dispenses with the necessity for a rapid rotation of the Talbot disk, and allows us to obtain at will a fine adjustment by screening one or the other light from the plate for a measured interval of time. In practice the *direct* light was thus partially cut off, a mechanically held screen being advanced a little above the plane of the revolving disk. The reader will not fail to observe that the incomplete coincidence of the times of exposure has the disadvantage of rendering the calculation dependent upon the assumption that the light is uniform over the duration of an experiment. Error that might otherwise enter is, however, in great degree obviated by the precaution of choosing the *middle* of the total period of exposure as the time for screening.

The above is a sufficient explanation of the general scheme, but there are many points of importance still to be described. With respect to the source of light, it was at first supposed that even if the radiation upwards and downwards could not be assumed to be equal, at any rate a reversal by rotation of the lamp through 180° in the plane of the diagram would suffice to eliminate error. On examination, however, it appeared that owing to veins in the glass bulb the radiation in various directions was very irregular, so much so that it was feared that mere reversal might prove an insufficient precaution. The difficulty thus arising was met by covering the bulb, or at least an equatorial belt of sufficient width, with thin tissue-paper, by which anything like sudden variations of radiation with direction would be prevented, and by causing the lamp to revolve slowly about its axis during the whole time of exposure. The diameter of the bulb was about 1¼ inch, and the illuminating-power rather less than that of one candle.

Another point of great importance is to secure that the light regularly reflected from the upper surface of the liquid, which we wish to measure, shall be free from admixture. It must be remembered that by far the greater part of the light incident upon the liquid penetrates into the interior, and must be annulled or at any rate diverted into a harmless direction. To this end it is necessary that the liquid be free from turbidity and that proper provision be made for the disposal of the light after its passage. It is not · sufficient merely to blacken the bottom of the dish in which the water is contained. But the desired object is attained by the insertion into the water of a piece of opaque glass, held at such a slight inclination to the horizon that the light from the lamp regularly reflected at its upper surface is thrown to one side. As additional precautions the disk and its mountings were blackened, as were also the walls and ceiling of the room in which the experiments were made.

The surface of water must be large enough to avoid curvature due to capillarity. Shortly before an experiment it is cleansed with the aid of a hoop of thin sheet-brass about 2 inches wide. The hoop is deposited upon the water so doubled up that it includes but an insensible area, and is then opened out into a circle. In this way not only is the greasy film usually present upon the surface greatly attenuated, but also dust is swept away. The avoidance of dust, especially of a fibrous character, is important. Otherwise the resulting deformation of the surface causes the field of the reflected light to become patchy and irregular.

We come now to the silvered glass reflectors, which are assumed to reflect the direct and reflected lights equally well. It seems safe to suppose that no appreciable error can enter depending upon the slightly differing angles at which the reflexion takes place in the two cases. But the mirrors are liable to tarnish, and, indeed, in the earlier experiments soon showed signs of being affected. The influence of this tarnish would be much greater in photographs done upon ordinary plates, sensitive principally to blue light, than in the estimation of the eye; and it was thought desirable to eliminate once for all any question of the effect of differential tarnishing by interchanging the mirrors in the middle of each exposure. For this purpose a somewhat elaborate mounting had to be contrived. It was executed by Mr Gordon and answered its purpose extremely well.

The mirrors are carried by a brass tube B (Fig. 2), which revolves in an

Fig. 2.

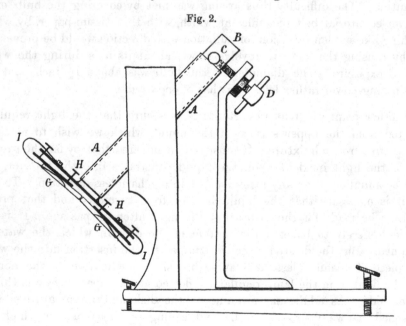

external tube AA rigidly attached to the stand of the apparatus. A lateral arm C, some inches in length, projects from B, and near its extremity bears against one or other of two screw-stops D. The lower end of B carries perpendicular to itself a brass plate *EE* (Fig. 3). The mirrors *GG* are of plate-glass and are fixed by cement to two brass plates *FF*. The latter

Fig. 3.

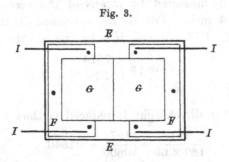

plates are attached by friction only to *EE*, being on the one hand pushed away by adjusting-screws *HH*, and on the other held up by four steel springs *I*. The edges of the reflecting surfaces meet accurately in a line passing through the axis of rotation, and the stops *D* are so adjusted that the transition from the one bearing to the other corresponds to a rotation through precisely 180°, so that on reversal the common edge of the reflectors recovers its position. The two mirrors were originally silvered in one piece, and the common edge corresponds to the division made by a diamond-cut at the back. These arrangements were so successful that in spite of the reversal between the two parts of the exposure the division-line appears sharp in the photographs and exhibits no appearance of duplicity.

When not in use the reflecting-surfaces are protected by a sort of cap of tin-plate, which fits loosely over them. The improvement thus obtained was very remarkable, the mirrors not suffering so much in a month as they formerly did in a day before the protection was provided.

The following are the measures of distances required for the calculation.

From the division-line *C* to the axis of rotation of the lamp *A* (Fig. 1),
$$AC = 82{\cdot}21 \text{ inches};$$
$$AB = 11{\cdot}28, \qquad BC = 93{\cdot}15,$$
so that
$$AB + BC = 104{\cdot}43.$$

The factor expressing the ratio of the squares of the distances is thus 1·6137.

The angle of incidence is best obtained from a measurement of the *horizontal* distance between *C* and *A*. This proved to be 11½ inches; so that
$$\sin \theta = \frac{11\frac{1}{2}}{104{\cdot}3} = {\cdot}11, \quad \text{and} \quad \theta = 6\tfrac{1}{2}°.$$

This applies to all the experiments referred to in the present paper.

The estimation of the angular opening in the disk used for the water experiments depended upon measurements of corresponding chord and diameter. The chord, measured by means of the screw of a travelling-microscope, was ·7574 inch. The radius, expressed in terms of the same unit, was found to be 7·79. Hence, if α be the angular opening,

$$2 \sin \tfrac{1}{2}\alpha = \frac{·7574}{7·79},$$

or $\tfrac{1}{2}\alpha = 2°\ 47' = 167'$.

The ratio in which the direct light is reduced is thus

$$\frac{167}{180 \times 60} = \frac{167}{10800} = ·01546.$$

It will now be necessary to give some details with respect to the actual matches as determined photographically. At first the intention was to employ ordinary plates (Ilford), which worked very satisfactorily. But when the attempt was made to compare the result with theory, the comparison was found to be embarrassed by uncertainty as to the effective wave-length of the light in operation. Moreover, as these plates are scarcely sensitive to yellow and green light, the effective wave-length is liable to considerable variation with the current used to ignite the lamp. Photographs were indeed taken of the spectrum of the lamp as actually employed, but the unsymmetrical character of the falling off at the two ends made it difficult to fix upon the centre of activity. Recourse was then had to Edwards' "isochromatic" plates. The spectrum of the lamp, as photographed upon these plates *after passing through a pale yellow glass*, was very well defined, lying with almost perfect symmetry between the sodium and the thallium lines. It was, therefore, determined to use these plates and the same yellow glass in the actual experiments, so that

$$\lambda = \tfrac{1}{2}(5892 + 5349) = 5620$$

could be taken as the representative wave-length.

The only disadvantage arising from this change was in the necessary prolongation of the exposure, which became somewhat tedious. Although no dense image is required or indeed desirable, the exposure should be such that the development does not need to be forced. Two photographs, with different times of screening, were usually taken upon the same plate, the object being to obtain a reversal of relative intensity, so that in one image the semicircle representing the direct light should be more intense and in the other image the semicircle representing the reflected light. The best way of examining the pictures depended somewhat upon circumstances. When the exposure and development had been suitable, the most effective view for the detection of a feeble difference was obtained by placing the dry picture, film downwards, upon a piece of opal glass. The light returned to the eye had then for the most part traversed the film *twice*, with the effect of doubling any feeble difference which would occur on simple transmission. Under favourable circumstances it was possible to detect a reversal between the two images when the difference amounted to $3\frac{1}{2}$ per cent. A few such experiments might therefore be expected to give the required result accurate to less than one per cent.

With the Edwards' plates an exposure of 12 minutes was found to be necessary. This was divided into two parts of 6 minutes each, with an interval of one minute during which the mirrors were reversed. About the middle of each period of 6 minutes the direct light was screened off for a time which varied from picture to picture. For example, on June 6, the time of screening for one picture was 71 seconds, and for the second picture 48 seconds. This means that while in both pictures the exposure for the reflected light was 12 minutes or 720 seconds, the exposures for the direct light were respectively $720 - 2 \times 71 = 578$ seconds, and $720 - 2 \times 48 = 624$ seconds. The water was distilled, and its temperature was $17^{\circ}\cdot7$ C. The examination of the finished pictures showed that the contrast was reversed, so that the total exposure (to the direct light) required for a balance was intermediate between 578 and 624, and, further, that the first mentioned was the nearer to the mark.

The general conclusion derived from a large number of photographs was that the balance corresponded to a total screening of 121 seconds, viz., to an exposure of $720 - 121 = 599$ seconds. This is for the direct light, the exposure to the reflected light being always 720 seconds. The ratio of exposures required for a balance is thus

$$599 : 720;$$

and this may be considered to correspond to a temperature of 18° C.

We can now calculate the observed reflexion for $6\frac{1}{2}^{\circ}$ incidence, reckoned as a fraction of the incident light. We have

$$\frac{599}{720} \cdot \frac{167}{10800} \cdot \left(\frac{104\cdot43}{82\cdot21}\right)^{2} = \cdot02076.$$

The above relates to the impression upon Edwards' plates after the light had been transmitted through a yellow glass. When Ilford plates were substituted and the yellow glass omitted, the reflexion appeared decidedly more powerful, and the ratio of exposures necessary for a balance was about $425 : 480$, or $637 : 720$. It appears, therefore, that the reflexion of the light operative in this case is some 6 per cent. more than before, or about $\cdot0220$ of the incident light. As to a large increase of reflexion there was no doubt; but, owing perhaps to variations in the quality of the light, the agreement between individual results was not so good as before.

It now remains to calculate the reflexion as given by Fresnel's formulæ; and it appears from the discussion at the commencement of this paper that we may ignore the small angle of incidence ($6\frac{1}{2}°$) and take the formula in the simple form given by Young, viz.:—

$$R = (\mu - 1)^2/(\mu + 1)^2.$$

As to the value of μ for water, Wüllner* gives

$$\mu_\lambda = 1.326067 - .000099\,t + .30531\,\lambda^{-2},$$

t denoting the temperature in Centigrade degrees. Applied to 18° and to $\lambda = 5620$, this gives

$$\mu = 1.333951,$$

whence

$$(\mu - 1)^2/(\mu + 1)^2 = .02047.$$

The reflexion actually found is accordingly about $1\frac{1}{2}$ per cent. greater than that given by Fresnel's formulæ.

In order to estimate the effect, according to the formula, of a change in index, we may use

$$\frac{\delta R}{R} = \frac{4\,\delta\mu}{\mu^2 - 1},$$

or, in the case of water,

$$\delta R/R = 5\,\delta\mu \text{ nearly.}$$

To cause a variation of $1\frac{1}{2}$ per cent. in the reflexion, $\delta\mu$ would have to be .003, and to cause 6 per cent. $\delta\mu$ would have to be .012. The latter exceeds the variation of μ in passing between the lines D and H.

The agreement with Fresnel's formulæ is thus pretty good, but the question arises whether it ought not to be better. Apart from à priori ideas as to the result to be expected, I should have estimated the errors of experiment as not likely to exceed one-half per cent., and certainly no straining of judgment in respect of the photometric pictures would bring about agreement. On the other hand, it must be remembered that one per cent. is not a large error in photometry, and that in the present case a one per cent. error in the reflexion is but one in 5000 reckoned as a fraction of the incident light. While, therefore, the disagreement may be real, it is too small a foundation upon which to build with any confidence.

It only remains to record the results of some observations upon the reflexion from mercury. In these experiments the revolving disk was dispensed with, and the photographs were taken upon Edwards' plates through yellow glass. The angle of incidence and all the other arrangements remained as before. In order to obtain a balance it appeared that the direct light

* *Pogg. Ann.* Bd. cxxxiii.

required to be screened for 64 seconds out of 120 seconds. The reflexion is accordingly

$$\frac{56}{120}\left(\frac{104\cdot43}{82\cdot21}\right)^2 = \cdot753.$$

The mercury was of good quality, and was filtered into a glass vessel just before use. The level was adjusted to be the same as that adopted for the observations upon water. A surface thus obtained would not be free from a greasy layer, but it is not probable that this would sensibly influence the reflexion.

APPENDIX.

The calculation of the reflexion depends upon the assumption that the reflecting surface is plane; and a very moderate concavity would suffice to explain the small excess in the observed number for water over that calculated from Fresnel's formulæ. It is thus of importance to assure ourselves that the concavity due to capillarity is really small enough to be neglected. For this purpose an estimate founded upon the capillary surface applicable in two dimensions will suffice.

If θ be the inclination to the horizon at any point, x the horizontal and y the vertical coordinate, the equations to the surface are :—

$$x = -2a\cos\tfrac{1}{2}\theta + a\log\cot\tfrac{1}{4}\theta, \qquad y = 2a\sin\tfrac{1}{2}\theta,$$

where

$$a^2 = T/g\rho.$$

At a great distance from the edge,

$$\theta = 0, \qquad y = 0, \qquad x = \infty.$$

At the vertical edge of a wetted vessel, $\theta = \tfrac{1}{2}\pi$.

The origin of x corresponds to

$$\theta = \pi, \qquad y = 2a.$$

In the case of water $T = 74$, $\rho = 1$, and $g = 981$ c.g.s.; so that

$$a = \cdot274 \text{ centim.}$$

In the experiments upon reflexion the part of the surface in action was about 11 centim. away from the boundary, so that $x/a = 40$, and θ is very small.

For the curvature

$$1/\rho = y/a^2 = 2 \sin \tfrac{1}{2}\theta \,./a;$$

or for our present purpose

$$1/\rho = \theta/a.$$

To find θ we have approximately,

$$\cot \tfrac{1}{4}\theta = e^{38}, \quad \text{or} \quad \theta = 4e^{-38}.$$

Accordingly

$$\frac{1}{\rho} = \frac{4}{\cdot 274 \times e^{38}}.$$

This may be multiplied by 4 to represent the increase of effect in the actual circumstances as compared with what is supposed in the two-dimensional problem; but it remains absolutely insensible in comparison with the other curvatures involved.

199.

ON THE INTERFERENCE BANDS OF APPROXIMATELY HOMOGENEOUS LIGHT; IN A LETTER TO PROF. A. MICHELSON.

[*Philosophical Magazine*, XXXIV. pp. 407—411, 1892.]

WHEN we were discussing together the results of your interesting work upon high interference, you asked my opinion upon one or two questions connected therewith. I have delayed answering until I had the opportunity of seeing your paper in print (*Phil. Mag.* Sept. 1892), but now I may as well send you what I have to say.

First, as to the definiteness with which the character of the spectral line $\phi(x)$ can be deduced from the "visibility-curve." By Fourier's theorem,

$$\phi(x) = \frac{1}{\pi} \int_0^\infty du \left\{ \cos ux \int_{-\infty}^{+\infty} \cos uv\, \phi(v)\, dv + \sin ux \int_{-\infty}^{+\infty} \sin uv\, \phi(v)\, dv \right\};$$

or in your notation, if we identify u with $2\pi D$,

$$\phi(x) = \frac{1}{\pi} \int_0^\infty du \left\{ C \cos ux + S \sin ux \right\}.$$

Hence, if C and S are both given as functions of u, $\phi(x)$ is absolutely, and uniquely, determined. However, the visibility-curve by itself gives, not both C and S, but only $C^2 + S^2$; so that we must conclude that in general an indefinite variety of structures is consistent with a visibility-curve given in all its parts.

But if we may assume that the structure is symmetrical, $S = 0$; and ϕ is then determined by means of C. And, since $V^2 = C^2/P^2$, the visibility-curve determines C, or at least C^2. In practice, considerations of continuity would always fix the choice of the square root. Thus, in the case of a spectral band of uniform brightness, where

$$V^2 = \sin^2 \pi n / \pi^2 n^2,$$

we are to take

$$C = \sin \pi n / \pi n,$$

and not

$$C = + \sqrt{(\sin^2 \pi n / \pi^2 n^2)}.$$

In order to determine both C and S, observations would have to be made not only upon the visibility, but also upon the situation of the bands. You remark that "it is theoretically possible by this means to determine, in case of an unequal double, or a line unsymmetrically broadened, whether the brighter side is towards the blue or the red end of the spectrum." But I suppose that a complete determination of both C and S, though theoretically possible, would be an extremely difficult task.

If the spectral line has a given total width, the "visibility" begins to fall away from the maximum (unity) most rapidly when the brightness of the line is all concentrated at the edges, so as to constitute a double line.

It is interesting to note that in several simple cases the bands seen with ever increasing retardation represent the character of the luminous vibration itself. In the case of a mathematical spectral line, the waves are regular to infinity, and the bands are formed without limit and with maximum visibility throughout. Again, in the case of a double line (with equal components) the waves divide themselves into groups with intermediate evanescences, and this is also the character of the interference bands. Thirdly, if the spectral line be a band of uniform brightness, and if the waves at the origin be supposed to be all in one phase, the actual compound vibration will be accurately represented by the corresponding interference bands. But this law is not general for the reason that in one case we have to deal with *amplitudes* and in the other with *intensities*. The accuracy of correspondence thus requires that the finite amplitudes involved be all of one magnitude. A partial exception to this statement occurs in the case of a spectral line in which the distribution of brightness is exponential.

Another question related to the effect of the gradual loss of energy, from communication to the ether, upon the homogeneity of the light radiated from freely vibrating molecules. In illustration of this we may consider the analysis by Fourier's theorem of a vibration in which the amplitude follows the exponential law, rising from zero to a maximum, and afterwards falling again to zero. It is easily proved that

$$e^{-a^2x^2} \cos rx = \frac{1}{2a\sqrt{\pi}} \int_0^\infty du \cos ux \left\{ e^{-(u-r)^2/4a^2} + e^{-(u+r)^2/4a^2} \right\},$$

in which the second member expresses an aggregate of trains of waves, each individual train being absolutely homogeneous. If a be small in comparison with r, as will happen when the amplitude on the left varies but slowly, $e^{-(u+r)^2/4a^2}$ may be neglected, and $e^{-(u-r)^2/4a^2}$ is sensible only when u is very nearly equal to r.

As an example in which the departure from regularity consists only in an abrupt change of phase, let us suppose that

$$\psi(x) = \pm \sin(2\pi x/l),$$

the sign being reversed at every interval of ml, so that the positive sign applies from 0 to ml, $2\,ml$ to $3\,ml$, $4\,ml$ to $5\,ml$, &c., and the negative sign from ml to $2\,ml$, $3\,ml$ to $4\,ml$, &c.   As the analysis into simple waves we find

$$\psi\,(x) = \Sigma\,\frac{2\cos\,(2\pi nx/2ml)}{m\pi\,(1-n^2/4m^2)},$$

the summation extending to odd values 1, 3, 5, ... of n.   The fundamental component $\cos\,(2\pi x/2ml)$ and every odd harmonic occur, but not to the same extent.   When n is nearly equal to $2m$, the terms rise to great relative magnitude.   The most important are thus

$$\cos\frac{2\pi x}{l}\left(1\pm\frac{1}{2m}\right),\qquad \cos\frac{2\pi x}{l}\left(1\pm\frac{2}{2m}\right),\ \ \&c.;$$

and it is especially to be remarked that what might at first sight be regarded as the principal, if not the solitary, wave-length, viz. l, does not occur *at all*.

Besides communication of energy to the ether, and disturbance during encounters with neighbours, the motion of the molecule itself has to be considered as hostile to homogeneity of radiation.   The effect, according to Döppler's principle, of motion in the line of sight was calculated by me on a former occasion and is fully regarded in your paper.   But there is another, and perhaps more important, consequence of molecular motion, which does not appear to have been remarked.   Besides the motion of translation there is the motion of rotation to be reckoned with.   The effect of the latter will depend upon the law of radiation in various directions from a stationary molecule.   As to this we do not know much, but enough to exclude the case of radiation alike in all directions, as from an ideal source of sound.   Such a symmetry is indeed inconsistent with the law of transverse vibrations.   The simplest supposition is that the radiation is like that generated in an elastic solid, at one point of which there acts a periodic force in a given direction. In this case the amplitude in any direction varies as the sine of the angle between the ray and the force, and the direction of (transverse) vibration lies in the plane containing these two lines.   A complete investigation of the radiation from such molecules vibrating and rotating about all possible axes would be rather complicated, but from one or two particular cases it is easy to recognize the general character of the effect produced.   Suppose, for example, that the axis of rotation is perpendicular to the axis of vibration, and consider the radiation in a direction perpendicular to the former axis. If ω be the angular velocity, the amplitude varies as $\cos\omega t$, and the vibration may be represented by

$$2\cos\omega t\,.\,\cos nt = \cos\,(n+\omega)\,t + \cos\,(n-\omega)\,t.$$

The spectrum would thus show a *double* line, whose components are separated by a distance proportional to ω.

Again, if the ray be parallel to the axis of rotation, the amplitude is indeed constant in magnitude, but its direction rotates. The plane-polarized rays into which the vibration may be resolved are represented as before by $\cos \omega t \cdot \cos nt$. There is of course one case in which these complications fail to occur, *i.e.* when the axis of rotation coincides with the axis of vibration; but with axes distributed at random we must expect vibrations $(n \pm \omega)$ to be almost as important as the vibration n. The law of distribution of brightness in the spectral line would probably be exponential, as when the widening is due to motion of molecules as wholes in the line of sight.

It will be of interest to compare the magnitudes of the two effects. If v be the linear velocity of a molecule and V that of light, the comparison is between ω and nv/V, or between ω and v/λ. If r be the radius of a molecule, the circumferential velocity of rotation is ωr, and we may compare ωr with vr/λ. Now, according to Boltzmann's theorem, $r\omega$ would be of the same order of magnitude as v, so that the importance of the rotatory and linear effects would be somewhat as $\lambda : r$. There is every reason to suppose that λ is much greater than r, and thus (if Boltzmann's relation held good) to expect that the disturbance of homogeneity due to rotation would largely outweigh that due to translation.

Your results seem already to interpose serious obstacles in the way of accepting such a conclusion; and the fact that light may thus be thrown upon a much controverted question in molecular physics is only another proof of the importance of the research upon which you are engaged.

200.

ON THE INFLUENCE OF OBSTACLES ARRANGED IN RECTANGULAR ORDER UPON THE PROPERTIES OF A MEDIUM.

[*Philosophical Magazine*, XXXIV. pp. 481—502, 1892.]

THE remarkable formula, arrived at almost simultaneously by L. Lorenz[*] and H. A. Lorentz[†], and expressing the relation between refractive index and density, is well known; but the demonstrations are rather difficult to follow, and the limits of application are far from obvious. Indeed, in some discussions the necessity for any limitation at all is ignored. I have thought that it might be worth while to consider the problem in the more definite form which it assumes when the obstacles are supposed to be arranged in rectangular or square order, and to show how the approximation may be pursued when the dimensions of the obstacles are no longer very small in comparison with the distances between them.

Taking, first, the case of two dimensions, let us investigate the conductivity for heat, or electricity, of an otherwise uniform medium interrupted by cylindrical obstacles which are arranged in rectangular order. The sides of the rectangle will be denoted by α, β, and the radius of the cylinders by a. The simplest cases would be obtained by supposing the material composing the cylinders to be either non-conducting or perfectly conducting; but it will be sufficient to suppose that it has a definite conductivity different from that of the remainder of the medium.

By the principle of superposition the conductivity of the interrupted medium for a current in any direction can be deduced from its conductivities in the three principal directions. Since conduction parallel to the axes of the cylinders presents nothing special for our consideration, we may limit

* *Wied. Ann.* XI. p. 70 (1880).
† *Wied. Ann.* IX. p. 641 (1880).

our attention to conduction parallel to one of the sides (α) of the rectangular structure.   In this case lines parallel to α, symmetrically situated between

Fig. 1.

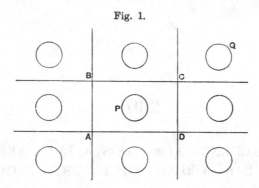

the cylinders, such as AD, BC, are lines of flow, and the perpendicular lines AB, CD are equipotential.

If we take the centre of one of the cylinders P as origin of polar coordinates, the potential external to the cylinder may be expanded in the series

$$V = A_0 + (A_1 r + B_1 r^{-1}) \cos\theta + (A_3 r^3 + B_3 r^{-3}) \cos 3\theta + \ldots, \quad \ldots\ldots(1)$$

and at points within the cylinder in the series

$$V' = C_0 + C_1 r \cos\theta + C_3 r^3 \cos 3\theta + \ldots, \quad \ldots\ldots\ldots\ldots(2)$$

θ being measured from the direction of α.   The sines of θ and its multiples are excluded by the symmetry with respect to $\theta = 0$, and the cosines of the even multiples by the symmetry with respect to $\theta = \frac{1}{2}\pi$.   At the bounding surface, where $r = a$, we have the conditions

$$V = V', \qquad \nu \, dV'/dr = dV/dr,$$

ν denoting the conductivity of the material composing the cylinders in terms of that of the remainder reckoned as unity.   The application of these conditions to the term in $\cos n\theta$ gives

$$B_n = \frac{1-\nu}{1+\nu} a^{2n} A_n. \quad \ldots\ldots\ldots\ldots\ldots\ldots\ldots\ldots\ldots(3)$$

In the case where the cylinders are perfectly conducting, $\nu = \infty$.   If they are non-conducting, $\nu = 0$.

The values of the coefficients A_1, B_1, A_3, $B_3 \ldots$ are necessarily the same for all the cylinders, and each may be regarded as a similar multiple source of potential.   The first term A_0, however, varies from cylinder to cylinder, as we pass up or down the stream.

Let us now apply Green's theorem,

$$\int \left(U \frac{dV}{dn} - V \frac{dU}{dn} \right) ds = 0 \quad \ldots\ldots\ldots\ldots\ldots(4)$$

to the contour of the region between the rectangle $ABCD$ and the cylinder P. Within this region V satisfies Laplace's equation, as also will U, if we assume

$$U = x = r \cos \theta. \quad\dots\dots\dots\dots\dots\dots\dots\dots\dots(5)$$

Over the sides BC, AD, dU/dn, dV/dn both vanish. On CD, $\int dV/dn \,.\, ds$ represents the total current across the rectangle, which we may denote by C. The value of this part of the integral over CD, AB is thus αC. The value of the remainder of the integral over the same lines is $-V_1\beta$, where V_1 is the fall in potential corresponding to one rectangle, as between CD and AB.

On the circular part of the contour,

$$U = a \cos \theta, \quad dU/dn = - dU/dr = - \cos \theta ;$$

and thus the only terms in (1) which will contribute to the result are those in $\cos \theta$. Thus we may write

$$V = (A_1 a + B_1 a^{-1}) \cos \theta,$$
$$dV/dn = -(A_1 - B_1 a^{-2}) \cos \theta ;$$

so that this part of the integral is $2\pi B_1$. The final result from the application of (4) is thus

$$\alpha C - \beta V_1 + 2\pi B_1 = 0. \quad\dots\dots\dots\dots\dots\dots\dots(6)$$

If $B_1 = 0$, we fall back upon the uninterrupted medium of which the conductivity is unity. For the case of the actual medium we require a further relation between B_1 and V_1.

The potential V at any point may be regarded as due to external sources at infinity (by which the flow is caused) and to multiple sources situated on the axes of the cylinders. The first part may be denoted by Hx. In considering the second it will conduce to clearness if we imagine the (infinite) region occupied by the cylinders to have a rectangular boundary parallel to α and β. Even then the manner in which the infinite system of sources is to be taken into account will depend upon the shape of the rectangle. The simplest case, which suffices for our purpose, is when we suppose the rectangular boundary to be infinitely more extended parallel to α than parallel to β. It is then evident that the periodic difference V_1 may be reckoned as due entirely to Hx, and equated to $H\alpha$. For the difference due to the sources upon the axes will be equivalent to the addition of one extra column at $+\infty$, and the removal of one at $-\infty$, and in the case supposed such a transference is immaterial*. Thus

$$V_1 = H\alpha \quad\dots\dots\dots\dots\dots\dots\dots\dots\dots(7)$$

simply, and it remains to connect H with B_1.

---

* It would be otherwise if the infinite rectangle were supposed to be of another shape, e.g. to be square.

This we may do by equating two forms of the expression for the potential at a point x, y near P. The part of the potential due to Hx and to the multiple sources Q (P not included) is

$$A_0 + A_1 r \cos \theta + A_3 r^3 \cos 3\theta + \dots;$$

or, if we subtract Hx, we may say that the potential at x, y due to the multiple sources at Q is the real part of

$$A_0 + (A_1 - H)(x + iy) + A_3(x + iy)^3 + A_5(x + iy)^5 + \dots \quad \dots\dots(8)$$

But if x', y' are the coordinates of the same point when referred to the centre of one of the Q's, the same potential may be expressed by

$$\Sigma \{B_1(x' + iy')^{-1} + B_3(x' + iy')^{-3} + \dots\}, \quad \dots\dots\dots\dots(9)$$

the summation being extended over all the Q's. If ξ, η be the coordinates of a Q referred to P,

$$x' = x - \xi, \quad y' = y - \eta;$$

so that

$$B_n(x' + iy')^{-n} = B_n(x + iy - \xi - i\eta)^{-n}.$$

Since (8) is the expansion of (9) in rising powers of $(x + iy)$, we obtain, equating term to term,

$$\left.\begin{aligned}
H - A_1 &= B_1 \Sigma_2 + 3B_3 \Sigma_4 + 5B_5 \Sigma_6 + \dots \\
-1.2.3 A_3 &= 1.2.3 B_1 \Sigma_4 + 3.4.5 B_3 \Sigma_6 + \dots \\
-1.2.3.4.5 A_5 &= 1.2.3.4.5 B_1 \Sigma_6 + 3.4.5.6.7 B_5 \Sigma_8 + \dots
\end{aligned}\right\} \dots(10)$$

and so on, where

$$\Sigma_{2n} = \Sigma (\xi + i\eta)^{-2n}, \quad \dots\dots\dots\dots\dots\dots\dots(11)$$

the summation extending over all the Q's.

By (3) each B can be expressed in terms of the corresponding A. For brevity, we will write

$$A_n = \nu' a^{-2n} B_n, \quad \dots\dots\dots\dots\dots\dots\dots(12)$$

where

$$\nu' = (1 + \nu)/(1 - \nu). \quad \dots\dots\dots\dots\dots\dots(13)$$

We are now prepared to find the approximate value of the conductivity. From (6) the conductivity of the rectangle is

$$\frac{C}{V_1} = \frac{\beta}{\alpha}\left\{1 - \frac{2\pi B_1}{\beta V_1}\right\} = \frac{\beta}{\alpha}\left\{1 - \frac{2\pi B_1}{\alpha \beta H}\right\};$$

so that the specific conductivity of the actual medium for currents parallel to α is

$$1 - \frac{2\pi B_1}{\alpha \beta H}, \quad \dots\dots\dots\dots\dots\dots\dots(14)$$

and the ratio of H to B_1 is given approximately by (10) and (12).

In the first approximation we neglect Σ_4, $\Sigma_6 \dots$, so that $A_3, A_5 \dots B_3, B_5 \dots$ vanish. In this case

$$H = A_1 + B_1 \Sigma_2 = B_1(\nu' a^{-2} + \Sigma_2), \quad \dots\dots\dots\dots(15)$$

and the conductivity is

$$1 - \frac{2\pi a^2}{\alpha\beta(\nu' + a^2\Sigma_2)}. \quad\dots\dots\dots\dots\dots(16)$$

The second approximation gives

$$\frac{Ha^2}{B_1} = \nu' + a^2\Sigma_2 - \frac{3}{\nu'}a^3\Sigma_4^2, \quad\dots\dots\dots\dots(17)$$

and the series may be continued as far as desired.

The problem is thus reduced to the evaluation of the quantities $\Sigma_2, \Sigma_4, \dots$. We will consider first the important particular case which arises when the cylinders are in *square* order, that is when $\beta = \alpha$. ξ and η in (11) are then both multiples of α, and we may write

$$\Sigma_n = \alpha^{-n}S_n, \quad\dots\dots\dots\dots\dots\dots(18)$$

where

$$S_n = \Sigma(m' + im)^{-n}; \quad\dots\dots\dots\dots\dots(19)$$

the summation being extended to all integral values of m, m', positive or negative, except the pair $m = 0$, $m' = 0$. The quantities S are thus purely numerical, and real.

The next thing to be remarked is that, since m, m' are as much positive as negative, S_n vanishes for every odd value of n. This holds even when α and β are unequal.

Again,

$$S_{2n} = \Sigma(m' + im)^{-2n} = i^{-2n}\Sigma(-im' + m)^{-2n}$$
$$= (-1)^n \Sigma(-im' + m)^{-2n} = (-1)^n S_{2n}.$$

Whenever n is odd, $S_{2n} = -S_{2n}$, or S_{2n} vanishes. Thus for square order,

$$S_6 = S_{10} = S_{14} = \dots\dots = 0. \quad\dots\dots\dots\dots(20)$$

This argument does not, without reservation, apply to S_2. In that case the sum is not convergent; and the symmetry between m and m', essential to the proof of evanescence, only holds under the restriction that the infinite region over which the summation takes place is symmetrical with respect to the two directions α and β—is, for example, square or circular. On the contrary, we have supposed, and must of course continue to suppose, that the region in question is infinitely elongated in the direction of α.

The question of convergency may be tested by replacing the parts of the sum relating to a great distance by the corresponding integral. This is

$$\iint \frac{dx\,dy}{(x + iy)^{2n}} = \iint \frac{\cos 2n\theta\, r\,dr\,d\theta}{r^{2n}};$$

and herein

$$\int r^{-2n+1}dr = r^{-2n+2}/(-2n + 2);$$

so that if $n > 1$ there is convergency, but if $n = 1$ the integral contains an infinite logarithm.

We have now to investigate the value of S_2 appropriate to our purpose; that is, when the summation extends over the region bounded by $x = \pm u$, $y = \pm v$, where u and v are both infinite, but so that $v/u = 0$. If we suppose that the region of summation is that bounded by $x = \pm v$, $y = \pm v$, the sum vanishes by symmetry. We may therefore regard the summation as extending over the region bounded externally by $x = \pm \infty$, $y = \pm v$, and internally

Fig. 2.

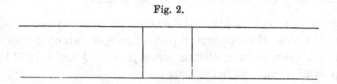

by $x = \pm v$ (Fig. 2). When v is very great, the sum may be replaced by the corresponding integral. Hence

$$S_2 = 2 \iint \frac{dx\,dy}{(x + iy)^2}, \quad \dots\dots\dots\dots\dots\dots(21)$$

the limits for y being $\pm v$, and those for x being v and ∞. Ultimately v is to be made infinite.

We have

$$\int_{-v}^{+v} \frac{dy}{(x + iy)^2} = \frac{i}{x + iv} - \frac{i}{x - iv} = \frac{2v}{x^2 + v^2};$$

and

$$\int_{v}^{\infty} \frac{2v\,dx}{x^2 + v^2} = 2 \tan^{-1} \infty - 2 \tan^{-1} 1 = \tfrac{1}{2}\pi.$$

Accordingly

$$S_2 = \pi. \quad \dots\dots\dots\dots\dots\dots\dots\dots\dots\dots\dots\dots\dots(22)$$

In the case of square order, equations (10), (12) give

$$\frac{Ha^2}{B_1} = \nu' + a^2 \Sigma_2 - \frac{3}{\nu'} a^8 \Sigma_4{}^2 - \frac{7}{\nu'} a^{16} \Sigma_8{}^2 - \dots$$

$$= \nu' + \frac{\pi a^2}{\alpha^2} - \frac{3}{\nu'} \frac{a^8}{\alpha^8} S_4{}^2 - \frac{7}{\nu'} \frac{a^{16}}{\alpha^{16}} S_8{}^2 - \dots; \quad \dots\dots\dots(23)$$

and by (14)

$$\text{Conductivity} = 1 - \frac{2\pi a^2}{\alpha^2} \cdot \frac{B_1}{Ha^2}. \quad \dots\dots\dots\dots\dots\dots(24)$$

If p denote the proportional space occupied by the cylinders,

$$p = \pi a^2 / \alpha^2; \quad \dots\dots\dots\dots\dots\dots\dots\dots(25)$$

and

$$\text{Conductivity} = 1 - \frac{2p}{\nu' + p - \frac{3p^4}{\nu' \pi^4} S_4{}^2 - \frac{7p^8}{\nu' \pi^8} S_8{}^2}. \quad \dots\dots\dots\dots(26)$$

Of the double summation indicated in (19) one part can be effected without difficulty. Consider the roots of

$$\sin(\xi - im\pi) = 0.$$

They are all included in the form

$$\xi = m'\pi + im\pi,$$

where m' is any integer, positive, negative, or zero. Hence we see that $\sin(\xi - im\pi)$ may be written in the form

$$A\left(1 - \frac{\xi}{im\pi}\right)\left(1 - \frac{\xi}{im\pi + \pi}\right)\left(1 - \frac{\xi}{im\pi - \pi}\right)\left(1 - \frac{\xi}{im\pi + 2\pi}\right)\dots,$$

in which

$$A = -\sin im\pi.$$

Thus

$$\log\{\cos\xi - \cot im\pi \sin\xi\} = \log\left(1 - \frac{\xi}{im\pi}\right) + \log\left(1 - \frac{\xi}{im\pi + \pi}\right) + \dots.$$

If we change the sign of m, and add the two equations, we get

$$\log\left\{1 - \frac{\sin^2\xi}{\sin^2 im\pi}\right\} = \log\left\{1 - \frac{\xi^2}{(im\pi)^2}\right\}$$

$$+ \log\left\{1 - \frac{\xi^2}{(im\pi + \pi)^2}\right\} + \log\left\{1 - \frac{\xi^2}{(im\pi - \pi)^2}\right\} + \dots;$$

whence, on expansion of the logarithms,

$$\frac{\sin^2\xi}{\sin^2 im\pi} + \frac{\sin^4\xi}{2\sin^4 im\pi} + \frac{\sin^6\xi}{3\sin^6 im\pi} + \dots$$

$$= \xi^2\left\{\frac{1}{(im\pi)^2} + \frac{1}{(im\pi + \pi)^2} + \frac{1}{(im\pi - \pi)^2} + \dots\right\}$$

$$+ \tfrac{1}{2}\xi^4\left\{\frac{1}{(im\pi)^4} + \frac{1}{(im\pi + \pi)^4} + \frac{1}{(im\pi - \pi)^4} + \dots\right\}$$

$$+ \tfrac{1}{3}\xi^6\left\{\frac{1}{(im\pi)^6} + \frac{1}{(im\pi + \pi)^6} + \frac{1}{(im\pi - \pi)^6} + \dots\right\} + \dots.$$

By expanding the sines on the left and equating the corresponding powers of ξ, we find

$$\frac{1}{(im)^2} + \frac{1}{(im+1)^2} + \frac{1}{(im-1)^2} + \frac{1}{(im+2)^2} + \dots\dots = \frac{\pi^2}{\sin^2 im\pi}, \dots\dots(27)$$

$$\frac{1}{(im)^4} + \frac{1}{(im+1)^4} + \dots\dots = -\frac{2\pi^4}{3\sin^2 im\pi} + \frac{\pi^4}{\sin^4 im\pi}, \dots\dots\dots\dots(28)$$

$$\frac{1}{(im)^6} + \frac{1}{(im+1)^6} + \dots\dots = \frac{2\pi^6}{15\sin^2 im\pi} - \frac{\pi^6}{\sin^4 im\pi} + \frac{\pi^6}{\sin^6 im\pi}. \dots(29)$$

In the summation with respect to m, required in (19), we are to take all positive and negative integral values. But in the case of $m = 0$ we are to leave out the first term, corresponding to $m' = 0$. When $m = 0$,

$$\frac{\pi^2}{\sin^2 im\pi} - \frac{1}{(im)^2} = \frac{\pi^2}{3},$$

which, as is well known, is the value of

$$\frac{1}{1^2} + \frac{1}{(-1)^2} + \frac{1}{2^2} + \frac{1}{(-2)^2} + \dots.$$

Hence

$$S_2 = 2\pi^2 \sum_{m=1}^{m=\infty} \sin^{-2} im\pi + \tfrac{1}{3}\pi^2; \quad \dots\dots\dots\dots\dots(30)$$

and in like manner

$$S_4 = \frac{\pi^4}{45} + 2\pi^4 \sum_{m=1}^{m=\infty} \{ -\tfrac{2}{3}\sin^{-2} im\pi + \sin^{-4} im\pi \}, \quad \dots\dots\dots\dots(31)$$

$$S_6 = \frac{2\pi^6}{27 \cdot 35} + 2\pi^6 \sum_{m=1}^{m=\infty} \{ \tfrac{2}{15}\sin^{-2} im\pi - \sin^{-4} im\pi + \sin^{-6} im\pi \}. \quad \dots\dots(32)$$

We have seen already that $S_6 = 0$, and that $S_2 = \pi$. The comparison of the latter with (30) gives

$$\sum_{m=1}^{m=\infty} \sin^{-2} im\pi = \frac{1}{2\pi} - \frac{1}{6}. \quad \dots\dots\dots\dots\dots(33)$$

We will now apply (31) to the numerical calculation of S_4. We find:

m	$-\sin^{-2} im\pi$	$\sin^{-4} im\pi$
1	·00749767	·0000562150
2	· 1395	· 2
3	· 3	
Sum	·00751165	·0000562152

so that

$$S_4 = \pi^4 \times ·03235020. \quad \dots\dots\dots\dots\dots\dots(34)$$

In the same way we may verify (33), and that (32) = 0.

If we introduce this value into (26), taking for example the case where the cylinders are non-conductive $(\nu' = 1)$, we get

$$1 - \frac{2p}{1 + p - ·3058p^4}. \quad \dots\dots\dots\dots\dots\dots(35)$$

From the above example it appears that in the summation with respect to m there is a high degree of convergency. The reason for this will appear more clearly if we consider the nature of the first summation (with respect to m'). In (19) we have to deal with the sum of $(x + iy)^{-n}$, where y is for the moment regarded as constant, while x receives the values $x = m'$. If, instead of being concentrated at equidistant points, the values of x were uniformly distributed, the sum would become

$$\int_{-\infty}^{+\infty} \frac{dx}{(x + iy)^n}.$$

Now, n being greater than 1, the value of this integral is zero. We see, then, that the finite value of the sum depends entirely upon the discontinuity of its formation, and thus a high degree of convergency when y increases may be expected.

The same mode of calculation may be applied without difficulty to any particular case of a rectangular arrangement. For example, in (11)

$$\Sigma_2 = \Sigma (m'\alpha + im\beta)^{-2} = \alpha^{-2}\Sigma (m' + im\beta/\alpha)^{-2}.$$

If m be given, the summation with respect to m' leads, as before, to

$$\Sigma (m' + im\beta/\alpha)^{-2} = \frac{\pi^2}{\sin^2(im\pi\beta/\alpha)};$$

and thus

$$\alpha^2\Sigma_2 = 2\pi^2 \sum_{m=1}^{m=\infty} \sin^{-2}(im\pi\beta/\alpha) + \tfrac{1}{3}\pi^2. \quad\dots\dots\dots\dots\dots(36)$$

The numerical calculation would now proceed as before, and the final approximate result for the conductivity is given by (16). Since (36) is not symmetrical with respect to α and β, the conductivity of the medium is different in the two principal directions.

When $\beta = \alpha$, we know that $\alpha^{-2}\Sigma_2 = \pi$. And since this does not differ much from $\tfrac{1}{3}\pi^2$, it follows that the series on the right of (36) contributes but little to the total. The same will be true, even though β be not equal to α, provided the ratio of the two quantities be moderate. We may then identify $\alpha^{-2}\Sigma_2$ with π, or with $\tfrac{1}{3}\pi^2$, if we are content with a very rough approximation.

The question of the values of the sums denoted by Σ_{2n} is intimately connected with the theory of the θ-functions*, inasmuch as the roots of $H(u)$, or $\theta_1(\pi u/2K)$, are of the form

$$2mK + 2m'iK'.$$

The analytical question is accordingly that of the expansion of $\log \theta_1(x)$ in ascending powers of x. Now, Jacobi† has himself investigated the expansion in powers of x of

$$\theta_1(x) = 2 \{q^{1/4} \sin x - q^{9/4} \sin 3x + q^{25/4} \sin 5x - \dots\}, \quad\dots\dots\dots(37)$$

---

* Cayley's *Elliptic Functions*, p. 300. The notation is that of Jacobi.
† *Crelle*, Bd. LIV. p. 82.

where
$$q = e^{-\pi K'/K}. \quad\dots\dots\dots\dots\dots\dots(38)$$
So far as the cube of x the result is
$$\frac{\theta_1(x)}{D\,.\,x} = 1 - \frac{2x^2}{3\pi^2}\left\{3KE - (2-k^2)K^2\right\} + \dots, \quad\dots\dots(39)$$

D being a constant which it is not necessary further to specify. K and E are the elliptic functions of k usually so denoted. By what has been stated above the roots of $\theta_1(x)$ are of the form
$$\pi\,(m + m'iK'/K); \quad\dots\dots\dots\dots\dots\dots(40)$$
so that
$$\Sigma\,\{m + im'K'/K\}^{-2} = \tfrac{4}{3}\{3KE - (2-k^2)K^2\}, \quad\dots\dots\dots(41)$$

the summation on the left being extended to all integral values of m and m', except $m = 0$, $m' = 0$.

This is the general solution for Σ_2. If $K' = K$, $k^2 = \tfrac{1}{2}$, and
$$\Sigma\,\{m + im'\}^{-2} = 2\{2KE - K^2\} = \pi,$$
since in general*,
$$EK' + E'K - KK' = \tfrac{1}{2}\pi.$$

In proceeding further it is convenient to use the form in which an exponential factor is removed from the series. This is
$$\theta_1(x) = A^{\frac{1}{2}}e^{-\frac{1}{2}ABx^2}\left\{s_0 Ax - s_1\frac{A^3x^3}{3!} + s_2\frac{A^5x^5}{5!} - s_3\frac{A^7x^7}{7!} + \dots\right\}, \quad\dots(42)$$
in which
$$A = \frac{2K}{\pi}, \qquad B = \frac{2E}{\pi} - k'^2\frac{2K}{\pi}, \quad\dots\dots\dots\dots(43)$$
$$s_0 = \beta, \quad s_1 = \alpha\beta, \quad s_2 = \beta(\alpha^2 - 2\beta^4), \quad s_3 = \alpha\beta(\alpha^2 - 6\beta^4),$$
the law of formation of s being
$$s_{m+1} = 2m\,(2m+1)\,\beta^4 s_{m-1} + \alpha\beta\,ds_m/d\beta - 8\beta^4 ds_m/d\alpha, \quad\dots\dots(44)$$
while
$$\alpha = k'^2 - k^2, \qquad \beta = \sqrt{(kk')}. \quad\dots\dots\dots\dots(45)$$

I have thought it worth while to quote these expressions, as they do not seem to be easily accessible; but I propose to apply them only to the case of square order, $K' = K$, $k'^2 = k^2 = \tfrac{1}{2}$. Thus
$$AB = 1/\pi, \quad \alpha = 0, \quad \beta = 1/\sqrt{2}; \quad\dots\dots\dots(46)$$
$$s_0 = \beta, \quad s_1 = 0, \quad s_2 = -2\beta^5, \quad s_3 = 0, \quad s_4 = -36\beta^9,$$
and
$$\theta_1(x) = \frac{A^{\frac{3}{2}}x}{\sqrt{2}}\,e^{-x^2/2\pi}\left\{1 - \frac{A^4x^4}{2.5!} - \frac{A^8x^8}{4.8!} + \dots\right\}. \quad\dots\dots\dots(47)$$

* Cayley's *Elliptic Functions*, p. 49.

Hence

$$\log \frac{\theta_1(x)}{D \cdot x} = -\frac{x^2}{2\pi} - \frac{A^4 x^4}{2 \cdot 5!} - \frac{A^8 x^8}{16 \cdot 35 \cdot 5!} - \dots \quad \dots\dots\dots(48)$$

If $\pm \lambda_1, \pm \lambda_2, \dots$ are the roots of $\theta_1(x)/x = 0$, we have

$$\Sigma \lambda^{-2} = \frac{1}{2\pi}, \quad \Sigma \lambda^{-4} = \frac{A^4}{5!}, \quad \Sigma \lambda^{-6} = 0, \quad \Sigma \lambda^{-8} = \frac{A^8}{7 \cdot 5 \cdot 4 \cdot 5!}.$$

Now by (40) the roots in question are $\pi(m + im')$, and thus

$$S_2 = \pi, \quad S_4 = \frac{\pi^4}{60} A^4, \quad S_8 = \frac{\pi^8 A^8}{70 \cdot 5!}, \quad \dots\dots\dots\dots(49)$$

in which

$$A = \frac{2}{\pi} K = 1 + \frac{1^2}{2^2} \cdot \frac{1}{2} + \frac{1^2 \cdot 3^2}{2^2 \cdot 4^2} \cdot \frac{1}{4} + \frac{1^2 \cdot 3^2 \cdot 5^2}{2^2 \cdot 4^2 \cdot 6^2} \cdot \frac{1}{8} + \dots = 1 \cdot 18034.$$

Leaving the two-dimensional problem, I will now pass on to the case of a medium interrupted by *spherical* obstacles arranged in rectangular order. As before, we may suppose that the side of the rectangle in the direction of flow is α, the two others being β and γ. The radius of the sphere is a.

The course of the investigation runs so nearly parallel to that already given, that it will suffice to indicate some of the steps with brevity. In place of (1) and (2) we have the expansions

$$V = A_0 + (A_1 r + B_1 r^{-2}) Y_1 + \dots$$
$$+ (A_n r^n + B_n r^{-n-1}) Y_n + \dots, \quad \dots\dots\dots(50)$$
$$V' = C_0 + C_1 Y_1 r + \dots + C_n Y_n r^n + \dots, \quad \dots\dots\dots(51)$$

Y_n denoting the spherical surface harmonic of order n. And from the surface conditions

$$V = V', \quad \nu dV'/dn = dV/dn,$$

we find

$$B_n = \frac{1 - \nu}{1 + \nu + 1/n} a^{2n+1} A_n. \quad \dots\dots\dots\dots(52)$$

We must now consider the limitations to be imposed upon Y_n. In general,

$$Y_n = \sum_{s=0}^{s=n} \Theta_n^{(s)} (H_s \cos s\phi + K_s \sin s\phi), \quad \dots\dots\dots(53)$$

where

$$\Theta_n^{(s)} = \sin^s \theta \left(\cos^{n-s}\theta - \frac{(n-s)(n-s-1)}{2(2n-1)} \cos^{n-s-2}\theta + \dots\dots \right), \quad \dots(54)$$

θ being supposed to be measured from the axis of x (parallel to α), and ϕ from the plane of xz. In the present application symmetry requires that s should be even, and that Y_n (except when $n = 0$) should be reversed when

$(\pi - \theta)$ is written for θ. Hence even values of n are to be excluded altogether. Further, no sines of $s\phi$ are admissible. Thus we may take

$$Y_1 = \cos\theta, \quad\dots\dots\dots\dots\dots\dots\dots\dots\dots\dots\dots\dots\dots\dots\dots(55)$$

$$Y_3 = \cos^3\theta - \tfrac{3}{5}\cos\theta + H_2\sin^2\theta\cos\theta\cos 2\phi, \quad\dots\dots\dots\dots(56)$$

$$Y_5 = \cos^5\theta - \tfrac{10}{9}\cos^3\theta + \tfrac{5}{21}\cos\theta$$

$$\qquad + L_2\sin^2\theta\,(\cos^3\theta - \tfrac{1}{3}\cos\theta)\cos 2\phi$$

$$\qquad + L_4\sin^4\theta\cos\theta\cos 4\phi. \quad\dots\dots\dots\dots\dots\dots(57)$$

In the case where $\beta = \gamma$ symmetry further requires that

$$H_2 = 0, \quad L_2 = 0. \quad\dots\dots\dots\dots\dots\dots\dots\dots(58)$$

In applying Green's theorem (4) the only difference is that we must now understand by s the area of the *surface* bounding the region of integration. If C denote the total current flowing across the faces $\beta\gamma$, V_1 the periodic difference of potential, the analogue of (6) is

$$\alpha C - \beta\gamma V_1 + 4\pi B_1 = 0. \quad\dots\dots\dots\dots\dots\dots(59)$$

We suppose, as before, that the system of obstacles, extended without limit in every direction, is yet infinitely more extended in the direction of α than in the directions of β and γ. Then, if Hx be the potential due to the sources at infinity other than the spheres, $V_1 = H\alpha$, and

$$\frac{C}{V_1} = \frac{\beta\gamma}{\alpha}\left\{1 - \frac{4\pi B_1}{\alpha\beta\gamma H}\right\};$$

so that the specific conductivity of the compound medium parallel to α is

$$1 - \frac{4\pi B_1}{\alpha\beta\gamma H}. \quad\dots\dots\dots\dots\dots\dots\dots(60)$$

We will now show how the ratio B_1/H is to be calculated approximately, limiting ourselves, however, for the sake of simplicity to the case of cubic order, where $\alpha = \beta = \gamma$. The potential round P, viz.

$$A_0 + A_1 r\,.\,Y_1 + A_3 r^3 Y_3 + \dots,$$

may be regarded as due to Hx and to the other spheres Q acting as sources of potential. Thus, if we revert to rectangular coordinates and denote the coordinates of a point relatively to P by x, y, z, and relatively to one of the Q's by x', y', z', we have

$$A_0 + (A_1 - H)x + A_3(x^3 - \tfrac{3}{5}xr^2) + \dots$$

$$= B_1\Sigma\frac{x'}{r'^3} + B_3\Sigma\frac{x'^3 - \tfrac{3}{5}x'r'^2}{r'^7} + \dots, \quad\dots\dots\dots(61)$$

in which

$$x' = x - \xi, \quad y' = y - \eta, \quad z' = z - \zeta,$$

if ξ, η, ζ be the coordinates of Q referred to P. The left side of (61) is thus the expansion of the right in ascending powers of x, y, z. Accordingly,

$A_1 - H$ is found by taking d/dx of the right-hand member and then making x, y, z vanish. In like manner $6A_3$ will be found from the third differential coefficient. Now, at the origin,

$$\frac{d}{dx}\frac{x'}{r'^3} = -\frac{d}{d\xi}\frac{x'}{r'^3} = -\frac{d}{d\xi}\frac{-\xi}{\rho^3} = \frac{\rho^2 - 3\xi^2}{\rho^5},$$

in which

$$\rho^2 = \xi^2 + \eta^2 + \zeta^2.$$

It will be observed that we start with a harmonic of order 1 and that the differentiation raises the order to 2. The law that each differentiation raises the order by unity is general; and, so far as we shall proceed, the harmonics are all zonal, and may be expressed in the usual way as functions $P_n(\mu)$ of μ where $\mu = \xi/\rho$. Thus

$$\frac{d}{dx}\frac{x'}{r'^3} = -2\rho^{-3}P_2(\mu).$$

In like manner,

$$\frac{d}{dx}\frac{x'^0 - \tfrac{3}{5}x'r'^2}{r'^7} = \frac{d}{d\xi}\frac{\xi^3 - \tfrac{3}{5}\xi\rho^2}{\rho^7} = -\frac{8}{5}\rho^{-5}P_4(\mu),$$

and

$$\frac{d^3}{dx^3}\frac{x'}{r'^3} = \frac{d^3}{d\xi^3}\frac{\xi}{\rho^3} = -24\rho^{-5}P_4(\mu).$$

The comparison of terms in (61) thus gives

$$\left.\begin{array}{l} A_1 - H = -2B_1\Sigma\rho^{-3}P_2 - \tfrac{8}{5}B_3\Sigma\rho^{-5}P_4 + \ldots \\ A_3 = -4B_1\Sigma\rho^{-5}P_4 + \ldots \\ \ldots = \ldots \end{array}\right\} \quad \ldots\ldots\ldots\ldots(62)$$

In each of the quantities, such as $\Sigma\rho^{-3}P_2$, the summation is to be extended to all the points whose coordinates are of the form $l\alpha, m\alpha, n\alpha$, where l, m, n are any set of integers, positive or negative, except $0, 0, 0$. If we take $\alpha = 1$, and denote the corresponding sums by $S_2, S_4, \ldots$, these quantities will be purely numerical, and

$$\Sigma\rho^{-n-1}P_n = \alpha^{-n-1}S_n. \quad\ldots\ldots\ldots\ldots\ldots\ldots(63)$$

From (52), (62) we now obtain

$$\frac{Ha^3}{B_1} = \frac{2+\nu}{1-\nu} + 2S_2\frac{a^3}{\alpha^3} - \frac{32}{5}\frac{1-\nu}{\tfrac{1}{3}+\nu}S_4^2\frac{a^{10}}{\alpha^{10}} + \ldots, \quad\ldots\ldots\ldots(64)$$

which with (60) gives the desired result for the conductivity of the medium.

We now proceed to the calculation of S_2. We have

$$S_2 = \Sigma\frac{3\mu^2 - 1}{2\rho^3} = \Sigma\frac{2\xi^2 - \eta^2 - \zeta^2}{2\rho^5} = -\tfrac{1}{2}\Sigma\frac{d}{d\xi}\left(\frac{\xi}{\rho^3}\right).$$

By the symmetry of a cubical arrangement, it follows that

$$\Sigma(\xi^2/\rho^5) = \Sigma(\eta^2/\rho^5) = \Sigma(\zeta^2/\rho^5);$$

so that if S were calculated for a space bounded by a cube, it would necessarily vanish. But for our purpose S_2 is to be calculated over the space bounded by $\xi = \pm \infty$, $\eta = \pm v$, $\zeta = \pm v$, where v is finally to be made infinite; and, as we have just seen, we may exclude the space bounded by

$$\xi = \pm v, \quad \eta = \pm v, \quad \zeta = \pm v;$$

so that $\frac{1}{2} S_2$ will be obtained from the space bounded by

$$\xi = v, \quad \xi = \infty, \quad \eta = \pm v, \quad \zeta = \pm v.$$

Now when ρ is sufficiently great, the summation may be replaced by an integration; thus

$$S_2 = - \iiint \frac{d}{d\xi} \left(\frac{\xi}{\rho^3} \right) d\xi \, d\eta \, d\zeta.$$

In this,

$$\int_v^\infty \frac{d}{d\xi} \frac{\xi}{\rho^3} \, d\xi = - \frac{v}{(v^2 + \eta^2 + \zeta^2)^{\frac{3}{2}}},$$

$$\int_{-v}^{+v} \frac{v \, d\eta}{(v^2 + \eta^2 + \zeta^2)^{\frac{3}{2}}} = \frac{2v^2}{(v^2 + \zeta^2)(2v^2 + \zeta^2)^{\frac{1}{2}}},$$

and finally

$$\int_{-v}^{+v} \frac{v^2 \, d\zeta}{(v^2 + \zeta^2)(2v^2 + \zeta^2)^{\frac{1}{2}}} = \int_0^1 \frac{2 \, d\theta}{\sqrt{(2 + \tan^2 \theta)}} = \int_0^{1/\sqrt{2}} \frac{2 \, ds}{\sqrt{(2 - s^2)}} = \frac{\pi}{3}.$$

Thus

$$S_2 = \frac{2\pi}{3}. \quad \dots\dots\dots\dots\dots\dots\dots\dots\dots\dots\dots(65)$$

If we neglect a^{10}/α^{10}, and write p for the ratio of volumes, viz.

$$p = \frac{4\pi a^3}{3\alpha^3}, \quad \dots\dots\dots\dots\dots\dots\dots\dots\dots(66)$$

we have by (60) for the conductivity

$$\frac{(2+\nu)/(1-\nu) - 2p}{(2+\nu)/(1-\nu) + p}, \quad \dots\dots\dots\dots\dots\dots(67)^*$$

or in the particular case of non-conducting obstacles ($\nu = 0$)

$$\frac{1-p}{1+\frac{1}{2}p}. \quad \dots\dots\dots\dots\dots\dots\dots\dots\dots(68)$$

In order to carry on the approximation we must calculate S_4 &c. Not seeing any general analytical method, such as was available in the former problem, I have calculated an approximate value of S_4 by direct summation from the formula

$$S_4 = \Sigma \, \frac{35 \, \xi^4 - 30 \, \xi^2 \rho^2 + 3 \rho^4}{8 \, \rho^9}.$$

---

* Compare Maxwell's *Electricity*, § 314.

We may limit ourselves to the consideration of positive and zero values of ξ, η, ζ. Every term for which ξ, η, ζ are finite is repeated in each octant, that is 8 times. If one of the three coordinates vanish, the repetition is fourfold, and if two vanish, twofold.

The following table contains the result for all points which lie within $\rho^2 = 18$. This repetition in the case, for example, of $\rho^2 = 9$ represents two kinds of composition. In the first

$$\rho^2 = 2^2 + 2^2 + 1^2 = 9,$$

and in the second

$$\rho^2 = 3^2 + 0^2 + 0^2 = 9.$$

The success of the approximation is favoured by the fact that P vanishes when integrated over the complete sphere, so that the sum required is only a kind of residue depending upon the discontinuity of the summation.

The result is

$$S_4 = 3 \cdot 11. \quad \dots\dots\dots\dots\dots\dots\dots\dots\dots(69)$$

	ρ^2			ρ^2	
0, 0, 1	1	$+ 3 \cdot 5000$	0, 0, 3	9	$+ \cdot 0144$
0, 1, 1	2	$- \cdot 3094$	0, 1, 3	10	$+ \cdot 0243$
1, 1, 1	3	$- \cdot 1996$	1, 1, 3	11	$+ \cdot 0075$
0, 0, 2	4	$+ \cdot 1094$	2, 2, 2	12	$- \cdot 0062$
0, 1, 2	5	$+ \cdot 0501$	0, 2, 3	13	$- \cdot 0015$
1, 1, 2	6	$- \cdot 0397$	1, 2, 3	14	$- \cdot 0095$
0, 2, 2	8	$- \cdot 0097$	0, 0, 4	16	$+ \cdot 0034$
1, 2, 2	9	$- \cdot 0277$	2, 2, 3	17	$- \cdot 0061$
			0, 1, 4	17	$+ \cdot 0085$

The results of our investigation have been expressed for the sake of simplicity in electrical language as the conductivity of a compound medium, but they may now be applied to certain problems of vibration. The simplest of these is the problem of wave-motion in a gaseous medium obstructed by rigid and fixed cylinders or spheres. It is assumed that the wave-length is very great in comparison with the period (α, β, γ) of the structure. Under these circumstances the flow of gas round the obstacles follows the same law as that of electricity, and the kinetic energy of the motion is at once given by the expressions already obtained. In fact the kinetic energy corresponding to a given total flow is increased by the obstacles in the same proportion as the electrical resistances of the original problem, so that the influence of the obstacles is taken into account if we suppose that the density of the gas is increased in the above ratio of resistances. In the case of cylinders in square order (35), the ratio is approximately

$$(1 + p)/(1 - p),$$

and in the case of spheres in cubic order by (68) it is approximately

$$(1 + \tfrac{1}{2}p)/(1 - p).$$

But this is not the only effect of the obstacles which we must take into account in considering the velocity of propagation. The potential energy also undergoes a change. The space available for compression or rarefaction is now $(1 - p)$ only instead of 1; and in this proportion is increased the potential energy corresponding to a given accumulation of gas*. For cylindrical obstruction the square of the velocity of propagation is thus altered in the ratio

$$\frac{1 - p}{1 + p} \div (1 - p) = \frac{1}{1 + p};$$

so that if μ denote the refractive index, referred to that of the unobstructed medium as unity, we find

$$\mu^2 = 1 + p,$$

or

$$(\mu^2 - 1)/p = \text{constant}, \quad \dots\dots\dots\dots\dots\dots(70)$$

which shows that a medium thus constituted would follow Newton's law as to the relation between refraction and density of obstructing matter. The same law (70) obtains also in the case of spherical obstacles; but reckoned absolutely the effect of spheres is only that of cylinders of halved density. It must be remembered, however, that while the velocity in the last case is the same in all directions, in the case of cylinders it is otherwise. For waves propagated parallel to the cylinders the velocity is uninfluenced by their presence. The medium containing the cylinders has therefore some of the properties which we are accustomed to associate with double refraction, although here the refraction is necessarily single. To this point we shall presently return, but in the meantime it may be well to apply the formulæ to the more general case where the obstacles have the properties of fluid, with finite density and compressibility.

To deduce the formula for the kinetic energy we have only to bear in mind that density corresponds to electrical *resistance*. Hence, by (26), if σ denote the density of the cylindrical obstacle, that of the remainder of the medium being unity, the kinetic energy is altered by the obstacles in the approximate ratio

$$\frac{(\sigma + 1)/(\sigma - 1) + p}{(\sigma + 1)/(\sigma - 1) - p} \dots\dots\dots\dots\dots\dots\dots(71)$$

* *Theory of Sound*, § 303.

The effect of this is the same as if the density of the whole medium were increased in the like ratio.

The change in the potential energy depends upon the "compressibility" of the obstacles. If the material composing them resists compression m times as much as the remainder of the medium, the volume p counts only as p/m, and the whole space available may be reckoned as $1 - p + p/m$ instead of 1. In this proportion is the potential energy of a given accumulation reduced. Accordingly, if μ be the refractive index as altered by the obstacles,

$$\mu^2 = (71) \times (1 - p + p/m). \quad \dots\dots\dots\dots\dots(72)$$

The compressibilities of all actual gases are nearly the same, so that if we suppose ourselves to be thus limited, we may set $m = 1$, and

$$\mu^2 = \frac{(\sigma+1)/(\sigma-1)+p}{(\sigma+1)/(\sigma-1)-p}; \quad \dots\dots\dots\dots(73)$$

or, as it may also be written,

$$\frac{\mu^2-1}{\mu^2+1}\frac{1}{p} = \text{constant}. \quad \dots\dots\dots\dots(74)$$

In the case of spherical obstacles of density σ we obtain in like manner $(m=1)$,

$$\mu^2 = \frac{(2\sigma+1)/(\sigma-1)+p}{(2\sigma+1)/(\sigma-1)-2p}, \quad \dots\dots\dots\dots(75)$$

or

$$\frac{\mu^2-1}{\mu^2+\frac{1}{2}}\frac{1}{p} = \text{constant}. \quad \dots\dots\dots\dots(76)$$

In the general case, where m is arbitrary, the equation expressing p in terms of μ^2 is a quadratic, and there are no simple formulæ analogous to (74) and (76).

It must not be forgotten that the application of these formulæ is limited to moderately small values of p. If it be desired to push the application as far as possible, we must employ closer approximations to (26), &c. It may be remarked that however far we may go in this direction, the final formula will always give μ^2 explicitly as a function of p. For example, in the case of rigid cylindrical obstacles, we have from (35)

$$\mu^2 = (1-p)\frac{1+p-\cdot3058p^4+\dots}{1-p-\cdot3058p^4+\dots}. \quad \dots\dots\dots\dots(77)$$

It will be evident that results such as these afford no foundation for a theory by which the refractive properties of a mixture are to be deduced by addition from the corresponding properties of the components. Such theories require formulæ in which p occurs in the first power only, as in (76).

If the obstacles are themselves elongated, or even, though their form be spherical, if they are disposed in a rectangular order which is not cubic, the velocity of wave-propagation becomes a function of the direction of the wave-normal. As in Optics, we may regard the character of the refraction as determined by the form of the *wave-surface*.

The æolotropy of the structure will not introduce any corresponding property into the potential energy, which depends only upon the volumes and compressibilities concerned. The present question, therefore, reduces itself to the consideration of the kinetic energy as influenced by the direction of wave-propagation. And this, as we have seen, is a matter of the electrical resistance of certain compound conductors, on the supposition, which we continue to make, that the wave-length is very large in comparison with the periods of the structure. The theory of electrical conduction in general has been treated by Maxwell (*Electricity*, § 297). A parallel treatment of the present question shows that in all cases it is possible to assign a system of principal axes, having the property that if the wave-normal coincide with any one of them the direction of flow will also lie in the same direction, whereas in general there would be a divergence. To each principal axis corresponds an efficient "density," and the equations of motion, applicable to the medium in the gross, take the form

$$\sigma_x \frac{d^2\xi}{dt^2} = m_1 \frac{d\delta}{dx}, \quad \sigma_y \frac{d^2\eta}{dt^2} = m_1 \frac{d\delta}{dy}, \quad \sigma_z \frac{d^2\zeta}{dt^2} = m_1 \frac{d\delta}{dz},$$

where ξ, η, ζ are the displacements parallel to the axes, m_1 is the compressibility, and

$$\delta = \frac{d\xi}{dx} + \frac{d\eta}{dy} + \frac{d\zeta}{dz}.$$

If λ, μ, ν are the direction-cosines of the displacement, l, m, n of the wave-normal, we may take

$$\xi = \lambda\theta, \quad \eta = \mu\theta, \quad \zeta = \nu\theta,$$

where

$$\theta = e^{i(lx+my+nz - Vt)}.$$

Thus

$$d\delta/dx = -l\theta(l\lambda + m\mu + n\nu), \quad \&c.$$

and the equations become

$$\sigma_x \lambda V^2 = m_1 l(l\lambda + m\mu + n\nu),$$
$$\sigma_y \mu V^2 = m_1 m(l\lambda + m\mu + n\nu),$$
$$\sigma_z \nu V^2 = m_1 n(l\lambda + m\mu + n\nu),$$

from which, on elimination of $\lambda : \mu : \nu$, we get

$$V^2 = m_1 \left(\frac{l^2}{\sigma_x} + \frac{m^2}{\sigma_y} + \frac{n^2}{\sigma_z}\right) = a^2 l^2 + b^2 m^2 + c^2 n^2, \quad \dots\dots\dots(78)$$

if a, b, c denote the velocities in the principal directions x, y, z.

The wave-surface after unit time is accordingly the ellipsoid whose axes are a, b, c.

As an example, if the medium, otherwise uniform, be obstructed by rigid cylinders occupying a moderate fraction (p) of the whole space, the velocity in the direction z, parallel to the cylinders, is unaltered; so that

$$c^2 = 1, \qquad a^2 - b^2 = 1/(1+p).$$

In the application of our results to the electric theory of light we contemplate a medium interrupted by spherical, or cylindrical, obstacles, whose inductive capacity is different from that of the undisturbed medium. On the other hand, the magnetic constant is supposed to retain its value unbroken. This being so, the kinetic energy of the electric currents for the same total flux is the same as if there were no obstacles, at least if we regard the wave-length as infinitely great*. And the potential energy of electric displacement is subject to the same mathematical laws as the resistance of our compound electrical conductor, specific inductive capacity in the one question corresponding to electrical conductivity in the other.

Accordingly, if ν denote the inductive capacity of the material composing the spherical obstacles, that of the undisturbed medium being unity, then the approximate value of μ^2 is given at once by (67). The equation may also be written in the form given by Lorentz,

$$\frac{\mu^2 - 1}{\mu^2 + 2} \frac{1}{p} = \frac{\nu - 1}{\nu + 2} = \text{constant}; \quad \dots\dots\dots\dots\dots(79)$$

and, indeed, it appears to have been by the above argument that (79) was originally discovered.

The above formula applies in strictness only when the spheres are arranged in cubic order†, and, further, when p is moderate. The next approximation is

$$\mu^2 = 1 + \frac{3p}{\dfrac{\nu + 2}{\nu - 1} - p - 1{\cdot}65 \dfrac{\nu - 1}{\nu + 4/3} p^{10/3}} \quad . \quad \dots\dots\dots\dots(80)$$

If the obstacles be cylindrical, and arranged in square order, the compound medium is doubly refracting, as in the usual electric theory of light, in which the medium is supposed to have an inductive capacity variable with the direction of displacement, independently of any discontinuity in its structure. The double refraction is of course of the uniaxal kind, and the wave-surface is the sphere and ellipsoid of Huygens.

* See Prof. Willard Gibbs's "Comparison of the Elastic and Electric Theories of Light," *Am. Journ. Sci.* xxxv. (1888).

† An irregular isotropic arrangement would, doubtless, give the same result.

For displacements parallel to the cylinders the resultant inductive capacity (analogous to conductivity in the conduction problem) is clearly $1 - p + \nu p$; so that the value of μ^2 for the principal extraordinary index is

$$\mu^2 = 1 + (\nu - 1)\,p, \quad \dots\dots\dots\dots\dots\dots\dots\dots\dots(81)$$

giving Newton's law for the relation between index and density.

For the ordinary index we have

$$\mu^2 = (26),$$

in which $\nu' = (1 + \nu)/(1 - \nu)$, while S_4, S_8 ... have the values given by (49). If we omit p^4, &c. we get

$$\mu^2 = \frac{\nu' - p}{\nu' + p}, \quad \dots\dots\dots\dots\dots\dots\dots\dots(82)$$

or

$$\frac{\mu^2 - 1}{\mu^2 + 1}\frac{1}{p} = -\frac{1}{\nu'} = \frac{\nu - 1}{\nu + 1}. \quad \dots\dots\dots\dots\dots\dots(83)$$

The general conclusion as regards the optical application is that, even if we may neglect dispersion, we must not expect such formulæ as (79) to be more than approximately correct in the case of dense fluid and solid bodies.

201.

ON THE DENSITIES OF THE PRINCIPAL GASES.

[*Proceedings of the Royal Society*, LIII. pp. 134—149, 1893.]

IN former communications* I have described the arrangements by which I determined the ratio of densities of oxygen and hydrogen (15·882). For the purpose of that work it was not necessary to know with precision the actual volume of gas weighed, nor even the pressure at which the containing vessel was filled. But I was desirous, before leaving the subject, of ascertaining not merely the relative, but also the absolute, densities of the more important gases, that is, of comparing their weights with that of an equal volume of water. To effect this it was necessary to weigh the globe, used to contain the gases, when charged with water, an operation not quite so simple as at first sight it appears. And, further, in the corresponding work upon the gases, a precise absolute specification is required of the temperature and pressure at which a filling takes place. To render the former weighings available for this purpose, it would be necessary to determine the errors of the barometers then employed. There would, perhaps, be no great difficulty in doing this; but I was of opinion that it would be an improvement to use a manometer in direct connexion with the globe, without the intervention of the atmosphere. In the latter manner of working, there is a doubt as to the time required for full establishment of equilibrium of pressure, especially when the passages through the taps are partially obstructed by grease. When the directly connected manometer is employed, there is no temptation to hurry from fear of the entrance of air by diffusion, and, moreover (Note A), the time actually required for the establishment of equilibrium is greatly diminished. With respect to temperature, also, it was thought better to avoid all further questions by surrounding the globe with ice, as in Regnault's original determinations. It is true that this procedure involves a subsequent cleaning and wiping of the globe, by which the errors of weighing are considerably augmented; but, as it was not proposed to experiment further with hydrogen, the objection was of less force. In the case of the heavier gases, unsystematic errors of weighing are less to be feared than doubts as to the actual temperature.

* *Roy. Soc. Proc.* February, 1888 [Vol. III. p. 37]; February, 1892 [Vol. III. p. 524].

In order to secure the unsystematic character of these errors, it is necessary to wash and wipe the working globe after an exhaustion in the same manner as after a filling. The dummy globe (of equal external volume, as required in Regnault's method of weighing gases) need not be wiped merely to secure symmetry, but it was thought desirable to do so before each weighing. In this way there would be no tendency to a progressive change. In wiping the globes the utmost care is required to avoid removing any loosely attached grease in the neighbourhood of the tap. The results to be given later will show that, whether the working globe be full or empty, the relative weights of the two globes can usually be recovered to an accuracy of about 0·3 milligramme. As in the former papers, the results were usually calculated by comparison of each "full" weight with the mean of the immediately preceding and following empty weights. The balance and the arrangements for weighing remained as already described.

The Manometer.

The arrangements adopted for the measurement of pressure must be described in some detail, as they offer several points of novelty. The apparatus actually used would, indeed, be more accurately spoken of as a manometric gauge, but it would be easy so to modify it as to fit it for measurements extending over a small range.

The object in view was to avoid certain defects to which ordinary barometers are liable, when applied to absolute measurements. Of these three especially may be formulated :—

a. It is difficult to be sure that the vacuum at the top of the mercury is suitable for the purpose.

b. No measurements of a length can be regarded as satisfactory in which different methods of reading are used for the two extremities.

c. There is necessarily some uncertainty due to irregular refraction by the walls of the tube. The apparent level of the mercury may deviate from the real position.

d. To the above may be added that the accurate observation of the barometer, as used by Regnault and most of his successors, requires the use of a cathetometer, an expensive and not always satisfactory instrument.

The guiding idea of the present apparatus is the actual application of a measuring rod to the upper and lower mercury surfaces, arranged so as to be vertically superposed. The rod AA, Fig. 1, is of iron (7 mm. in diameter), pointed below, B. At the upper end, C, it divides at the level of the mercury into a sort of fork, and terminates in a point similar to that at B, and, like it, directed downwards. The coincidence of these points with their images reflected in the mercury surfaces, is observed with the aid of lenses of about 30 mm. focus, held in position upon the wooden framework of the apparatus. It is, of course, independent of any irregular refraction which the tube may exercise. The verticality of the line joining the points is tested without difficulty by a plumb-line.

Fig. 1.

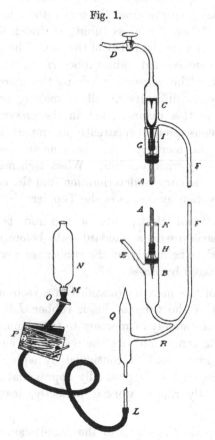

The upper and lower chambers C, B are formed from tubing of the same diameter (about 21 mm. internal). The upper communicates through a tap, D, with the Töppler, by means of which a suitable vacuum can at any time be established and tested. In ordinary use, D stands permanently open, but its introduction was found useful in the preliminary arrangements and in testing for leaks. The connexion between the lower chamber B and the vessel in which the pressure is to be verified takes place through a side tube, E.

The greater part of the column of mercury to which the pressure is due is contained in the connecting tube FF, of about 3 mm. internal diameter. The temperature is taken by a thermometer whose bulb is situated near the middle of *FF*. Towards the close of operations the more sensitive parts are protected by a packing of tow or cotton-wool, held in position between two wooden boards. The anterior board is provided with a suitable glass window, through which the thermometer may be read.

It is an essential requirement of a manometer on the present plan that the measuring rod pass air-tight from the upper and lower chambers into the atmosphere. To effect this the glass tubing is drawn out until its internal diameter is not much greater than that of the rod. The joints are then made by short lengths of thick-walled india-rubber *H, G*, wired on and drowned externally in mercury. The vessels for holding the mercury are shown at *I, K*. There is usually no difficulty at all in making perfectly tight joints between glass tubes in this manner ; but in the present case some trouble was experienced in consequence apparently of imperfect approximation between the *iron* and the mercury. At one time it was found necessary to supplement the mercury with vaseline. When tightness is once obtained, there seems to be no tendency to deterioration, and the condition of things is under constant observation by means of the Töppler.

The distance between the points of the rod is determined under microscopes by comparison with a standard scale, before the apparatus is put together. As the rod is held only by the rubber connexions, there is no fear of its length being altered by stress.

The adjustment of the mercury (distilled in a vacuum) to the right level is effected by means of the tube of black rubber *LM*, terminating in the reservoir *N*. When the supply of mercury to the manometer is a little short of what is needed, the connexion with the reservoir is cut off by a pinch-cock at *O*, and the fine adjustment is continued by squeezing the tube at *P* between a pair of hinged boards, gradually approximated by a screw. This plan, though apparently rough, worked perfectly, leaving nothing to be desired.

It remains to explain the object of the vessel shown at *Q*. In the early trials, when the rubber tube was connected directly to *R*, the gradual fouling of the mercury surface, which it seems impossible to avoid, threatened to interfere with the setting at *B*. By means of *Q*, the mercury can be discharged from the measuring chambers, and a fresh surface constituted at *B* as well as at *C*.

The manometer above described was constructed by my assistant, Mr Gordon, at a nominal cost for materials ; and it is thought that the same principle may be applied with advantage in other investigations. In cases where a certain latitude in respect of pressure is necessary, the measuring rod might be constructed in two portions, sliding upon one another. Probably a range of a few millimetres could be obtained without interfering with the india-rubber connexions.

The length of the iron rod was obtained by comparison under microscopes with a standard bar R divided into millimetres. In terms of R the length at 15° C. is 762·248 mm. It remains to reduce to standard millimetres. Mr Chaney has been good enough to make a comparison between R and the iridio-platinum standard metre, 1890, of the Board of Trade. From this it appears that the metre bar R is at 15° C. 0·3454 mm. too long; so that the true distance between the measuring points of the iron rod is at 15° C.

$$762·248 \times 1·0003454 = 762·511 \text{ mm.}$$

Connexions with Pump and Manometer.

Some of the details of the process of filling the globe with gas under standard conditions will be best described later under the head of the particular gas; but the general arrangement and the connexions with the pump and the manometer are common to all. They are sketched in Fig. 2, in

Fig. 2.

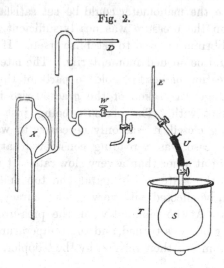

which S represents the globe, T the inverted bell-glass employed to contain the enveloping ice. The connexion with the rest of the apparatus is by a short tube U of thick rubber, carefully wired on. The tightness of these joints was always tested with the aid of the Töppler X, the tap V leading to the gas-generating apparatus being closed. The side tube at D leads to the vacuum chamber of the manometer, while that at E leads to the pressure chamber B. The wash-out of the tubes, and in some cases of the generator, was aided by the Töppler. When this operation was judged to be complete, V was again closed, and a good vacuum made in the parts still connected to the pump. W would then be closed, and the actual filling commenced by opening V, and finally the tap of the globe. The lower chamber of the manometer was now in connexion with the globe, and through a regulating tap (not shown) with the gas-generating apparatus. By means of the Töppler the vacuum in the manometer could be carried to any desired point. But with respect to this a remark must be made. It is a feature of the method employed* that the exhaustions of the globe are carried to such a point that the weight of the residual gas may be neglected, thus eliminating errors due to a second manometer reading. There is no difficulty in attaining this result, but the delicacy of the Töppler employed as a gauge is so great that the residual gas still admits of tolerably accurate measurement. Now in exhausting the head of the manometer it would be easy to carry the process to a point much in excess of what is necessary in the case of the globe, but there is evidently no advantage in so doing. The best results will be obtained by carrying both exhaustions to the *same* degree of perfection.

At the close of the filling the pressure has to be adjusted to an exact value, and it might appear that the double adjustment required (of pressure and of mercury) would be troublesome. Such was not found to be the case. After a little practice the manometer could be set satisfactorily without too great a delay. When the pressure was nearly sufficient, the regulating tap was closed, and equilibrium allowed to establish itself. If more gas was then required, the tap could be opened momentarily. The later adjustments were effected by the application of heat or cold to parts of the connecting tubes. At the close, advantage was taken of the gradual rise in the temperature which was usually met with. The pressure being just short of what was required, and V being closed, it was only necessary to wait until the point was reached. In no case was a reading considered satisfactory when the pressure was changing at other than a very slow rate. It is believed that the comparison between the state of things at the top and at the bottom of the manometer could be effected with very great accuracy, and this is all that the method requires. At the moment when the pressure was judged to be right, the tap of the globe was turned, and the temperature of the manometer was read. The vacuum was then verified by the Töppler.

The Weights.

The object of the investigation being to ascertain the *ratio* of densities of water and of certain gases under given conditions, the absolute values of the weights employed is evidently a matter of indifference. This is a point which I think it desirable to emphasise, because v. Jolly, in his, in many respects, excellent work upon this subject†, attributes a discrepancy between his final result for oxygen and that of Regnault to a possible variation in the standard of weight. On the same ground we may omit to allow for the buoyancy of the weights as used in air, since only the *variations* of buoyancy,

* Due to von Jolly.
† *Munich Acad. Trans.* Vol. XIII. Part II. p. 49, 1880.

due, for example, to changing barometer, could enter; and these affect the result so little that they may safely be neglected*.

But, while the absolute values of the weights are of no consequence, their relative values must be known with great precision. The investigation of these over the large range required (from a kilogramme to a centigramme) is a laborious matter, but it presents nothing special for remark. The weights quoted in this paper are, in all cases, corrected, so as to give the results as they would have been obtained from a perfectly adjusted system.

The Water Contents of the Globe.

The globe, packed in finely-divided ice, was filled with boiled distilled water up to the level of the top of the channel through the plug of the tap, that is, being itself at 0°, was filled with water also at 0°. Thus charged the globe had now to be weighed; but this was a matter of some difficulty, owing to the very small capacity available above the tap. At about 9° there would be a risk of overflow. Of course the water could be retained by the addition of extra tubing, but this was a complication that it was desired to avoid. In February, 1892, during a frost, an opportunity was found to effect the weighing in a cold cellar at a temperature ranging from 4° to 7°. The weights required (on the same side of the balance as the globe and its supports) amounted to 0·1822 gram. On the other side were other weights whose values did not require to be known so long as they remained unmoved during the whole series of operations. Barometer (corrected) 758·9 mm.; temperature 6·3°.

A few days later the globe was discharged, dried, and replaced in the balance with tap open. 1834·1701 grams had now to be associated with it in order to obtain equilibrium. The difference,

$$1834·170 - 0·182 = 1833·988,$$

represents the weight of the water less that of the air displaced by it. The difference of atmospheric conditions was sufficiently small to allow the neglect of the *variation* in the buoyancy of the glass globe and of the brass counter-poises.

It remains to estimate the actual weight of the air displaced by the water under the above mentioned atmospheric conditions. It appears that, on this account, we are to add 2·314, thus obtaining

$$1836·30$$

as the weight of the water at 0° which fills the globe at 0°.

* In v. Jolly's calculations the buoyancy of the weights seems to be allowed for in dealing with the water, and neglected in dealing with the gases. If this be so, the result would be affected with a slight error, which, however, far exceeds any that could arise from neglecting buoyancy altogether.

A further small correction is required to take account of the fact that the usual standard density is that of water at 4° and not at 0°. According to Broch (Everett's *C. G. S. System of Units*), the factor required is 0·99988, so that we have

$$\frac{1836 \cdot 30}{0 \cdot 99988} = 1836 \cdot 52$$

as the weight of water at 4° which would fill the globe at 0°.

Air.

Air drawn from outside (in the country) was passed through a solution of potash. On leaving the regulating tap it traversed tubes filled with fragments of potash, and a long length of phosphoric anhydride, followed by a filter of glass wool. The arrangements beyond the regulating tap were the same for all the gases experimented upon. At the close of the filling it was necessary to use a condensing syringe in order to force the pressure up to the required point, but the air thus introduced would not reach the globe. It may be well to give the results for air in some detail, so as to enable the reader to form a judgment as to the degree of accuracy attained in the manipulations.

Date	Globe empty	Globe full	Temp. of manometer	Correction to 15°	Corrected to 15°
1892					
September 24.........	2·90941				
„ 27.........	...	0·53327	17·8	−0·00112	0·53219
„ 28.........	2·90867	...			
„ 29.........	...	0·53271	15·7	−0·00028	0·53243
October 1.........	2·90923	...			
„ 3.........	...	0·53151	12·7	+0·00093	0·53244
„ 4.........	2·90872				
		Tap regreased			
„ 7.........	2·91036				
„ 8.........	...	0·53296	12·4	+0·00105	0·53401
„ 10.........	2·91056	...			
„ 11.........	...	0·53251	11·8	+0·00129	0·53380
„ 12.........	2·91039				
„ 13.........	...	0·53201	11·0	+0·00161	0·53362
„ 14.........	2·91043				
„ 15.........	...	0·53219	10·6	+0·00177	0·53396

The column headed "globe empty" gives the (corrected) weights, on the side of the working globe, required for balance. The third column gives the corresponding weights when the globe was full of air, having been charged at 0° and up to the pressure required to bring the mercury in the manometer into contact with the two points of the measuring rod.

This pressure was not quite the same on different occasions, being subject to a temperature correction for the density of mercury and for the expansion of the iron rod. The correction is given in the fifth column, and the weights that would have been required, had the temperature been 15°, in the sixth. The numbers in the second and sixth columns should agree, but they are liable to a discontinuity when the tap is regreased.

In deducing the weight of the gas we compare each weighing " full" with the mean of the preceding and following weights " empty," except in the case of October 15, when there was no subsequent weighing empty. The results are

September 27 .... 2·37686

„   29 .... 2·37651

October   3 .... 2·37653

„   8 .... 2·37646

„   11 .... 2·37668

„   13 .... 2·37679

„   15 .... 2·37647

Mean .... 2·37661

There is here no evidence of the variation in the density of air suspected by Regnault and v. Jolly. Even if we include the result for September 27th, obviously affected by irregularity in the weights of the globe empty, the extreme difference is only 0·4 milligram, or about 1/6000th part.

To allow for the contraction of the globe (No. 14) when weighed empty, discussed in my former papers, we are to add 0·00056 to the apparent weight, so that the result for air becomes

2·37717.

This is the weight of the contents at 0° and under the pressure defined by the manometer gauge at 15° of the thermometer. The reduction to standard conditions is, for the present, postponed.

Oxygen.

This gas has been prepared by three distinct methods: (a) from chlorates, (b) from permanganate of potash, (c) by electrolysis.

In the first method mixed chlorates of potash and soda were employed, as recommended by Shenstone, the advantage lying in the readier fusibility. The fused mass was contained in a Florence flask, and during the wash-out was allowed slowly to liberate gas into a vacuum. After all air had been expelled, the regulating tap was closed, and the pressure allowed gradually to rise to that of the atmosphere. The temperature could then be pushed without fear of distorting the glass, and the gas was drawn off through the regulating tap. A very close watch over the temperature was necessary to prevent the evolution of gas from becoming too rapid. In case of excess, the superfluous gas was caused to blow off into the atmosphere, rather than risk imperfect action of the potash and phosphoric anhydride. Two sets of five fillings were effected with this oxygen. In the first set (May, 1892) the highest result was 2·6272, and the lowest 2·6266, mean 2·62691. In the second set (June, July, 1892) the highest result was 2·6273 and the lowest 2·6267, mean 2·62693.

The second method (b) proved very convenient, the evolution of gas being under much better control than in the case of chlorates. The recrystallised salt was heated in a Florence flask, the wash-out, in this case also, being facilitated by a vacuum. Three fillings gave satisfactory results, the highest being 2·6273, the lowest 2·6270, and the mean 2·62714. The gas was quite free from smell.

By the third method I have not as many results as I could have wished, operations having been interrupted by the breakage of the electrolytic generator. This was, however, of less importance, as I had evidence from former work that there is no material difference between the oxygen from chlorates and that obtained by electrolysis. The gas was passed over hot copper [oxide], as detailed in previous papers. The result of one filling, with the apparatus as here described, was 2·6271. To this may be added the result of two fillings obtained at an earlier stage of the work, when the head of the manometer was exhausted by an independent Sprengel pump, instead of by the Töppler. The value then obtained was 2·6272. The results stand thus :—

Electrolysis (2), May, 1892	2·6272
„ (1) „	2·6271
Chlorates (5), May, 1892	2·6269
„ (5), June, 1892	2·6269
Permanganate (3), January, 1893 ...	2·6271
Mean............................	2·62704
Correction for contraction ...	0·00056
	2·62760

It will be seen that the agreement between the different methods is very good, the differences, such as they are, having all the appearance of being accidental. Oxygen prepared by electrolysis is perhaps most in danger of being light (from contamination with hydrogen), and that from chlorates of being abnormally heavy.

Nitrogen.

This gas was prepared, in the usual manner, from air by removal of oxygen with heated copper. Precautions are required, in the first place, to secure a sufficient action of the reduced copper, and, secondly, as was shown by v. Jolly, and later by Leduc, to avoid contamination with hydrogen which may be liberated from the copper. I have followed the plan, recommended by v. Jolly, of causing the gas to pass finally over a length of unreduced copper. The arrangements were as follows :—

Air drawn through solution of potash was deprived of its oxygen by reduced copper, contained in a tube of hard glass heated by a large flame. It then traversed a **U**-tube, in which was deposited most of the water of combustion. The gas, practically free, as the event proved, from oxygen, was passed, as a further precaution, over a length of copper heated in a combustion furnace, then through strong sulphuric acid*, and afterwards back through the furnace over a length of oxide of copper. It then passed on to the regulating tap, and thence through the remainder of the apparatus, as already described. In no case did the copper in the furnace, even at the end where the gas entered, show any sign of losing its metallic appearance.

Three results, obtained in August, 1892, were—

$$\begin{array}{ll} \text{August} \ 8 \ \dots\dots\dots\dots & 2\cdot31035 \\ \text{,,} \quad 10 \ \dots\dots\dots\dots & 2\cdot31026 \\ \text{,,} \quad 15 \ \dots\dots\dots\dots & 2\cdot31024 \\ \qquad \text{Mean}\dots\dots\dots\dots & 2\cdot31028 \end{array}$$

To these may be added the results of two special experiments made to test the removal of hydrogen by the copper oxide. For this purpose a small hydrogen generator, which could be set in action by closing an external contact, was included between the two tubes of reduced copper, the gas being caused to bubble through the electrolytic liquid. The quantity of hydrogen liberated was calculated from the deflection of a galvanometer included in the circuit, and was sufficient, if retained, to alter the density very materially. Care was taken that the small stream of hydrogen should be uniform during the whole time (about $2\frac{1}{2}$ hours) occupied by the filling, but, as will be seen, the impurity was effectually removed by the copper oxide†. Two experiments gave—

$$\begin{array}{ll} \text{September} \ 17 \ \dots\dots\dots\dots & 2\cdot31012 \\ \text{,,} \quad \ 20 \ \dots\dots\dots\dots & 2\cdot31027 \\ \qquad \text{Mean}\dots\dots\dots\dots & 2\cdot31020 \end{array}$$

We may take as the number for nitrogen—

$$\begin{array}{lr} & 2\cdot31026 \\ \text{Correction for contraction}\dots & 56 \\ \hline & 2\cdot31082 \end{array}$$

* There was no need for this, but the acid was in position for another purpose.

† Much larger quantities of hydrogen, sufficient to reduce the oxide over several centimetres, have been introduced without appreciably altering the weight of the gas.

Although the subject is not yet ripe for discussion, I cannot omit to notice here that nitrogen prepared from ammonia, and expected to be pure, turned out to be decidedly lighter than the above. When the oxygen of air is burned by excess of ammonia, the deficiency is about 1/1000th part*. When oxygen is substituted for air, so that all (instead of about one-seventh part) of the nitrogen is derived from ammonia, the deficiency of weight may amount to $\frac{1}{2}$ per cent. It seems certain that the abnormal lightness cannot be explained by contamination with hydrogen, or with ammonia, or with water, and everything suggests that the explanation is to be sought in a dissociated state of the nitrogen itself. Until the questions arising out of these observations are thoroughly cleared up, the above number for nitrogen must be received with a certain reserve. But it has not been thought necessary, on this account, to delay the presentation of the present paper, more especially as the method employed in preparing the nitrogen for which the results are recorded is that used by previous experimenters.

Reduction to Standard Pressure.

The pressure to which the numbers so far given relate is that due to 762·511 mm. of mercury at a temperature of 14·85°†, and under the gravity operative in my laboratory in latitude 51° 47′. In order to compare the results with those of other experimenters, it will be convenient to reduce them not only to 760 mm. of mercury pressure at 0°, but also to the value of gravity at Paris. The corrective factor for length is 760/762·511. In order to correct for temperature, we will employ the formula‡

$$1 + 0·0001818\, t + 0·00000000017\, t^2$$

for the volume of mercury at $t°$. The factor of correction for temperature is thus 1·002700. For gravity we may employ the formula—

$$g = 980·6056 - 2·5028 \cos 2\lambda,$$

λ being the latitude. Thus, for my laboratory—

$$g = 981·193,$$

and for Paris—

$$g = 980·939,$$

the difference of elevation being negligible. The factor of correction is thus 0·99974.

The product of the three factors, corrective for length, for temperature, and for gravity, is accordingly 0·99914. Thus multiplied, the numbers are as follows :—

Air	Oxygen	Nitrogen
2·37512	2·62534	2·30883

* *Nature*, Vol. xlvi. p. 512. [Vol. iv. p. 1.]
† The thermometer employed with the manometer read 0·15° too high.
‡ Everett, p. 142.

and these may now be compared with the water contents of the globe, viz., 1836·52.

The densities of the various gases under standard conditions, referred to that of distilled water at 4°, are thus:—

Air	Oxygen	Nitrogen
0·00129327	0·00142952	0·00125718

With regard to hydrogen, we may calculate its density by means of the ratio of densities of oxygen and hydrogen formerly given by me, viz., 15·882. Hence

Hydrogen
0·000090009

The following table shows the results arrived at by various experimenters. Von Jolly did not examine hydrogen. The numbers are multiplied by 1000 so as to exhibit the weights in grams per litre:—

	Air	Oxygen	Nitrogen	Hydrogen
Regnault, 1847	1·29319	1·42980	1·25617	0·08958
Corrected by Crafts	1·29349	1·43011	1·25647	0·08988
Von Jolly, 1880	1·29351	1·42939	1·25787	...
Ditto corrected	1·29383	1·42971	1·25819	...
Leduc, 1891*	1·29330	1·42910	1·25709	0·08985
Rayleigh, 1893	1·29327	1·42952	1·25718	0·09001

The correction of Regnault by Crafts† represents allowance for the contraction of Regnault's globe when exhausted, but the data were not obtained from the identical globe used by Regnault. In the fourth row I have introduced a similar correction to the results of von Jolly. This is merely an estimate founded upon the probability that the proportional contraction would be about the same as in my own case and in that of M. Leduc.

In taking a mean we may omit the uncorrected numbers, and also that obtained by Regnault for nitrogen, as there is reason to suppose that his gas was contaminated with hydrogen. Thus

Mean Numbers.

Air	Oxygen	Nitrogen	Hydrogen
1·29347	1·42961	1·25749	0·08991

The evaluation of the densities as compared with water is exposed to many sources of error which do not affect the comparison of one gas with

* Bulletin des Séances de la Société de Physique.
† Comptes Rendus, Vol. cvi. p. 1664.

another. It may therefore be instructive to exhibit the results of various
workers referred to air as unity*.

	Oxygen	Nitrogen	Hydrogen
Regnault (corrected)	1·10562	0·97138	0·06949
v. Jolly (corrected)	1·10502	0·97245	...
Leduc	1·1050	0·9720	0·06947
Rayleigh...........................	1·10535	0·97209	0·06960
Mean	1·10525	0·97218	0·06952

As usually happens in such cases, the concordance of the numbers
obtained by various experimenters is not so good as might be expected from
the work of each taken separately. The most serious discrepancy is in the
difficult case of hydrogen. M. Leduc suggests† that my number is too high
on account of penetration of air through the blow-off tube (used to establish
equilibrium of pressure with the atmosphere), which he reckons at 1 m. long
and 1 cm. in diameter. In reality the length was about double, and the
diameter one-half of these estimates; and the explanation is difficult to
maintain, in view of the fact, recorded in my paper, that a prolongation of
the time of contact from 4^m to 30^m had no appreciable ill effect. It must be
admitted, however, that there is a certain presumption in favour of a lower
number, unless it can be explained as due to an insufficient estimate of the
correction for contraction. On account of the doubt as to the appropriate
value of this correction, no great weight can be assigned to Regnault's
number for hydrogen. If the atomic weight of oxygen be indeed 15·88, and
the ratio of densities of oxygen and hydrogen be 15·90, as M. Leduc makes
them, we should have to accept a much higher number for the ratio of
volumes than that (2·0002) resulting from the very elaborate measurements
of Morley. But while I write the information reaches me that Mr A. Scott's
recent work upon the volume ratio leads him to just such a higher ratio,
viz., 2·00245, a number à priori more probable than 2·0002. Under the
circumstances both the volume ratio and the density of hydrogen must be
regarded as still uncertain to the 1/1000th part.

* [1902.   Cooke's value for hydrogen, viz. ·06958, of date 1889, should have been included in
the above.]
    † Comptes Rendus, July, 1892.

NOTE A.

On the Establishment of Equilibrium of Pressure in Two Vessels connected by a Constricted Channel.

It may be worth while to give explicitly the theory of this process, supposing that the difference of pressures is small throughout, and that the capacity of the channel may be neglected. If v_1, p_1 denote the volume and pressure of the gas in the first vessel at time t; v_2, p_2 the corresponding quantities for the second vessel, we have

$$v_1 dp_1/dt + c(p_1 - p_2) = 0,$$
$$v_2 dp_2/dt + c(p_2 - p_1) = 0,$$

where c is a constant which we may regard as the *conductivity* of the channel. In these equations inertia is neglected, only resistances of a viscous nature being regarded, as amply suffices for the practical problem. From the above we may at once deduce

$$\frac{d(p_1 - p_2)}{dt} + \left(\frac{c}{v_1} + \frac{c}{v_2}\right)(p_1 - p_2) = 0;$$

showing that $(p_1 - p_2)$ varies as e^{-qt}, where

$$q = \frac{c}{v_1} + \frac{c}{v_2} = \frac{1}{\tau},$$

if τ be the time in which the difference of pressures is reduced in the ratio of $e : 1$.

Let us now apply this result (a) to the case where the globe of volume v_1 communicates with the atmosphere, (b) to the case where the globe is connected with a manometer of relatively small volume v_2. For (a) we have

$$1/\tau = c/v_1,$$

and for (b) $$1/\tau = c/v_2;$$

so that $$\tau/\tau' = v_1/v_2.$$

For such a manometer as is described in the text, the ratio v_1/v_2 is at least as high as 30; and in this proportion is diminished the time required for the establishment of equilibrium up to any standard of perfection that may be fixed upon.

[1902. The question of the weight of nitrogen is further treated in Arts. 210, 214. It will be understood that the results given in the present paper relate to the atmospheric mixture of nitrogen and argon.]

202.

INTERFERENCE BANDS AND THEIR APPLICATIONS.

[Proceedings of the Royal Institution, XIV. pp. 72—78, 1893; *Nature*, XLVIII. pp. 212—214, 1893.]

THE formation of the interference bands, known as Newton's Rings, when two slightly curved glass plates are pressed into contact, was illustrated by an acoustical analogue. A high-pressure flame B (Fig. 1) is sensitive to sounds which reach it in the direction EB, but is insensitive to similar sounds which reach it in the nearly perpendicular direction AB. A is a "bird-call," giving a pure sound (inaudible) of wavelength (λ) equal to about 1 cm.; C and D are reflectors of perforated zinc. If C acts alone, the flame is visibly excited by the waves reflected from it, though by far the greater part of the energy is transmitted. If D, held parallel to C, be then brought into action, the result depends upon the interval between the two partial reflectors. The reflected sounds may co-operate, in which case the flame flares vigorously; or they may interfere, so that the flame recovers, and behaves as if no sound at all were falling upon it. The first effect occurs when the reflectors are close together, or are separated by any multiple of $\frac{1}{2}\sqrt{2}.\lambda$; the second when the interval is midway between those of the above-mentioned series, that is, when it coincides with an odd multiple of $\frac{1}{4}\sqrt{2}.\lambda$. The factor $\sqrt{2}$ depends upon the obliquity of the reflection.

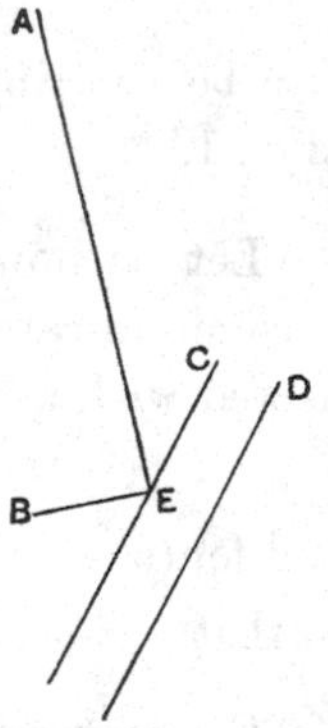

The coloured rings, as usually formed between glass plates, lose a good deal of their richness by contamination with white light reflected from the exterior surfaces. The reflection from the hindermost surface is easily got rid of by employing an opaque glass, but the reflection from the first surface is less easy to deal with. One plan, used in the lecture, depends upon the use of slightly wedge-shaped glasses (2°) so combined that the exterior surfaces are parallel to one another, but inclined to the interior operative surfaces. In this arrangement the false light is thrown somewhat to one

Fig. 1.

side, and can be stopped by a screen suitably held at the place where the image of the electric arc is formed.

The formation of colour and the ultimate disappearance of the bands as the interval between the surfaces increases, depends upon the mixed character of white light. For each colour the bands are upon a scale proportional to the wave-length for that colour. If we wish to observe the bands when the interval is considerable—bands of high interference as they are called—the most natural course is to employ approximately homogeneous light, such as that afforded by a soda flame. Unfortunately, this light is hardly bright enough for projection upon a large scale.

A partial escape from this difficulty is afforded by Newton's observations as to what occurs when a ring system is regarded through a prism. In this case the bands upon one side may become approximately achromatic, and are thus visible to a tolerably high order, in spite of the whiteness of the light. Under these circumstances there is, of course, no difficulty in obtaining sufficient illumination; and bands formed in this way were projected upon the screen*.

The bands seen when light from a soda flame falls upon nearly parallel surfaces have often been employed as a test of flatness. Two flat surfaces can be made to fit, and then the bands are few and broad, if not entirely absent; and, however the surfaces may be presented to one another, the bands should be straight, parallel, and equidistant. If this condition be violated, one or other of the surfaces deviates from flatness. In Fig. 2, A and B represent the glasses to be tested, and C is a lens of 2 or 3 feet focal length. Rays diverging from a soda flame at E are rendered parallel by the lens, and after reflection from the surfaces are recombined by the lens at E. To make an observation, the coincidence of the radiant point and its image must be somewhat disturbed, the one being displaced to a position a little beyond, and the other to a position a little in front of, the diagram.

The eye, protected from the flame by a suitable screen, is placed at the image, and being focused upon AB, sees the field traversed by bands. The reflector D is introduced as a matter of convenience to make the line of vision horizontal.

These bands may be photographed. The lens of the camera takes the place of the eye, and should be as close to the flame as possible. With suitable plates, sensitised by cyanin, the exposure required may vary from ten minutes to an hour. To get the best results, the hinder surface of A should be blackened, and the front surface of B should be thrown out of action by the superposition of a wedge-shaped plate of glass, the intervening space being filled with oil of turpentine or other fluid having nearly the same

* The theory is given in a paper upon " Achromatic Interference Bands," *Phil. Mag.* Aug. 1889. [Vol. III. p. 288.]

refraction as glass. Moreover, the light should be purified from blue rays by a trough containing solution of bichromate of potash. With these precautions the dark parts of the bands are very black, and the exposure may be prolonged much beyond what would otherwise be admissible.

The lantern slides exhibited showed the elliptical rings indicative of a curvature of the same sign in both directions, the hyperbolic bands corresponding to a saddle-shaped surface, and the approximately parallel system due to the juxtaposition of two telescopic "flats," kindly lent by Mr Common. On other plates were seen grooves due to rubbing with rouge along a defined track, and depressions, some of considerable regularity, obtained by the action of diluted hydrofluoric acid, which was allowed to stand for some minutes as a drop upon the surface of the glass.

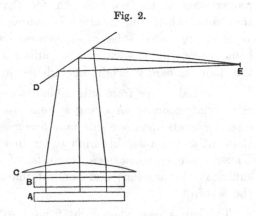

Fig. 2.

By this method it is easy to compare one flat with another, and thus, if the first be known to be free from error, to determine the errors of the second. But how are we to obtain and verify a standard? The plan usually followed is to bring *three* surfaces into comparison. The fact that two surfaces can be made to fit another in all azimuths proves that they are spherical and of equal curvatures, but one convex and the other concave, the case of perfect flatness not being excluded. If A and B fit one another, and also A and C, it follows that B and C must be similar. Hence, if B and C also fit one another, all three surfaces must be flat. By an extension of this process the errors of three surfaces which are not flat can be found from a consideration of the interference bands which they present when combined in three pairs.

But although the method just referred to is theoretically complete, its application in practice is extremely tedious, especially when the surfaces are not of revolution. A very simple solution of the difficulty has been found in the use of a free surface of water, which, when protected from tremors and motes, is as flat as can be desired*. In order to avoid all trace of capillary curvature it is desirable to allow a margin of about $1\frac{1}{2}$ inch. The surface to be tested is supported horizontally at a short distance ($\frac{1}{10}$ or $\frac{1}{20}$ inch) below that of the water, and the whole is carried upon a large and massive levelling stand. By the aid of screws the glass surface is brought into approximate

---

* The diameter would need to be 4 feet in order that the depression at the circumference, due to the general curvature of the earth, should amount to $\frac{1}{20}\lambda$.

parallelism with the water. In practice the principal trouble is in the avoidance of tremors and motes. When the apparatus is set up on the floor of a cellar in the country, the tremors are sufficiently excluded, but care must be taken to protect the surface from the slightest draught. To this end the space over the water must be enclosed almost air-tight. In towns, during the hours of traffic, it would probably require great precaution to avoid the disturbing effects of tremors. In this respect it is advantageous to diminish the thickness of the layer of water; but if the thinning be carried too far, the subsidence of the water surface to equilibrium becomes surprisingly slow, and a doubt may be felt whether after all there may not remain some deviation from flatness due to irregularities of temperature.

Fig. 3.

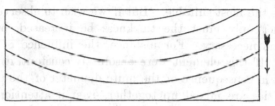

With the aid of the levelling screws the bands may be made as broad as the nature of the surface admits; but it is usually better so to adjust the level that the field is traversed by five or six approximately parallel bands. Fig. 3 represents bands actually observed from the face of a prism. That these are not straight, parallel, and equidistant is a proof that the surface deviates from flatness. The question next arising is to determine the direction of the deviation. This may be effected by observing the displacement of the bands due to a known motion of the levelling screws; but a simpler process is open to us. It is evident that if the surface under test were to be moved downwards parallel to itself, so as to increase the thickness of the layer of water, every band would move in a certain direction, viz. *towards* the side where the layer is thinnest. What amounts to the same, the retardation may be increased, without touching the apparatus, by so moving the eye as to *diminish* the obliquity of the reflection. Suppose, for example, in Fig. 3, that the movement in question causes the bands to travel downwards, as indicated by the arrow. The inference is that the surface is concave. More glass must be removed at the ends of the bands than in the middle in order to straighten them. If the object be to correct the errors by local polishing operations upon the surface, the rule is that *the bands, or any parts of them, may be rubbed in the direction of the arrow.*

A good many surfaces have thus been operated upon; and although a fair amount of success has been attained, further experiment is required in order to determine the best procedure. There is a tendency to leave the marginal parts behind; so that the bands, though straight over the greater part of their length, remain curved at their extremities. In some cases hydro-fluoric acid has been resorted to, but it appears to be rather difficult to control.

The delicacy of the test is sufficient for every optical purpose. A deviation from straightness amounting to $\frac{1}{10}$ of a band interval could hardly escape the eye, even on simple inspection. This corresponds to a departure from flatness of $\frac{1}{20}$ of a wave-length in water, or about $\frac{1}{30}$ of the wave-length in air. Probably a deviation of $\frac{1}{100}\lambda$ could be made apparent.

For practical purposes a layer of moderate thickness, adjusted so that the two systems of bands corresponding to the duplicity of the soda line do not interfere, is the most suitable. But if we wish to observe bands of high interference, not only must the thickness be increased, but certain pre-cautions become necessary. For instance, the influence of obliquity must be considered. If this element were absolutely constant, it would entail no ill effect. But in consequence of the finite diameter of the pupil of the eye, various obliquities are mixed up together, even if attention be confined to one part of the field. When the thickness of the layer is increased, it becomes necessary to reduce the obliquity to a minimum, and further to diminish the aperture of the eye by the interposition of a suitable slit. The effect of obliquity is shown by the formula

$$[2\mu t \cos \theta' = n\lambda].$$

The necessary parallelism of the operative surfaces may be obtained, as in the above described apparatus, by the aid of levelling. But a much simpler device may be employed, by which the experimental difficulties are greatly reduced. If we superpose a layer of water upon a surface of mercury, the flatness and parallelism of the surfaces take care of themselves. The objection that the two surfaces would reflect very unequally may be obviated by the addition of so much dissolved colouring matter, *e.g.* soluble aniline blue, to the water as shall equalise the intensities of the two reflected lights. If the adjustments are properly made, the whole field, with the exception of a margin near the sides of the containing vessel, may be brought to one degree of brightness, being in fact all included within a fraction of a band. The width of the margin, within which rings appear, is about one inch, in agreement with calculation founded upon the known values of the capillary constants. During the establishment of equilibrium after a disturbance, bands are seen due to variable thickness, and when the layer is thin, they persist for a considerable time.

When the thickness of the layer is increased beyond a certain point, the difficulty above discussed, depending upon obliquity, becomes excessive, and it is advisable to change the manner of observation to that adopted by

Michelson*. In this case the eye is focused, not, as before, upon the operative surfaces, but upon the flame, or rather upon its image at E (Fig. 2). For this purpose it is only necessary to introduce an eye-piece of low power, which with the lens C (in its second operation) may be regarded as a telescope. The bands now seen depend entirely upon obliquity according to the formula above written, and therefore take the form of circular arcs. Since the thickness of the layer is absolutely constant, there is nothing to interfere with the perfection of the bands except want of homogeneity in the light.

But, as Fizeau found many years ago, the latter difficulty soon becomes serious. At a very moderate thickness it becomes necessary to reduce the supply of soda, and even with a very feeble flame a limit is soon reached. When the thickness was pushed as far as possible, the retardation, calculated from the volume of liquid and the diameter of the vessel, was found to be 50,000 wave-lengths, almost exactly the limit fixed by Fizeau.

To carry the experiment further requires still more homogeneous sources of light. It is well known that Michelson has recently observed interference with retardations previously unheard of, and with the aid of an instrument of ingenious construction has obtained most interesting information with respect to the structure of various spectral lines.

A curious observation respecting the action of hydrofluoric acid upon polished glass surfaces was mentioned in conclusion. After the operation of the acid the surfaces appear to be covered with fine scratches, in a manner which at first suggested the idea that the glass had been left in a specially tender condition, and had become scratched during the subsequent wiping. But it soon appeared that the effect was a *development* of scratches previously existent in a latent state. Thus parallel lines ruled with a knife-edge, at first invisible even in a favourable light, became conspicuous after treatment with acid. Perhaps the simplest way of regarding the matter is to consider the case of a furrow with perpendicular sides and a flat bottom. If the acid may be supposed to eat in equally in all directions, the effect will be to *broaden* the furrow, while the depth remains unaltered. It is possible that this method might be employed with advantage to *intensify* (if a photographic term may be permitted) gratings ruled upon glass for the formation of spectra.

* [1902. The influence of the diameter of the pupil of the eye in lessening the visibility of fringes dependent primarily upon variable thickness, seems to have been first pointed out by Lummer (*Wied. Ann.* XXIII. p. 49, 1884), who also emphasised the advantages attending the use of a plate of uniform thickness and of rings dependent solely upon obliquity, whether the object be the investigation of high interference itself, or the examination for uniformity of plates intended to be plane-parallel.

The circular ring system dependent upon obliquity was first observed by Haidinger (*Pogg. Ann.* LXXVII. p. 219, 1849) and explained by Mascart (*Ann. de Chim.* XXIII. p. 116, 1871).]

203.

ON THE THEORY OF STELLAR SCINTILLATION.

[*Philosophical Magazine*, XXXVI. pp. 129—142, 1893.]

ARAGO'S theory of this phenomenon is still perhaps the most familiar, although I believe it may be regarded as abandoned by the best authorities. According to it the momentary disappearance of the light of the star is due to accidental interference between the rays which pass the two halves of the pupil of the eye or the object-glass of the telescope. When the relative retardation amounts to an odd multiple of the half wave-length of any kind of light, such light, it is argued, vanishes from the spectrum of the star. But this theory is based upon a complete misconception. "It is as far as possible from being true that a body emitting homogeneous light would disappear on merely covering half the aperture of vision with a half wave plate. Such a conclusion would be in the face of the principle of energy, which teaches plainly that the retardation in question would leave the aggregate brightness unaltered*." It follows indeed from the principle of interference that there will be darkness at the precise point which before the introduction of the half wave plate formed the centre of the image, but the light missing there is to be found in a slightly displaced position†.

---

* *Enc. Brit.*, "Wave Theory," p. 441. [Vol. III. p. 123.]

† Since the remarks in the text were written I have read the version of Arago's theory given by Mascart (*Traité d'Optique*, t. III. p. 348). From this some of the most objectionable features have been eliminated. But there can be no doubt as to Arago's meaning. " Supposons que les rayons qui tombent à gauche du centre de l'objectif aient rencontré, depuis les limites supérieures de l'atmosphère, des couches qui, à cause de leur densité, de leur température, ou de leur état hygrométrique, étaient douées d'une réfringence différente de celle que possédaient les couches traversées par les rayons de droite; il pourra arriver, qu'à raison de cette différence de réfringence, les rayons rouges de droite détruisent *en totalité* les rayons rouges de gauche, et que le foyer passe du blanc, son état normal, au vert............ Voilà donc le résultat théorique parfaitement d'accord avec les observations; voilà le phénomène de la scintillation dans une lunette rattaché d'une manière intime à la doctrine des interférences" (*l'Annuaire du Bureau des Longitudes pour* 1852, pp. 423, 425).

That the difference between Arago's theory and that followed in the present paper is fundamental will be recognized when it is noticed that, according to the former, the colour effects of scintillation would be nearly independent of atmospheric *dispersion*. Arago gives an interesting summary of the views held by early writers.

The older view that scintillation is due to the actual diversion of light from the aperture of vision by atmospheric irregularities was powerfully supported by Montigny*, to whom we owe also a leading feature of the true theory, that is, the explanation of the chromatic effects by reference to the different paths pursued by rays of different colours in virtue of *regular* atmospheric dispersion.   The path of the violet ray lies higher than that of the red ray which reaches the eye of the observer from the same star, and the separation may be sufficient to allow the one to escape the influence of an atmospheric irregularity which operates upon the other.   In Montigny's view the diversion of the light is caused by total reflexion at strata of varying density.

But the most important work upon this subject is undoubtedly that of Respighi†, who, following in the steps of Montigny and Wolf, applied the spectroscope to the investigation of stellar scintillation.   The results of these observations are summed up under thirteen heads, which it will be convenient to give almost at full length.

(I.)   In spectra of stars near the horizon we may observe dark or bright bands, transversal or perpendicular to the length of the spectrum, which more or less quickly travel from the red to the violet or from the violet to the red, or oscillate from one to the other colour; and this however the spectrum may be directed from the horizontal to the vertical.

(II.)   In normal atmospheric conditions the motion of the bands proceeds regularly from red to violet for stars in the west, and from violet to red for stars in the east; while in the neighbourhood of the meridian the movement is usually oscillatory, or even limited to one part of the spectrum.

(III.)   In observing the horizontal spectra of stars more and more elevated above the horizon, the bands are seen sensibly parallel to one another, but more or less inclined to the axis of the spectrum, passing from red to violet or reversely according as the star is in the west or the east.

(IV.)   The inclination of the bands, or the angle formed by them with the axis (? transversal) of the spectrum, depends upon the height of the star; it reduces to 0° at the horizon and increases rapidly with the altitude so as to reach 90° at an elevation of 30° or 40°, so that at this elevation the bands become longitudinal.

(V.)   The inclination of the bands, reckoned downwards, is towards the more refrangible end of the spectrum.

(VI.)   The bands are most marked and distinct when the altitude of the star is *least*.   At an altitude of more than 40° the longitudinal bands are reduced to mere shaded streaks, and often can only be observed upon the spectrum as slight general variations of brightness.

* *Mém. de l'Acad. d. Bruxelles*, t. xxviii. (1856).

† *Roma, Atti Nuovi Lincei*, xxi. (1868); *Assoc. Française, Compt. Rend.* i. (1872), p. 169.

(VII.) As the altitude increases, the movement of the bands becomes quicker and less regular.

(VIII.) As the prism is turned so as to bring the spectrum from the horizontal to the vertical position, the inclination of the bands to the transversal of the spectrum continually diminishes until it becomes zero when the spectrum is nearly vertical; but the bands then become less marked, retaining, however, the movement in the direction indicated above (III.).

(IX.) Luminous bands are less frequent and less regular than dark bands, and occur well marked only in the spectra of stars near the horizon.

(X.) In the midst of this general and violent movement of bright and dark masses in the spectra of stars, the black spectral lines proper to the light of each star remain sensibly quiescent or undergo very slight oscillations.

(XI.) Under abnormal atmospheric conditions the bands are fainter and less regular in shape and movement.

(XII.) When strong winds prevail the bands are usually rather faint and ill defined, and then the spectrum exhibits mere changes of brightness, even in the case of stars near the horizon.

(XIII.) Good definition and regular movement of the bands seems to be a sign of the probable continuance of fine weather, and, on the other hand, irregularity in these phenomena indicates probable change.

These results show plainly that the changes of intensity and colour in the images of stars are produced by a momentary real diversion of the luminous rays from the object-glass of the telescope; that in the neighbourhood of the horizon rays of different colours are affected separately and successively, and that all the rays of a given colour are momentarily withdrawn from the whole of the object-glass.

Most of his conclusions from observation were readily explained by Respighi as due to irregular refractions, not necessarily or usually amounting (as Montigny supposed) to total reflexions, taking place at a sufficient distance from the observer. The progress of the bands in one direction along the spectrum (II.) is attributed to the diurnal motion. In the case of a setting star, for instance, the blue rays by which it is seen, pursuing a higher course through the atmosphere, encounter an obstacle somewhat later than do the red rays. Hence the band travels towards the violet end of the spectrum. In the neighbourhood of the meridian this cause of a progressive movement ceases to operate.

The observations recorded in (III.) are of special interest as establishing a connexion between the rates with which various parts of the object-glass and of the spectrum are affected. Since the spectrum is horizontal, various parts of its width correspond to various horizontal sections of the objective, and the existence of bands at a definite inclination shows that at the moment when the shadow of the obstacle thrown by blue rays reaches the bottom of the glass the shadow at the top is that thrown by green, yellow, or red rays of less refrangibility.   When the altitude of the star reaches 30° or 40°, the difference of path due to atmospheric dispersion is insufficient to differentiate the various parts of the spectrum.   The bands then appear longitudinal.

The *definite* obliquity of the bands at moderate altitudes, reported by Respighi, leads to a conclusion of some interest, which does not appear to have been noticed.   In the case of a given star, observed at a given altitude, the linear separation at the telescope of the shadows of the same obstacle thrown by rays of various colours will of necessity depend upon the distance of the obstacle.   But the definiteness of the obliquity of the bands requires that this separation shall not vary, and therefore that the obstacles to which the effects are due are sensibly at one distance only.   It would seem to follow from this that, under "normal atmospheric conditions," scintillation depends upon irregularities limited to a comparatively narrow horizontal stratum situated overhead.   A further consequence will be that the distance of the obstacles increases as the altitude of the star diminishes, and this according to a definite law.

The principal object of the present communication is to exhibit some of the consequences of the theory of scintillation in a definite mathematical form.   The investigation may be conducted by simple methods, if, as suffices for most purposes, we regard the whole refraction as small, and neglect the influence of the earth's curvature.   When the object is to calculate with accuracy the refraction itself, further approximations are necessary, but even in this case the required result can be obtained with more ease than is generally supposed.

The foundation upon which it is most convenient to build is the idea of James Thomson*, which establishes instantaneously the connexion between the curvature of a ray travelling in a medium of varying optical constitution and the rate at which the index changes at the point in question.   The following is from Everett's memoir:—

"Draw normal planes to a ray at two consecutive points of its path. Then the distance of their intersection from either point will be ρ, the radius of curvature.   But these normal planes are tangential to the wave-front in its two consecutive positions.   Hence it is easily shown by similar triangles that a very short line dN drawn from either of the points towards the centre of curvature is to the whole length ρ, of which it forms part, as dv the

* *Brit. Assoc. Rep.* 1872.   Everett, *Phil. Mag.* March 1873.

difference of the velocities of light at its two ends is to v the velocity at either end.   That is

$$dN/\rho = -dv/v,$$

the negative sign being used because the velocity diminishes in approaching the centre of curvature.   But, since v varies inversely as μ, we have

$$-dv/v = d\mu/\mu.$$

Hence the curvature $1/\rho$ is given by any of the four following expressions:—

$$\frac{1}{\rho} = -\frac{1}{v}\frac{dv}{dN} = -\frac{d\log v}{dN} = \frac{1}{\mu}\frac{d\mu}{dN} = \frac{d\log\mu}{dN}.\quad\text{..............(1)}$$

"The curvatures of different rays at the same point are directly as the rates of increase of μ in travelling along their respective normals." If θ denote the angle which the ray makes with the direction of most rapid increase of index, the curvatures will be directly as the values of $\sin\theta$.   In fact, if $d\mu/dr$ denote the rate at which μ increases in a direction normal to the surfaces of equal index, we have

$$\frac{d\mu}{dN} = \frac{d\mu}{dr}\sin\theta,$$

and therefore

$$\pm\frac{1}{\rho} = \frac{1}{\mu}\frac{d\mu}{dr}\sin\theta = \frac{d\log\mu}{dr}\sin\theta.\quad\text{....................(2)}$$

Everett shows how the well-known equation

$$\mu p = \text{const}.\text{....................................(3)}$$

can be deduced from (2), p being the perpendicular upon the ray from the centre of *spherical* surfaces of equal index.   In general,

$$\frac{1}{\rho} = \frac{1}{r}\frac{dp}{dr},\qquad \sin\theta = \frac{p}{r},$$

and thus

$$-\frac{1}{r}\frac{dp}{dr} = \frac{p}{r}\frac{d\log\mu}{dr},$$

giving (3) on integration.

At a first application of (2) we may find by means of it a first approximation to the law of atmospheric refraction, on the supposition that the whole refraction is small and that the curvature of the earth may be neglected.   Under these limitations θ in (2) may be treated as constant along the whole path of the ray; and if $d\psi$ be the angle through which the ray turns in describing the element of arc ds, we have

$$d\psi = \frac{d\log\mu}{dr}\sin\theta\,ds = \tan\theta \,.\, d\log\mu.$$

If we integrate this along the whole course of the ray through the atmosphere, that is from $\mu = 1$ to $\mu = \mu_0$, we get, as the whole refraction,

$$\psi = \log \mu_0 \tan \theta = (\mu_0 - 1) \tan \theta, \quad \dots\dots\dots\dots\dots(4)$$

for to the order of approximation in question $\log \mu_0$ may be identified with $(\mu_0 - 1)$.

If $\delta\psi$ denote the chromatic variation of ψ corresponding to $\delta\mu_0$, we have from (4)

$$\delta\psi/\psi = \delta\mu_0/(\mu_0 - 1). \quad \dots\dots\dots\dots\dots(5)$$

According to Mascart* the value of the right-hand member of (5) in the case of air and of the lines B and H is

$$\delta\mu_0/(\mu_0 - 1) = \cdot024. \quad \dots\dots\dots\dots\dots(6)$$

We will now take a step further and calculate the linear deviation of a ray from a straight course, still upon the supposition that the whole refraction is small. If η denote the linear deviation (reckoned perpendicularly) at any point defined by the length s measured along the ray θ, we have

$$\frac{d^2\eta}{ds^2} = \frac{1}{\rho} = \tan \theta \frac{d \log \mu}{ds},$$

so that

$$\frac{d\eta}{ds} = \int \tan \theta \, d \log \mu = \tan \theta \, (\mu - 1) + \alpha,$$

α being a constant of integration. A second integration now gives

$$\eta = \tan \theta \int (\mu - 1) \, ds + \alpha s + \beta, \quad \dots\dots\dots\dots\dots(7)$$

which determines the path of the ray. If y be the height of any point above the surface of the earth, $ds = dy \sec \theta$; so that (7) may also be written

$$\eta = \frac{\sin \theta}{\cos^2 \theta} \int (\mu - 1) \, dy + \alpha s + \beta. \quad \dots\dots\dots\dots\dots(8)$$

The origin of s is arbitrary, but we may conveniently take it at the point (A) where the ray strikes the earth's surface.

We will now consider also a second ray, of another colour, deviating from the line θ by the distance $\eta + \delta\eta$, and corresponding to a change of μ to $\mu + \delta\mu$. The distance between the two rays at any point y is

$$\delta\eta = \frac{\sin \theta}{\cos^2 \theta} \int_0^y \delta\mu \, dy + \delta\alpha \cdot s + \delta\beta. \quad \dots\dots\dots\dots\dots(9)$$

In this equation $\delta\beta$ denotes the separation of the rays at A, where $y = 0$,

* Everett's *C. G. S. System of Units.*

$s = 0$. And $\delta\alpha$ denotes the angle between the rays when outside the atmosphere.

Equation (9) may be applied at once to Montigny's problem, that is to determine the separation of two rays of different colours, both coming from the same star, and both arriving at the same point A. The first condition gives $\delta\alpha = 0$, and the second gives $\delta\beta = 0$. Accordingly,

$$\delta\eta = \frac{\sin\theta}{\cos^2\theta}\int_0^y \delta\mu\, dy \dots\dots\dots\dots\dots\dots(10)$$

is the solution of the question.

The integral in (10) may be otherwise expressed by means of the principle that $(\mu - 1)$ and $\delta\mu$ are proportional to the density. Thus, if l denote the "height of the homogeneous atmosphere," and h the elevation in such an atmosphere determined by the condition that there shall be as much air below it as actually exists below y,

$$\int_0^y \delta\mu\, dy = \delta\mu_0 h, \dots\dots\dots\dots\dots\dots(11)$$

$\delta\mu_0$ being the value of $\delta\mu$ at the surface of the earth. Equation (10) thus becomes

$$\delta\eta = \frac{\delta\mu_0 h \sin\theta}{\cos^2\theta}. \dots\dots\dots\dots\dots(12)$$

At the limits of the atmosphere and beyond, $h = l$, and the separation there is

$$\delta\eta = \frac{\delta\mu_0 l \sin\theta}{\cos^2\theta}. \dots\dots\dots\dots\dots(13)$$

These results are applicable to all altitudes higher than about 10°.

The formulæ given by Montigny (*loc. cit.*) are quite different from the above. That corresponding to (13) is

$$\delta\eta = \delta\mu_0\, a \sin\theta, \dots\dots\dots\dots\dots\dots(14)$$

a being the radius of the earth! The substitution of a for l increases the calculated result some 800 times. But this is in a large measure compensated by the factor $\sec^2\theta$ in (13), for at low altitudes $\sec\theta$ is large. According to Montigny the separation at moderately low altitudes would be nearly independent of the altitude, a conclusion entirely wide of the truth.

The value of $(\mu_0 - 1)$ for air at 0° and 760 millim. at Paris is ·0002927, so that $\delta\mu_0$ (for the lines B and H) is ·000007025. The height of the homogeneous atmosphere is $7·990 \times 10^5$ centim., and thus $\delta\eta$ reckoned in centim. is

$$\delta\eta = 5·612\, \frac{h \sin\theta}{l \cos^2\theta}. \dots\dots\dots\dots\dots(15)$$

The following are a few corresponding values of θ and $\sin\theta/\cos^2\theta$:—

θ	$\sin\theta/\cos^2\theta$	θ	$\sin\theta/\cos^2\theta$
$0°$	0·000	$60°$	3·46
20	0·387	70	8·03
40	1·095	80	32·66

Thus at the limit of the atmosphere the separation of rays which reach the observer at an apparent altitude of 10° is 185 centim. Nearer the horizon the separation would be still greater, but its value cannot well be found from (15). Although these estimates are considerably less than those of Montigny, the separation near the horizon seems to be sufficient to explain the vertical position of the bands in the spectrum, recorded by Respighi (I.). The fact that the margin is not very great suggests that the obstacles to which scintillation is due may often be situated at a considerable elevation.

We have now to consider the effect of an obstacle situated at a given point B at level y on the course of the ray. And the first desideratum will be the estimation of the separation at A, the object-glass of the telescope, of rays of various colours coming from the same star, which all pass through the given point B. It will appear at once that no fresh question is raised. For, since the rays come from the same star at the same time, $\delta\alpha = 0$, and thus by (9) $\delta\eta_A = \delta\beta$. The value of $\delta\beta$ is given at once by the condition that $\delta\eta_B = 0$. Thus

$$- \delta\eta_A = \frac{\sin\theta}{\cos^2\theta} \int_0^y \delta\mu\, dy = \frac{\delta\mu_0 h \sin\theta}{\cos^2\theta}, \quad \text{..............(16)}$$

as before. The discussion, already given of (15), is thus immediately applicable.

Equation (16) solves the problem of determining the inclination of the bands seen in the spectra of stars not very low (III.). It is only necessary to equate $-\delta\eta_A$ to the aperture of the telescope. $\delta\mu_0$ then gives the range of refrangibility covered by the bands as inclined. In practice h would not be known beforehand; but from the observed inclination of the bands it would be possible to determine it.

In a given state of the atmosphere h, so far as it is definite, must be constant and then $\delta\mu_0$ must be proportional to $\cos^2\theta/\sin\theta$. This gives the relation between the altitude of the star and the inclination of the bands.

When θ is small, $\delta\mu_0$ is large; that is, the bands become longitudinal.

As a numerical example, let us suppose that the aperture of the telescope is 10 centim., and that at an altitude of 10° the obliquity of the bands is such that the vertical diameter of the object-glass corresponds to the entire range from B to H.   In this case (15) gives

$$h = \frac{10\,l}{185} = \cdot054\,l,$$

indicating that the obstacles to which the bands are due are situated at such a level that about $\frac{1}{20}$ of the whole mass of the atmosphere is below them.

The next question to which (9) may be applied is to find the angle $\delta\alpha$ outside the atmosphere between two rays of different colours which pass through the *two* points A and B.   Here $\delta\eta_A = 0$, and thus $\delta\beta = 0$.   And further, since $\delta\eta_B = 0$, we get

$$-\,\delta\alpha = \frac{\sin\theta}{s\cos^2\theta} \int_0^y \delta\mu\,dy = \frac{\delta\mu_0 h \tan\theta}{y}. \quad \dots\dots\dots\dots(17)$$

If the height of the obstacle above the ground be so small that the density of the air below it is sensibly uniform, then $h = y$, and

$$-\,\delta\alpha = \delta\mu_0 \tan\theta. \quad \dots\dots\dots\dots\dots\dots\dots\dots(18)$$

In this case the angle is the same as that of the spectrum of the star observed at A, as appears from (4) and (5).   In general, y is greater than h, so that $\delta\alpha$ is somewhat less than the value given by (18).

The interest of (18) lies in the application of it to find the time occupied by a band in traversing the spectrum in virtue of the diurnal motion, according to Respighi's observation (II.).   The time required is that necessary for the star to rise or fall through the angle of its dispersion-spectrum at the altitude in question.   At an altitude of 10°, this angle will be 8″, being always about $\frac{1}{40}$ of the whole refraction.   The rate at which a star rises or falls depends of course upon the declination of the star and upon the latitude of the observer, and may vary from zero to 15° per hour.   At the latter maximum rate the star would describe 8″ in about one half of a second, which would therefore be the time occupied by a band in crossing the spectrum under the circumstances supposed.   In the case of a star quite close to the horizon, the progress of the band would be a good deal slower.

The fact that the larger planets scintillate but little, even under favourable conditions, is readily explained by their sensible apparent magnitude. The separation of rays of one colour thus arising during their passage through the atmosphere is usually far greater than the already calculated separation, due to chromatic dispersion; so that if a fixed star of no apparent magnitude scintillates in colours, the different parts of the area of a planet must *à fortiori* scintillate independently.   But under these circumstances the eye perceives only an average effect, and there is no scintillation visible.

The non-scintillation of small stars situated near the horizon may be referred to the failure of the eye to appreciate colour when the light is faint.

In the case of stars higher up, the whole spectrum is affected simultaneously. A momentary accession of illumination, due to the passage of an atmospheric irregularity, may thus render visible a star which on account of its faintness could not be steadily seen through an undisturbed atmosphere*.

In the preceding discussion the refracting obstacles have for the sake of brevity been spoken of as throwing sharp shadows. This of course cannot happen, if only in consequence of diffraction; and it is of some interest to inquire into the magnitude of the necessary diffusion. The theory of diffraction shows that even in the case of an opaque screen with a definite straight boundary, the transition of illumination at the edge of the shadow occupies a space such as $\sqrt{(b\lambda)}$, where λ is the wave-length of the light, and b is the distance across which the shadow is thrown. We may take λ at 6×10^{-5} centim., and if b be reckoned in kilometres, we have as the space of transition, $\sqrt{(6b)}$. Thus if b were 4 kilometres, the space of transition would amount to about 5 centim. The inference is that the various parts of the aperture of a small telescope cannot be very differently affected unless the obstacles to which the scintillation is due are at a less distance than 4 kilometres.

One of the principal outstanding difficulties in the theory of scintillation is to see how the transition from one index to another in an atmospheric irregularity can be sufficiently sudden. The fact that the various parts of a not too small object-glass are diversely affected seems to prove that the transitions in question do not occupy many centimetres. Now, whether the irregularity be due to temperature or to moisture, we should expect that a transition, however abrupt at first, would after a few minutes or hours be eased off to a greater degree than would accord with the above estimate. Perhaps the abruptness of transition is, as it were, continually renewed by the coming into contact of fresh portions of light and dense air as the ascending and descending streams proceed in their courses. The speculations and experiments of Jevons on the Cirrus form of Cloud† may find some application here. A preliminary question requiring attention is as to the origin of the irregularities which cause scintillation. Is it always at the ground, and mainly under the influence of sunshine? Or may irregular absorption of solar heat in the atmosphere, due to varying proportions of moisture, give rise to transitions of the necessary abruptness? Again, we may ask how many obstacles are to be supposed operative upon the same

* The theory of Arago leads him to a directly opposite conclusion (*loc. cit.* p. 381).
† *Phil. Mag.* xiv. p. 22, 1857. For a mathematical investigation, by the author, see *Math. Soc. Proc.* xiv. April 1883. [Vol. ii. p. 200.]

ray? Is the ultimate effect only a small residue from many causes in the main neutralizing one another? It does not appear that in the present state of meteorological science satisfactory answers can be given to these questions.

A complete investigation of atmospheric refraction can only be made upon the basis of some hypothesis as to the distribution of temperature; but, as has already been hinted, a second approximation to the value of the refraction can be obtained independently of such knowledge and without difficulty. In Laplace's elaborate investigation it is very insufficiently recognized, if indeed it be recognized at all, that the whole difficulty of the problem depends upon the curvature of the earth. If this be neglected, that is if the strata are supposed to be plane, the desired result follows at once from the law of refraction, without the necessity of knowing anything more than the condition of affairs at the surface. For in virtue of the law of refraction,

$$\mu \sin \theta = \text{constant};$$

so that if θ be the apparent zenith distance of a star seen at the earth's surface, and $\delta\theta$ the refraction, we have at once

$$\mu_0 \sin \theta = \sin (\theta + \delta\theta), \dots\dots\dots\dots\dots\dots(19)$$

from which the refraction can be rigorously calculated. If an expansion be desired,

$$\delta\theta = \sin \delta\theta = \tan \theta \, (\mu_0 - \cos \delta\theta)$$

$$= (\mu_0 - 1) \tan \theta \, \{1 + \tfrac{1}{2} (\mu_0 - 1) \tan^2\theta\} \dots\dots\dots\dots(20)$$

is the second approximation.

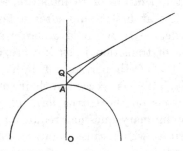

When the curvature of the earth is retained, so that the atmospheric strata are supposed to be spheres described round O the centre of the earth, the appropriate form of the law of refraction is

$$\mu p = \text{constant}.$$

Thus, if A be the point of observation at the earth's surface where the apparent zenith distance is θ, and if the original direction of the ray outside the atmosphere meet the vertical OA at the point Q,

$$\mu_0 . OA . \sin \theta = OQ . \sin (\theta + \delta\theta);$$

or if $OA = a$, $AQ = c$,

$$\mu_0 a \sin \theta = (a + c) \sin (\theta + \delta\theta). \quad\dots\dots\dots\dots\dots(21)$$

If c be neglected altogether, we fall back upon the former equations (19), (20). For the purposes of a second approximation c, though it cannot be neglected, may be calculated as if the refraction were small, and the curvature of the strata negligible. If η be the whole linear deviation of the ray due to the refraction,

$$c = \eta / \sin \theta, \quad\dots\dots\dots\dots\dots\dots(22)$$

and, as in (16),

$$\eta = (\mu_0 - 1) \, l \sin \theta / \cos^2 \theta, \quad\dots\dots\dots\dots\dots(23)$$

so that

$$c = \frac{(\mu_0 - 1) \, l}{\cos^2 \theta}. \quad\dots\dots\dots\dots\dots(24)$$

By equations (21), (24) the value of $\delta\theta$ may be calculated from the trigonometrical tables without further approximation.

To obtain an expansion, we have

$$\delta\theta = \sin \delta\theta = \frac{\mu_0 \tan \theta}{1 + c/a} - \tan \theta \cos \delta\theta$$

$$= \tan \theta \left\{ \frac{\mu_0}{1 + c/a} - 1 + \tfrac{1}{2} (\delta\theta)^2 \right\}$$

$$= (\mu_0 - 1) \tan \theta \left\{ 1 - \frac{\mu_0 c}{(\mu_0 - 1) a} + \tfrac{1}{2} (\mu_0 - 1) \tan^2 \theta \right\}$$

$$= (\mu_0 - 1) \left(1 - \frac{l}{a} \right) \tan \theta - (\mu_0 - 1) \left(\frac{l}{a} - \frac{\mu_0 - 1}{2} \right) \tan^3 \theta. \quad\dots\dots(25)$$

To this order of approximation the refraction can be expressed in terms of the condition of things at the earth's surface, and (25) is equivalent to an expression deduced at great length by Laplace.

From the value of l already quoted, and $a = 6{\cdot}3709 \times 10^8$ centim., we get

$$l/a = {\cdot}0012541. \quad\dots\dots\dots\dots\dots(26)$$

If further we take as the value under standard conditions for the line D

$$\mu_0 - 1 = {\cdot}0002927, \quad\dots\dots\dots\dots(27)$$

we find as the refraction expressed in seconds of arc

$$\delta\theta = 60''{\cdot}29 \tan \theta - 0''{\cdot}06688 \tan^3 \theta. \quad\dots\dots\dots(28)$$

In (28) θ is the apparent zenith distance, and it should be understood that the application of the formula must not be pushed too close to the horizon. If the density of the air at the surface of the earth differ from the standard density (0° ånd 760 millim.) the numbers in (28) must be altered proportionally. It will be observed that the result has been deduced entirely *à priori* on the basis of data obtained in laboratory experiments.

It may be convenient for reference to give a few values calculated from (28) of the refraction, and of the dispersion, reckoned at $\frac{1}{40}$ of the refraction.

Apparent zenith distance	Refraction	Dispersion (B to H)
0	0·0	0·0
20	21·9	·5
40	50·5	1·3
45	1 0·2	1·5
60	1 40·1	2·5
70	2 44·2	4·1
75	3 41·5	5·5
80	5 29·7	8·2
85	9 49·2	14·7

The results of the formula (28) agree with the best tables up to a zenith distance of 75°, at which point the value of the second term is 3″·5. For 85° the number usually given is about 10′ 16″, and for 90° about 36′; but at these low altitudes the refraction is necessarily uncertain on account of irregularities such as those concerned in the production of mirage.

[1902. When the above memoir was written I was unacquainted with the excellent work of K. Exner upon this subject (*Wien. Akad. Sitzber.* 84, p. 1038, 1881; Über die Scintillation. Eine Monographie. Wien, 1891). Exner's general conclusions are very similar to those at which I arrived independently, and his work includes also an interesting historical discussion.]

204.

ASTRONOMICAL PHOTOGRAPHY.

[*Nature*, XLVIII. p. 391, 1893.]

THE announcement (*Nature*, August 10) that it is in contemplation to raise a sum exceeding £2000 for the establishment of a special photographic telescope at the Cambridge Observatory, leads me to ask whether astronomers have duly considered the facilities afforded by modern photography. At the time of my early experience of the art, thirty-five years ago, it would have been thought a great feat to photograph the Fraunhofer lines in the yellow or red regions of the spectrum, although even then the statement so commonly made that chemical activity was limited to the blue and ultra-blue rays was quite unwarranted. With the earlier photographic processes the distinction was necessary between telescopes to be used with the eye or for photography. In the former case the focal length had to be a minimum for the yellow rays, in the latter for the blue rays of the spectrum.

But the situation is entirely changed. There is now no difficulty in preparing plates sensitive to all parts of the spectrum, witness the beautiful photographs of Rowland and Higgs. I have myself used "Orthochromatic" plates in experiments where it was desirable to work with the same rays as most influence the eye. The interference bands of sodium light may be photographed with the utmost facility on plates sensitised in a bath containing cyanin.

The question I wish to ask is whether the time has not come to accommodate the photographic plates to the telescopes rather than the telescopes to the plates. It is possible that plates already in the market may not exactly meet the requirements of the case, but I feel sure that a tithe of the sums lavished upon instruments would put us in possession of plates suitable for object-glasses that have been designed for visual purposes. There would be no difficulty even in studying the requirements of a particular instrument, over or under corrected as the case might be.

A doubt may arise whether plates so adjusted would be as sensitive as those now in use. Probably Captain Abney, or some other authority, could give the required information. For some astronomical purposes a moderate loss of sensitiveness could hardly be of much consequence; for others doubtless it would be a serious matter.

205.

GRINDING AND POLISHING OF GLASS SURFACES.

[British Association, Sept. 14, 1893, from a report in
Nature, XLVIII. p. 526, 1893.]

LORD RAYLEIGH stated that he had been investigating the nature of these processes, and gave a most interesting description of the results. He first pointed out that the process of grinding with emery is not, as is commonly supposed, a scratching process. The normal effect is the production of isolated detached pits—not scratches. The glass gives way under the emery; at the same time the emery gives way under the glass and suffers abrasion. An image seen through glass which has been finely ground (but not yet polished) has perfect definition. And so when the sun is viewed through a cloud the image is sharp as long as there is an image; even when the cloud thickens, the edge appears to be sharp until we lose the image altogether. A glass lens finely ground gives very good definition, but there is great loss of light by irregular reflection. To obviate this, the lens is polished, and examination under the microscope shows that in the process of polishing with pitch and rouge the polishing goes on entirely on the surface or plateau, the bottom of each pit being left untouched until the adjoining surface is entirely worked down to it. It appeared interesting to investigate the amount of glass removed during the process of polishing. This was done both by weighing and interference methods, and the amount removed was found to be surprisingly small. A sufficiently good polish was obtained when a thickness corresponding to $2\frac{1}{2}$ wave-lengths of sodium light was removed, and the polishing was complete when a thickness corresponding to 4 wave-lengths was removed. Lord Rayleigh is of opinion that the process of polishing is not continuous with that of grinding, but that it consists of a removal of molecular layers of the surface of the glass. Grinding is easy and rapid, whereas polishing is tedious and difficult. The action of hydrofluoric acid in dissolving glass was also investigated and was found to be much more regular than it has generally been assumed to be by chemists. It was found to be easy to remove a layer corresponding in thickness to half a wave-length of sodium light; and with due precautions as little as one-tenth of a wave-length. [1902. For a further discussion of this subject see *Nature*, LXIV. p. 385, 1901.]

206.

ON THE REFLECTION OF SOUND OR LIGHT FROM A CORRUGATED SURFACE.

[*British Association Report*, pp. 690, 691, 1893.]

THE angle of incidence is supposed to be zero, and the amplitude of the incident wave to be unity. If then

$$\zeta = c \cos px \quad \dots\dots\dots\dots\dots\dots\dots\dots(1)$$

be the equation of the surface, the problem of reflection is readily solved so long as p in (1) is small relatively to k or $2\pi/\lambda$; that is so long as the wave-length of the corrugation is large in comparison with that of the vibrations. The solution assumes a specially simple form when the second medium is impenetrable, so that the whole energy is thrown back either in the perpendicularly reflected wave or in the lateral spectra. Of this two cases are notable (α) when—in the application to sound—the second medium is gaseous and devoid of inertia, as in the theory of the 'open ends' of organ pipes. The amplitude A_0 of the perpendicularly reflected wave, so far as the fourth power of p/k inclusive, is then given by

$$- A_0 = J_0(2kc) + \frac{p^2}{k^2} . \tfrac{1}{2} kc . J_1(2kc) + \frac{p^4}{k^4} \{ \tfrac{1}{8} kc\, J_1(2kc) - \tfrac{1}{2} k^2 c^2 J_2(2kc) \}, \dots(2)$$

in which there is no limitation upon the value of kc, so that the corrugation may be as deep as we please in relation to λ. If p be very small, the result, viz. $- J_0(2kc)$, is the same as would be obtained by the methods usual in Optics; and it appears that these methods cease to be available when p cannot be neglected.

The second case (β) arises when sound is reflected from a rigid and fixed wall. We find, as far as p^2/k^2,

$$A_0 = J_0(2kc) - \frac{p^2}{2k^2} . kc . J_1(2kc). \quad\dots\dots\dots\dots\dots(3)$$

If p, instead of being relatively small, exceeds k in magnitude, there are no lateral spectra in the reflected vibrations; and if the second medium is impenetrable, the regular reflection is necessarily total. It thus appears that an extremely rough wall reflects sounds of medium pitch as well as if it were mathematically smooth.

The question arises whether, when the second medium is not impenetrable, the regular reflection from a rough wall ($p > k$) is the same as if $c = 0$. Reasons are given for concluding that the answer should be in the negative.

207.

ON A SIMPLE INTERFERENCE ARRANGEMENT.

[*British Association Report*, pp. 703, 704, 1893.]

IF a point, or line, of light be regarded through a telescope, the aperture of which is limited to two narrow parallel slits, interference bands are seen, of which the theory is given in treatises on Optics. The width of the bands is inversely proportional to the distance between the centres of the slits, and the width of the field, upon which the bands are seen, is inversely proportional to the width of the individual slits. If the latter element be given, it will usually be advantageous to approximate the slits until only a small number of bands are included. In this way not only are the bands rendered larger, but illumination may be gained by the then admissible widening of the original source.

Supposing, then, the proportions of the double slit to be given, we may inquire as to the effect of an alteration in scale. A diminution in ratio m will have the effect of magnifying m times the field and the bands (fixed in number) visible upon it. Since the total aperture is diminished m times, it might appear that the illumination would be diminished m^2 times, but the admissible widening of the original source m times reduces the loss, so that it stands at m times, instead of m^2 times.

It remains, and this is more particularly the object of the present note, to point out the effect of the telescope upon the angular magnitude and illumination of the bands. If the magnifying power of the telescope exceed the ratio of aperture of object-glass and pupil, its introduction is prejudicial. And even if the above limit be not exceeded, the use of the telescope is without advantage. The relation between the greatest brightness and the apparent magnitude of the bands is the same whether a telescope be used or not, the loss by reflections and absorptions being neglected. *The function of the telescope is merely to magnify the linear dimensions of the slit system.*

This magnification is sometimes important, especially when it is desirable to operate separately upon the interfering pencils. But when the object is merely to see the bands, the telescope may be abolished without loss. The only difficulty is to construct the very diminutive slit system then required. In the arrangement now exhibited the slits are very fine lines formed by ruling with a knife upon a silver film supported upon glass. This double slit is mounted at one end of a tube and at the other is placed a parallel slit. It then suffices to look through the tube at a candle or gas flame in order to see interference bands in a high degree of perfection.

It is suggested that this simple apparatus could be turned out very cheaply, and that its introduction into the market would tend to popularise acquaintance with interference phenomena.

208.

ON THE FLOW OF VISCOUS LIQUIDS, ESPECIALLY IN TWO DIMENSIONS.

[*Philosophical Magazine*, XXXVI. pp. 354—372, 1893.]

THE problems in fluid motion of which solutions have hitherto been given relate for the most part to two extreme conditions. In the first class the viscosity is supposed to be sensible, but the motion is assumed to be so slow that the terms involving the squares of the velocities may be omitted; in the second class the motion is not limited, but viscosity is supposed to be absent or negligible.

Special problems of the first class have been solved by Stokes and other mathematicians; and general theorems of importance have been established by v. Helmholtz* and by Korteweg†, relating to the laws of steady motion. Thus in the steady motion (M_0) of an incompressible fluid moving with velocities given at the boundary, less energy is dissipated than in the case of any other motion (M) consistent with the same conditions. And if the motion M be in progress, the rate of dissipation will constantly decrease until it reaches the minimum corresponding to M_0. It follows that the motion M_0 is always stable.

It is not necessary for our purpose to repeat the investigation of Korteweg; but it may be well to call attention to the fact that problems in viscous motion *in which the squares of the velocities are neglected*, fall under the general method of Lagrange, at least when this is extended by the introduction of a dissipation function‡. In the present application there is no potential energy to be considered, and everything depends upon the expressions for the kinetic energy T and the dissipation function F. The conditions to be satisfied may be expressed by ascribing given constant

---

* *Collected Works*, I. p. 223.
† *Phil. Mag.* XVI. p. 112, 1883.
‡ *Theory of Sound*, § 81. [See Vol. I. of the present collection, p. 176.]

values to some of the generalized velocities; but it is unnecessary to introduce more than one into the argument, inasmuch as any others may be
eliminated beforehand by means of the given relations. Suppose, then, that
$\dot{\psi}_r$ is given. The other coordinates ψ_1, ψ_2, ... may be so chosen that no
product of their velocities enters into the expressions for T and F, although
products with $\dot{\psi}_r$, such as $\dot{\psi}_1\dot{\psi}_r$, will enter. These coordinates are, in fact,
the *normal* coordinates of the system when $\dot{\psi}_r$ is constrained to vanish.
Thus simplified F becomes

$$F = \tfrac{1}{2}b_1\dot{\psi}_1{}^2 + \ldots + \tfrac{1}{2}b_s\dot{\psi}_s{}^2 + \ldots + b_{rs}\dot{\psi}_s\dot{\psi}_r + \ldots \quad\ldots\ldots\ldots(1)$$

and a similar expression applies to T with a written for b. Lagrange's
equation is now

$$a_s\ddot{\psi}_s + a_{rs}\ddot{\psi}_r + b_s\dot{\psi}_s + b_{rs}\dot{\psi}_r = 0,$$

ψ_s being any one of the coordinates ψ_1, ψ_2, ... In this equation $\ddot{\psi}_r = 0$,
and $\dot{\psi}_r$ has a prescribed value; so that

$$a_s\ddot{\psi}_s + b_s\dot{\psi}_s = - b_{rs}\dot{\psi}_r \quad\ldots\ldots\ldots\ldots\ldots\ldots\ldots(2)$$

is the equation giving $\dot{\psi}_s$. The solution of (2) is well known, and it appears
that $\dot{\psi}_s$ settles gradually down to the value given by

$$b_s\dot{\psi}_s = - b_{rs}\dot{\psi}_r, \quad\ldots\ldots\ldots\ldots\ldots\ldots\ldots(3)$$

since a_s, b_s are intrinsically positive. Further,

$$\frac{dF}{dt} = \Sigma\,\{b_s\dot{\psi}_s\ddot{\psi}_s + b_{rs}(\ddot{\psi}_s\dot{\psi}_r + \dot{\psi}_s\ddot{\psi}_r)\},$$

in which the summation extends to all values of s other than r. In this
$\ddot{\psi}_r = 0$, so that

$$\frac{dF}{dt} = \Sigma\,\ddot{\psi}_s\{b_s\dot{\psi}_s + b_{rs}\dot{\psi}_r\} = - \Sigma a_s\ddot{\psi}_s{}^2, \quad\ldots\ldots\ldots\ldots(4)$$

by (2). The last expression is intrinsically negative, proving that until the
steady motion is reached F continually decreases. Korteweg's theorem is
thus shown to be of general application to systems devoid of potential
energy for which T and F can be expressed as quadratic functions of the
velocities with constant coefficients.

It may be mentioned in passing that a similar theorem holds for systems
devoid of *kinetic* energy, for which, however, F and V (the potential energy)
are sensible, and may be proved in the same way. If such a system be
subjected to given displacements, it settles down into the configuration of
minimum V; and during the progress of the motion V continually decreases.

The theorem of Korteweg places in a clear light the general question of
the slow motion of a viscous liquid under given boundary conditions, and the
only remaining difficulty lies in finding the analytical expressions suitable
for special problems. It is proposed to consider a few simple cases relating
to motion in two dimensions.

Under the above restriction, as is well known, the motion may be expressed by means of Earnshaw's current function (ψ), which satisfies

$$\nabla^4 \psi = 0, \quad\ldots\ldots\ldots\ldots\ldots\ldots\ldots\ldots\ldots\ldots(5)$$

the same equation as governs the transverse displacement of an elastic plate, when in equilibrium*. Of this analogy we shall avail ourselves in the sequel. At a fixed wall ψ retains a constant value, and, further, in consequence of the friction $d\psi/dn$, representing the tangential velocity, is evanescent. The boundary conditions for a fixed wall in the fluid problem are therefore analogous to those of a clamped edge in the statical problem.

The motion within a simply connected area is determined by (5) and by the values of the component velocities over the boundary. If we suppose that two such motions are possible, their difference constitutes a motion also satisfying (5), and making ψ and $d\psi/dn$ zero over the boundary. Considerations respecting energy in this or in the analogous problem of the elastic plate are then sufficient to show that ψ must vanish throughout; and an analytical proof may readily be given by means of Green's theorem. For if ψ and χ are any two functions of x and y,

$$\int \left\{ \chi \frac{d\psi}{dn} - \psi \frac{d\chi}{dn} \right\} ds = \iint \{ \chi \nabla^2 \psi - \psi \nabla^2 \chi \} \, dx\, dy, \ldots\ldots\ldots(6)$$

the integrations being taken round and over the area in question. If we suppose that ψ and $d\psi/dn$ are zero over the boundary, the left-hand member vanishes. If, further, $\chi = \nabla^2 \psi$, we have

$$\iint (\nabla^2 \psi)^2 \, dx\, dy = \iint \psi \nabla^4 \psi \, dx\, dy, \quad\ldots\ldots\ldots\ldots\ldots(7)$$

of which the right-hand member vanishes by (5). Hence $\nabla^2 \psi$ vanishes all over the area, and by a known theorem, as ψ vanishes on the contour, this requires that ψ vanish throughout.

We will now investigate in detail the slow motion of viscous fluid within a circular boundary. In virtue of (5) $\nabla^2 \psi$, which represents the vorticity, satisfies Laplace's equation, and may therefore be expanded in positive and negative integral powers of r, each term such as r^n, or r^{-n}, being accompanied by the factor $\cos(n\theta + \alpha)$. But if, as we shall suppose, the vorticity be finite at the centre of the circle, where $r = 0$, the negative powers are excluded, and we have to consider only such terms as

$$\nabla^2 \psi = \left(\frac{d^2}{dr^2} + \frac{1}{r} \frac{d}{dr} + \frac{1}{r^2} \frac{d^2}{d\theta^2} \right) \psi = r^n \cos(n\theta + \alpha). \quad\ldots\ldots\ldots(8)$$

* [1902. If w be the displacement, parallel to z, at any point of a plane elastic plate in the plane of xy, the differential equation of equilibrium is $\nabla^4 w = 0$, impressed forces being absent.]

The solution of this is readily obtained. If we assume

$$\psi = r^m \cos(n\theta + \alpha), \quad \dots\dots\dots\dots\dots\dots(9)$$

we find $m = n + 2$. To this may be added, as satisfying $\nabla^2\psi = 0$, a term corresponding to $m = n$; so that the type of solution for $n\theta$ is

$$\psi = A_n r^{n+2} \cos(n\theta + \alpha) + B_n r^n \cos(n\theta + \beta). \quad \dots\dots\dots(10)$$

By differentiation,

$$\frac{d\psi}{dn} = (n+2)A_n r^{n+1} \cos(n\theta + \alpha) + nB_n r^{n-1} \cos(n\theta + \beta). \quad \dots\dots(11)$$

The first problem to which we will apply these equations is that of motion within the circle $r = 1$ under the condition that the tangential motion vanishes at every part of the circumference. By (11) $\beta = \alpha$, and

$$(n+2)A_n + nB_n = 0. \quad \dots\dots\dots\dots\dots\dots(12)$$

The normal velocity at the boundary is represented by $d\psi/d\theta$, and we might be tempted, in our search after simplicity, to suppose that this is sensible in the neighbourhood of one point only, for example $\theta = 0$. But in that case the condition of incompressibility would require that the total flow of fluid at the place in question should be zero. If the total quantity of fluid entering the enclosure at $\theta = 0$ is to be finite, provision must be made for its escape elsewhere. This might take the form of a *sink* at the centre of the circle; but it will come to much the same thing, and be more in harmony with our equations, as already laid down, to suppose that the escape takes place uniformly over the entire circumference. This state of things will be represented analytically by ascribing to ψ a sudden change of value from -1 to $+1$ at $\theta = 0$, with a gradual passage from the one value to the other as θ increases from 0 to 2π, or, as it may be more conveniently expressed for our present purpose, ψ is to be regarded as an *odd* function of θ such that from $\theta = 0$ to $\theta = \pi$ its value is

$$\theta = 1 - \theta/\pi . \quad \dots\dots\dots\dots\dots\dots(13)$$

The symmetry with respect to $\theta = 0$ shows that we are concerned in (10) only with the sines of multiples of θ, so that having regard to (12) we may take as the form of ψ applicable in the present problem,

$$\psi = \Sigma C_n \sin n\theta \{(n+2)r^n - nr^{n+2}\}, \quad \dots\dots\dots\dots(14)$$

in which n is any integer and C_n an arbitrary constant. It remains to determine the coefficients C in accordance with (13). When $r = 1$,

$$\psi = 2\Sigma C_n \sin n\theta = 1 - \theta/\pi;$$

and this must hold good for all values of θ from 0 to π. Multiplying by $\sin m\theta$ and integrating as usual, we find

$$C_n = \frac{1}{n\pi}; \quad \dots\dots\dots\dots\dots\dots(15)$$

so that

$$\pi \cdot \psi = \Sigma \sin n\theta \{(1 - 2/n)\, r^n - r^{n+2}\} \dots\dots\dots\dots(16)$$

is the value of ψ expressed in series.

These series may be summed. In the first place, $\Sigma r^n \sin n\theta$ is the real part of $-i\Sigma (re^{i\theta})^n$, or of

$$\frac{-ire^{i\theta}}{1 - re^{i\theta}}.$$

Thus
$$\Sigma r^n \sin n\theta = \frac{r \sin \theta}{1 - 2r \cos \theta + r^2}. \dots\dots\dots\dots(17)$$

Again, $\Sigma n^{-1} r^n \sin n\theta$ is the real part of $-i\Sigma n^{-1}(re^{i\theta})^n$, or of $i \log (1 - re^{i\theta})$; so that

$$\Sigma n^{-1} r^n \sin n\theta = \tan^{-1} \frac{r \sin \theta}{1 - r \cos \theta}. \dots\dots\dots\dots(18)$$

Thus, as the expression for ψ in finite terms, we have

$$\pi \cdot \psi = \frac{(1 - r^2)\, r \sin \theta}{1 - 2r \cos \theta + r^2} + 2 \tan^{-1} \frac{r \sin \theta}{1 - r \cos \theta}. \dots\dots\dots(19)$$

In (19) the separate parts admit of simple geometrical interpretation. The second represents simply twice the angle PAO, Fig. 1, which is known to constitute a solution of $\nabla^2 \psi = 0$. In the first term,

$$\frac{r \sin \theta}{1 - 2r \cos \theta + r^2} = \frac{PM}{AP^2} = \frac{\sin PAO}{AP},$$

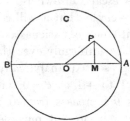

Fig. 1.

which is also obviously a solution of $\nabla^2 \psi = 0$. The remaining part of (19) is not a solution of $\nabla^2 \psi = 0$; but it satisfies $\nabla^4 \psi = 0$, as being derived from a solution of $\nabla^2 \psi = 0$ by multiplication with r^2.

On the foundation of (19) we may build up by simple integration the general expression for ψ, subject to the conditions that $d\psi/dr$ vanishes over the whole circumference, and that $d\psi/d\theta$ has any prescribed values consistent with the recurrence of ψ.

A simple example is afforded by the case of a source at A and an equal sink at B, where $\theta = \pi$ (Fig. 2). The fluid enters and leaves the enclosure by two perforations situated at opposite ends of a diameter, the walls being elsewhere impenetrable. The solution may be found independently, or from (19), by changing the sign of $\cos \theta$, and adding the equations together. Thus

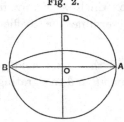

Fig. 2.

$$\tfrac{1}{2}\pi \cdot \psi = \frac{(1 - r^4)\, r \sin \theta}{1 - 2r^2 \cos 2\theta + r^4} + \tan^{-1} \frac{2r \sin \theta}{1 - r^2}. \quad \dots(20)$$

In this case the walls of the enclosure are of necessity stream-lines, the value of ψ being $+1$ from 0 to π, and -1 from 0 to $-\pi$.

When $\theta = \frac{1}{2}\pi$, that is along OD (Fig. 2),

$$\tfrac{1}{2}\pi \cdot \psi = \frac{r - r^3}{1 + r^2} + 2\tan^{-1} r, \quad\quad\quad\quad\quad\quad (21)$$

$$\tfrac{1}{2}\pi \frac{d\psi}{dr} = \frac{(3 + r^2)(1 - r^2)}{(1 + r^2)^2}. \quad\quad\quad\quad\quad\quad (22)$$

From (21) we obtain by interpolation the following corresponding values:—

ψ	·00	·25	·50	·75	1·00
r	·00	·1330	·2800	·4698	1·0000

In the neighbourhood of A or B, Fig. 1, (20) assumes a special form. Thus in the former case,

$$1 - 2r^2 \cos 2\theta + r^4 = (1 - r^2)^2 + 4r^2 \sin^2\theta = 4\left\{AM^2 + PM^2\right\},$$

$$(1 - r^4)\, r \sin\theta = 4\, PM \cdot AM,$$

$$\tan^{-1} \frac{2r \sin\theta}{1 - r^2} = \text{angle } PAO.$$

Thus if PAO be denoted by ϕ, the value of ψ in the neighbourhood of A is given by

$$\pi \cdot \psi = \sin 2\phi + 2\phi. \quad\quad\quad\quad\quad\quad (22')$$

That the functions of ϕ which occur in (22') satisfy the fundamental equation may be readily seen.

By calculation from (22') we get the following values for ϕ expressed as fractions of degrees:—

ψ	0	·25	·50	·75	1·00
ϕ	0	11°·40	23°·83	39°·40	90°·00

This example is of interest, from its bearing upon the laws of flow at a place where a channel is enlarged. In actual fluids there would be a tendency to shoot directly across from A to B, the region about U being occupied by an eddy, or backwater, such that the motion of the fluid near the wall is reversed. Nothing of the kind is indicated by the present solution. In (22) $d\psi/dr$ represents the velocity across the line $\theta = \frac{1}{2}\pi$, and we see that there is no change of sign. In fact the velocity decreases, as r increases, all the way from $r = 0$ to $r = 1$. The formation of a backwater may thus be connected with the terms involving the squares of the velocities, which are neglected in the present solution. And we may infer that if the motion were slow *enough*, or if the fluid were viscous enough, the backwater, usually observed in practice, would disappear.

Another particular case of some interest, included in the general solution already indicated, would be obtained by supposing similar *sources* to be situated at $\theta = 0$, $\theta = \pi$, and equal *sinks* at $\theta = \frac{1}{2}\pi$, $\theta = \frac{3}{4}\pi$.

We will now suppose that it is the *radial* velocity which vanishes at every point of the circumference $r = 1$, and that the tangential velocity also vanishes except in the neighbourhood of $\theta = 0$. In this case, by the symmetry, ψ in (10) reduces to a series of cosines. And

$$- d\psi/d\theta = \Sigma n \sin n\theta \, (A_n r^{n+2} + B_n r^n),$$

which is to vanish when $r = 1$ for all values of θ. Hence

$$A_n + B_n = 0; \qquad \dots\dots\dots\dots\dots\dots(23)$$

so that

$$\psi = (1 - r^2) \Sigma B_n r^n \cos n\theta, \qquad \dots\dots\dots\dots(24)$$

$$d\psi/dr = \Sigma B_n \cos n\theta \, \{n r^{n-1} - (n+2) r^{n+1}\}. \qquad \dots\dots\dots(25)$$

When $r = 1$,

$$d\psi/dr = -2\Sigma B_n \cos n\theta, \qquad \dots\dots\dots\dots(26)$$

and is to be made to vanish for all values of θ except in the neighbourhood of $\theta = 0$. If we suppose that the integral of $d\psi/dr$ with respect to θ over the whole region where $d\psi/dr$ is sensible, is 2, we find

$$B_0 = -1/2\pi, \qquad B_n = -1/\pi, \qquad \dots\dots\dots\dots(27)$$

the second equation applying to all values of n other than 0. Hence,

$$- \pi \cdot \psi = -\tfrac{1}{2}(1 - r^2) + (1 - r^2) \Sigma_0^\infty r^n \cos n\theta, \qquad \dots\dots\dots(28)$$

or in finite terms,

$$- \pi \cdot \psi = -\tfrac{1}{2}(1 - r^2) + (1 - r^2) \frac{1 - r \cos \theta}{1 - 2r \cos \theta + r^2} \cdot \qquad \dots\dots(29)$$

The equation may also be written

$$- 2\pi \cdot \psi = \frac{(1 - r^2)^2}{1 - 2r \cos \theta + r^2} \cdot \qquad \dots\dots\dots(30)$$

In (29),

$$\frac{1 - r \cos \theta}{1 - 2r \cos \theta + r^2} = \frac{AM}{AP^2} = \frac{\cos PAO}{AP},$$

which is a solution of $\nabla^2 \psi = 0$. When multiplied by r^2, or by $(1 - r^2)$, it remains a solution of $\nabla^4 \psi = 0$.

In (30) we may write x for $r \cos \theta$, and if the point under consideration lie upon the axis, $x^2 = r^2$. Hence on the axis,

$$- 2\pi \cdot \psi = (1 + x)^2, \qquad \dots\dots\dots\dots(31)$$

$$- \pi d\psi/dx = (1 + x), \qquad \dots\dots\dots\dots(32)$$

equations which may be applied at all points except near $x = 1$. It appears from (32) that the velocity transverse to the axis increases continuously from $x = -1$ to the neighbourhood of $x = +1$.

The lines of flow are readily constructed from (30), which we may write in the form

$$AP = \frac{1 - OP^2}{\sqrt{(-2\pi\psi)}}, \quad \dots\dots\dots\dots\dots\dots(33)$$

showing how P may be determined by the intersection of circles struck from O and A. A few of the lines of flow are shown in Fig. 3. The external circle AB corresponds to $\psi = 0$; AC, AO, AD correspond respectively to $-2\pi\psi = \frac{1}{4}$, 1, 2. As appears from (31), the highest value of $-2\pi\psi$ is 4, and gives a curve at A of infinitely small area.

Fig. 3.

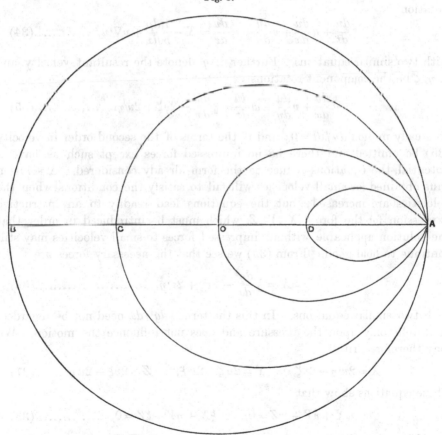

In the neighbourhood of A (Fig. 1), (30) reduces to a simpler form. Thus

$$-\pi \cdot \psi = \frac{(1 - OP^2)^2}{2AP^2} = 1 + \cos 2\phi, \quad \dots\dots\dots\dots(33')$$

where $\phi = PAO$. The second term here satisfies the fundamental equation as being derived by multiplication with AP^2 from a solution, $AP^{-2}\cos 2\phi$, of $\nabla^2\psi = 0$.

Equations (19), (30) give the means of expressing the stream-function subject to the conditions that both ψ and $d\psi/dr$ shall have values arbitrarily given at all points of the circumference of the circle. It is not necessary actually to write down the formulæ; but it may be well to notice that the same solution applies to the question of determining the transverse displacement w of a thin circular plate when w and dw/dr have arbitrarily prescribed values on the boundary.

As a preliminary to further questions, it will be desirable to consider for a moment the form of the general equations of viscous motion. In the usual notation,

$$\frac{du}{dt} + u\frac{du}{dx} + v\frac{du}{dy} + w\frac{du}{dz} = X - \frac{1}{\rho}\frac{dp}{dx} + \nu\nabla^2 u, \quad \ldots\ldots\ldots\ldots(34)$$

with two similar equations. Further, if q^2 denote the resultant velocity, and ξ, η, ζ be the component rotations,

$$u\frac{du}{dx} + v\frac{du}{dy} + w\frac{du}{dz} = \tfrac{1}{2}\frac{dq^2}{dx} - 2v\zeta + 2w\eta. \quad \ldots\ldots\ldots\ldots(35)$$

In steady motion $du/dt = 0$; and if the terms of the second order in velocity (35) be omitted and there be no impressed forces except such as have a potential, the equations reduce to the form already considered. A solution thus obtained for small velocities will fail to satisfy the conditions when the velocities are increased; but the equations lead readily to an instructive expression for the forces X, Y, Z, which must be introduced in order that the solution applicable without impressed forces to small velocities may still continue to hold good. From (35) we see that the necessary forces are

$$X = \tfrac{1}{2}\frac{dq^2}{dx} - 2v\zeta + 2w\eta, \quad \ldots\ldots\ldots\ldots\ldots\ldots(36)$$

with two similar equations. In this the term $\tfrac{1}{2}dq^2/dx$ need not be regarded, as it tells only upon the pressure and does not influence the motion. We may therefore write

$$X = 2w\eta - 2v\zeta, \qquad Y = 2u\zeta - 2w\xi, \qquad Z = 2v\xi - 2u\eta. \ldots\ldots(37)$$

These equations show that

$$uX + vY + wZ = 0, \qquad \xi X + \eta Y + \zeta Z = 0, \quad \ldots\ldots\ldots\ldots(38)$$

signifying that the force whose components are X, Y, Z, acts at every point in a direction perpendicular both to the velocity and to the axis of rotation. As regards its magnitude,

$$\tfrac{1}{4}(X^2 + Y^2 + Z^2) = (u^2 + v^2 + w^2)(\xi^2 + \eta^2 + \zeta^2) - (u\xi - v\eta - w\zeta)^2. \ldots(39)$$

If the motion take place in two dimensions, $w = 0$, $\xi = \eta = 0$, and

$$\tfrac{1}{4}(X^2 + Y^2) = (u^2 + v^2)\zeta^2. \quad \ldots\ldots\ldots\ldots\ldots\ldots(40)$$

In the case of symmetry round an axis,

$$u\xi + v\eta + w\zeta = 0,$$

and (39) reduces to

$$\tfrac{1}{4}(X^2 + Y^2 + Z^2) = (u^2 + v^2 + w^2)(\xi^2 + \eta^2 + \zeta^2). \quad \ldots\ldots\ldots(41)$$

These expressions for the forces necessary to the maintenance of a motion similar to the infinitely small motion give us in simple cases an idea of the direction in which the law is first departed from as the motion increases.

There are very few cases in which the problem of the rapid motion of a viscous fluid has been dealt with. When the motion is in one dimension, the troublesome terms do not present themselves, and the same solution holds good mathematically for the steady motion at all velocities. When the motion is so small that the laws appropriate to infinitely small motion hold good as a first approximation, a correction may be calculated. This has been effected by Whitehead*, and in an unpublished paper by Rowland, for the problem, first investigated by Stokes, of a sphere moving with velocity V through viscous liquid. For infinitely small motion the velocity of the fluid in the neighbourhood of the sphere is of order V. It follows from the solution referred to, or may be proved independently by considerations of dimensions, that in the second approximation involving V^2, the terms are of the order $V^2 a/\nu$, a being the radius of the sphere, and ν, equal to μ/ρ, the kinematic coefficient of viscosity. This method of approximation is thus only legitimate when Va/ν is small, a condition of a very restricting character. In the case of water $\nu = \cdot01$ c.g.s., and if $Va/\nu = \cdot1$, it is required that $Va = \cdot001$.

Thus even if a were as small as one millimetre ($\cdot1$), V should not exceed $\cdot01$ centimetre per second. With such diameters and velocities as often occur in practice, Va/ν would be a large, instead of a small, quantity; and a solution founded upon the type of infinitely slow motion is wholly inapplicable.

We will now recur to the suppositions that the motion is steady, is in two dimensions, and that its square may be neglected. Thus, writing as usual

$$u = d\psi/dy, \qquad v = -\,d\psi/dx,$$

we get from (34)

$$\nu\nabla^4\psi = \frac{dY}{dx} - \frac{dX}{dy}. \quad \ldots\ldots\ldots\ldots\ldots\ldots\ldots(42)$$

Forces derivable from a potential do not disturb the equation $\nabla^4\psi = 0$. In the analogy with a thin elastic plate, already referred to, a place where $dY/dx - dX/dy$ assumes a finite value in the fluid problem corresponds to a place where transverse force acts upon the plate.

---

* *Quart. Journ. of Math.* Vol. XXIII. p. 153 (1889).

The simplest example of the finiteness of the second member of (42) occurs when it is sensible at one point only. This is the case of forces derivable from a potential θ, where θ denotes the angle measured round the point in question. It is to be observed that in the fluid problem the forces themselves are not limited to the one point, but they have no "circulation" except round that point. In the elastic problem, on the other hand, the transverse force is limited to the one point.

The circumstance last mentioned renders the elastic problem the easier of the two to deal with in thought and expression, and we will accordingly avail ourselves of the analogy in the investigation which follows. It is proposed to examine the infinitely slow motion of fluid within an enclosure, which is maintained by forces having circulation at one point only, with the view of determining whether a contrary flow, or backwater, is possible. In the analogous elastic problem we have to consider a plate, subject at the boundary to the conditions that w (the transverse displacement) and dw/dn shall everywhere vanish, and disturbed from its original plane condition by a force acting transversely at a single point P. For distinctness we may suppose that the plane is horizontal and that the force at P acts downwards, in which direction the displacements are reckoned positive. At the point P itself the principle of energy shows that the displacement is positive, and it might appear probable that the displacement would be also positive at all other points of the plate. A similar conclusion is readily proved to be true in the case of a stretched membrane of any shape subjected to transverse force at any point, and also in one dimension for a bar resisting flexure by its stiffness. But a consideration of particular cases suffices to show that the theorem cannot be generally true in the present case.

For suppose that the plate (Fig. 4) is almost divided into two independent parts by a straight partition CD extending across, but perforated by an aperture AB; and that the force is applied at a distance from CD on the left. If the partition were complete, w and dw/dn would be zero over the whole, and the displacement in the neighbourhood on the left would be simple one-dimensional bending, with w positive throughout. On the right w would vanish throughout. In order to maintain this condition of things a certain couple acts upon the plate in virtue of the supposed constraints along CD.

Fig. 4.

C

A

B

D

Along the perforated portion AB the couple required to produce the one-dimensional bending fails. The actual deformation accordingly differs from the one-dimensional bending by the deformation that would be produced by a couple over AB acting upon the plate as clamped along CA, BD, but otherwise free from force. This deformation is evidently symmetrical with change of sign upon the two sides of CD, w being positive on the left, *negative* on the right, and vanishing on AB itself. Thus upon the whole a downward force acting on the left gives rise to an upward motion on the right, in opposition to the general law proposed for examination.

In the application to the hydrodynamical problem we see that the fluid moving on the left from D to B passes on in a straight course to A, and thence along AC, and that on the right an eddy, or backwater, is formed. At distances from the aperture large in comparison with AB the supplementary motion is of the character expressed in (33′).

Fig. 5.

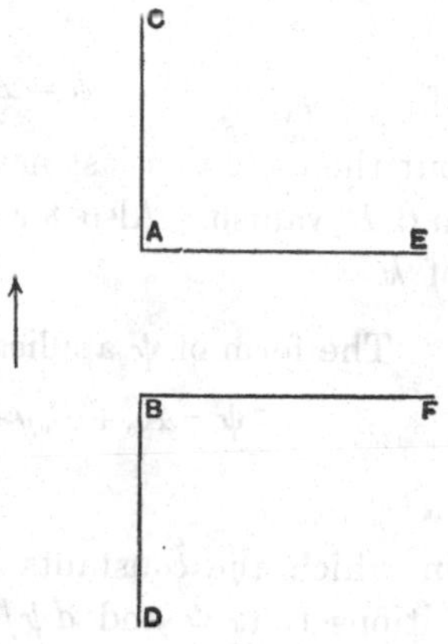

A similar argument may be applied to the case (Fig. 5) where fluid moves along a wall DC into which a channel AF opens, and it leads to the conclusion that the fluid on arrival at B will refuse to follow the wall BF, but will rather shoot across towards A.

These examples are of some interest as establishing that the formation of eddies observed in practice is not wholly due to the influence of the terms involving the squares of the velocities, but would persist in certain cases even though the motion were made infinitely slow.

We will now investigate the motion in two dimensions of a viscous

Fig. 6.

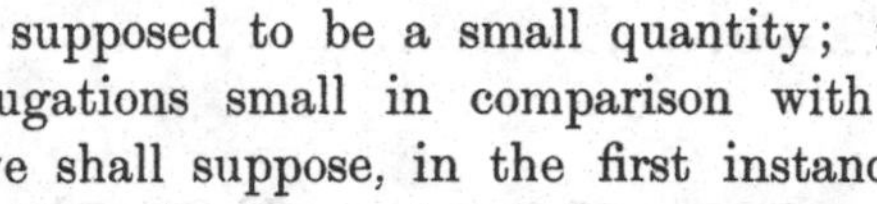

incompressible fluid past a corrugated wall AB (Fig. 6), whose equation may be taken to be

$$y = \beta \cos kx. \quad \dots\dots\dots\dots\dots\dots\dots\dots\dots(43)$$

In this $k\beta$ will be supposed to be a small quantity; in other words, the depth of the corrugations small in comparison with their wave-length $(2\pi/k)$. Further we shall suppose, in the first instance, that the motion is slow enough to allow the terms involving squares of the velocities to be neglected; in which case the equation for the stream-function may be written

$$\nabla^4\psi = 0. \quad \dots\dots\dots\dots\dots\dots\dots\dots\dots(44)$$

At a distance from the wall we suppose the motion to take place in plane strata, as defined by

$$\psi = Ly^2. \quad \dots\dots\dots\dots\dots\dots\dots\dots\dots(45)$$

In the absence of corrugations this value of ψ might hold good throughout, up to the wall at $y=0$. The effect of the corrugations will be to introduce terms periodic with respect to x; but the influence of these will be confined to the neighbourhood of the wall. For any term in ψ, proportional to $\cos mx$, (44) gives

$$\left(\frac{d^2}{dy^2} - m^2\right)^2 \psi = 0, \dots\dots\dots\dots(46)$$

or $$\psi = Ae^{-my} + Bye^{-my} + Ce^{my} + Dye^{my};$$

but the condition last named requires that of the four arbitrary constants C and D vanish. Also for our present purpose m is limited to be a multiple of k.

The form of ψ applicable to our present purpose is accordingly

$$\psi = A_0 + B_0 y + Ly^2 + \cos kx\,(A_1 e^{-ky} + B_1 y e^{-ky})$$
$$+ \cos 2kx\,(A_2 e^{-2ky} + B_2 y e^{-2ky}) + \dots, \dots\dots(47)$$

in which the constants A_0, B_0, A_1, ... are to be determined from the conditions that ψ and $d\psi/dy$ vanish when $y = \beta \cos kx$. It may be observed that the problem is mathematically identical with that of an elastic plate clamped at a sinuous edge, and deformed in such a manner that if there were no sinuosity the bending would be one-dimensional.

The boundary conditions are

$$A_0 + B_0\beta \cos kx + L\beta^2 \cos^2 kx$$
$$+ \cos kx\,(A_1 + B_1\beta \cos kx)\,e^{-k\beta \cos kx}$$
$$+ \cos 2kx\,(A_2 + B_2\beta \cos kx)\,e^{-2k\beta \cos kx}$$
$$+ \dots\dots = 0 \dots\dots\dots\dots(48)$$

and

$$B_0 + 2L\beta \cos kx$$
$$+ \cos kx\,(B_1 - kA_1 - B_1 k\beta \cos kx)\,e^{-k\beta \cos kx}$$
$$+ \cos 2kx\,(B_2 - 2kA_2 - 2B_2 k\beta \cos kx)\,e^{-2k\beta \cos kx}$$
$$+ \dots\dots = 0; \dots\dots\dots\dots(49)$$

or, with use of (48),

$$kA_0 + B_0 + (B_0 k\beta + 2L\beta)\cos kx + Lk\beta^2 \cos^2 kx$$
$$+ B_1 \cos kx\,e^{-k\beta \cos kx}$$
$$+ (B_2 - kA_2 - B_2 k\beta \cos kx)\,e^{-2k\beta \cos kx}$$
$$+ \dots\dots = 0. \dots\dots\dots\dots(50)$$

The exponentials in (48), (50) could be expanded in Fourier's series by means of Bessel's functions of an imaginary argument, and the complete equations formed which express the evanescence of the various Fourier terms. But the results are too complicated to be useful in the general case; and, if we regard $k\beta$ as small, it is hardly worth while to introduce the Bessel's functions at all. The first approximation, in which β^2 is neglected in (48), (50), gives

$$\left.\begin{aligned}A_0 &= 0, & A_1 &= 0, & A_3 &= 0, \ldots\\ B_0 &= 0, & B_1 &= -2L\beta, & B_3 &= 0, \ldots\end{aligned}\right\}; \quad \ldots\ldots\ldots\ldots(51)$$

and the second approximation, in which β^2 is retained, gives

$$\left.\begin{aligned}A_0 &= \tfrac{1}{2}\beta^2 L, & A_1 &= 0, & A_2 &= \tfrac{1}{2}\beta^2 L, \ldots\\ B_0 &= -2k\beta^2 L, & B_1 &= -2\beta L, & B_2 &= -k\beta^2 L, \ldots\end{aligned}\right\}, \quad \ldots\ldots(52)$$

the coefficients with higher suffixes than 2 vanishing to this order of approximation. Thus

$$\psi/L = \beta^2(\tfrac{1}{2} - 2ky) + y^2 - 2\beta y e^{-ky}\cos kx$$
$$+ \beta^2(\tfrac{1}{4} - ky)e^{-2ky}\cos 2kx, \ldots\ldots\ldots(53)$$

$$\frac{1}{L}\frac{d\psi}{dy} = -2k\beta^2 + 2y - 2\beta(1 - ky)e^{-ky}\cos kx$$
$$- 2k\beta^2(1 - ky)e^{-2ky}\cos 2kx, \ldots\ldots\ldots\ldots\ldots\ldots(54)$$

solutions applicable also to the problem of the elastic plate, if ψ be understood to mean the transverse displacement.

In the above investigation, so far as it applies to the hydrodynamical question, L^2 has been supposed to be negligible. We will now retain the square of L, but simplify the problem in another direction by neglecting the square of β, so that the first approximation is

$$\psi = Ly^2 - 2L\beta y e^{-ky}\cos kx. \quad \ldots\ldots\ldots\ldots\ldots(55)$$

The exact equation (derivable from (34)) for the motion of a viscous fluid in two dimensions is

$$\nabla^4\psi = \frac{u}{\nu}\frac{d\nabla^2\psi}{dx} + \frac{v}{\nu}\frac{d\nabla^2\psi}{dy}. \quad \ldots\ldots\ldots\ldots\ldots(56)$$

From (55),

$$\nabla^2\psi = 2L + 4Lk\beta e^{-ky}\cos kx,$$

$$u\frac{d\nabla^2\psi}{dx} + v\frac{d\nabla^2\psi}{dy} = -8L^2k^2\beta y e^{-ky}\sin kx. \quad \ldots\ldots\ldots\ldots(57)$$

Using this in (56) we have

$$\nabla^4\psi = -\frac{8k^2\beta L^2}{\nu}y e^{-ky}\sin kx. \quad \ldots\ldots\ldots\ldots\ldots(58)$$

The solution of

$$\left(\frac{d^2}{dy^2} - k^2\right)^2\psi = y e^{-ky} \quad \ldots\ldots\ldots\ldots\ldots\ldots(59)$$

is

$$\psi = \frac{y^2 e^{-ky}}{8k^3} + \frac{y^3 e^{-ky}}{24k^2}; \quad \ldots\ldots\ldots\ldots\ldots(60)$$

so that the required solution of (58), correct as far as the term involving L^2, is

$$\psi = Ly^2 - 2L\beta y e^{-ky} \cos kx - \frac{\beta L^2}{k\nu}(y^2 + \tfrac{1}{3}ky^3)\, e^{-ky} \sin kx. \quad \ldots(61)$$

It may be well to repeat that, though L^2 is retained, β^2 is neglected in (61); that is, the depth of the corrugations is supposed to be infinitely small.

The part of the motion proportional to L^2 is, of course, independent of the direction of the principal motion of the fluid, and is thus in a manner applicable even when the principal motion is alternating. With regard to the relative importance of the third and second terms in (61), we have to consider the value of

$$\frac{L}{2k\nu}(y + \tfrac{1}{3}ky^2),$$

and the conclusion will depend upon the value of y. If we suppose that $ky = 1$, the ratio is $2L : 3k^2\nu$; or, if we denote by V the undisturbed velocity of the fluid when $ky = 1$, $V/3k\nu$, or $V\lambda/6\pi\nu$, λ being the wave-length of the corrugation. With ordinary liquids and moderate values of λ, V would have to be very small in order to permit the success of the method of approximation.

The character of the motion proportional to L^2 is easily seen from the value of v. We have

$$v = -\frac{d\psi}{dx} = \frac{\beta L^2}{\nu}(y^2 + \tfrac{1}{3}ky^3)\, e^{-ky} \cos kx, \quad \ldots\ldots\ldots\ldots(62)$$

indicating a motion directed outwards from the wall over the places where the sinuosities encroach upon the fluid, and an inward motion where the sinuosities recede.

The application of the results towards the explanation of such phenomena as ripple-mark and wave-formation requires a calculation of the forces operative upon the boundary. We will confine ourselves to the first term in β and L, as expressed in (55).

The normal stress, parallel to y, is given by

$$Q = -p + 2\mu\frac{dv}{dy} = -p - 2\mu\frac{d^2\psi}{dx\,dy}; \quad \ldots\ldots\ldots\ldots(63)$$

and the tangential stress, parallel to x, is

$$U = \mu\left(\frac{dv}{dx} + \frac{du}{dy}\right) = \mu\left(\frac{d^2\psi}{dy^2} - \frac{d^2\psi}{dx^2}\right) \quad \ldots\ldots\ldots\ldots(64)$$

From (34), (55) we find

$$p = - 4\mu k\beta\, e^{-ky} \sin kx,$$

or when $y = 0$,

$$p = - 4k\beta \sin x, \quad \text{simply.}$$

Also, when $y = 0$,

$$2\mu\, \frac{d^2\psi}{dx\,dy} = - 4k\beta \sin kx;$$

so that
$$Q = 0. \dots\dots\dots\dots\dots\dots\dots\dots\dots\dots(65)$$

In like manner, when $y = 0$,

$$U = 2\mu \left\{1 + 2k\beta \cos kx\right\}. \dots\dots\dots\dots\dots(66)$$

So far as the first power of β the action upon the boundary is thus purely tangential, and of magnitude given by (66). The periodic part has the same sign as the constant part at the places where the boundary encroaches upon the fluid.

This result finds immediate application to the question of wave-formation under the action of wind, especially if we suppose that the waves move very slowly, as they would do if gravity (and cohesion) were small. The maintenance or augmentation of the waves requires that the forces operative at the surface be of suitable phase. Thus pressures acting upon the *retreating* shoulders are favourable, as are also tangential forces acting forwards at the *crests* of the waves, where the internal motion is itself in the forward direction. Equation (65) shows that the pressures produce no effect, and that we have only to consider the action of the tangential stress. We see from (66) that when the waves move in the same direction as the wind, the effect of the latter is to favour the development of the former. Whether the waves will actually increase depends upon whether the supply of energy, proportional to β^2, is greater or less than the loss from internal dissipation, itself proportional to the same quantity. If the waves are moving against the wind, the tendency is to a more rapid subsidence than would occur in a calm.

209.

THE SCIENTIFIC WORK OF TYNDALL.

[*Proceedings of the Royal Institution*, xiv. pp. 216—224, 1894.]

IT is fitting that the present season should not pass without a reference on these evenings to the work of him whose tragic death a few months since was felt as a personal grief and loss by every member of the Royal Institution. With much diffidence I have undertaken the task to-night, wishing that it had fallen to one better qualified by long and intimate acquaintance to do justice to the theme. For Tyndall was a personality of exceeding interest. He exercised an often magical charm upon those with whom he was closely associated, but when his opposition was aroused he showed himself a keen controversialist. My subject of to-night is but half the story.

Even the strictest devotion of the time at my disposal to survey of the scientific work of Tyndall will not allow of more than a very imperfect and fragmentary treatment. During his thirty years of labour within these walls he ranged over a vast field, and accumulated results of a very varied character, important not only to the cultivators of the physical sciences, but also to the biologist. All that I can hope to do is to bring back to your recollection the more salient points of his work, and to illustrate them where possible by experiments of his own devising.

In looking through the catalogue of scientific papers issued by the Royal Society, one of the first entries under the name of Tyndall relates to a matter comparatively simple, but still of some interest. It has been noticed that when a jet of liquid is allowed to play into a receiving vessel, a good deal of air is sometimes carried down with it, while at other times this does not happen. The matter was examined experimentally by Tyndall, and he found that it was closely connected with the peculiar transformation undergone by a jet of liquid which had been previously investigated by Savart. A jet as it issues from the nozzle is at first cylindrical, but after a time it becomes what the physiologists call *varicose*; it swells in some places and contracts in others. This effect becomes more exaggerated as the jet descends, until the swellings separate into distinct drops, which follow one another in single file. Savart showed that under the influence of vibration the resolution into drops takes place more rapidly, so that the place of resolution travels up closer to the nozzle.

Tyndall's observation was that the carrying down of air required a jet already resolved into drops when it strikes the liquid. I hope to be able to show you the experiment by projection upon the screen. At the present moment the jet is striking the water in the tank previously to resolution into drops, and is therefore carrying down no air. If I operate on the nozzle with a vibrating tuning-fork, the resolution occurs earlier, and the drops now carry down with them a considerable quantity of air.

Among the earlier of Tyndall's papers are some relating to ice, a subject which attracted him much, probably from his mountaineering experiences. About the time of which I am speaking Faraday made interesting observations upon a peculiar behaviour of ice, afterwards called by the name of regelation. He found that if two pieces of ice were brought into contact they stuck or froze together. The pressure required to produce this effect need not be more than exceedingly small. Tyndall found that if fragments of ice are squeezed they pack themselves into a continuous mass. We have here some small ice in a mould, where it can be subjected to a powerful squeeze. The ice under this operation will be regelated, and a mass obtained which may appear almost transparent, and as if it had never been fractured at all. The flow of glaciers has been attributed to this action, the fractures which the stresses produce being mended again by regelation. I should say, perhaps, that the question of glacier motion presents difficulties not yet wholly explained. There can be no doubt, however, that regelation plays an important part.

Another question treated by Tyndall is the manner in which ice first begins to melt under the action of a beam of light passing into it from an electric lamp. Ice usually melts by conducted heat, which reaches first the outside layers. But if we employ a beam from an electric lamp, the heat will reach the ice not only outside but internally, and the melting will begin at certain points in the interior. Here we have a slab of ice which we project upon the screen. We see that the melting begins at certain points, which develop a crystallised appearance resembling flowers. They are points in the interior of the ice, not upon the surface. Tyndall found that when the ice gives way at these internal points there is a formation of apparently empty space. He carefully melted under water such a piece of ice, and found that when the cavity was melted out there was no escape of air, proving that the cavity was really vacuous.

Various speculations have been made as to the cause of this internal melting at definite points, but here again I am not sure if the difficulty has been altogether removed. One point of importance brought out by Tyndall relates to the plane of the flowers. It is parallel to the direction in which the ice originally froze, that is, parallel to the original surface of the water from which it was formed.

I must not dwell further upon isolated questions, however interesting; but will pass on at once to our main subject, which may be divided into three distinct parts, relating namely to heat, especially dark radiation, sound, and the behaviour of small particles, such as compose dust, whether of living or dead matter.

The earlier publications of Tyndall on the subject of heat are for the most part embodied in his work entitled *Heat as a Mode of Motion*. This book has fascinated many readers. I could name more than one now distinguished physicist who drew his first scientific nutriment from it. At the time of its appearance the law of the equivalence of heat and work was quite recently established by the labours of Mayer and Joule, and had taken firm hold of the minds of scientific men; and a great part of Tyndall's book may be considered to be inspired by and founded upon this first law of thermodynamics. At the time of publication of Joule's labours, however, there seems to have been a considerable body of hostile opinion, favourable to the now obsolete notion that heat is a distinct entity called caloric. Looking back, it is a little difficult to find out who were responsible for this reception of the theory of caloric. Perhaps it was rather the popular writers of the time than the first scientific authorities. A scientific worker, especially if he devotes himself to original work, has not time to examine for himself all questions, even those relating to his own department, but must take something on trust from others whom he regards as authorities. One might say that a knowledge of science, like a knowledge of law, consists in knowing where to look for it. But even this kind of knowledge is not always easy to obtain. It is only by experience that one can find out who are most entitled to confidence. It is difficult now to understand the hesitation that was shown in fully accepting the doctrine that heat is a mode of motion, for all the great authorities, especially in England, seem to have favoured it. Not to mention Newton and Cavendish, we have Rumford making almost conclusive experiments in its support, Davy accepting it, and Young, who was hardly ever wrong, speaking of the antagonistic theory almost with contempt. On the Continent perhaps, and especially among the French school of chemists and physicists, caloric had more influential support.

As has been said, a great part, though not the whole of Tyndall's work was devoted to the new doctrine. Much relates to other matters, such as radiant heat. Objection has been taken to this phrase, not altogether without reason; for it may be said that when heat it is not radiant, and while radiant it is not heat. The term dark radiation, or dark radiance as Newcomb calls it, is preferable, and was often used by Tyndall. If we analyse, as Newton did, the components of light, we find that only certain parts are visible. The invisible parts produce, however, as great, or greater, effects in other ways than do the visible parts. The heating effect, for example, is vastly greater in the invisible region than in the visible. One of the experiments that Tyndall devised in order to illustrate this fact I hope now to repeat. He found that it was possible by means of a solution of iodine in bisulphide of carbon to isolate the invisible rays. This solution is opaque to light; even the sun could not be seen through it; but it is very fairly transparent to the invisible ultra-red radiation. By means of a concave reflector I concentrate the rays from an arc lamp. In their path is inserted the opaque solution, but in the focus of invisible radiation the heat developed is sufficient to cause the inflammation of a piece of gun-cotton.

Tyndall varied this beautiful experiment in many ways. By raising to incandescence a piece of platinum foil, he illustrated the transformation of invisible into visible radiation.

The most important work, however, that we owe to Tyndall in connexion with heat is the investigation of the absorption of invisible radiation by gaseous bodies. Melloni had examined the behaviour of solid and liquid bodies, but not of gases. He found that transparent bodies like glass might be very opaque to invisible radiation. Thus, as we all know, a glass screen will keep off the heat of a fire, while if we wish to protect ourselves from the sun, the glass screen would be useless. On the other hand rock-salt freely transmitted invisible radiation. But nothing had been done on the subject of gaseous absorption, when Tyndall attacked this very difficult problem. Some of his results are shown in the accompanying table. The absorption of the ordinary non-condensable, or rather, not easily condensable gases—for we must not talk of non-condensable gases now, least of all in this place—the absorption of these gases is very small; but when we pass to the more compound gases, such as nitric oxide, we find the absorption much greater—and in the case of olefiant gas we see that the absorbing power is as much as 6000 times that of the ordinary gases.

	Relative Absorption at 1 inch Pressure
Air	1
Oxygen	1
Nitrogen...	1
Hydrogen	1
Carbonic acid...	972
Nitric oxide	1590
Ammonia	5460
Olefiant gas	6030

There is one substance as to which there has been a great diversity of opinion—aqueous vapour. Tyndall found that aqueous vapour exercises a strong power of absorption—strong relatively to that of the air in which it is contained. This is of course a question of great importance, especially in relation to meteorology. Tyndall's conclusions were vehemently contested by many of the authorities of the time, among whom was Magnus, the celebrated physicist of Berlin. With a view to this lecture I have gone somewhat carefully into this question, and I have been greatly impressed by the care and skill showed by Tyndall, even in his earlier experiments upon this subject. He was at once sanguine and sceptical—a combination necessary for success in any branch of science. The experimentalist who is not sceptical will be led away on a false tack and accept conclusions which he would find it necessary to reject were he to pursue the matter further; if not sanguine, he will be discouraged altogether by the difficulties encountered in his earlier efforts, and so arrive at no conclusion at all. One criticism, however, may be made. Tyndall did not at first describe with sufficient detail the method and the precautions which he used. There was a want of that precise information necessary to allow another to follow in his steps. Perhaps this may have been due to his literary instinct, which made him averse from overloading his pages with technical experimental details.

The controversy above referred to I think we may now consider to be closed. Nobody now doubts the absorbing power of aqueous vapour. Indeed the question seems to have entered upon a new phase; for in a recent number of Wiedemann's *Annalen*, Paschen investigates the precise position in the spectrum of the rays which are absorbed by aqueous vapour.

I cannot attempt to show you here any of the early experiments on the absorption of vapours. But some years later Tyndall contrived an experiment, which will allow of reproduction. It is founded on some observations of Graham Bell, who discovered that various bodies become sonorous when exposed to intermittent radiation.

The radiation is supplied from incandescent lime, and is focused by a concave reflector. In the path of the rays is a revolving wheel provided with projecting teeth. When a tooth intervenes, the radiation is stopped; but in the interval between the teeth the radiation passes through, and falls upon any object held at the focus. The object in this case is a small glass bulb containing a few drops of ether, and communicating with the ear by a rubber tube. Under the operation of the intermittent radiation, the ether vapour expands and contracts; in other words a vibration is established, and a sound is heard by the observer. But if the vapour were absolutely diathermanous, no sound would be heard.

I have repeated the experiment of Tyndall which allowed him to distinguish between the behaviour of ordinary air and dry air. If, dispensing with ether, we fill the bulb with air in the ordinary moist state, a sound is heard with perfect distinctness, but if we drop in a little sulphuric acid, so as to dry the air, the sound disappears.

According to the law of exchanges, absorption is connected with radiation; so that while hydrogen or oxygen do not radiate, from ammonia we might expect to get considerable radiation. In the following experiment I aim at showing that the radiation of hot coal gas exceeds the radiation of equally hot air.

The face of the thermopile, protected by screens from the ball itself, is exposed to the radiation from the heated air which rises from a hot copper ball. The effect is manifested by the [spot of] light reflected from a galvanometer mirror. When we replace the air by a stream of coal gas, the galvanometer indicates an augmentation of heat, so that we have before us a demonstration that coal gas when heated does radiate more than equally hot air, from which we conclude that it would exercise more absorption than air.

I come now to the second division of my subject, that relating to Sound. Tyndall, as you know, wrote a book on Sound, founded on lectures delivered in this place. Many interesting and original discoveries are there embodied. One that I have been especially interested in myself, is on the subject of sensitive flames. Professor Leconte in America made the first observations at an amateur concert, but it was Tyndall who introduced the remarkable high-pressure flame now before you. It issues from a pin-hole burner, and the sensitiveness is entirely a question of the pressure at which the gas is supplied. Tyndall describes the phenomenon by saying that the flame under the influence of a high pressure is like something on the edge of a precipice. If left alone, it will maintain itself; but under the slightest touch it will be pushed over. The gas at high pressure will, if undisturbed, burn steadily and erect, but if a hiss is made in its neighbourhood it becomes at once unsteady, and ducks down. A very high sound is necessary. Even a whistle, as you see, does not act. Smooth pure sounds are practically without effect unless of very high pitch.

I will illustrate the importance of the flame as a means of investigation by an experiment in the diffraction of sound. I have here a source of sound, but of pitch so high as to be inaudible. The waves impinge perpendicularly upon a circular disc of plate glass. Behind the disc there is a sound shadow, and you might expect that the shadow would be most complete at the centre. But it is not so. When the burner occupies this [central] position the flame flares; but when by a slight motion of the disc the position of the flame is made eccentric, the existence of the shadow is manifested by the recovery of the flame. At the centre the intensity of sound is the same as if no obstacle were interposed.

The optical analogue of the above experiment was made at the suggestion of Poisson, who had deduced the result theoretically, but considered it so unlikely that he regarded it as an objection to the undulatory theory of light. Now, I need hardly say, it is regarded as a beautiful confirmation.

It is of importance to prove that the flame is not of the essence of the matter, that there is no need to have a flame, or to ignite it at the burner. Thus, it is quite possible to have a jet of gas so arranged that ignition does not occur until the jet has lost its sensitiveness. The sensitive part is that quite close to the nozzle, and the flame is only an indicator. But it is not necessary to have any kind of flame at all. Tyndall made observations on smoke-jets, showing that a jet of air can be made sensitive to sound. The difficulty is to see it, and to operate successfully upon it; because, as Tyndall soon found, a smoke-jet is much more difficult to deal with than flames, and is sensitive to much graver sounds. I doubt whether I am wise in trying to exhibit smoke-jets to an audience, but I have a special means of projection by which I ought at least to succeed in making them visible. It consists in a device by which the main part of the light from the lamp is stopped at the image of the arc, so that the only light which can reach the screen is light which by diffusion has been diverted out of its course. Thus we shall get an exhibition of a jet of smoke upon the screen, showing bright on a dark ground. The jet issues near the mouth of a resonator of pitch 256. When undisturbed it pursues a straight course, and remains cylindrical. But if a fork of suitable pitch be sounded in the neighbourhood, the jet spreads out into a sort of fan, or even bifurcates, as you see upon the screen. The real motion of the jet cannot of course be ascertained by mere inspection. It consists in a continuously increasing *sinuosity*, leading after a while to complete disruption. If two forks slightly out of unison are sounded together, the jet expands and re-collects itself, synchronously with the audible beats. I should say that my jet is a very coarse imitation of Tyndall's. The nozzle that I am using is much too large. With a proper nozzle, and in a perfectly undisturbed atmosphere—undisturbed not only by sounds, but free from all draughts—the sensitiveness is wonderful. The slightest noise is seen to act instantly and to bring the jet down to a fraction of its former height.

Another important part of Tyndall's work on Sound was carried out as adviser of the Trinity House. When in thick weather the ordinary lights fail, an attempt is made to replace them with sound signals. These are found to vary much in their action, sometimes being heard to a very great distance, and at other times failing to make themselves audible even at a moderate distance. Two explanations have been suggested, depending upon acoustic refraction and acoustic reflection.

Under the influence of variations of temperature refraction occurs in the atmosphere. For example, sound travels more quickly in warm than in cold air. If, as often happens, it is colder above, the upper part of the sound wave tends to lag behind, and the wave is liable to be tilted upwards and so to be carried over the head of the would-be observer on the surface of the ground. This explanation of acoustic refraction by variation of temperature was given by Prof. Osborne Reynolds. As Sir G. Stokes showed, refraction is also caused by wind. The difference between refraction by wind and by temperature variations is that in one case everything turns upon the direction in which the sound is going, while in the second case this consideration is immaterial. The sound is heard by an observer down wind, and not so well by an observer up wind. The explanation by refraction of the frequent failure of sound signals was that adopted by Prof. Henry in America, a distinguished worker upon this subject. Tyndall's investigations, however, led him to favour another explanation. His view was that sound was actually reflected by atmospheric irregularities. He observed, what appears to be amply sufficient to establish his case, that prolonged signals from fog sirens give rise to echoes audible after the signal has stopped. This echo was heard from the air over the sea, and lasted in many cases a long time, up to 15 seconds. There seems here no alternative but to suppose that reflection must have occurred internally in the atmosphere. In some cases the explanation of the occasional diminished penetration of sound seems to be rather by refraction, and in others by reflection.

Tyndall proved that a single layer of hot air is sufficient to cause reflection, and I propose to repeat his experiment. The source of sound, a toy reed, is placed at one end of one metallic tube, and a sensitive flame at one end of a second. The opposite ends of these tubes are placed near each other, but in a position which does not permit the sound waves issuing from the one to enter the other directly. Accordingly the flame shows no response. If, however, a pane of glass be held suitably, the waves are reflected back and the flame is excited. Tyndall's experiment consists in the demonstration that a flat gas flame is competent to act the part of a reflector. When I hold the gas flame in the proper position, the percipient flame flares; when the flat flame is removed or held at an unsuitable angle, there is almost complete recovery.

It is true that in the atmosphere no such violent transitions of density can occur as are met with in a flame; but, on the other hand, the interruptions may be very numerous, as is indeed rendered probable by the phenomena of stellar scintillation.

The third portion of my subject must be treated very briefly. The guiding idea of much of Tyndall's work on atmospheric particles was the application of an intense illumination to render them evident. Fine particles of mastic, precipitated on admixture of varnish with a large quantity of water, had already been examined by Brücke. Chemically precipitated sulphur is convenient, and allows the influence of size to be watched as the particles grow. But the most interesting observations of Tyndall relate to precipitates in gases caused by the chemical action of the light itself. This may be illustrated by causing the concentrated rays of the electric lamp to pass through a flask containing vapour of peroxide of chlorine. Within a few seconds dense clouds are produced.

When the particles are very small in comparison with the wave-length, the laws governing the dispersion of the light are simple. Tyndall pursued the investigation to the case where the particles have grown beyond the limit above indicated, and found that the polarisation of the dispersed light was affected in a peculiar and interesting manner.

Atmospheric dust, especially in London, is largely organic. If, following Tyndall, we hold a spirit lamp under the track of the light from the electric lamp, the dark spaces, resulting from the combustion of the dust, have all the appearance of smoke.

In confined and undisturbed spaces the dust settles out. I have here a large flask which has been closed for some days. If I hold it to the lamp, the track of the light, plainly visible before entering and after leaving the flask, is there interrupted. This, it will be evident, is a matter of considerable importance in connexion with organic germs.

The question of the spontaneous generation of life occupied Tyndall for several years. He brought to bear upon it untiring perseverance and refined experimental skill, and his results are those now generally accepted. Guarding himself from too absolute statements as to other times and other conditions, he concluded that under the circumstances of our experiments life is always founded upon life. The putrefaction of vegetable and animal infusions, even when initially sterilised, is to be attributed to the intrusion of organic germs from the atmosphere.

The universal presence of such germs is often regarded as a hypothesis difficult of acceptance. It may be illustrated by an experiment from the inorganic world. I have here, and can project upon the screen, glass pots, each containing a shallow layer of a supersaturated solution of sulphate of soda. Protected by glass covers, they have stood without crystallising for forty-eight hours. But if I remove the cover, a few seconds or minutes will see the crystallisation commence. It has begun, and long needles are invading the field of view. Here it must be understood that, with a few exceptions, the crystalline germ required to start the action must be of the same nature as the dissolved salt; and the conclusion is that small crystals of sulphate of soda are universally present in the atmosphere.

I have now completed my task. With more or less success I have laid before you the substance of some of Tyndall's contributions to knowledge. What I could not hope to recall was the brilliant and often poetic exposition by which his vivid imagination illumined the dry facts of science. Some reminiscence of this may still be recovered by the reader of his treatises and memoirs; but much survives only as an influence exerted upon the minds of his contemporaries, and manifested in subsequent advances due to his inspiration.

210.

ON AN ANOMALY ENCOUNTERED IN DETERMINATIONS OF THE DENSITY OF NITROGEN GAS.

[*Proceedings of the Royal Society*, LV. pp. 340—344, *April*, 1894.]

IN a former communication* I have described how nitrogen, prepared by Lupton's† method, proved to be lighter by about 1/1000 part than that derived from air in the usual manner. In both cases a red-hot tube containing copper is employed, but with this difference. In the latter method the atmospheric oxygen is removed by oxidation of the copper itself, while in [Harcourt's] method it combines with the hydrogen of ammonia, through which the air is caused to pass on its way to the furnace, the copper remaining unaltered. In order to exaggerate the effect, the air was subsequently replaced by oxygen. Under these conditions the whole, instead of only about one-seventh part of the nitrogen is derived from ammonia, and the discrepancy was found to be exalted to about one-half per cent.

Upon the assumption that similar gas should be obtained by both methods, we may explain the discrepancy by supposing either that the atmospheric nitrogen was too heavy on account of imperfect removal of oxygen, or that the ammonia nitrogen was too light on account of contamination with gases lighter than pure nitrogen. Independently of the fact that the action of the copper in the first case was pushed to great lengths, there are two arguments which appeared to exclude the supposition that oxygen was still present in the prepared gas. One of these depends upon the large quantity of oxygen that would be required, in view of the small difference between the weights of the two gases. As much as 1/30th part of oxygen would be necessary to raise the density by 1/200, or about one-sixth of all the oxygen originally present. This seemed to be out of the question. But even if so high a degree of imperfection in the action of the copper could be admitted,

---

* "On the Densities of the Principal Gases," *Roy. Soc. Proc.* Vol. LIII. p. 146, 1893. [Vol. IV. p. 39. See also p. 1.]

† [1902. The use of ammonia to burn atmospheric oxygen is due to Mr Vernon Harcourt.]

the large alteration caused by the substitution of oxygen for air in [Harcourt's] process would remain unexplained. Moreover, as has been described in the former paper, the introduction of hydrogen into the gas made no difference, such hydrogen being removed by the hot oxide of copper subsequently traversed. It is surely impossible that the supposed residual oxygen could have survived such treatment.

Another argument may be founded upon more recent results, presently to be given, from which it appears that almost exactly the same density is found when the oxygen of air is removed by hot iron reduced with hydrogen, instead of by copper, or in the cold by ferrous hydrate.

But the difficulties in the way of accepting the second alternative are hardly less formidable. For the question at once arises, of what gas, lighter than nitrogen, does the contamination consist? In order that the reader may the better judge, it may be well to specify more fully what were the arrangements adopted. The gas, whether air or oxygen, after passing through potash was charged with ammonia as it traversed a small wash-bottle, and thence proceeded to the furnace. The first passage through the furnace was in a tube packed with metallic copper, in the form of fine wire. Then followed a wash-bottle of sulphuric acid by which the greater part of the excess of ammonia would be arrested, and a second passage through the furnace in a tube containing copper oxide. The gas then traversed a long length of pumice charged with sulphuric acid, and a small wash-bottle containing Nessler solution. On the other side of the regulating tap the arrangements were always as formerly described, and included tubes of finely divided potash and of phosphoric anhydride. The rate of passage was usually about half a litre per hour.

Of the possible impurities, lighter than nitrogen, those most demanding consideration are hydrogen, ammonia, and water vapour. The last may be dismissed at once, and the absence of ammonia is almost equally certain. The question of hydrogen appears the most important. But this gas, and hydrocarbons, such as CH_4, could they be present, should be burnt by the copper oxide; and the experiments already referred to, in which hydrogen was purposely introduced into atmospheric nitrogen, seem to prove conclusively that the burning would really take place. Some further experiments of the same kind will presently be given.

The gas from ammonia and oxygen was sometimes odourless, but at other times smelt strongly of nitrous fumes, and, after mixture with moist air, reddened litmus paper. On one occasion the oxidation of the nitrogen went so far that the gas showed colour in the blow-off tube of the Töppler, although the thickness of the layer was only about half an inch. But the presence of nitric oxide is, of course, no explanation of the abnormal lightness. The conditions under which the oxidation takes place proved to be difficult of control, and it was thought desirable to examine nitrogen derived by *reduction* from nitric and nitrous oxides.

The former source was the first experimented upon. The gas was evolved from copper and diluted nitric acid in the usual way, and, after passing through potash, was reduced by *iron*, copper not being sufficiently active, at least without a very high temperature. The iron was prepared from blacksmith's scale. In order to get quit of carbon, it was first treated with a current of oxygen at a red heat, and afterwards reduced by hydrogen, the reduction being repeated after each employment. The greater part of the work of reducing the gas was performed outside the furnace, in a tube heated locally with a Bunsen flame. In the passage through the furnace in a tube containing similar iron the work would be completed, if necessary. Next followed washing with sulphuric acid (as required in the ammonia process), a second passage through the furnace over copper oxide, and further washing with sulphuric acid. In order to obtain an indication of any unreduced nitric oxide, a wash-bottle containing ferrous sulphate was introduced, after which followed the Nessler test and drying tubes, as already described. As thus arranged, the apparatus could be employed without alteration, whether the nitrogen to be collected was derived from air, from ammonia, from nitric oxide, from nitrous oxide, or from ammonium nitrite.

The numbers which follow are the weights of the gas contained by the globe at zero, at the pressure defined by the manometer when the temperature is 15°. They are corrected for the errors in the weights, but not for the shrinkage of the globe when exhausted, and thus correspond to the number 2·31026, as formerly given for nitrogen.

Nitrogen from NO by Hot Iron.

November 29, 1893............2·30143 ⎫
December   2, 1893........ ...2·29890 ⎪ Mean, 2·30008
December   5, 1893............2·29816 ⎬
December   6, 1893............2·30182 ⎭

Nitrogen from N_2O by Hot Iron*.

December 26, 1893 ....2·29869 ⎫ Mean, 2·29904
December 28, 1893 ....2·29940 ⎭

Nitrogen from Ammonium Nitrite passed over Hot Iron.

January   9, 1894............2·29849 ⎫ Mean, 2·29869
January 13, 1894............2·29889 ⎭

* The N_2O was prepared from zinc and very dilute nitric acid.

With these are to be compared the weights of nitrogen derived from the atmosphere.

Nitrogen from Air by Hot Iron.

December 12, 1893............2·31017
December 14, 1893............2·30986 (H)
December 19, 1893............2·31010 (H)   Mean, 2·31003
December 22, 1893............2·31001

Nitrogen from Air by Ferrous Hydrate.

January 27, 1894 ....2·31024
January 30, 1894 ....2·31010   Mean, 2·31020
February 1, 1894 ....2·31028

In the last case a large volume of air was confined for several hours in a glass reservoir with a mixture of slaked lime and ferrous sulphate. The gas was displaced by deoxygenated water, and further purified by passage through a tube packed with a similar mixture. The hot tubes were not used.

If we bring together the means for atmospheric nitrogen obtained by various methods, the agreement is seen to be good, and may be regarded as inconsistent with the supposition of residual oxygen in quantity sufficient to influence the weights.

Atmospheric Nitrogen.

By hot copper, 1892......................2·31026
By hot iron, 1893.......................2·31003
By ferrous hydrate, 1894...............2·31020

Two of the results relating to hot iron, those of December 14 and December 19, were obtained from nitrogen, into which hydrogen had been purposely introduced. An electrolytic generator was inserted between the two tubes containing hot iron, as formerly described. The generator worked under its own electromotive force, and the current was measured by a tangent galvanometer. Thus, on December 19, the deflection throughout the time of filling was 3°, representing about 1/15 ampère. In two hours and a half the hydrogen introduced into the gas would be about 70 c.c., sufficient, if retained, to reduce the weight by about 4 per cent. The fact that there was no sensible reduction proves that the hydrogen was effectively removed by the copper oxide.

The nitrogen, obtained altogether in four ways from chemical compounds, is materially lighter than the above, the difference amounting to about 11 mg., or about 1/200 part of the whole. It is also to be observed that the agreement of individual results is less close in the case of chemical nitrogen than of atmospheric nitrogen.

I have made some experiments to try whether the densities were influenced by exposing the gas to the silent electric discharge. A Siemens tube, as used for generating ozone, was inserted in the path of the gas after desiccation with phosphoric anhydride. The following were the results :—

Nitrogen from Air by Hot Iron, Electrified.

January 1, 1894............2·31163 ⎫
                                    ⎬ Mean, 2·31059
January 4, 1894............2·30956 ⎭

Nitrogen from N_2O by Hot Iron, Electrified.

January 2, 1894............2·30074 ⎫
                                    ⎬ Mean, 2·30064
January 5, 1894............2·30054 ⎭

The somewhat anomalous result of January 1 is partly explained by the failure to obtain a subsequent weighing of the globe empty, and there is no indication that any effect was produced by the electrification.

One more observation I will bring forward in conclusion. Nitrogen prepared from oxygen and ammonia, and about one-half per cent. lighter than ordinary atmospheric nitrogen, was stored in the globe for eight months. The globe was then connected to the apparatus, and the pressure was readjusted in the usual manner to the standard conditions. On re-weighing no change was observed, so that the abnormally light nitrogen did not become dense by keeping.

[1902. For the explanation of the discrepancy here set forth, as due to a previously unrecognised constituent of the atmosphere, see the memoir by Rayleigh and Ramsay, Art. 214 below.]

211.

ON THE MINIMUM CURRENT AUDIBLE IN THE TELEPHONE.

[*Philosophical Magazine*, XXXVIII. pp. 285—295, 1894*.]

THE estimates which have been put forward of the minimum current perceptible in the Bell telephone vary largely. Mr Preece gives 6×10^{-13} ampere†; Prof. Tait, for a current reversed 500 times per second, 2×10^{-12} ampere‡. De la Rue gives 1×10^{-8} ampere, and the same figure is recorded by Brough§ as applicable to the *strongest* current with which the instrument is worked. Various methods, more or less worthy of confidence, have been employed, but the only experimenter who has described his procedure with detail sufficient to allow of criticism is Prof. Ferraris‖, whose results may be thus expressed :—

Pitch	Frequency	Minimum current in amperes
Do₃	264	23×10^{-9}
Fa₃	352	17×10^{-9}
La₃	440	10×10^{-9}
Do₄	528	7×10^{-9}
Re₄	594	5×10^{-9}

The currents were from a make-and-break apparatus, and in each case are reckoned as if only the first periodic term of the Fourier series, representative of the actual current, were effective. On this account the quantities in the third column should probably be increased, for the presence of overtones could hardly fail to favour audibility.

Although a considerable margin must be allowed for varying pitch, varying acuteness of audition, and varying construction of the instruments, it is scarcely possible to suppose that all the results above mentioned can be

---

* Read at the Oxford Meeting of the British Association.

† *Brit. Assoc. Report*, Manchester, 1887, p. 611.

‡ *Edin. Proc.* Vol. IX. p. 551 (1878). Prof. Tait speaks of a *billion* B.A. units, and, as he kindly informs me, a billion here means 10^{12}.

§ *Proceedings of the Asiatic Society of Bengal*, 1877, p. 255.

‖ *Atti della R. Accad. d. Sci. di Torino*, Vol. XIII. p. 1024 (1877).

correct, even in the roughest sense. The question is of considerable interest in connexion with the theory of the telephone. For it appears that *à priori* calculations of the possible efficiency of the instrument are difficult to reconcile with numbers such as those of Tait and of Preece, at least without attributing to the ear a degree of sensitiveness to aerial vibration far surpassing even the marvellous estimates that have hitherto been given*.

Under these circumstances it appeared to be desirable to undertake fresh observations, in which regard should be paid to various sources of error that may have escaped attention in the earlier days of telephony. The importance of defining the resistance of the instruments and of employing pure tones of various pitch need not be insisted upon.

As regards resistance, a low-resistance telephone, although suitable in certain cases, must not be expected to show the same sensitiveness to current as an instrument of higher resistance. If we suppose that the total space available for the windings is given, and that the proportion of it occupied by the copper is also given, a simple relation obtains between the resistance and the minimum current. For if γ be the current, n be the number of convolutions, and r the resistance, we have, as in the theory of galvanometers, $n\gamma = \text{const.}$, $n^{-2}r = \text{const.}$, so that $\gamma\sqrt{r} = \text{const.}$, or the minimum current is inversely *as the square root of the resistance.*

The telephones employed in the experiments about to be narrated were two, of which one (T_1) is a very efficient instrument of 70-ohms resistance. The other (T_2), of less finished workmanship, was rewound in the laboratory with comparatively thick wire. The interior diameter of the windings is 9 mm., and the exterior diameter is 26 mm. The width of the groove, or the axial dimension of the coil, is 8 mm., the number of windings is 160, and the resistance is ·8 ohm. Since the dimensions of the coils are about the same in the two cases, we should expect, according to the above law, that about 10 times as much current would be required in T_2 as in T_1. Both instruments are of the Bell (unipolar) type, and comparison with other specimens shows that there is nothing exceptional in their sensibility.

In view of the immense discrepancies above recorded, it is evident that what is required is not so much accuracy of measurement as assured soundness in method. It appeared to me that electromotive forces of the necessary harmonic type would be best secured by the employment of a revolving magnet in the proximity of an inductor-coil of known construction. The electromotive force thus generated operates in a circuit of known resistance; and, if the self-induction can be neglected, the calculation of the current presents no difficulty. The sound as heard in the telephone may be reduced

---

* *Proc. Roy. Soc.* Vol. xxvi. p. 248 (1877). [Vol. i. p. 328; see also Art. 213 below.] Also Wien, *Wied. Ann.* Vol. xxxvi. p. 834 (1889).

to the required point either by varying the distance (B) between the magnet and the inductor, or by increasing the resistance (R) of the circuit. In fact both these quantities may be varied; and the agreement of results obtained with widely different values of R constitutes an effective test of the legitimacy of neglecting self-induction. When R is too much reduced, the time-constant of the circuit becomes comparable with the period of vibration, and the current is no longer increased in proportion to the reduction of R. This complication is most likely to occur when the pitch is high.

In order to keep as clear as possible of the complication due to self-induction, I employed in the earlier experiments a resistance-coil of 100,000 ohms, constructed as usual of wire doubled upon itself. But it soon appeared that in avoiding Scylla I had fallen upon Charybdis. The first suspicion of something wrong arose from the observation that the sound was nearly as loud when the 100,000 ohms was included as when a 10,000-ohm coil was substituted for it. The first explanation that suggested itself was that the sound was being conveyed mechanically instead of electrically, as is indeed quite possible under certain conditions of experiment. But a careful observation of the effect of breaking the continuity of the leads, one at a time, proved that the propagation was really electrical. Subsequent inquiry showed that the anomaly was due to a condenser, or leyden, like action of the doubled wire of the 100,000-ohm coil. When the junction at the middle was unsoldered, so as to interrupt the metallic continuity, the sounds heard in the telephone were nearly as loud as before. In this condition the resistance should have been enormous, and was in fact about 12 megohms* as indicated by a galvanometer. It was evident that the coil was acting principally as a leyden rather than as a resistance, and that any calculation founded upon results obtained with it would be entirely fallacious.

It is easy to form an estimate of the point at which the complication due to capacity would begin to manifest itself. Consider the case of a simple resistance R in parallel with a leyden of capacity C, and let the currents in the two branches be x and y respectively. If V be the difference of potential at the common terminals, proportional to e^{ipt}, we have

$$x = V/R, \qquad y = CdV/dt = ipVC;$$

so that

$$\frac{x+y}{V} = \frac{1+ipRC}{R}.$$

The amplitude of the total current is increased by the leyden in the ratio $\sqrt{(1+p^2R^2C^2)} : 1$; and the action of the leyden becomes important when $pRC = 1$. With a frequency of 640, $p = 4020$; so that, if $R = 10^{14}$ c.g.s., the critical value of C is $\frac{1}{402} \times 10^{-15}$ c.g.s., or about $\frac{1}{400}$ of a microfarad.

* Doubtless the insulation between the wires should have been much higher.

It will be seen that even if the capacity remained unaltered, a reduction of resistance in the ratio say of 10 to 1 would greatly diminish the complication due to condenser-like action; but perhaps the best evidence that the results obtained are not prejudiced in this manner is afforded by the experiments in which the principal resistance was a column of plumbago.

The revolving magnet was of clock-spring, about $2\frac{1}{2}$ cm. long, and so bent as to be driven directly, windmill fashion, from an organ bellows. It was mounted transversely upon a portion of a sewing-needle, the terminals of which were carried in slight indentations at the ends of a U-shaped piece of brass. As fitted to the wind-trunk, the axis of rotation was horizontal.

The inductor-coil, with its plane horizontal, was situated so that its centre was vertically below that of the magnet at distance B. Thus, if A be the mean radius of the coil, n the number of convolutions, the galvanometer-constant G of the coil at the place occupied by the magnet is given by

$$G = \frac{2\pi n A^2}{C^3}, \dots\dots\dots\dots\dots\dots\dots\dots\dots(1)$$

where $C^2 = A^2 + B^2$; and if m be the magnetic moment of the magnet, and ϕ the angle of rotation, the mutual potential M may be represented by*

$$M = Gm \sin \phi. \dots\dots\dots\dots\dots\dots\dots(2)$$

If the frequency of revolution be $p/2\pi$, $\phi = pt$; and then

$$dM/dt = Gmp \cos pt. \dots\dots\dots\dots\dots(3)$$

The expression (3) represents the electromotive force operative in the circuit. If the inductance can be neglected, the corresponding current is obtained on division of (3) by R, the total resistance of the circuit.

The moment m is deduced by observation of the deflection of a magnetometer-needle from the position which it assumes under the operation of the earth's horizontal force H. If the magnet be situated to the east at distance r, and be itself directed east and west, the angular deflection θ from equilibrium is given by

$$\tan \theta = \frac{2m/r^3}{H}.$$

The relation between the angle θ and the double deflection d in scale-divisions, obtained on *reversal* of m, is approximately $\theta = d/4D$, where D is the distance between mirror and scale; so that we may take

$$m = \frac{Hr^3 d}{8D}. \dots\dots\dots\dots\dots\dots\dots(4)$$

---

* Maxwell, *Electricity and Magnetism*, Vol. II. § 700.

The amplitude of the oscillatory current, generated under these conditions, is accordingly

$$\frac{n\pi pHA^2r^3d}{4C^3RD} \dots\dots\dots\dots\dots\dots\dots\dots\dots\dots\dots(5)$$

If c.g.s. units are employed, $H = \cdot18$. A must of course be measured in centimetres; but any units that are convenient may be used for r and C, and for d and D. The current will then be given in terms of the c.g.s. unit, which is equal to 10 amperes.

The inductor-coil used in most of the experiments is wound upon an ebonite ring, and is the one that was employed as the "suspended coil" in the determination of the electro-chemical equivalent of silver*. The number of convolutions (n) is 242. The axial dimension of the section is 1·4 cm. and the radial dimension is ·97 cm. The mean radius A is 10·25 cm., and the resistance is about 10½ ohms.

In making the observations the current from the inductor-coil was led to a distant part of the house by leads of doubled wire, and was there connected to the telephone and resistances. Among the latter was a plumbago resistance on Prof. F. J. Smith's plan† of about 84,000 ohms; but in most of the experiments a resistance-box going up to 10,000 ohms was employed, with the advantage of allowing the adjustment of sound to be made by the observer at the telephone. The attempt to hit off the least possible sound was found to be very fatiguing and unsatisfactory; and in all the results here recorded the sounds were adjusted so as to be *easily* audible after attention for a few seconds. Experiment showed that the resistances could then be *doubled* without losing the sound, although perhaps it would not be caught at once by an unprepared ear. But it must not be supposed that the observation admits of precision, at least without greater precautions than could well be taken. Much depends upon the state of the ear as regards fatigue, and upon freedom from external disturbance.

The pitch was determined before and after an observation by removing the added resistance and comparing the loud sound then heard with a harmonium. The *octave* thus estimated might be a little uncertain. It was verified by listening to the beats of the sound from the telephone and from a nearly unisonant tuning-fork, both sounds being nearly pure tones.

When the magnet was driven at full speed the frequency was found to be 307, and at this pitch a series of observations was made with various values of C and of R. Thus when $B = 7\cdot75$ inches, or $C = 8\cdot7$ inches, the resistance from the box required to produce the standard sound in telephone T_1 was

* *Phil. Trans.* Part II. 1884, p. 421. [Vol. II. p. 290.]
† *Phil. Mag.* Vol. xxxv. p. 210 (1893).

8000 ohms, so that $R = 8100 \times 10^9$. The quantities required for the calculation of (5) are as follows :—

$$n = 242, \qquad p = 2\pi \times 307, \qquad H = \cdot 18,$$
$$A = 10\cdot25, \qquad r = 8\cdot25, \qquad d = 140,$$
$$C = 8\cdot7, \qquad R = 81 \times 10^{11}, \qquad D = 1370,$$

r and C being reckoned in inches, d and D in scale-divisions of about $\frac{1}{40}$ inch. From these data the current required to produce the standard sound is found to be $7\cdot4 \times 10^{-8}$ c.g.s., or $7\cdot4 \times 10^{-7}$ amperes, for telephone T_1.

The results obtained by the method of the revolving magnet are collected into the accompanying table. The "wooden coil" is of smaller dimensions than the "ebonite coil," the mean radius being only $3\cdot5$ cm. The number of convolutions is 370.

Frequency $= 307$.   Ebonite coil.

Telephone	R in ohms	Current in amperes	Sound
T_1	84100 Plumbago	$3\cdot8 \times 10^{-7}$	Below standard
T_1	8100 Box	$7\cdot4 \times 10^{-7}$	Standard
T_1	4100 Box	$5\cdot2 \times 10^{-7}$	....
T_2	500 Box	$1\cdot2 \times 10^{-5}$	....
T_2	200 Box	$1\cdot0 \times 10^{-5}$	....

Frequency $= 307$.   Wooden coil.

T_1	84100 Plumbago	$3\cdot6 \times 10^{-7}$	Standard
T_1	10100 Box	$3\cdot7 \times 10^{-7}$	....
T_1	1600 Box	$5\cdot4 \times 10^{-7}$	....
T_2	350 Box	$1\cdot1 \times 10^{-5}$	....

Frequency $= 192$.   Ebonite coil.

T_1	3100 Box	$2\cdot5 \times 10^{-6}$	Standard

The method of the revolving magnet seemed to be quite satisfactory so far as it went, but it was desirable to extend the determinations to frequencies higher than could well be reached in this manner. For this purpose recourse was had to magnetized tuning-forks, vibrating with known amplitudes. If, for the moment, we suppose the magnetic poles to be concentrated at the extremities of the prongs, a vibrating-fork may be regarded as a simple magnet, fixed in position and direction, but of moment proportional to the instantaneous distance between the poles. Thus, if the magnetic axis pass perpendicularly through the centre of the mean plane of the inductor-coil, the situation is very similar to that obtaining in the case of the revolving magnet. The angle ϕ in (2) is no longer variable, but such that $\sin \phi = 1$ throughout. On the other hand m varies harmonically. If l be the mean distance between the poles, 2β the extreme arc from rest to rest traversed by each pole during the vibration, m_0 the mean magnetic moment,

$$m/m_0 = 1 + 2\beta/l \,.\, \sin pt,$$

and

$$dM/dt = Gm_0p \,.\, 2\beta/l \,.\, \cos pt. \quad\quad\quad\quad\quad\quad(6)$$

The formula corresponding to (5) is thus derived from it by simple introduction of the factor $2\beta/l$.

The forks were excited by bowing, and the observation of amplitude was effected by comparison with a finely divided scale under a magnifying-glass. It was convenient to observe the extreme end of a prong where the motion is greatest, but the double amplitude thus measured must be distinguished from 2β. In order to allow for the distance between the resultant poles and the extremities of the prongs, the measured amplitude was reduced in the ratio of 2 to 3. The observation of the magnetic moment at the magnetometer is not embarrassed by the diffusion of the free polarity.

In order to explain the determination more completely, I will give full details of an observation with a fork c' of frequency 256. The distance l between the middles of the prongs was ·875 inch, and the double amplitude of the vibration at the end of one of the prongs was ·09 inch. Thus 2β is reckoned as ·06 inch. The inductor-coil was the ebonite coil already described, and the sound was judged to be of the standard distinctness when, for example, $B = 15$ inches, or $C = 15·5$ inches, and the added resistance was 1000 ohms, so that $R = 1100 \times 10^9$. The quantities required for the computation of (5) as extended are

$$n = 242, \quad\quad p = 2\pi \times 256, \quad\quad H = ·18,$$
$$A = 10·25, \quad\quad r = 15, \quad\quad d = 410,$$
$$C = 15·5, \quad\quad R = 11 \times 10^{11}, \quad\quad D = 1370,$$
$$2\beta = ·06, \quad\quad l = ·875\,;$$

and they give for the current corresponding to the standard sound $9·8 \times 10^{-8}$ C.G.S., or $9·8 \times 10^{-7}$ amperes.

A summary of the results obtained with forks of pitch $c,\ c',\ e',\ g',\ c'',\ e'',\ g''$ is annexed. As the pitch rose, the difficulties of observation increased, both

Telephone	R in ohms	Current in amperes	
	$c = 128.$		
T_1	1100	$2·8 \times 10^{-5}$	
	$c' = 256.$		
T_1	8100 Box	$6·8 \times 10^{-7}$	
T_1	1100	$9·8 \times 10^{-7}$	
T_2	500	....	$1·1 \times 10^{-5}$

Telephone	R in ohms	Current in amperes	

$$e' = 320.$$

Telephone	R in ohms	Current in amperes	
T_1	84000 Plumbago	$3 \cdot 8 \times 10^{-7}$	
T_1	6100 Box	$2 \cdot 6 \times 10^{-7}$	
T_1	1600	$3 \cdot 1 \times 10^{-7}$	

$$g' = 384.$$

Telephone	R in ohms	Current in amperes	
T_1	84000 Plumbago	$1 \cdot 4 \times 10^{-7}$	
T_1	9500 Box	$1 \cdot 6 \times 10^{-7}$	
T_1	2100	$1 \cdot 4 \times 10^{-7}$	
T_1	900	$1 \cdot 7 \times 10^{-7}$	
T_2	600	....	$1 \cdot 9 \times 10^{-6}$
T_2	300	....	$2 \cdot 2 \times 10^{-6}$

$$c'' = 512.$$

Telephone	R in ohms	Current in amperes	
T_1	84000 Plumbago	$8 \cdot 9 \times 10^{-8}$	
T_1	9000 Box	$4 \cdot 8 \times 10^{-8}$	
T_1	3600	$5 \cdot 2 \times 10^{-8}$	
T_1	700	$8 \cdot 2 \times 10^{-8}$	
T_2	11 ?	....	$5 \cdot 2 \times 10^{-6}$?
T_2	100 Box	....	$1 \cdot 9 \times 10^{-6}$
T_2	300	....	$1 \cdot 4 \times 10^{-6}$
T_2	500	....	$2 \cdot 5 \times 10^{-6}$
T_2	900	....	$2 \cdot 4 \times 10^{-6}$

$$e'' = 640.$$

Telephone	R in ohms	Current in amperes	
T_1	84000 Plumbago	$3 \cdot 8 \times 10^{-8}$	
T_1	5100 Box	$3 \cdot 8 \times 10^{-8}$	
T_1	1100	$5 \cdot 5 \times 10^{-8}$	

$$g'' = 768.$$

Telephone	R in ohms	Current in amperes	
T_1	84000 Plumbago	$1 \cdot 1 \times 10^{-7}$	
T_1	7100 Box	$\cdot 9 \times 10^{-7}$	
T_1	2100	$1 \cdot 1 \times 10^{-7}$	

on account of the less duration of the sound and of the smaller amplitudes available for measurement. In one observation with telephone T_2 at pitch c'', the resistance, estimated at 11 ohms, was that of the coil, telephone, and leads only. No trustworthy result was to be expected under such conditions, but the number is included in order to show how small was the influence of self-induction, even where it had every opportunity of manifesting itself. If we bring together the numbers* derived with the revolving magnet and with the forks, we obtain in the case of T_1 :—

---

* The observations recorded were made with my own ears. Mr Gordon obtained very similar numbers when he took my place.

Pitch	Source	Current in 10^{-8} amperes
128	Fork..........................	2800
192	Revolving magnet	250
256	Fork..........................	83
307	Revolving magnet	49
320	Fork..........................	32
384	....	15
512	....	7
640	....	4·4
768	....	10

It would appear that the maximum sensitiveness to current occurs in the region of frequency 640; but observations at still higher frequencies would be needed to establish this conclusion beyond doubt. Attention must be paid to the fact that the sounds were not the least that could be heard, and that before a comparison is made with the numbers given by other experimenters there should be a division by 2, if not by 3. But this consideration does not fully explain the difference between the above table and that of Ferraris already quoted, from which it appears that in his experiments a current of 5×10^{-9} amperes was audible.

It is interesting to note that the sensitiveness of the telephone to periodic currents is of the same order as that of the galvanometer of equal resistance to steady currents*, viz. that the currents (at pitch 512) just audible in the telephone would, on commutation, be just easily visible by a deflection in the latter instrument. But there is probably more room for further refinements in the galvanometer than in the telephone.

If we compare the performances of the two telephones T_1 and T_2, we find ratios of sensitiveness to current ranging from 13 to 30; so that T_2 shows itself inferior in a degree beyond what may be accounted for by the resistances. It is singular that an experiment of another kind led to the opposite conclusion. The circuit of a Daniell cell A was permanently closed through resistance-coils of 5 ohms and of 1000 ohms. The two telephones in series with one another and with a resistance-box C were placed in a derived circuit where was also a scraping contact-apparatus B, as indicated in the figure.

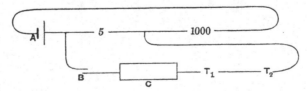

The adjustment was made by varying the resistance in C until the sound was just easily audible in the telephone under trial. Experiments conducted

* See, for example, Ayrton, Mather, and Sumpner, *Phil. Mag.* Vol. xxx. p. 90, 1890, " On Galvanometers."

upon this plan showed that T_1 was only about five times as sensitive to current as T_2. It was noticed, however, that the sounds, though as equal as could be estimated, were not of the same quality, and in this probably lies the explanation of the discrepancy between the two methods of experimenting. In the latter the original sound is composite, and the telephone selects the most favourable elements—that is, those nearly in agreement with the natural pitch of its own plate. In this way the loudness of the selected sound becomes a question of the freedom of vibration of the plate, an element which is almost without influence when the sound is of pitch far removed from that of the proper tone of the telephone. There was independent reason for the suspicion that T_1 had not so well defined a proper pitch as was met with in the case of some other telephones.

P.S.—Measurements with the electro-dynamometer have been made by Cross and Page* of the currents used in practical telephony. The experiments were varied by the employment of several transmitters, and various vowel sounds were investigated. The currents found were of the order 2×10^{-4} amperes.

* *Electrical Review*, Nov. 14, 1885. I owe this reference to Mr Swinburne.

212.

AN ATTEMPT AT A QUANTITATIVE THEORY OF THE TELEPHONE.

[*Philosophical Magazine*, XXXVIII. pp. 295—301, 1894*.]

THE theory of the telephone cannot be said to be understood, in any but the most general manner, until it is possible to estimate from the data of construction what its sensitiveness should be, at least so far as to connect the magnitude of the vibratory current with the resulting condensations and rarefactions in the external ear-passage. Unfortunately such an estimate is a matter of extreme difficulty, partly on account of imperfection in our knowledge of the magnetic properties of iron, and partly from mathematical difficulties arising from the particular forms employed in actual construction; and indeed the problem does not appear to have been attacked hitherto. In view, however, of the doubts that have been expressed as to theory, and of the highly discrepant estimates of actual sensitiveness which have been put forward, it appears desirable to make the attempt. It will be understood that at present the question is as to the order of magnitude only, and that the result will not be without value should it prove to be 10 or even 100 times in error.

One of the elements required to be known, the number (n) of convolutions, cannot be directly observed in the case of a finished instrument; but it may be inferred with sufficient accuracy for the present purpose from the dimensions and the *resistance* of the coil. Denote the axial dimension by ξ, the inner and outer radii by η_1 and η_2, the section of the wire by σ and its total length by l, so that $l\sigma$ is the total volume of copper. The area of section of the coil by an axial plane is $\xi(\eta_2 - \eta_1)$, and of this the area $n\sigma$ is occupied by

---

* Read at the Oxford Meeting of the British Association.

copper. If we suppose the latter to be *half* the former, we shall not be far from the mark. Thus

$$n\sigma = \tfrac{1}{2}\xi\,(\eta_2 - \eta_1). \quad \dots\dots\dots\dots\dots\dots\dots(1)$$

On the same assumption,

$$l\sigma = \tfrac{1}{2}\,\pi\xi\,(\eta_2{}^2 - \eta_1{}^2). \quad \dots\dots\dots\dots\dots\dots(2)$$

Accordingly, if R be the whole resistance of the coil, and r the specific resistance of copper,

$$R = r\frac{l}{\sigma} = \frac{2\pi r n^2(\eta_2 + \eta_1)}{\xi\,(\eta_2 - \eta_1)}. \quad \dots\dots\dots\dots\dots(3)$$

As applicable to actual telephones we may take $\xi = 1$ centim., $\eta_2 = 3\eta_1$; and then $R = 4\pi r n^2$. In c.g.s. measure $r = 1600$, and thus

$$n^2 = \frac{R}{4\pi \times 1600}. \quad \dots\dots\dots\dots\dots\dots\dots(4)$$

If the resistance be 100 ohms, $R = 10^{11}$, and $n = 2230$.

When the resistance varies, other circumstances remaining the same,

$$n \propto \sqrt{R}.$$

We have now to connect the periodic force upon the telephone-plate with the periodic current in the coil. As has already been stated, only a very rough estimate is possible *à priori*. We will commence by considering the case of an unlimited cylindrical core, divided by a transverse fracture into two parts, and encompassed by an infinite cylindrical magnetizing coil containing n turns to the centimetre. If γ be the current, the magnetizing force δH due to it is

$$\delta H = 4\pi n\gamma. \quad \dots\dots\dots\dots\dots\dots\dots(5)$$

If we regard the core as composed of soft iron, magnetized strongly by a constant force H, the mechanical force with which the two parts attract one another per unit of area is in the usual notation

$$I(H + 2\pi I);$$

and what we require is the variation of this quantity, when H becomes $H + \delta H$. This may be written

$$\delta H \left\{ I + (H + 4\pi I)\frac{dI}{dH} \right\}. \quad \dots\dots\dots\dots(6)$$

The value of dI/dH to be here employed is that appropriate to small cyclical changes. It is greatest when I is small, and then [*] amounts to about $100/4\pi$. As I increases, dI/dH diminishes, and finally approaches to zero in the state of saturation. In order to increase (6) it is thus advisable to augment I up to a certain point, but not to approach saturation so nearly as to

---

* *Phil. Mag.* xxiii. p. 225 (1887). [Vol. ii. p. 579.]

bring about a great diminution in the value of dI/dH. In the absence of precise information we may estimate that the maximum of (6) will be reached when I is about half the saturation value, or equal to 800*; and that dI/dH also has half its maximum value, or $50/4\pi$. At this rate the force due to δH is about $40,000\,\delta H$, reckoned per unit of area of the divided core, or by (5)

$$40,000 \times 4\pi n\gamma. \quad\quad\quad\quad (7)$$

But before (7) can be applied to the core of a telephone electromagnet it must be subjected to large deductions. For in the telephone the total number of windings n is limited to about one centimetre measured parallel to the axis, whereas in (7) the electromagnet is supposed to be infinitely long, and n denotes the number of windings *per* centimetre. If we are to suppose in (7) that the windings are really limited to one centimetre, lying immediately on one side of the division, there must be a loss of effect which I estimate at 5 times. We have now further to imagine the second part of the divided cylinder to be replaced by the plate of the telephone, and that not in actual contact with the remaining cylindrical part. The reduction of effect on this account I estimate at 4 times†. The force on the telephone-plate per unit area of core is thus

$$2000 \times 4\pi n\gamma; \quad\quad\quad\quad (8)$$

or if, as for the telephone of 100-ohms resistance, $n = 2200$, and area of section $= \cdot 31$ sq. cm.,

$$\text{force} = 1\cdot 7 \times 10^7 \gamma. \quad\quad\quad\quad (9)$$

In (9) the force is in dynes, and the current γ is in c.g.s. measure. If Γ denote the current reckoned in amperes,

$$\text{force} = 1\cdot 7 \times 10^6 \Gamma, \quad\quad\quad\quad (10)$$

and this must be supposed to be operative at the centre of the plate.

We shall presently consider what effect such a force may be expected to produce; but before proceeding to this I may record the result of some experiments directed to check the applicability of (10), and made subsequently to the theoretical estimates. A Bell telephone, similar to T_1, was mounted vertically, mouth downwards, having attached to the centre of its plate a slender strip of glass. This strip was also vertical and carried at its lower end a small scale-pan. The whole weight of the attachments was only $\cdot 44$ gram. The movement of the glass strip in the direction of its length was observed through a reading-microscope focused upon accidental markings. The telephone, itself of 70-ohms resistance, was connected through a reversing-key with a Daniell cell and with an external resistance varied from time to time. In taking an observation the current was first sent in such a direction as to depress the plate, and the web was adjusted upon the mark. The

---

* Ewing, *Magnetic Induction*, 1891, p. 136.

† I should say that these estimates were all made in ignorance of the result to which they would lead.

current was then reversed, by which the plate was drawn up, but by addition of weights in the pan it was brought back again to the *same* position as before. The force due to the current is thus measured by the *half* of the weight applied.

The results were as follows:—

External resistance in ohms...... 100   200   500
Weight in grams   8   4   2

When 1000 ohms were included, the displacement on reversal was still just visible. We may conclude that a force of 1 gram weight corresponds to a current of about $\frac{1}{300}$ of an ampere. Now, 1 gram weight is equal to 981 dynes, so that for comparison with (10)

$$\text{force} = \cdot 6 \times 10^6 \Gamma. \quad\dots\dots\dots\dots\dots\dots\dots\dots\dots(11)$$

The force observed is thus about the third part of that which had been estimated, and the agreement is sufficient.

Although not needed for the above comparison, we shall presently require to know the linear displacement of the centre of the telephone-plate due to a given force.   Observations with the aid of a micrometer-eyepiece showed that a force of 5 grams weight gave a displacement of $10^{-4} \times 6 \cdot 62$ centim., or $10^{-4} \times 1 \cdot 32$ for each gram, viz. $10^{-7} \times 1 \cdot 34$ centim. per dyne.   Thus by (11) the displacement x due to a current Γ expressed in amperes is

$$x = \cdot 080 \Gamma. \quad\dots\dots\dots\dots\dots\dots\dots\dots\dots(12)$$

We have now to estimate what motion of the telephone-plate may be expected to result from a given periodic force operating at its centre.   The effect depends largely upon the relation between the frequency of the imposed vibration and those natural to the plate regarded as a freely vibrating body. If we attempt to calculate the natural frequencies *à priori*, we are met by uncertainty as to the precise mechanical conditions.   From the manner in which a telephone-plate is supported we should naturally regard the ideal condition as one in which the whole of the circular boundary is *clamped*.   On this basis a calculation may be made, and it appears* that the frequency of the gravest symmetrical mode should be about 991 in the case of the telephone in question.   But it may well be doubted whether we are justified in assuming that the clamping is complete, and any relaxation tells in the direction of a lowered frequency.   A more trustworthy conclusion may perhaps be founded upon the observed connexion between displacement and force of restitution, coupled with an estimate of the inertia of the moving parts.   The total weight of the plate is 3·4 grams; the outside diameter is 5·7 centim., and the inside diameter, corresponding to the free portion of

---

* *Theory of Sound,* 2nd ed. § 221 *a.*

the plate, is 4·5. The effective mass, supposed to be situated at the centre, I estimate to be that corresponding to a diameter of 2·5 centim., viz. ·65 gram. A force of restitution per unit displacement equal to $(10^{-7} \times 1·34)^{-1}$, or $10^6 \times 7·5$, is supposed to urge the above mass to its position of equilibrium. The frequency of the resulting vibration is

$$\frac{1}{2\pi} \sqrt{\left\{\frac{10^6 \times 7·5}{·65}\right\}} = 541.$$

With the aid of a special electric maintenance the plate may be made to *speak* on its own account. The frequency so found, viz. 896, corresponds undoubtedly to a free vibration, but it does not follow that the vibration is the gravest of which the plate is capable; and there were indications pointing to the opposite conclusion.

As it is almost impossible to form an *à priori* estimate of the amplitude of vibration (x) when the frequency of the force is in the neighbourhood of any of the free frequencies, I will take for calculation the case of frequency 256, which is presumably much lower than any of them. Under these circumstances an "equilibrium theory" may be employed, the displacement coexisting with any applied force being the same as if the force were permanent. At this pitch the minimum current recorded in the table* is $8·3 \times 10^{-7}$ amperes; so that by (12) the maximum excursion corresponding thereto is given by $x = ·080 \times 8·3 \times 10^{-7} = 6·8 \times 10^{-8}$ centim.

The excursion thus found must not be compared with that calculated formerly † for free progressive waves. The proper comparison is rather between the *condensations* s in the two cases. In a progressive wave the connexion between s and v, the maximum velocity, is $v = as$, where a is the velocity of propagation. But in the present case the excursion x takes effect upon a very small volume. If A be the effective area of the plate, and S the whole volume included between the plate and the tympanum of the ear, we may take $s = Ax/S$. This relation assumes that the condensations and rarefactions are uniform throughout the space in question, an assumption justified by the smallness of its dimensions in comparison with the wave-length, and further that the behaviour is the same as if the space were closed air-tight. It would seem that a slight deficiency in the latter respect would not be material.

For the numerical application I estimate that $A = 4$ sq. cm., $S = 20$ cub cm.; so that with the above value of x

$$s = 1·4 \times 10^{-8}, \quad \dots\dots\dots\dots\dots\dots\dots\dots(13)$$

s being reckoned in atmospheres.

---

* *Supra*, p. 294. [Vol. IV. p. 117.]

† *Proc. Roy. Soc.* Vol. XXVI. p. 248 (1877). [Vol. I. p. 328.]

The value of s corresponding to but just audible progressive waves of frequency 256 was found to be $5·9 \times 10^{-9}$, in sufficiently good agreement with (13)*.

But if the equilibrium theory be applied to the notes of higher pitch, such as 512, we find the actual sensitiveness of the telephone greater than according to the calculation. In this case† $\Gamma = 7 \times 10^{-8}$; so that by (12)

$$x = 5·6 \times 10^{-9},$$

and

$$s = Ax/S = 1·1 \times 10^{-9}, \quad \dots\dots\dots\dots\dots(14)$$

decidedly smaller than that $(4·5 \times 10^{-9})$ deduced from the observations upon progressive waves. The conclusion seems to be that for these frequencies the equilibrium theory of the telephone-plate fails, and that in virtue of resonance the sensitiveness of the instrument is specially exalted.

I will not dwell further upon these calculations, which involve too much guesswork to be very satisfactory. They suffice, however, to show that the "push and pull" theory is capable of giving an adequate account of the action of the telephone, so far at least as my own observations are concerned. But it is doubtful, to say the least, whether it could be reconciled with estimates of sensitiveness such as those of Tait and of Preece.

* I hope shortly to publish an account of the observations upon which this statement is founded. [See following Art. 213.]

† *Supra*, p. 294. [Vol. IV. p. 117.]

213.

ON THE AMPLITUDE OF AERIAL WAVES WHICH ARE BUT JUST AUDIBLE.

[*Philosophical Magazine*, XXXVIII. pp. 365—370, 1894*.]

THE problem of determining the absolute value of the amplitude, or particle velocity, of a sound which is but just audible to the ear, is one of considerable difficulty. In a short paper published seventeen years ago† I explained a method by which it was easy to demonstrate a superior limit. A whistle, blown under given conditions, consumes a known amount of energy per second. Upon the assumption that the whole of this energy is converted into sound, that the sound is conveyed without loss, and that it is uniformly distributed over the surface of a hemisphere, it is easy to calculate the amplitude at any distance; and the result is necessarily a superior limit to the actual amplitude. In the case of the whistle experimented on, of frequency 2730, the superior limit so arrived at for a sound just easily audible was 8.1×10^{-8} cm. The maximum particle velocity v and the maximum condensation s are the quantities more immediately determined by the observations, and they are related by the well-known equation $v = as$, in which a denotes the velocity of propagation. In the experiment above referred to the superior limit for v was $.0014$ cm. per second, and that for s was 4.1×10^{-8}. I estimated that on a still night an amplitude, or velocity, one-tenth of the above would probably be audible. A very similar number has been arrived at by Wien‡, who used an entirely different method§.

In connexion with calculations respecting the sensitiveness of telephones, I was desirous of checking the above estimates, and made some attempts to do so by the former method. In order to avoid possible complications of

---

* Read at the Oxford Meeting of the British Association.

† *Proc. Roy. Soc.* Vol. XXVI. p. 248 (1878). [Vol. I. p. 328.]

‡ *Wied. Ann.* XXXVI. p. 834 (1889).

§ The first estimate of the amplitude of but just audible sounds, with which I have only recently become acquainted, is that of Töpler and Boltzmann (*Pogg. Ann.* CXLI. p. 321 (1870)). It depends upon an ingenious application of v. Helmholtz's theory of the open organ-pipe to data relating to the maximum condensation within the pipe as obtained by the authors experimentally. The value of s was found to be 6.5×10^{-8} for a pitch of 181.—August 21.

atmospheric refraction which may occur when large distances are in question, I sought to construct pipes which should generate sound of given pitch upon a much smaller scale, but with the usual economy of wind. In this I did not succeed, and it seems as if there is some obstacle to the desired reduction of scale.

The experiments here to be recorded were conducted with tuning-forks. A fork of known dimensions, vibrating with a known amplitude, may be regarded as a store of energy of which the amount may readily be calculated. This energy is gradually consumed by internal friction and by generation of sound. When a resonator is employed the latter element is the more important, and in some cases we may regard the dying down of the amplitude as sufficiently accounted for by the emission of sound. Adopting this view for the present, we may deduce the rate of emission of sonorous energy from the observed amplitude of the fork at the moment in question and from the rate at which the amplitude decreases. Thus if the law of decrease be $e^{-\frac{1}{2}kt}$ for the amplitude of the fork, or e^{-kt} for the energy, and if E be the total energy at time t, the rate at which energy is emitted at that time is $-dE/dt$, or kE. The value of k is deducible from observations of the rate of decay, *e.g.* of the time during which the amplitude is halved. With these arrangements there is no difficulty in converting energy into sound upon a small scale, and thus in reducing the distance of audibility to such a figure as 30 metres. Under these circumstances the observations are much more manageable than when the operators are separated by half a mile, and there is no reason to fear disturbance from atmospheric refraction.

The fork is mounted upon a stand to which is also firmly attached the observing-microscope. Suitable points of light are obtained from starch grains, and the line of light into which each point is extended by the vibration is determined with the aid of an eyepiece-micrometer. Each division of the micrometer-scale represents 001 centim. The resonator, when in use, is situated in the position of maximum effect, with its mouth under the free ends of the vibrating prongs.

The course of an experiment was as follows:—In the first place the rates of dying down were observed, with and without the resonator, the stand being situated upon the ground in the middle of a lawn. The fork was set in vibration with a bow, and the time required for the double amplitude to fall to half its original value was determined. Thus in the case of a fork of frequency 256, the time during which the vibration fell from 20 micrometer-divisions to 10 micrometer-divisions was 16^s without the resonator, and $\cdot9^s$ when the resonator was in position. These times of halving were, as far as could be observed, independent of the initial amplitude. To determine the minimum audible, one observer (myself) took up a position 30 yards (27·4 metres) from the fork, and a second (Mr Gordon) communicated a large vibration to the fork. At the moment when the double amplitude measured 20 micrometer-divisions the second observer gave a signal, and immediately afterwards withdrew to a distance. The business of the first observer was to estimate for how many seconds after the signal the sound still remained audible. In the case referred to the time was 12^s. When the distance was reduced to 15 yards (13·7 metres), an initial double amplitude of 10 micrometer-divisions was audible for almost exactly the same time.

These estimates of audibility are not made without some difficulty. There are usually 2 or 3 seconds during which the observer is in doubt whether he hears or only imagines, and different individuals decide the question in opposite ways. There is also of course room for a real difference of hearing, but this has not obtruded itself much. A given observer on a given day will often agree with himself surprisingly well, but the accuracy thus suggested is, I think, illusory. Much depends upon freedom from disturbing noises. The wind in the trees or the twittering of birds embarrasses the observer, and interferes more or less with the accuracy of results.

The equality of emission of sound in various horizontal directions was tested, but no difference could be found. The sound issues almost entirely from the resonator, and this may be expected to act as a simple source.

When the time of audibility is regarded as known, it is easy to deduce the amplitude of the vibration of the fork at the moment when the sound ceases to impress the observer. From this the rate of emission of sonorous energy and the amplitude of the aerial vibration as it reaches the observer are to be calculated.

The first step in the calculation is the expression of the total energy of the fork as a function of the amplitude of vibration measured at the extremity of one of the prongs. This problem is considered in § 164 of my *Theory of Sound*. If l be the length, ρ the density, and ω the sectional area of a rod clamped at one end and free at the other, the kinetic energy T is connected with the displacement η at the free end by the equation (10)

$$T = \tfrac{1}{6}\rho l\omega (d\eta/dt)^2.$$

At the moment of passage through the position of equilibrium $\eta = 0$ and $d\eta/dt$ has its maximum value, the whole energy being then kinetic. The maximum value of $d\eta/dt$ is connected with the maximum value of η by the equation

$$(d\eta/dt)_{\text{max.}} = 2\pi/\tau \cdot (\eta)_{\text{max.}};$$

so that if we now denote the double amplitude by 2η, the whole energy of the vibrating bar is

$$\tfrac{1}{8}\rho\omega l\pi^2/\tau^2 \cdot (2\eta)^2,$$

or for the two bars composing the fork

$$E = \tfrac{1}{4}\rho\omega l\pi^2/\tau^2 \cdot (2\eta)^2, \quad \dots\dots\dots\dots\dots(A)$$

where $\rho\omega l$ is the mass of each prong.

The application of (A) to the 256-fork, vibrating with a double amplitude of 20 micrometer-divisions, is as follows.   We have

$$l = 14 \cdot 0 \text{ cm.}, \qquad \omega = \cdot 6 \times 1 \cdot 1 = \cdot 66 \text{ sq. cm.},$$
$$1/\tau = 256, \qquad \rho = 7 \cdot 8, \qquad 2\eta = \cdot 050 \text{ cm.};$$

and thus

$$E = 4 \cdot 06 \times 10^3 \text{ ergs.}$$

This is the whole energy of the fork when the actual double amplitude at the ends of the prongs is ·050 centim.

As has already been shown, the energy lost per second is kE, if the amplitude vary as $e^{-\frac{1}{2}kt}$. For the present purpose k must be regarded as made up of two parts, one k_1 representing the dissipation which occurs in the absence of the resonator, the other k_2 due to the resonator. It is the latter part only which is effective towards the production of sound. For when the resonator is out of use the fork is practically silent; and, indeed, even if it were worth while to make a correction on account of the residual sound, its phase would only accidentally agree with that of the sound issuing from the resonator.

The values of k_1 and k are conveniently derived from the times, t_1 and t, during which the amplitude falls to *one-half*. Thus

$$k = 2 \log_e 2 . / t, \qquad k_1 = 2 \log_e 2 . / t_1;$$

so that

$$k_2 = 2 \log_e 2 . (1/t - 1/t_1) = 1 \cdot 386 \, (1/t - 1/t_1).$$

And the energy converted into sound per second is $k_2 E$.

We may now apply these formulæ to the case, already quoted, of the 256-fork, for which $t = 9$, $t_1 = 16$. Thus t_2, the time which would be occupied in halving the amplitude were the dissipation due entirely to the resonator, is 20·6; and $k_2 = \cdot 0674$. Accordingly,

$$k_2 E = 267 \text{ ergs per second,}$$

corresponding to a double amplitude represented by 20 micrometer-divisions. In the experiment quoted the duration of audibility was 12 seconds, during which the amplitude would fall in the ratio $2^{12/9} : 1$, and the energy in the ratio $4^{12/9} : 1$. Hence at the moment when the sound was just becoming inaudible the energy emitted as sound was 42·1 ergs per second[*].

* It is of interest to compare with the energy-emission of a source of light. An incandescent electric-lamp of 200 candles absorbs about a horse-power, or say 10^{10} ergs per second. Of the total radiation only about $\frac{1}{100}$ part acts effectively upon the eye; so that radiation of suitable quality consuming 5×10^5 ergs per second corresponds to a candle-power. This is about 10^4 times that emitted as sound by the fork in the experiment described above. At a distance of $10^2 \times 30$, or 3000 metres the stream of energy from the ideal candle would be about equal to the stream of energy just audible to the ear. It appears that the streams of energy required to influence the eye and the ear are of the same order of magnitude, a conclusion already drawn by Töpler and Boltzmann.—August 21.

The question now remains, What is the corresponding amplitude or condensation in the progressive aerial waves at 27·4 metres from the source? If we suppose, as in my former calculations, that the ground reflects well, we are to treat the waves as hemispherical. On the whole this seems to be the best supposition to make, although the reflexion is doubtless imperfect. The area S covered at the distance of the observer is thus $2\pi \times 2740^2$ sq. centim., and since*

$$S \cdot \tfrac{1}{2} a \rho v^2 = S \cdot \tfrac{1}{2} \rho a^3 s^2 = 42 \cdot 1,$$

we find

$$s^2 = \frac{42 \cdot 1}{\pi \times 2740^2 \times \cdot 00125 \times 34100^3},$$

and

$$s = 6 \cdot 0 \times 10^{-9}.$$

The condensation s is here reckoned in atmospheres; and the result shows that the ear is able to recognize the addition and subtraction of densities far less than those to be found in our highest vacua.

The amplitude of aerial vibration is given by $as\tau/2\pi$, where $1/\tau = 256$, and is thus equal to $1 \cdot 27 \times 10^{-7}$ cm.

It is to be observed that the numbers thus obtained are still somewhat of the nature of superior limits, for they depend upon the assumption that all the dissipation *due to the resonator* represents production of sound. This may not be strictly the case even with the moderate amplitudes here in question, but the uncertainty under this head is far less than in the case of resonators or organ-pipes caused to speak by wind. From the nature of the calculation by which the amplitude or condensation in the aerial waves is deduced, a considerable loss of energy does not largely influence the final numbers.

Similar experiments have been tried at various times with forks of pitch 384 and 512. The results were not quite so accordant as was at first hoped might be the case, but they suffice to fix with some approximation the condensation necessary for audibility. The mean results are as follows:—

c',   frequency $= 256$, $s = 6 \cdot 0 \times 10^{-9}$,

g',   „ $= 384$, $s = 4 \cdot 6 \times 10^{-9}$,

c'',   „ $= 512$, $s = 4 \cdot 6 \times 10^{-9}$,

no reliable distinction appearing between the two last numbers. Even the distinction between 6·0 and 4·6 should be accepted with reserve; so that the comparison must not be taken to prove much more than that the condensation necessary for audibility varies but slowly in the singly dashed octave.

* *Theory of Sound*, § 245.

214.

ARGON, A NEW CONSTITUENT OF THE ATMOSPHERE*.

BY LORD RAYLEIGH, SEC. R.S., AND PROFESSOR WILLIAM RAMSAY, F.R.S.

[*Philosophical Transactions*, 186 (A), pp. 187—241, 1895.]

"Modern discoveries have not been made by large collections of facts, with subsequent discussion, separation, and resulting deduction of a truth thus rendered perceptible. A few facts have suggested an *hypothesis*, which means a *supposition*, proper to explain them. The necessary results of this supposition are worked out, and then, and not till then, other facts are examined to see if their ulterior results are found in Nature."—De Morgan, *A Budget of Paradoxes*, Ed. 1872, p. 55.

1. *Density of Nitrogen from Various Sources.*

IN a former paper† it has been shown that nitrogen extracted from chemical compounds is about one-half per cent. lighter than "atmospheric nitrogen."

The mean numbers for the weights of gas contained in the globe used were as follows :—

	grams.
From nitric oxide	2·3001
From nitrous oxide	2·2990
From ammonium nitrite	2·2987

while for "atmospheric" nitrogen there was found—

By hot copper, 1892	2·3103
By hot iron, 1893	2·3100
By ferrous hydrate, 1894	2·3102

At the suggestion of Professor Thorpe, experiments were subsequently tried with nitrogen liberated from *urea* by the action of sodium hypobromite.

---

* This memoir is included in the present collection by kind permission of Prof. Ramsay.

† Rayleigh, "On an Anomaly encountered in Determinations of the Density of Nitrogen Gas," *Proc. Roy. Soc.* Vol. LV. p. 340, 1894. [Vol. IV. p. 104.]

The carbon and hydrogen of the urea are supposed to be oxidized by the reaction to CO_2 and H_2O, the former of which would be retained by the large excess of alkali employed. It was accordingly hoped that the gas would require no further purification than drying. If it proved to be light, it would at any rate be free from the suspicion of containing hydrogen.

The hypobromite was prepared from commercial materials in the proportions recommended for the analysis of urea—100 grams. caustic soda, 250 cub. centims. water, and 25 cub. centims. of bromine. For our purpose about one and a half times the above quantities were required. The gas was liberated in a bottle of about 900 cub. centims. capacity, in which a vacuum was first established. The full quantity of hypobromite solution was allowed to run in slowly, so that any dissolved gas might be at once disengaged. The urea was then fed in, at first in a dilute condition, but, as the pressure rose, in a 10 per cent. solution. The washing out of the apparatus, being effected with gas in a highly rarefied state, made but a slight demand upon the materials. The reaction was well under control, and the gas could be liberated as slowly as desired.

In the first experiment, the gas was submitted to no other treatment than slow passage through potash and phosphoric anhydride, but it soon became apparent that the nitrogen was contaminated. The "inert and inodorous" gas attacked vigorously the mercury of the Töpler pump, and was described as smelling like a dead rat. As to the weight, it proved to be in excess even of the weight of atmospheric nitrogen.

The corrosion of the mercury and the evil smell were in great degree obviated by passing the gas over hot metals. For the fillings of June 6, 9, 13, the gas passed through a short length of tube containing copper in the form of fine wire, heated by a flat Bunsen burner, then through the furnace over red-hot iron, and back over copper oxide. On June 19 the furnace tubes were omitted, the gas being treated with the red-hot copper only. The results, reduced so as to correspond with those above quoted, were—

June 6	2·2978
„ 9	2·2987
„ 13	2·2982
„ 19	2·2994
Mean	2·2985

Without using heat it has not been found possible to prevent the corrosion of the mercury. Even when no urea is employed, and air simply bubbled through the hypobromite solution is allowed to pass with constant shaking over mercury contained in a U-tube, the surface of the metal was soon fouled. When *hypochlorite* was substituted for *hypobromite* in the last experiment there was a decided improvement, and it was thought desirable to try whether the gas prepared from hypochlorite and urea would be pure on simple desiccation. A filling on June 25 gave as the weight 2·3343, showing an excess of 36 mgs., as compared with other chemical nitrogen, and of about 25 mgs. as compared with atmospheric nitrogen. A test with alkaline pyrogallate appeared to prove the absence from this gas of free oxygen, and only a trace of carbon could be detected when a considerable quantity of the gas was passed over red-hot cupric oxide into solution of baryta.

Although the results relating to urea nitrogen are interesting for comparison with that obtained from other nitrogen compounds, the original object was not attained on account of the necessity of retaining the treatment with hot metals. We have found, however, that nitrogen from ammonium nitrite may be prepared without the employment of hot tubes, whose weight agrees with that above quoted. It is true that the gas smells slightly of ammonia, easily removable by sulphuric acid, and apparently also of oxides of nitrogen. The solution of potassium nitrite and ammonium chloride was heated in a water-bath, of which the temperature rose to the boiling-point only towards the close of operations. In the earlier stages the temperature required careful watching in order to prevent the decomposition taking place too rapidly. The gas was washed with sulphuric acid, and after passing a Nessler test, was finally treated with potash and phosphoric anhydride in the usual way. The following results have been obtained :—

$$
\begin{array}{ll}
\text{July } 4 & 2\cdot2983 \\
\text{,, } 9 & 2\cdot2989 \\
\text{,, } 13 & 2\cdot2990 \\
\hline
\text{Mean} & 2\cdot2987
\end{array}
$$

It will be seen that in spite of the slight nitrous smell there is no appreciable difference in the densities of gas prepared from ammonium nitrite with and without the treatment by hot metals. The result is interesting, as showing that the agreement of numbers obtained for chemical nitrogen does not depend upon the use of a red heat in the process of purification.

The five results obtained in more or less distinct ways for chemical nitrogen stand thus :—

$$
\begin{array}{ll}
\text{From nitric oxide} & 2\cdot3001 \\
\text{From nitrous oxide} & 2\cdot2990 \\
\text{From ammonium nitrite purified at a red heat} & 2\cdot2987 \\
\text{From urea} & 2\cdot2985 \\
\text{From ammonium nitrite purified in the cold} & 2\cdot2987 \\
\hline
\text{Mean} & 2\cdot2990
\end{array}
$$

These numbers, as well as those above quoted for "atmospheric nitrogen," are subject to a correction (additive)* of ·0006 for the shrinkage of the globe when exhausted†. If they are then multiplied in the ratio of 2·3108 : 1·2572, they will express the weights of the gas in grams. per litre. Thus, as regards the mean numbers, we find as the weight per litre under standard conditions of chemical nitrogen 1·2511, that of atmospheric nitrogen being 1·2572.

It is of interest to compare the density of nitrogen obtained from chemical compounds with that of oxygen. We have $N_2 : O_2 = 2·2996 : 2·6276 = 0·87517$; so that if $O_2 = 16$, $N_2 = 14·003$. Thus, when the comparison is with chemical nitrogen, the ratio is very nearly that of 16 : 14. But if "atmospheric nitrogen" be substituted, the ratio of small integers is widely departed from.

The determination by Stas of the atomic weight of nitrogen from synthesis of silver nitrate is probably the most trustworthy, inasmuch as the atomic weight of silver was determined with reference to oxygen with the greatest care, and oxygen is assumed to have the atomic weight 16. If, as found by Stas, $AgNO_3 : Ag = 1·57490 : 1$, and $Ag : O = 107·930 : 16$, then

$$N : O = 14·049 : 16.$$

To the above list may be added nitrogen, prepared in yet another manner, whose weight has been determined subsequently to the isolation of the new dense constituent of the atmosphere. In this case nitrogen was actually extracted from air by means of magnesium. The nitrogen thus separated was then converted into ammonia by action of water upon the magnesium nitride, and afterwards liberated in the free state by means of calcium hypo-chlorite. The purification was conducted in the usual way, and included passage over red-hot copper and copper oxide. The following was the result:—

Globe empty, October 30, November 5 . . 2·82313
Globe full, October 31 . . . . ·52395
Weight of gas . . . . 2·29918

It differs inappreciably from the mean of other results, viz., 2·2990, and is of special interest as relating to gas which, at one stage of its history, formed part of the atmosphere.

Another determination with a different apparatus of the density of "chemical" nitrogen from the same source, magnesium nitride, which had been prepared by passing "atmospheric" nitrogen over ignited magnesium, may here be recorded. The sample differed from that previously mentioned, inasmuch as it had not been subjected to treatment with red-hot copper.

[* In the Abstract of this paper (*Proc. Roy. Soc.* Vol. LVII. p. 265) the correction of ·0006 was erroneously treated as a deduction.—April, 1895.]

† Rayleigh, "On the Densities of the Principal Gases," *Proc. Roy. Soc.* Vol. LIII. p. 134, 1893. [Vol. IV. p. 39.]

After treating the nitride with water, the resulting ammonia was distilled off, and collected in hydrochloric acid; the solution was evaporated to dryness; the dry ammonium chloride was dissolved in water, and its concentrated solution added to a freshly prepared solution of sodium hypobromite. The nitrogen was collected in a gas-holder over water which had previously been boiled, so as at all events partially to expel air. The nitrogen passed into the vacuous globe through a solution of potassium hydroxide, and through two drying-tubes, one containing soda-lime, and the other phosphoric anhydride.

At 18·38° C. and 754·4 mgs. pressure, 162·843 cub. centims. of this nitrogen weighed 0·18963 gram. Hence:—

Weight of 1 litre at 0° C. and 760 millims. pressure ... 1·2521 gram.

The mean result of the weight of 1 litre of "chemical" nitrogen has been found to equal 1·2511. It is therefore seen that "chemical" nitrogen, derived from "atmospheric" nitrogen, without any exposure to red-hot copper, possesses the usual density.

Experiments were also made, which had for their object to prove that the ammonia, produced from the magnesium nitride, is identical with ordinary ammonia, and contains no other compound of a basic character. For this purpose, the ammonia was converted into ammonium chloride, and the percentage of chloride determined by titration with a solution of silver nitrate which had been standardized by titrating a specimen of pure sublimed ammonium chloride. The silver solution was of such a strength that 1 cub. centim. precipitated the chlorine from 0·001701 gram. of ammonium chloride.

1. Ammonium chloride from orange-coloured sample of magnesium nitride.

0·1106 gram. required 43·10 cub. centims. of silver nitrate = 66·35 per cent. of chlorine.

2. Ammonium chloride from blackish magnesium nitride.

0·1118 gram. required 43·6 cub. centims. of silver nitrate = 66·35 per cent. of chlorine.

3. Ammonium chloride from nitride containing a large amount of unattacked magnesium.

0·0630 gram. required 24·55 cub. centims. of silver nitrate = 66·30 per cent. of chlorine.

Taking for the atomic weights of hydrogen, H = 1·0032, of nitrogen, N = 14·04, and of chlorine, Cl = 35·46, the theoretical amount of chlorine in ammonium chloride is 66·27 per cent.

From these results—that nitrogen prepared from magnesium nitride obtained by passing "atmospheric" nitrogen over red-hot magnesium has the density of "chemical" nitrogen, and that ammonium chloride prepared from magnesium nitride contains practically the same percentage of chlorine as pure ammonium chloride—it may be concluded that red-hot magnesium withdraws from "atmospheric" nitrogen no substance other than nitrogen capable of forming a basic compound with hydrogen.

In a subsequent part of this paper, attention will again be called to this statement. (See addendum, p. 240.)

2. *Reasons for Suspecting a hitherto Undiscovered Constituent in Air.*

When the discrepancy of weights was first encountered, attempts were naturally made to explain it by contamination with known impurities. Of these the most likely appeared to be hydrogen, present in the lighter gas, in spite of the passage over red-hot cupric oxide. But, inasmuch as the intentional introduction of hydrogen into the heavier gas, afterwards treated in the same way with cupric oxide, had no effect upon its weight, this explanation had to be abandoned; and, finally, it became clear that the difference could not be accounted for by the presence of any known impurity. At this stage it seemed not improbable that the lightness of the gas extracted from chemical compounds was to be explained by partial dissociation of nitrogen molecules N_2 into detached atoms. In order to test this suggestion, both kinds of gas were submitted to the action of the silent electric discharge, with the result that both retained their weights unaltered. This was discouraging, and a further experiment pointed still more markedly in the negative direction. The chemical behaviour of nitrogen is such as to suggest that dissociated atoms would possess a higher degree of activity, and that, even though they might be formed in the first instance, their life would probably be short. On standing, they might be expected to disappear, in partial analogy with the known behaviour of ozone. With this idea in view, a sample of chemically-prepared nitrogen was stored for eight months. But, at the end of this time, the density showed no sign of increase, remaining exactly as at first*.

Regarding it as established that one or other of the gases must be a mixture, containing, as the case might be, an ingredient much heavier or much lighter than ordinary nitrogen, we had to consider the relative probabilities of the various possible interpretations. Except upon the already discredited hypothesis of dissociation, it was difficult to see how the gas of chemical origin could be a mixture. To suppose this would be to admit two kinds of nitric acid, hardly reconcilable with the work of Stas and others

---

* Rayleigh, *Proc. Roy. Soc.* Vol. LV. p. 344, 1894. [Vol. IV. p. 108.]

upon the atomic weight of that substance. The simplest explanation in many respects was to admit the existence of a second ingredient in air from which oxygen, moisture, and carbonic anhydride had already been removed. The proportional amount required was not great. If the density of the supposed gas were double that of nitrogen, one-half per cent. only by volume would be needed; or, if the density were but half as much again as that of nitrogen, then one per cent. would still suffice. But in accepting this explanation, even provisionally, we had to face the improbability that a gas surrounding us on all sides, and present in enormous quantities, could have remained so long unsuspected.

The method of most universal application by which to test whether a gas is pure or a mixture of components of different densities is that of diffusion. By this means Graham succeeded in effecting a partial separation of the nitrogen and oxygen of the air, in spite of the comparatively small difference of densities. If the atmosphere contain an unknown gas of anything like the density supposed, it should be possible to prove the fact by operations conducted upon air which had undergone atmolysis. If, for example, the parts least disposed to penetrate porous walls were retained, the "nitrogen" derived from it by the usual processes should be heavier than that derived in like manner from unprepared air. This experiment, although in view from the first, was not executed until a later stage of the inquiry (§ 6), when results were obtained sufficient of themselves to prove that the atmosphere contains a previously unknown gas.

But although the method of diffusion was capable of deciding the main, or at any rate the first question, it held out no prospect of isolating the new constituent of the atmosphere, and we therefore turned our attention in the first instance to the consideration of methods more strictly chemical. And here the question forced itself upon us as to what really was the evidence in favour of the prevalent doctrine that the inert residue from air after withdrawal of oxygen, water, and carbonic anhydride, is all of one kind.

The identification of "phlogisticated air" with the constituent of nitric acid is due to Cavendish, whose method consisted in operating with electric sparks upon a short column of gas confined with potash over mercury at the upper end of an inverted $\mathbf{U}$-tube*. This tube (M) was only about $\frac{1}{10}$ inch in diameter, and the column of gas was usually about 1 inch in length. After describing some preliminary trials, Cavendish proceeds:—
"I introduced into the tube a little soap-lees (potash), and then let up some dephlogisticated† and common air, mixed in the above-mentioned proportions

---

* "Experiments on Air," *Phil. Trans.* Vol. lxxv. p. 372, 1785.

[† The explanation of combustion in Cavendish's day was still vague. It was generally imagined that substances capable of burning contained an unknown principle, to which the name "phlogiston" was applied, and which escaped during combustion. Thus, metals and hydrogen and other gases were said to be "phlogisticated" if they were capable of burning in air. Oxygen

which rising to the top of the tube M, divided the soap-lees into its two legs. As fast as the air was diminished by the electric spark, I continued adding more of the same kind, till no further diminution took place: after which a little pure dephlogisticated air, and after that a little common air, were added, in order to see whether the cessation of diminution was not owing to some imperfection in the proportion of the two kinds of air to each other; but without effect. The soap-lees being then poured out of the tube, and separated from the quicksilver, seemed to be perfectly neutralised, and they did not at all discolour paper tinged with the juice of blue flowers. Being evaporated to dryness, they left a small quantity of salt, which was evidently nitre, as appeared by the manner in which paper, impregnated with a solution of it, burned."

Attempts to repeat Cavendish's experiment in Cavendish's manner have only increased the admiration with which we regard this wonderful investigation. Working on almost microscopical quantities of material, and by operations extending over days and weeks, he thus established one of the most important facts in chemistry. And what is still more to the purpose, he raises as distinctly as we could do, and to a certain extent resolves, the question above suggested. The passage is so important that it will be desirable to quote it at full length.

"As far as the experiments hitherto published extend, we scarcely know more of the phlogisticated part of our atmosphere than that it is not diminished by lime-water, caustic alkalies, or nitrous air; that it is unfit to support fire or maintain life in animals; and that its specific gravity is not much less than that of common air; so that, though the nitrous acid, by being united to phlogiston, is converted into air possessed of these properties, and consequently, though it was reasonable to suppose, that part at least of the phlogisticated air of the atmosphere consists of this acid united to phlogiston, yet it was fairly to be doubted whether the whole is of this kind, or whether there are not in reality many different substances confounded together by us under the name of phlogisticated air. I therefore made an experiment to determine whether the whole of a given portion of the phlogisticated air of the atmosphere could be reduced to nitrous acid, or whether there was not a part of a different nature to the rest which would refuse to undergo that change. The foregoing experiments indeed in some measure decided this point, as much the greatest part of the air let up into the tube lost its elasticity; yet as some remained unabsorbed it did not appear for certain whether that was of the same nature as the rest or not.

being non-inflammable was named " dephlogisticated air," and nitrogen, because it was incapable of supporting combustion or life was named by Priestley " phlogisticated air," although up till Cavendish's time it had not been made to unite with oxygen.

The term used for oxygen by Cavendish is " dephlogisticated air," and for nitrogen, " phlogisticated air."—April, 1895.]

For this purpose I diminished a similar mixture of dephlogisticated and common air, in the same manner as before, till it was reduced to a small part of its original bulk. I then, in order to decompound as much as I could of the phlogisticated air which remained in the tube, added some dephlogisticated air to it and continued the spark until no further diminution took place. Having by these means condensed as much as I could of the phlogisticated air, I let up some solution of liver of sulphur to absorb the dephlogisticated air; after which only a small bubble of air remained unabsorbed, which certainly was not more than $\frac{1}{120}$ of the bulk of the phlogisticated air let up into the tube; so that, if there is any part of the phlogisticated air of our atmosphere which differs from the rest, and cannot be reduced to nitrous acid, we may safely conclude that it is not more than $\frac{1}{120}$ part of the whole."

Although Cavendish was satisfied with his result, and does not decide whether the small residue was genuine, our experiments about to be related render it not improbable that his residue was really of a different kind from the main bulk of the "phlogisticated air," and contained the gas now called argon.

Cavendish gives data* from which it is possible to determine the rate of absorption of the mixed gases in his experiment. The electrical machine used "was one of Mr Nairne's patent machines, the cylinder of which is $12\frac{1}{2}$ inches long and 7 in diameter. A conductor, 5 feet long and 6 inches in diameter, was adapted to it, and the ball which received the spark was placed two or three inches from another ball, fixed to the end of the conductor. Now, when the machine worked well, Mr Gilpin supposes he got about two or three hundred sparks a minute, and the diminution of the air during the half hour which he continued working at a time varied in general from 40 to 120 measures, but was usually greatest when there was most air in the tube, provided the quantity was not so great as to prevent the spark from passing readily." The "measure" spoken of represents the volume of one grain of quicksilver, or ·0048 cub. centim., so that an absorption of one cub. centim. of mixed gas per hour was about the most favourable rate. Of the mixed gas about two-fifths would be nitrogen.

3. *Methods of Causing Free Nitrogen to Combine.*

The concord between the determinations of density of nitrogen obtained from sources other than the atmosphere, having made it at least probable that some heavier gas exists in the atmosphere, hitherto undetected, it became necessary to submit atmospheric nitrogen to examination, with a view of isolating, if possible, the unknown and overlooked constituent, or it might be constituents.

* *Phil. Trans.* Vol. LXXVIII. p. 271, 1788.

Nitrogen, however, is an element which does not easily enter into direct combination with other elements; but with certain elements, and under certain conditions, combination may be induced. The elements which have been directly united to nitrogen are (a) boron, (b) silicon, (c) titanium, (d) lithium, (e) strontium and barium, (f) magnesium, (g) aluminium, (h) mercury, (i) manganese, (j) hydrogen, and (k) oxygen, the last two by help of an electrical discharge.

(a) *Nitride of boron* was prepared by Wöhler and Deville* by heating amorphous boron to a white heat in a current of nitrogen. Experiments were made to test whether the reaction would take place in a tube of difficultly fusible glass; but it was found that the combination took place at a bright red heat to only a small extent, and that the boron, which had been prepared by heating powdered boron oxide with magnesium dust, was only superficially attacked. Boron is, therefore, not a convenient absorbent for nitrogen. [M. Moissan informs us that the reputation it possesses is due to the fact that early experiments were made with boron which had been obtained by means of sodium, and which probably contained a boride of that metal.—April, 1895.]

(b) *Nitride of silicon†* also requires for its formation a white heat, and complete union is difficult to bring about. Moreover, it is not easy to obtain large quantities of silicon. This method was therefore not attempted.

(c) *Nitride of titanium* is said to have been formed by Deville and Caron‡, by heating titanium to whiteness in a current of nitrogen. This process was not tried by us. As titanium has an unusual tendency to unite with nitrogen, it might, perhaps, be worth while to set the element free in presence of atmospheric nitrogen, with a view to the absorption of the nitrogen. This has, in effect, been already done by Wöhler and Deville§; they passed a mixture of the vapour of titanium chloride and nitrogen over red-hot aluminium, and obtained a large yield of nitride. It is possible that a mixture of the precipitated oxide of titanium with magnesium dust might be an effective absorbing agent at a comparatively low temperature. [Since writing the above we have been informed by M. Moissan that titanium, heated to 800°, burns brilliantly in a current of nitrogen. It might therefore be used with advantage to remove nitrogen from air, inasmuch as we have found that it does not combine with argon.—April, 1895.]

(d), (e) Lithium at a dull red heat absorbs nitrogen‖, but the difficulty of obtaining the metal in quantity precludes its application. On the other

* *Annales de Chimie*, (3), LII. p. 82.
† Schutzenberger, *Comptes Rendus*, LXXXIX. 644.
‡ *Annalen der Chemie u. Pharmacie*, CI. 360.
§ *Annalen der Chemie u. Pharmacie*, LXXIII. 34.
‖ Ouvrard, *Comptes Rendus*, CXIV. 120.

hand, strontium and barium, prepared by electrolysing solutions of their chlorides in contact with mercury, and subsequently removing the mercury by distillation, are said by Maquenne* to absorb nitrogen with readiness. Although we have not tried these metals for removing nitrogen, still our experience with their amalgams has led us to doubt their efficacy, for it is extremely difficult to free them from mercury by distillation, and the product is a fused ingot, exposing very little surface to the action of the gas. The process might, however, be worth a trial.

Barium is the efficient absorbent for nitrogen when a mixture of barium carbonate and carbon is ignited in a current of nitrogen, yielding cyanide. Experiments have shown, however, that the formation of cyanides takes place much more readily and abundantly at a high temperature, a temperature not easily reached with laboratory appliances. Should the process ever come to be worked on a large scale, the gas rejected by the barium will undoubtedly prove a most convenient source of argon.

(*f*) *Nitride of magnesium* was prepared by Deville and Caron (*loc. cit.*) during the distillation of impure magnesium. It has been more carefully investigated by Briegleb and Geuther†, who obtained it by igniting metallic magnesium in a current of nitrogen. It forms an orange-brown, friable substance, very porous, and it is easily produced at a bright red heat. When magnesium, preferably in the form of thin turnings, is heated in a combustion tube in a current of nitrogen, the tube is attacked superficially, a coating of magnesium silicide being formed. As the temperature rises to bright redness, the magnesium begins to glow brightly, and combustion takes place, beginning at that end of the tube through which the gas is introduced. The combustion proceeds regularly, the glow extending down the tube, until all the metal has united with nitrogen. The heat developed by the combination is considerable, and the glass softens; but by careful attention and regulation of the rate of the current, the tube lasts out an operation. A piece of combustion tubing of the usual length for organic analysis packed tightly with magnesium turnings, and containing about 30 grams., absorbs between seven and eight litres of nitrogen. It is essential that oxygen be excluded from the tube, otherwise a fusible substance is produced, possibly nitrate, which blocks the tube. With the precaution of excluding oxygen, the nitride is loose and porous, and can easily be removed from the tube with a rod; but it is not possible to use a tube twice, for the glass is generally softened and deformed.

(*g*) *Nitride of aluminium* has been investigated by Mallet‡. He obtained it in crystals by heating the metal to whiteness in a carbon crucible.

* Ouvrard, *Comptes Rendus*, cxiv. 25, and 220.
† *Annalen der Chemie u. Pharmacie*, cxxiii. 228.
‡ *Journ. Chem. Soc.* 1876, Vol. ii. p. 349.

But aluminium shows no tendency to unite with nitrogen at a red heat, and cannot be used as an absorbent for the gas.

(h) Gerresheim * states that he has induced combination between nitrogen and mercury; but the affinity between these elements is of the slightest, for the compound is explosive.

(i) In addition to these, metallic manganese in a finely divided state has been shown to absorb nitrogen at a not very elevated temperature, forming a nitride of the formula Mn_5N_2†.

(j) [A mixture of nitrogen with hydrogen, standing over acid, is absorbed at a fair rate under the influence of electric sparks. But with an apparatus such as that shown in Fig. 1, the efficiency is but a fraction (perhaps $\frac{1}{3}$) of that obtainable when oxygen is substituted for hydrogen and alkali for acid. —April, 1895.]

4. *Early Experiments on sparking Nitrogen with Oxygen in presence of Alkali.*

In our earliest attempts to isolate the suspected gas by the method of Cavendish, we used a Ruhmkorff coil of medium size actuated by a battery of five Grove cells. The gases were contained in a test-tube A, Fig. 1, standing over a large quantity of weak alkali B, and the current was conveyed in wires insulated by U-shaped glass tubes CC passing through the liquid round the mouth of the test-tube. The inner platinum ends DD of the wires were sealed into the glass insulating tubes, but reliance was not placed upon these sealings. In order to secure tightness in spite of cracks, mercury was placed in the bends. This disposition of the electrodes complicates the apparatus somewhat and entails the use of a large depth of liquid in order to render possible the withdrawal of the tubes, but it has the great advantage of dispensing with sealing electrodes of platinum into the principal vessel, which might give way and cause the loss of the experiment at the most inconvenient moment. With the given battery and coil a somewhat short spark, or arc, of about 5 millims. was found to be more favourable than a longer one. When the mixed gases were in the right proportion, the rate of absorption was about 30 cub. centims. per hour, or 30 times as fast as Cavendish could work with the electrical machine of his day.

To take an example, one experiment of this kind started with 50 cub. centims. of air. To this, oxygen was gradually added until, oxygen being in excess, there was no perceptible contraction during an hour's sparking. The remaining gas was then transferred at the pneumatic trough to a small

* *Annalen der Chemie u. Pharmacie*, cxcv. 373.
† O. Prehlinger, *Monatsh. f. Chemie*, xv. 391.

measuring vessel, sealed by mercury, in which the volume was found to be
1·0 cub. centim.   On treatment with alkaline pyrogallate, the gas shrank
to ·32 cub. centim.   That this small residue could not be nitrogen was
argued from the fact that it had withstood the prolonged action of the
spark, although mixed with oxygen in nearly the most favourable proportion.

Fig. 1.

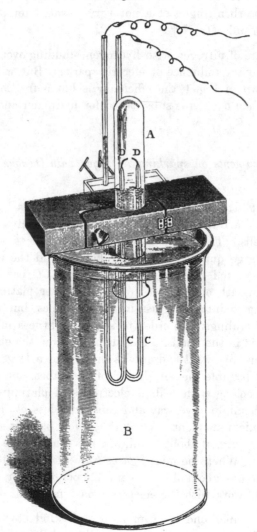

    The residue was then transferred to the test-tube with an addition of
another 50 cub. centims. of air, and the whole worked up with oxygen as
before.   The residue was now 2·2 cub. centims., and, after removal of oxygen,
·76 cub. centim.

Although it seemed almost impossible that these residues could be either nitrogen or hydrogen, some anxiety was not unnatural, seeing that the final sparking took place under somewhat abnormal conditions. The space was very restricted, and the temperature (and with it the proportion of aqueous vapour) was unduly high. But any doubts that were felt upon this score were removed by comparison experiments in which the whole quantity of air operated on was very small. Thus, when a mixture of 5 cub. centims. of air with 7 cub. centims. of oxygen was sparked for one hour and a quarter, the residue was ·47 cub. centim., and, after removal of oxygen, ·06 cub. centim. Several repetitions having given similar results, it became clear that the final residue did not depend upon anything that might happen when sparks passed through a greatly reduced volume, *but was in proportion to the amount of air operated upon.*

No satisfactory examination of the residue which refused to be oxidised could be made without the accumulation of a larger quantity. This, however, was difficult of attainment at the time in question. The gas seemed to rebel against the law of addition. It was thought that the cause probably lay in the solubility of the gas in water, a suspicion since confirmed. At length, however, a sufficiency was collected to allow of sparking in a specially constructed tube, when a comparison with the air spectrum taken under similar conditions proved that, at any rate, the gas was not nitrogen. At first scarcely a trace of the principal nitrogen lines could be seen, but after standing over water for an hour or two these lines became apparent.

[The apparatus shown in Fig. 1 has proved to be convenient for the purification of small quantities of argon, and for determinations of the amount of argon present in various samples of gas, *e.g.*, in the gases expelled from solution in water. To set it in action an alternating current is much to be preferred to a battery and break. At the Royal Institution the primary of a small Ruhmkorff was fed from the 100-volt alternating current supply, controlled by two large incandescent lamps in series with the coil. With this arrangement the voltage at the terminals of the secondary, available for starting the sparks, was about 2000, and could be raised to 4000 by plugging out one of the lamps. With both lamps in use the rate of absorption of mixed gases was 80 cub. centims. per hour, and this was about as much as could well be carried out in a test-tube. Even with this amount of power it was found better to abandon the sealings at *D*. No inconvenience arises from the open ends, if the tubes are wide enough to ensure the liberation of any gas included over the mercury when they are sunk below the liquid.

The power actually expended upon the coil is very small. When the apparatus is at work the current taken is only 2·4 amperes. As regards the voltage, by far the greater part is consumed in the lamps. The efficient voltage at the terminals of the primary coil is best found indirectly. Thus, if

A be the current in amperes, V the total voltage, V_1 the voltage at the terminals of the coil, V_2 that at the terminals of the lamps, the watts used are[*]

$$W = \frac{A}{2V_2}(V^2 - V_2{}^2 - V_1{}^2).$$

In the present case a Cardew voltmeter gave $V = 90\frac{1}{2}$, $V_2 = 88$; and $V_1{}^2$ in the formula may be neglected. Thus,

$$W = \frac{A}{2V_2}(V + V_2)(V - V_2) = A(V - V_2)$$

$$= 2\cdot4 \times 2\cdot5 = 6\cdot0 \text{ approximately.}$$

The work consumed by the coil when the sparks are passing is, thus, less than $\frac{1}{100}$ of a horse-power; but, in designing an apparatus, it must further be remembered that in order to maintain the arc, a pretty high voltage is required at the terminals of the secondary when no current is passing in it.— April, 1895.]

5. *Early Experiments on Withdrawal of Nitrogen from Air by means of Red-hot Magnesium.*

It having been proved that nitrogen, at a bright red heat, was easily absorbed by magnesium, best in the form of turnings, an attempt was successfully made to remove that gas from the residue left after eliminating oxygen from air by means of red-hot copper.

Fig. 2.

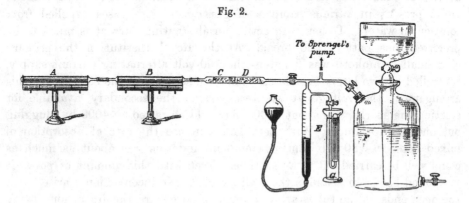

The preliminary experiment was made in the following manner:— A combustion tube, A, was filled with magnesium turnings, packed tightly by pushing them in with a rod. This tube was connected with a second piece of combustion tubing, B, by means of thick-walled india-rubber tubing,

[*] Ayrton and Sumpner, *Proc. Roy. Soc.* Vol. XLIX. p. 427, 1891.

carefully wired; B contained copper oxide, and, in its turn, was connected with the tube CD, one-half of which contained soda-lime previously ignited to expel moisture, while the other half was filled with phosphoric anhydride. E is a measuring vessel, and F is a gas-holder containing "atmospheric nitrogen."

In beginning an experiment, the tubes were heated with long-flame burners, and pumped empty; a little hydrogen was formed by the action of the moisture on the metallic magnesium; it was oxidised by the copper oxide and absorbed by the phosphoric pentoxide. A gauge attached to the Sprengel's pump, connected with the apparatus, showed when a vacuum had been reached. A quantity of nitrogen was then measured in E, and admitted into contact with the red-hot magnesium. Absorption took place, rapidly at first and then slowly, as shown by the gauge on the Sprengel's pump. A fresh quantity was then measured and admitted, and these operations were repeated until no more could be absorbed. The system of tubes was then pumped empty by means of the Sprengel's pump, and the gas was collected. The magnesium tube was then detached and replaced by another. The unabsorbed gas was returned to the measuring-tube by a device shown in the figure (G) and the absorption recommenced. After 1094 cub. centims. of gas had been thus treated, there was left about 50 cub. centims. of gas, which resisted rapid absorption. It still contained nitrogen, however, judging by the diminution of volume which it experienced when allowed to stand in contact with red-hot magnesium. Its density was, nevertheless, determined by weighing a small bulb of about 40 cub. centims. capacity, first with air, and afterwards with the gas. The data are these:—

grm.

(a) Weight of bulb and air — that of glass counterpoise . . 0·8094

   „   „   alone — that of glass counterpoise . . . 0·7588

   „   air . . . . 0·0506

(b) Weight of bulb and gas — that of glass counterpoise . . 0·8108

   „   „   alone — that of glass counterpoise . . . 0·7588

   „   gas . . . . 0·0520

Taking as the weight of a litre of air, 1·29347 grms., the mean of the latest results, and of oxygen ($= 16$) 1·42961 grms.[*], the density of the residual gas is 14·88.

This result was encouraging, although weighted with the unavoidable error attaching to the weighing of a very small amount. Still the fact remains that the supposed nitrogen was heavier than air. It would hardly have been possible to make a mistake of 2·7 milligrams.

* For note see foot of p. 146.

It is right here to place on record the fact that this first experiment was to a great extent carried out by Mr Percy Williams, to whose skill in manipulation and great care its success is due, and to whom we desire here to express our thanks.

Experiments were now begun on a larger scale, the apparatus employed being shown in Figs. 3 and 4.

Fig. 3.

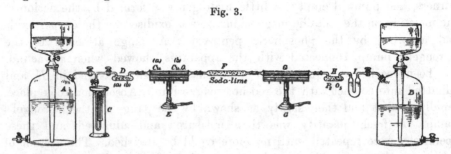

A and *B* are large glass gas-holders of about 10 litres capacity. *C* is an arrangement by which gas could be introduced at will into the gas-holder *A*, either by means of an india-rubber tube slipped over the open end of the U-tube, or, as shown in the figure, from a test-tube. The tube *D* was half

* The results on which this and the subsequent calculations are based are as follows (the weights are those of 1 litre) :—

	Air	Oxygen	Nitrogen	Hydrogen
Regnault............	1·29349	1·43011	1·25647	0·08988
Von Jolly	1·29383	1·42971	1·25819	
Leduc	1·29330	1·42910	1·25709	0·08985
Rayleigh............	1·29327	1·42952	1·25718	0·09001

Regnault's numbers have an approximate correction applied to them by Crafts. The mean of these numbers is taken, that of Regnault for nitrogen being omitted, as there is reason to believe that his specimen was contaminated with hydrogen.

Air	Oxygen	Nitrogen	Hydrogen
1·29347	1·42961	1·25749	0·08991

This ratio gives for air the composition by volume—

Oxygen	20·91 per cent.
Nitrogen	79·09 „

a result verified by experiment.

It is, of course, to be understood that these densities of nitrogen refer to atmospheric nitrogen, that is, to air from which oxygen, water vapour carbon dioxide, and ammonia have been removed.

filled with soda-lime (*a*), half with phosphoric anhydride (*b*). Similarly, the tube *E*, which was kept at a red heat by means of the long-flame burner, was filled half with very porous copper (*a*), reduced from dusty oxide by heating in hydrogen, half with copper oxide in a granular form (*b*). The next tube, *F*, contained granular soda-lime, while *G* contained magnesium turnings, also heated to bright redness by means of a long-flame burner. *H* contained phosphoric anhydride, and *I* soda-lime. All joints were sealed, excepting those connecting the hard-glass tubes *E* and *G* to the tubes next them.

The gas-holder *A* having been filled with nitrogen, prepared by passing air over red-hot copper, and introduced at *C*, the gas was slowly passed through the system of tubes into the gas-holder *B*, and back again. The magnesium in the tube *G* having then ceased to absorb was quickly removed and replaced by a fresh tube. This tube was of course full of air, and before the tube *G* was heated, the air was carried back from *B* towards *A* by passing a little nitrogen from right to left. The oxygen in the air was removed by the metallic copper, and the nitrogen passed into the gas-holder *A*, to be returned in the opposite direction to *B*.

Fig. 4.

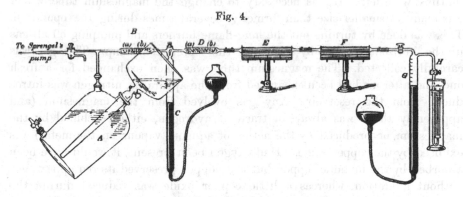

In the course of about ten days most of the nitrogen had been absorbed. The magnesium was not always completely exhausted; usually the nitride presented the appearance of a blackish-yellow mass, easily shaken out of the tube. It is needless to say that the tube was always somewhat attacked, becoming black with a coating of magnesium silicide. The nitride of magnesium, whether blackish or orange, if left for a few hours exposed to moist air, was completely converted into white, dusty hydroxide, and during exposure it gave off a strong odour of ammonia. If kept in a stoppered bottle, however, it was quite stable.

It was then necessary, in order to continue the absorption, to carry on operations on a smaller scale, with precautions to exclude atmospheric air as completely as possible. There was at this stage a residue of 1500 cub. centims.

The apparatus was therefore altered to that shown in Fig. 4, so as to make it possible to withdraw all the gas out of the gas-holder A.

The left-hand exit led to the Sprengel's pump; the compartment (a) of the drying-tube B was filled with soda-lime, and (b) with phosphoric anhydride. C is a tube into which the gas could be drawn from the gas-holder A. The stop-cock, as shown, allows gas to pass through the horizontal tubes, and does not communicate with A; but a vertical groove allows it to be placed in communication either with the gas-holder, or with the apparatus to the right. The compartment (a) of the second drying-tube D contained soda-lime, and (b) phosphoric anhydride. The tube D communicated with a hard-glass tube E, heated over a long-flame burner; it was partly filled with metallic copper, and partly with copper oxide. This tube, as well as the tube F filled with magnesium turnings, was connected to the drying-tube with india-rubber. The gas then entered G, a graduated reservoir, and the arrangement H permitted the removal or introduction of gas from or into the apparatus. The gas was gradually transferred from the gas-holder to the tube C, and passed backwards and forwards over the red-hot magnesium until only about 200 cub. centims. were left. It was necessary to change the magnesium tube, which was made of smaller size than formerly, several times during the operation. This was done by turning out the long-flame burners and pumping off all gas in the horizontal tubes by means of the Sprengel's pump. This gas was carefully collected. The magnesium tube was then exchanged for a fresh one, and after air had been exhausted from the apparatus, nitrogen was introduced from the reservoir. Any gas evolved from the magnesium (and apparently there was always a trace of hydrogen, either occluded by the magnesium, or produced by the action of aqueous vapour on the metal) was oxidised by the copper oxide. Had oxygen been present, it would have been absorbed by the metallic copper, but the copper preserved its red appearance without alteration, whereas a little copper oxide was reduced during the series of operations. The gas, which had been removed by pumping, was reintroduced at H, and the absorption continued.

The volume of the gas was thus, as has been said, reduced to about 200 cub. centims. It would have been advisable to take exact measurements, but, unfortunately, some of the original nitrogen had been lost through leakage; and a natural anxiety to see if there was any unknown gas led to pushing on operations as quickly as possible.

The density of the gas was next determined. The bulb or globe in which the gas was weighed was sealed to a two-way stop-cock, and the weight of distilled and air-free water filling it at 17·15° was 162·654 grms., corresponding to a capacity of 162·843 cub. centims. The shrinkage on removing air completely was 0·0212 cub. centim. Its weight, when empty, should therefore be increased by the weight of that volume of air, which may be taken as

0·000026 grm. This correction, however, is perhaps hardly worth applying in the present case.

The counterpoise was an exactly similar bulb of equal capacity, and weighing about 0·2 grm. heavier than the empty globe. The balance was a very sensitive one by Oertling, which easily registered one-tenth of a milligrm. By the process of swinging, one-hundredth of a milligrm. could be determined with fair accuracy.

In weighing the empty globe, 0·2 grm. was placed on the same pan as that which hung from the end of the beam to which it was suspended, and the final weight was adjusted by means of a rider, or by small weights on the other pan. This process practically leads to weighing by substitution of gas for weights. The bulb was always handled with gloves, to avoid moisture or grease from the fingers.

Three experiments, of which it is unnecessary to give details, were made to test the degree of accuracy with which a gas could be weighed, the gas being dried air, freed from carbon dioxide. The mean result gave for the weight of one litre of air at 0° and 760 millims. pressure, 1·2935 grm. Regnault found 1·29340, a correction having been applied by Crafts to allow for the estimated alteration of volume caused by the contraction of his vacuous bulb. The mean result of determinations by several observers is 1·29347; while one of us found 1·29327.

The globe was then filled with the carefully dried gas.

Temperature, 18·80°.   Pressure, 759·3 millims.
Weight of 162·843 cub. centims. of gas . . . . . .   0·21897 grm.
Weight of 1 litre gas at 0° and 760 millims. . . . .   1·4386 ,,
Density, that of air compared with O, = 16, being 14·476   16·100 grms.

It is evident from these numbers that the dense constituent of the air was being concentrated. As a check, the bulb was pumped empty and again weighed; its weight was 0·21903 grm. This makes the density 16·105.

It appeared advisable to continue to absorb nitrogen from this gas. The first tube of magnesium removed a considerable quantity of gas; the nitride was converted into ammonium chloride, and the sample contained 66·30 per cent. of chlorine, showing, as has before been remarked, that if any of the heavier constituent of the atmosphere had been absorbed, it formed no basic compound with hydrogen. The second tube of magnesium was hardly attacked; most of the magnesium had melted, and formed a layer at the lower part of the tube. That which was still left in the body of the tube was black on the surface, but had evidently not been much attacked. The ammonium chloride which it yielded weighed only 0·0035 grm.

The density of the remaining gas was then determined. But as its volume was only a little over 100 cub. centims., the bulb, the capacity of which was 162 cub. centims., had to be filled at reduced pressure. This was easily done by replacing the pear-shaped reservoir of the mercury gas-holder by a straight tube, and noting the level of the mercury in the gas-holder and in the tube which served as a mercury reservoir against a graduated mirror-scale by help of a cathetometer at the moment of closing the stop-cock of the density bulb.

The details of the experiment are these :—

Temperature, 19·12° C.   Barometric pressure, 749·8 millims. (corr.).
Difference read on gas-holder and tube, 225·25 millims. (corr.).
Actual pressure, 524·55 millims.
Weight of 162·843 cub. centims. of gas . . . .   0·17913 grm.
Weight of 1 litre at 0° and 760 millims. pressure .   1·7054   „
Density . . . .   19·086   grms.

This gas is accordingly at least 19 times as heavy as hydrogen.

A portion of the gas was then mixed with oxygen, and submitted to a rapid discharge of sparks for four hours in presence of caustic potash. It contracted, and on absorbing the excess of oxygen with pyrogallate of potassium the contraction amounted to 15·4 per cent. of the original volume. The question then arises, if the gas contain 15·4 per cent. of nitrogen, of density 14·014, and 84·6 per cent. of other gas, and if the density of the mixture were 19·086, what would be the density of the other gas ? Calculation leads to the number 20·0.

A vacuum-tube was filled with a specimen of the gas of density 19·086, and it could not be doubted that it contained nitrogen, the bands of which were distinctly visible. It was probable, therefore, that the true density of the pure gas lay not far from 20 times that of hydrogen. At the same time many lines were seen which could not be recognised as belonging to the spectrum of any known substance.

Such were the preliminary experiments made with the aid of magnesium to separate from atmospheric nitrogen its dense constituent. The methods adopted in preparing large quantities will be subsequently described.

6.   *Proof of the Presence of Argon in Air, by means of Atmolysis.*

It has already (§ 2) been suggested that if " atmospheric nitrogen " contains two gases of different densities, it should be possible to obtain direct evidence of the fact by the method of atmolysis. The present section contains an account of carefully conducted experiments directed to this end.

The atmolyser was prepared (after Graham) by combining a number of "churchwarden" tobacco pipes. At first twelve pipes were used in three groups, each group including four pipes connected in series. The three groups were then connected in parallel, and placed in a large glass tube closed in such a way that a partial vacuum could be maintained in the space outside the pipes by a water-pump. One end of the combination of pipes was open to the atmosphere, or rather was connected with the interior of an open bottle containing sticks of caustic alkali, the object being mainly to dry the air. The other end of the combination was connected to a bottle aspirator, initially full of water, and so arranged as to draw about two per cent. of the air which entered the other end of the pipes. The gas collected was thus a very small proportion of that which leaked through the pores of the pipes, and should be relatively rich in the heavier constituents of the atmosphere. The flow of water from the aspirator could not be maintained very constant, but the rate of two per cent. was never much exceeded. The necessary four litres took about sixteen hours to collect.

The air thus obtained was treated exactly as ordinary air had been treated in determinations of the density of atmospheric nitrogen. Oxygen was removed by red-hot copper followed by cupric oxide, ammonia by sulphuric acid, carbonic anhydride and moisture by potash and phosphoric anhydride.

The following are the results:—

Globe empty July 10, 14	2·81789
Globe full September 15 (twelve pipes) . .	·50286
Weight of gas	2·31503
Ordinary atmospheric nitrogen	2·31016
Difference	+ ·00487

Globe empty September 17	2·81345
Globe full September 18 (twelve pipes) . .	·50191
Weight of gas	2·31154
Ordinary atmospheric nitrogen	2·31016
Difference	+ ·00138

Globe empty September 21	2·82320
Globe full September 20 (twelve pipes) . .	·51031
Weight of gas	2·31289
Ordinary atmospheric nitrogen	2·31016
Difference	+ ·00273

```
Globe empty September 21, October 30  .   .    2·82306
Globe full September 22 (twelve pipes)  .   .    ·51140
                                                ─────────
Weight of gas . . . . . . . . . . .    2·31166
Ordinary atmospheric nitrogen . . . . .    2·31016
                                                ─────────
          Difference  . . . . . . . . + ·00150
```

The mean excess of the four determinations is ·00262 gram., or if we omit the first, which depended upon a vacuum weighing of two months old, ·00187 gram.

The gas from prepared air was thus in every case denser than from unprepared air, and to an extent much beyond the possible errors of experiment. The excess was, however, less than had been expected, and it was thought that the arrangement of the pipes could be improved. The final delivery of gas from each of the groups in parallel being so small in comparison with the whole streams concerned, it seemed possible that each group was not contributing its proper share, and even that there might be a flow in the wrong direction at the delivery end of one or two of them. To meet this objection, the arrangement in parallel had to be abandoned, and for the remaining experiments eight pipes were connected in simple series. The porous surface in operation was thus reduced, but this was partly compensated for by an improved vacuum. Two experiments were made under the new conditions :—

```
Globe empty, October 30, November 5  .   .    2·82313
Globe full, November 3 (eight pipes) .   .   .    ·50930
                                                ─────────
Weight of gas . . . . . . . . . .    2·31383
Ordinary atmospheric nitrogen . . . . .    2·31016
                                                ─────────
          Difference  . . . . . . . . + ·00367

Globe empty, November 5, 8  . . . . .    2·82355
Globe full, November 6 (eight pipes) . . .    ·51011
                                                ─────────
Weight of gas . . . . . . . . . . .    2·31344
Ordinary atmospheric nitrogen . . . . .    2·31016
                                                ─────────
          Difference  . . . . . . . . + ·00328
```

The excess being larger than before is doubtless due to the greater efficiency of the atmolysing apparatus. It should be mentioned that the above recorded experiments include all that have been tried, and the conclusion seems inevitable that "atmospheric nitrogen" is a mixture and not a simple body.

It was hoped that the concentration of the heavier constituent would be sufficient to facilitate its preparation in a pure state by the use of prepared air in substitution for ordinary air in the oxygen apparatus. The advance of $3\frac{1}{2}$ mg. on the 11 mg., by which atmospheric nitrogen is heavier than chemical nitrogen, is indeed not to be despised, and the use of prepared air would be convenient if the diffusion apparatus could be set up on a large scale and be made thoroughly self-acting.

7. *Negative Experiments to prove that Argon is not derived from Nitrogen or from Chemical Sources.*

Although the evidence of the existence of argon in the atmosphere, derived from the comparison of densities of atmospheric and chemical nitrogen and from the diffusion experiments (§ 6), appeared overwhelming, we have thought it undesirable to shrink from any labour that would tend to complete the verification. With this object in view, an experiment was undertaken and carried to a conclusion on November 13, in which 3 litres of chemical nitrogen, prepared from ammonium nitrite, were treated with oxygen in precisely the manner in which atmospheric nitrogen had been found to yield a residue of argon. In the course of operations an accident occurred, by which no gas could have been lost, but of such a nature that from 100 to 200 cub. centims. of air must have entered the working vessel. The gas remaining at the close of the large scale operations was worked up as usual with battery and coil until the spectrum showed only slight traces of the nitrogen lines. When cold, the residue measured 4 cub. centims. This was transferred, and after treatment with alkaline pyrogallate to remove oxygen, measured 3·3 cub. centims. If atmospheric nitrogen had been employed, the final residue should have been about 30 cub. centims. Of the 3·3 cub. centims. actually left, a part is accounted for by the accident alluded to, and the result of the experiment is to show that argon is not formed by sparking a mixture of oxygen and chemical nitrogen.

In a second experiment of the same kind 5660 cub. centims. of nitrogen from ammonium nitrite were treated with oxygen in the large apparatus (Fig. 7, § 8). The final residue was 3·5 cub. centims.; and, as evidenced by the spectrum, it consisted mainly of argon.

The source of the residual argon is to be found in the water used for the manipulation of the large quantities of gas (6 litres of nitrogen and 11 litres of oxygen) employed. Unfortunately the gases had been collected by allowing them to bubble up into aspirators charged with ordinary water, and they were displaced by ordinary water. In order to obtain information with respect to the contamination that may be acquired in this way, a parallel experiment was tried with carbonic anhydride. Eleven litres of the gas, prepared from marble and hydrochloric acid with ordinary precautions for the exclusion of air, were collected exactly as oxygen was commonly collected. It was then transferred by displacement with water to a gas pipette charged with a solution containing 100 grms. of caustic soda. The residue which refused absorption measured as much as 110 cub. centims. In another experiment where the water employed had been partially de-aerated, the residue left amounted to 71 cub. centims., of which 26 cub. centims. were oxygen. The quantities of dissolved gases thus extracted from water during the collection of oxygen and nitrogen suffice to explain the residual argon of the negative experiments.

It may perhaps be objected that the impurity was contained in the carbonic anhydride itself as it issued from the generating vessel, and was not derived from the water in the gas-holder; and indeed there seems to be a general impression that it is difficult to obtain carbonic anhydride in a state of purity. To test this question, 18 litres of the gas, made in the same generator and from the same materials, were passed directly into the absorption pipette. Under these conditions, the residue was only $6\frac{1}{2}$ cub. centims., corresponding to 4 cub. centims. from 11 litres. The quantity of gas employed was determined by decomposing the resulting sodium carbonate with hydrochloric acid, allowance being made for a little carbonic anhydride contained in the soda as taken from the stock bottle. It will be seen that there is no difficulty in reducing the impurity to $\frac{1}{3000}$th, even when india-rubber connections are freely used, and no extraordinary precautions are taken. The large amount of impurity found in the gas when collected over water must therefore have been extracted from the water.

A similar set of experiments was carried out with magnesium. The nitrogen, of which three litres were used, was prepared by the action of bleaching-powder on ammonium chloride. It was circulated in the usual apparatus over red-hot magnesium, until its volume had been reduced to about 100 cub. centims. An equal volume of hydrogen was then added, owing to the impossibility of circulating a vacuum. The circulation then proceeded until all absorption had apparently stopped. The remaining gas was then passed over red-hot copper oxide into the Sprengel's pump, and collected. As it appeared still to contain hydrogen, which had escaped oxidation, owing to its great rarefaction, it was passed over copper oxide for a second and a third time. As there was still a residue, measuring 12·5 cub. centims., the gas was left in contact with red-hot magnesium for several hours, and then pumped out; its volume was then 4·5 cub. centims. Absorption was, however, still proceeding, when the experiment terminated, for at a low pressure, the rate is exceedingly slow. This gas, after being sparked with oxygen contracted to 3·0 cub. centims., and on examination was seen to consist mainly of argon. The amount of residue obtainable from three litres of atmospheric nitrogen should have amounted to a large multiple of this quantity.

In another experiment, 15 litres of nitrogen prepared from a mixture of ammonium chloride and sodium nitrite by warming in a flask (some nitrogen having first been drawn off by a vacuum-pump, in order to expel all air from the flask and from the contained liquid) were collected over water in a large gas-holder. The nitrogen was not bubbled through the water, but was admitted from above, while the water escaped below. This nitrogen was absorbed by red-hot magnesium, contained in tubes heated in a combustion-furnace. The unabsorbed gas was circulated over red-hot magnesium in a special small apparatus, by which its volume was reduced to 15 cub. centims. As it was impracticable further to reduce the volume by means of magnesium, the residual 15 cub. centims. were transferred to a tube, mixed with oxygen, and submitted to sparking over caustic soda. The residue after absorption of oxygen, which undoubtedly consisted of pure argon, amounted to 3·5 cub. centims. This is one-fortieth of the quantity which would have been obtained from atmospheric nitrogen, and its presence can be accounted for, we venture to think, first from the water in the gas-holder, which had not been freed from dissolved gas by boiling *in vacuo* (it has already been shown that a consider-able gain may ensue from this source), and second, from leakage of air which accidentally took place, owing to the breaking of a tube. The leakage may have amounted to 200 cub. centims., but it could not be accurately ascertained. Quantitative negative experiments of this nature are exceedingly difficult, and require a long time to carry them to a successful conclusion.

8. *Separation of Argon on a Large Scale.*

To separate nitrogen from "atmospheric nitrogen" on a large scale, by help of magnesium, several devices were tried. It is not necessary to describe them all in detail. Suffice it to say that an attempt was made to cause a store of "atmospheric nitrogen" to circulate by means of a fan, driven by a water-motor. The difficulty encountered here was leakage at the bearing of the fan, and the introduced air produced a cake which blocked the tube on coming into contact with the magnesium. It might have been possible to remove oxygen by metallic copper; but instead of thus complicating the apparatus, a water-injector was made use of to induce circulation. Here also it is unnecessary to enter into details. For, though the plan worked well, and although about 120 litres of "atmospheric nitrogen" were absorbed, the yield of argon was not large, about 600 cub. centims. having been collected. This loss was subsequently discovered to be due partially, at least, to the rela-tively high solubility of argon in water. In order to propel the gas over magnesium, through a long combustion-tube packed with turnings, a consider-able water-pressure, involving a large flow of water, was necessary. The gas was brought into intimate contact with this water, and presuming that several thousand litres of water ran through the injector, it is obvious that a not inconsiderable amount of argon must have been dissolved. Its proportion was increasing at each circulation, and consequently its partial pressure also increased. Hence, towards the end of the operation, at least, there is every reason to believe that a serious loss had occurred.

It was next attempted to pass " atmospheric nitrogen " from a gas-holder first through a combustion tube of the usual length packed with metallic copper reduced from the oxide; then through a small U-tube containing a little water, which was intended as an index of the rate of flow; the gas was then dried by passage through tubes filled with soda-lime and phosphoric anhydride; and it next passed through a long iron tube (gas-pipe) packed with magnesium turnings, and heated to bright redness in a second combustion-furnace.

After the iron tube followed a second small U-tube containing water, intended to indicate the rate at which the argon escaped into a small gas-holder placed to receive it. The nitrogen was absorbed rapidly, and argon entered the small gas-holder. But there was reason to suspect that the iron tube is permeable by argon at a red heat. The first tube-full allowed very little argon to pass. After it had been removed and replaced by a second, the same thing was noticed. The first tube was difficult to clean; the nitride of magnesium forms a cake on the interior of the tube, and it was very difficult to remove it; moreover this rendered the filling of the tube very troublesome, inasmuch as its interior was so rough that the magnesium turnings could only with difficulty be forced down. However, the permeability to argon, if such be the case, appeared to have decreased. The iron tube was coated internally with a skin of magnesium nitride, which appeared to diminish its permeability to argon. After all the magnesium in the tube had been converted into nitride (and this was easily known, because a bright glow proceeded gradually from one end of the tube to the other) the argon remaining in the iron tube was " washed " out by a current of nitrogen; so that, after a number of operations, the small gas-holder contained a mixture of argon with a considerable quantity of nitrogen.

On the whole, the use of iron tubes is not to be recommended, owing to the difficulty in cleaning them, and the possible loss through their permeability to argon. There is no such risk of loss with glass tubes, but each operation requires a new tube, and the cost of the glass is considerable if much nitrogen is to be absorbed. Tubes of porcelain were tried; but the glaze in the interior is destroyed by the action of the red-hot magnesium, and the tubes crack on cooling.

By these processes 157 litres of " atmospheric nitrogen " were reduced in volume to about 2·5 litres in all of a mixture of nitrogen and argon. This mixture was afterwards circulated over red-hot magnesium, in order to remove the last portion of nitrogen.

Fig. 5.

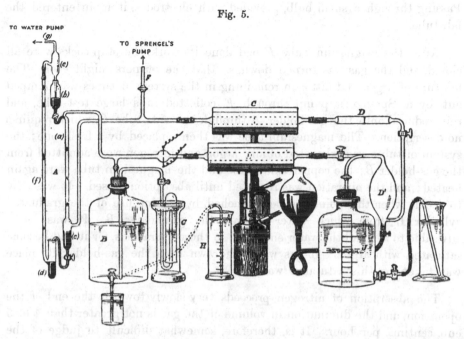

As the apparatus employed for this purpose proved very convenient, a full description of its construction is here given. A diagram is shown in Fig. 5, which sufficiently explains the arrangement of the apparatus. *A* is the circulator. It consists of a sort of Sprengel's pump (*a*) to which a supply of mercury is admitted from a small reservoir (*b*). This mercury is delivered into a gas-separator (*c*), and the mercury overflows into the reservoir (*d*). When its level rises, so that it blocks the tube (*f*), it ascends in pellets or pistons into (*e*), a reservoir which is connected through (*g*) with a water-pump. The mercury falls into (*b*), and again passes down the Sprengel tube (*a*). No attention is, therefore, required, for the apparatus works quite automatically. This form of apparatus was employed several years ago by Dr Collie.

The gas is drawn from the gas-holder *B*, and passes through a tube *C*, which is heated to redness by a long-flame burner, and which contains in one half metallic copper, and in the other half copper oxide. This precaution is taken in order to remove any oxygen which may possibly be present, and also any hydrogen or hydrocarbon. In practice, it was never found that the copper became oxidised, or the oxide reduced. It is, however, useful to guard against any possible contamination. The gas next traversed a drying-tube *D*, the anterior portion containing ignited soda-lime, and the posterior portion phosphoric anhydride. From this it passed a reservoir, *D′*, from which it could be transferred, when all absorption had ceased, into the small gas-holder.

It then passed through E, a piece of combustion-tube, drawn out at both ends, filled with magnesium turnings, and heated by a long-flame burner to redness. Passing through a small bulb, provided with electrodes, it again entered the fall-tube.

After the magnesium tube E had done its work, the stop-cocks were all closed, and the gas was turned down, so that the burners might cool. The mixture of argon and nitrogen remaining in the system of tubes was pumped out by a Sprengel's pump through F, collected in a large test-tube, and reintroduced into the gas-holder B through the side-tube G, which requires no description. The magnesium tube was then replaced by a fresh one; the system of tubes was exhausted of air; argon and nitrogen were admitted from the gas-holder B; the copper-oxide tube and the magnesium tube were again heated; and the operation was repeated until absorption ceased. It was easy to decide when this point had been reached, by making use of the graduated cylinder H, from which water entered the gas-holder B. It was found advisable to keep all the water employed in these operations, for it had become saturated with argon. If gas was withdrawn from the gas-holder, its place was taken by this saturated water.

The absorption of nitrogen proceeds very slowly towards the end of the operation, and the diminution in volume of the gas is not greater than 4 or 5 cub. centims. per hour. It is, therefore, somewhat difficult to judge of the end-point, as will be seen when experiments on the density of this gas are described. The magnesium tube, towards the end of the operations, was made so hot that the metal was melted in the lower part of the tube, and sublimed in the upper part. The argon and residual nitrogen had, therefore, been thoroughly mixed with gaseous magnesium during its passage through the tube E.

To avoid possible contamination with air in the Sprengel's pump, the last portion of gas collected from the system of tubes was not re-admitted to the gas-holder B, but was separately stored.

The crude argon was collected in two operations. First, the quantity made by absorption by magnesium in glass tubes with the water-pump circulator was purified. Later, after a second supply had been prepared by absorption in iron tubes, the mixture of argon and nitrogen was united with the first quantity and circulated by means of the mercury circulator, in the gas-holder B. Attention will be drawn to the particular sample of gas employed in describing further experiments made with the argon.

By means of magnesium, about 7 litres of nitrogen can be absorbed in an hour. The changing of the tubes of magnesium, however, takes some time; consequently, the largest amount absorbed in one day was nearly 30 litres.

At a later date a quantitative experiment was carried out on a large scale, the amount of argon from 100 litres of "atmospheric" nitrogen, measured at 20°, having been absorbed by magnesium, and the resulting argon measured at 12°. During the process of absorbing nitrogen in the combustion-furnace, however, one tube cracked, and it is estimated that about 4 litres of nitrogen escaped before the crack was noticed. With this deduction, and assuming that the nitrogen had been measured at 12°, 93·4 litres of atmospheric nitrogen were taken. The magnesium required for absorption weighed 409 grms. The amount required by theory should have been 285 grms.; but it must be remembered that in many cases the magnesium was by no means wholly converted into nitride. The first operation yielded about 3 litres of a mixture of nitrogen and argon, which was purified in the circulating apparatus. The total residue, after absorption of the nitrogen, amounted to 921 cub. centims. The yield is therefore 0·986 per cent.

At first no doubt the nitrogen gains a little argon from the water over which it stands. But, later, when the argon forms the greater portion of the gaseous mixture, its solubility in water must materially decrease its volume. It is difficult to estimate the loss from this cause. The gas-holder, from which the final circulation took place, held three litres of water. Taking the solubility of argon as 4 per cent., this would mean a loss of about 120 cub. centims. If this is not an over-estimate, the yield of argon would be increased to 1040 cub. centims., or 1·11 per cent. The truth probably lies between these two estimates.

It may be concluded, with probability, that the argon forms approximately 1 per cent. of the "atmospheric" nitrogen.

The principal objection to the oxygen method of isolating argon, as hitherto described, is the extreme slowness of the operation. An absorption of 30 cub. centims. of mixed gas means the removal of but 12 cub. centims. of nitrogen. At this rate 8 hours are required for the isolation of 1 cub. centim. of argon, supposed to be present in the proportion of 1 per cent.

In extending the scale of operations we had the great advantage of the advice of Mr Crookes, who a short time ago called attention to the flame rising from platinum terminals, which convey a high tension alternating electric discharge, and pointed out its dependence upon combustion of the nitrogen and oxygen of the air*. Mr Crookes was kind enough to arrange an impromptu demonstration at his own house with a small alternating current plant, in which it appeared that the absorption of mixed gas was at the rate of 500 cub. centims. per hour, or nearly 20 times as fast as with the battery.

* *Chemical News*, Vol. LXV. p. 301, 1892.

The arrangement is similar to that first described by Spottiswoode*. The primary of a Ruhmkorff coil is connected directly with the alternator, no break or condenser being required; so that, in fact, the coil acts simply as a high potential transformer. When the arc is established the platinum terminals may be separated much beyond the initial striking distance.

The plant with which the large scale operations have been made consists of a De Meritens alternator, kindly lent by Professor J. J. Thomson, and a gas engine. As transformer, one of Swinburne's hedgehog pattern has been employed with success, but the ratio of transformation (24 : 1) is scarcely sufficient. A higher potential, although, perhaps, not more efficient, is more convenient. The striking distance is greater, and the arc is not so liable to go out. Accordingly most of the work to be described has been performed with transformers of the Ruhmkorff type.

The apparatus has been varied greatly, and it cannot be regarded as having even yet assumed a final form. But it will give a sufficient idea of the method if we describe an experiment in which a tolerably good account was kept of the air and oxygen employed. The working vessel was a glass flask, *A* (Fig. 6), of about 1500 cub. centims. capacity, and stood, neck downwards, over a large jar of alkali, *B*. As in the small scale experiments, the leading-in wires were insulated by glass tubes, *DD*, suitably bent and carried through the liquid up the neck. For the greater part of the length iron wires were employed, but the internal extremities, *EE*, were of platinum, doubled upon itself at the terminals from which the discharge escaped. The glass protecting tubes must be carried up for some distance above the internal level of the liquid, but it is desirable that the arc itself should not be much raised above that level. A general idea of the disposition of the electrodes will be obtained from Fig. 6. To ensure gas tightness the bends were occupied by mercury. A tube, *C*, for the supply or withdrawal of gas was carried in the same way through the neck.

The Ruhmkorff employed in this operation was one of medium size. When the mixture was rightly proportioned and the arc of full length, the rate of absorption was about 700 cub. centims. per hour. A good deal of time is lost in starting, for, especially when there is soda on the platinums, the arc is liable to go out if lengthened prematurely. After seven days the total quantity of air let in amounted to 7925 cub. centims., and of oxygen (prepared from chlorate of potash) 9137 cub. centims. On the eighth and ninth days oxygen alone was added, of which about 500 cub. centims. was consumed, while there remained about 700 cub. centims. in the flask. Hence the proportion in which the air and oxygen combined was as 70 : 96. On the eighth day there was about three hours' work, and the absorption slackened off to

* "A Mode of Exciting an Induction-coil," *Phil. Mag.* Vol. VIII. p. 390, 1879.

about one quarter of the previous rate. On the ninth day (September 8) the rate fell off still more, and after three hours' work became very slow. The progress towards removal of nitrogen was examined from time to time with the spectroscope, the points being approximated and connected with a small

Fig. 6.

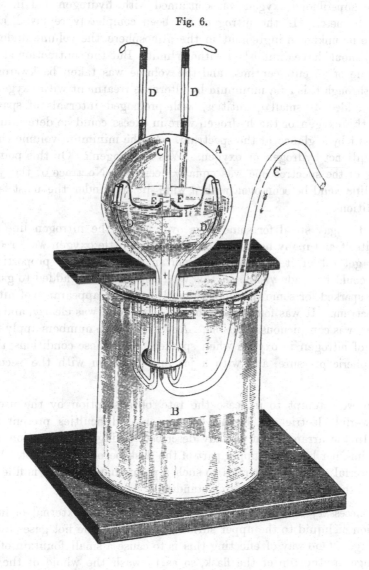

Leyden jar. At this stage the yellow nitrogen line was faint, but plainly visible. After about four hours' more work, the yellow line had disappeared, and for two hours there had been no visible contraction. It will be seen that the removal of the last part of the nitrogen was very slow, mainly on account of the large excess of oxygen present.

The final treatment of the residual 700 cub. centims. of gas was on the model of the small scale operations already described (§ 4). By means of a pipette the gas was gradually transferred to a large test-tube standing over alkali. Under the influence of sparks (from battery and coil) passing all the while, the superfluous oxygen was consumed with hydrogen fed in slowly from a voltameter. If the nitrogen had been completely removed, and if there were no unknown ingredient in the atmosphere, the volume under this treatment should have diminished without limit. But the contraction stopped at a volume of 65 cub. centims., and the volume was taken backwards and forwards through this as a minimum by alternate treatment with oxygen and hydrogen added in small quantities, with prolonged intervals of sparking. Whether the oxygen or the hydrogen were in excess could be determined at any moment by a glance at the spectrum. At the minimum volume the gas was certainly not hydrogen or oxygen. Was it nitrogen? On this point the testimony of the spectroscope was equally decisive. No trace of the yellow nitrogen line could be seen even with a wide slit and under the most favourable conditions.

When the gas stood for some days over water, the nitrogen line again asserted itself, and many hours of sparking with a little oxygen were required again to get rid of it. As it was important to know what proportions of nitrogen could be made visible in this way, a little air was added to gas that had been sparked for some time subsequently to the disappearance of nitrogen in its spectrum. It was found that about $1\frac{1}{2}$ per cent. was clearly, and about 3 per cent. was conspicuously, visible. About the same numbers apply to the visibility of nitrogen in oxygen when sparked under these conditions, that is, at atmospheric pressure, and with a jar in connection with the secondary terminals.

When we attempt to increase the rate of absorption by the use of a more powerful electric arc, further experimental difficulties present themselves. In the arrangement already described, giving an absorption of 700 cub. centims. per hour, the upper part of the flask becomes very hot. With a more powerful arc the heat rises to such a point that the flask is filled with steam and the operation comes to a standstill.

It is necessary to keep the vessel cool by either the external or internal application of liquid to the upper surface upon which the hot gases from the arc impinge. One way of effecting this is to cause a small fountain of alkali to impinge on the top of the flask, so as to wash the whole of the upper surface. This plan is very effective, but it is open to the objection that a breakdown would be disastrous, and it would involve special arrangements to avoid losing the argon by solution in the large quantity of alkali required. It is simpler in many respects to keep the vessel cool by immersing it in a large body of water, and the inverted flask arrangement (Fig. 6) has been applied in this manner. But, on the whole, it appears to be preferable to limit the application of the cooling water to the upper part of the external surface, building up for this purpose a suitable wall of sheet lead cemented round the glass. The most convenient apparatus for large-scale operations that has hitherto been tried is shown in the accompanying figure (Fig. 7).

Fig. 7.

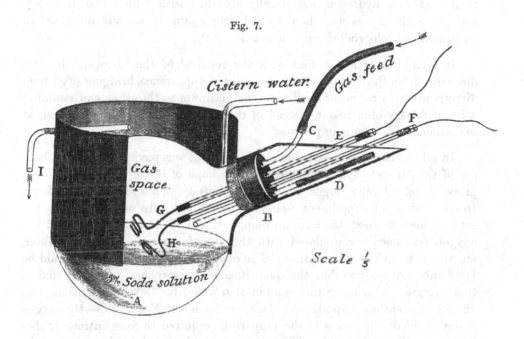

The vessel A is a large globe of about 6 litres capacity, intended for demonstrating the combustion of phosphorus in oxygen gas, and stands in an inclined position. It is about half filled with a solution of caustic soda. The neck is fitted with a rubber stopper, B, provided with four perforations. Two of these are fitted with tubes, C, D, suitable for the supply or withdrawal of gas or liquid. The other two allow the passage of the stout glass tubes, E, F, which contain the electrodes. For greater security against leakage, the interior of these tubes is charged with water, held in place by small corks, and the outer ends are cemented up. The electrodes are formed of stout iron wires terminated by thick platinums, G, H, triply folded together, and welded at the ends. The lead walls required to enclose the cooling water are partially shown at I. For greater security the india-rubber cork is also drowned in water, held in place with the aid of sheet-lead. The lower part of the globe is occupied by about 3 litres of a 5 per cent. solution of caustic soda, the solution rising to within about half-an-inch of the platinum terminals. With this apparatus an absorption of 3 litres of mixed gas per hour can be attained,—about 3000 times the rate at which Cavendish could work.

When it is desired to stop operations, the feed of air (or of chemical nitrogen in blank experiments) is cut off, oxygen alone being supplied as long as any visible absorption occurs. Thus at the close the gas space is occupied by argon and oxygen with such nitrogen as cannot readily be taken up in a condition of so great dilution. The oxygen, being too much for convenient treatment with hydrogen, was usually absorbed with copper and ammonia, and the residual gas was then worked over again as already described in an apparatus constructed upon a smaller scale.

It is worthy of notice that with the removal of the nitrogen, the arc-discharge from the dynamo changes greatly in appearance, bridging over more directly and in a narrower band from one platinum to the other, and assuming a beautiful sky-blue colour, instead of the greenish hue apparent so long as oxidation of nitrogen is in progress.

In all the large-scale experiments, an attempt was made to keep a reckoning of the air and oxygen employed, in the hope of obtaining data as to the proportional volume of argon in air, but various accidents too often interfered. In one successful experiment (January, 1895), specially undertaken for the sake of measurement, the total air employed was 9250 cub. centims., and the oxygen consumed, manipulated with the aid of partially de-aerated water, amounted to 10,820 cub. centims. The oxygen contained in the air would be 1942 cub. centims.; so that the quantities of "atmospheric nitrogen" and of total oxygen which enter into combination would be 7308 cub. centims., and 12,762 cub. centims. respectively. This corresponds to $N + 1.75\ O$—the oxygen being decidedly in excess of the proportion required to form nitrous acid—$2HNO_2$, or $H_2O + N_2 + 3\ O$. The argon ultimately found on absorption of the excess of oxygen was 75.0 cub. centims., reduced to conditions similar to those under which the air was measured, or a little more than 1 per cent. of the "atmospheric nitrogen" used. It is probable, however, that some of the argon was lost by solution during the protracted operations required in order to get quit of the last traces of nitrogen.

[In recent operations at the Royal Institution, where a public supply of alternating current at 100 volts is available, the scale of the apparatus has been still further increased.

The capacity of the working vessel is 20 litres, of which about one half is occupied by a strong solution of caustic soda. The platinum terminals are very massive, and the flame rising from them is prevented from impinging directly upon the glass by a plate of platinum held over it and supported by a wire which passes through the rubber cork. In the electrical arrangements we have had the advantage of Mr Swinburne's advice. The transformers are two of the "hedgehog" pattern, the thick wires being connected in parallel and the thin wires in series. In order to control the current taken when the arc is short or the platinums actually in contact, a choking-coil, provided with a movable core of fine iron wires, is inserted in the thick wire circuit. In normal working the current taken from the mains is about 22 amperes, so that some $2\frac{1}{2}$ h. p. is consumed. At the same time the actual voltage at the platinum terminals is 1500. When the discharge ceases, the voltage at the platinum rises to 3000*, which is the force actually available for re-starting the discharge if momentarily stopped.

With this discharge, the rate of absorption of mixed gases is about 7 litres per hour. When the argon has accumulated to a considerable extent, the rate falls off, and after several days' work, about 6 litres per hour becomes the maximum. In commencing operations it is advisable to introduce, first, the oxygen necessary to combine with the already included air, after which the feed of mixed gases should consist of about 11 parts of oxygen to 9 parts of air. The mixed gases may be contained in a large gas holder, and then, the feed being automatic, very little attention is required. When it is desired to determine the rate of absorption, auxiliary gas-holders of glass, graduated into litres, are called into play. If the rate is unsatisfactory, a determination may be made of the proportion of oxygen in the working vessel, and the necessary gas, air, or oxygen, as the case may be, introduced directly.

In re-starting the arc after a period of intermission, it is desirable to cut off the connection with the principal gas-holder. The gas (about two litres in amount) ejected from the working vessel by the expansion is then retained in the auxiliary holder, and no argon finds its way further back. The connection between the working vessel and the auxiliary holder should be made without india-rubber, which is liable to be attacked by the ozonized gases.

The apparatus has been kept in operation for fourteen hours continuously, and there should be no difficulty in working day and night. An electric signal could easily be arranged to give notice of the extinction of the arc, which sometimes occurs unexpectedly; or an automatic device for re-striking the arc could be contrived.—*April*, 1895.]

9. *Density of Argon prepared by means of Oxygen.*

A first estimate of the density of argon prepared by the oxygen method was founded upon the data recorded already respecting the volume present in air, on the assumption that the accurately known densities of "atmospheric" and of chemical nitrogen differ on account of the presence of argon in the former, and that during the treatment with oxygen nothing is oxidised except nitrogen. Thus, if

---

* A still higher voltage on open circuit would be preferable.

$D =$ density of chemical nitrogen,

$D' =$ „   atmospheric nitrogen,

$d =$ „   argon,

$\alpha =$ proportional volume of argon in atmospheric nitrogen,

the law of mixtures gives

$$\alpha d + (1 - \alpha) D = D'$$

or

$$d = D + (D' - D)/\alpha.$$

In this formula $D' - D$ and α are both small, but they are known with fair accuracy.   From the data already given for the experiment of September 8th

$$\alpha = \frac{65}{0 \cdot 79 \times 7925} = 0 \cdot 0104 \,;$$

whence, if on an arbitrary scale of reckoning $D = 2 \cdot 2990$, $D' = 2 \cdot 3102$, we find $d = 3 \cdot 378$.   Thus if N_2 be 14, or O_2 be 16, the density of argon is $20 \cdot 6$.

Again, from the January experiment,

$$\alpha = \frac{75 \cdot 0}{7308} = 0 \cdot 0103 \,;$$

whence, if $N = 14$, the density of argon is $20 \cdot 6$, as before.   There can be little doubt, however, that these numbers are too high, the true value of α being greater than is supposed in the above calculations.

A direct determination by weighing is desirable, but hitherto it has not been feasible to collect by this means sufficient to fill the large globe (§ 1) employed for other gases.   A *mixture* of about 400 cub. centims. of argon with pure oxygen, however, gave the weight $2 \cdot 7315$, $0 \cdot 1045$ in excess of the weight of oxygen, viz., $2 \cdot 6270$.   Thus, if α be the ratio of the volume of argon to the whole volume, the number for argon will be

$$2 \cdot 6270 + 0 \cdot 1045/\alpha.$$

The value of α, being involved only in the excess of weight above that of oxygen, does not require to be known very accurately.   Sufficiently concordant analyses by two methods gave $\alpha = 0 \cdot 1845$; whence, for the weight of the gas we get $3 \cdot 193$; so that if $O = 16$, the density of the gas would be $19 \cdot 45$.   An allowance for residual nitrogen, still visible in the gas before admixture of oxygen, raises this number to $19 \cdot 7$, which may be taken as the density of pure argon resulting from this determination*.

---

* [The proportion of nitrogen (4 or 5 per cent. of the volume) was estimated from the appearance of the nitrogen lines in the spectrum, these being somewhat more easily visible than when 3 per cent. of nitrogen was introduced into pure argon (§ 8).—*April*, 1895.]

10. *Density of Argon prepared by means of Magnesium*.*

It has already been stated that the density of the residual gas from the first and preliminary attempt to separate oxygen and nitrogen from air by means of magnesium was 19·086, and allowing for contraction on sparking with oxygen the density is calculable as 20·01. The following determinations of density were also made :—

(*a*) After absorption in glass tubes, the water circulator having been used, and subsequent circulation by means of mercury circulator until rate of contraction had become slow, 162·843 cub. centims., measured at 757·7 millims. (corr.) pressure, and 16·81° C., weighed 0·2683 grm.   Hence,

> Weight of 1 litre at 0° and 760 millims. . . . . . 1·7543 grms.
> Density compared with hydrogen (O = 16) . . . 19·63   „

This gas was again circulated over red-hot magnesium for two days. Before circulation it contained nitrogen as was evident from its spectrum ; after circulating, nitrogen appeared to be absent, and absorption had completely stopped. The density was again determined.

(*b*) 162·843 cub. centims., measured at 745·4 millims. (corr.) pressure, and 17·25° C., weighed 0·2735 grm.   Hence,

> Weight of 1 litre at 0° and 760 millims. . . . . . 1·8206 grms.
> Density compared with hydrogen (O = 16) . . . 20·38   „

Several portions of this gas, having been withdrawn for various purposes, were somewhat contaminated with air, owing to leakage, passage through the pump, &c. All these portions were united in the gas-holder with the main stock, and circulated for eight hours, during the last three of which no contraction occurred. The gas removed from the system of tubes by the mercury-pump was not restored to the gas-holder, but kept separate.

(*c*) 162·843 cub. centims., measured at 758·1 millims. (corr.) pressure, and 17·09° C., weighed 0·27705 grm.   Hence,

> Weight of 1 litre at 0° and 760 millims. . . . . . 1·8124 grms.
> Density compared with hydrogen (O = 16) . . . 20·28   „

The contents of the gas-holder were subsequently increased by a mixture of nitrogen and argon from 37 litres of atmospheric nitrogen, and after circulating, density was determined. The absorption was however not complete.

(*d*) 162·843 cub. centims., measured at 767·6 millims. (corr.) pressure, and 16·31° C., weighed 0·2703 grm.   Hence,

* See Addendum, p. 184.

Weight of 1 litre at 0° and 760 millims. . . . . 1·742 grms.

Density compared with hydrogen (O = 16) . . . 19·49   „

The gas was further circulated, until all absorption had ceased. This took about six hours. Density was again determined.

(*e*)   162·843 cub. centims., measured at 767·7 millims. (corr.) pressure, and 15·00° C., weighed 0·2773 grm. Hence,

Weight of 1 litre at 0° and 760 millims. . . . . 1·7784 grms.

Density compared with hydrogen (O = 16) . . . 19·90   „

(*f*)   A second determination was carried out, without further circulation.

162·843 cub. centims., measured at 769·0 millims. (corr.) pressure, and 16·00° C., weighed 0·2757 grm. Hence,

Weight of 1 litre at 0° and 760 millims. . . . . 1·7713 grms.

Density compared with hydrogen (O = 16) . . . 19·82   „

(*g*)   After various experiments had been made with the same sample of gas, it was again circulated until all absorption ceased. A vacuum-tube was filled with it, and showed no trace of nitrogen.

The density was again determined :—

162·843 cub. centims., measured at 750 millims. (corr.) pressure, and at 15·62° C., weighed 0·26915 grm.

Weight of 1 litre at 0° and 760 millims. . . . . 1·7707 grms.

Density compared with hydrogen (O = 16) . . . 19·82   „

These comprise all the determinations of density made. It should be stated that there was some uncertainty discovered later about the weight of the vacuous globe in (*b*) and (*c*). Rejecting these weighings, the mean of (*e*), (*f*), and (*g*) is 19·88. The density may be taken as 19·9, with approximate accuracy.

It is better to leave these results without comment at this point, and to return to them later.

11.   *Spectrum of Argon.*

Vacuum tubes were filled with argon prepared by means of magnesium at various stages in this work, and an examination of these tubes has been undertaken by Mr Crookes, to whom we wish to express our cordial thanks for his kindness in affording us helpful information with regard to its spectrum. The first tube was filled with the early preparation of density 19·09, which obviously contained some nitrogen. A photograph of the spectrum was taken, and compared with a photograph of the spectrum of nitrogen, and it was at once evident that a spectrum different from that of nitrogen had been registered.

Since that time many other samples have been examined.

The spectrum of argon, seen in a vacuum tube of about 3 millims. pressure, consists of a great number of lines, distributed over almost the whole visible field. Two lines are specially characteristic; they are less refrangible than the red lines of hydrogen or lithium, and serve well to identify the gas when examined in this way. Mr Crookes, who gives a full account of the spectrum in a separate communication, has kindly furnished us with the accurate wave-lengths of these lines as well as of some others next to be described; they are respectively $696\cdot56$ and $705\cdot64 \times 10^{-6}$ millim.

Besides these red lines, a bright yellow line, more refrangible than the sodium line, occurs at $603\cdot84$. A group of five bright green lines occurs next, besides a number of less intensity. Of this group of five, the second, which is perhaps the most brilliant, has the wave-length $561\cdot00$. There is next a blue, or blue-violet, line of wave-length $470\cdot2$ and last, in the less easily visible part of the spectrum, there are five strong violet lines, of which the fourth, which is the most brilliant, has the wave-length $420\cdot0$.

Unfortunately, the red lines, which are not to be mistaken for those of any other substance, are only to be seen at atmospheric pressure when a very powerful jar-discharge is passed through argon. The spectrum, seen under these conditions, has been examined by Professor Schuster. The most characteristic lines are perhaps those in the neighbourhood of F, and are very easily seen if there be not too much nitrogen, in spite of the presence of some oxygen and water-vapour. The approximate wave-lengths are:—

$487\cdot91$	Strong.
$(486\cdot07)$	F.
$484\cdot71$	Not quite so strong.
$480\cdot52$	Strong.
$476\cdot50$ $473\cdot53$ } . . . . $472\cdot56$	Fairly strong characteristic triplet.

It is necessary to anticipate Mr Crookes's communication, and to state that when the current is passed from the induction-coil in one direction, that end of the capillary tube next the positive pole appears of a redder, and that next the negative of a bluer hue. There are, in effect, two spectra, which Mr Crookes has succeeded in separating to a considerable extent. Mr E. C. C. Baly *, who has noticed a similar phenomenon, attributes it to the presence of two gases. The conclusion would follow that what we have termed " argon " is in reality a mixture of two gases which have as yet not been separated. This conclusion, if true, is of great importance, and experi-

* *Proc. Phys. Soc.* 1893, p. 147. He says: "When an electric current is passed through a mixture of two gases, one is separated from the other, and appears in the negative glow."

ments are now in progress to test it by the use of other physical methods. The full bearing of this possibility will appear later.

A comparison was made of the spectrum seen in a vacuum tube with the spectrum in a "plenum" tube, *i.e.*, one filled at atmospheric pressure. Both spectra were thrown into a field at the same time. It was evident that they were identical, although the relative strengths of the lines were not always the same. The seventeen most striking lines were absolutely coincident.

The presence of a small quantity of nitrogen interferes greatly with the argon spectrum. But we have found that in a tube with platinum electrodes, after the discharge has been passed for four hours, the spectrum of nitrogen disappears, and the argon spectrum manifests itself in full purity. A specially constructed tube, with magnesium electrodes, which we hoped would yield good results, removed all traces of nitrogen it is true, but hydrogen was evolved from the magnesium, and showed its characteristic lines very strongly. However, these are easily identified. The gas evolved on heating magnesium *in vacuo*, as proved by a separate experiment, consists entirely of hydrogen.

Mr Crookes has proved the identity of the chief lines of the spectrum of gas separated from air-nitrogen by aid of magnesium with that remaining after sparking air-nitrogen with oxygen, in presence of caustic soda solution.

Professor Schuster has also found the principal lines identical in the spectra of the two gases, when taken from the jar-discharge at atmospheric pressure.

12. *Solubility of Argon in Water.*

The tendency of the gas to disappear when manipulated over water in small quantities having suggested that it might be more than usually soluble in that liquid, special experiments were tried to determine the degree of solubility.

The most satisfactory measures relating to the gas isolated by means of oxygen were those of September 28. The sample contained a trace of oxygen, and (as judged by the spectrum) a residue of about 2 per cent. of nitrogen. The procedure and the calculations followed pretty closely the course marked out by Bunsen*, and it is scarcely necessary to record the details. The quantity of gas operated upon was about 4 cub. centims., of which about $1\frac{1}{2}$ cub. centims. were absorbed. The final result for the solubility was 3·94 per 100 of water at 12° C., about $2\frac{1}{2}$ times that of nitrogen. Similar results have been obtained with argon prepared by means of magnesium. At a temperature of 13·9°, 131 arbitrary measures of water absorbed

---

* *Gasometry*, p. 141.

5·3 of argon.   This corresponds to a solubility in distilled water, previously freed from dissolved gas by boiling *in vacuo* for a quarter of an hour, and admitted to the tube containing argon without contact with air, of 4·05 cub. centims. of argon per 100 of water.

The fact that the gas is more soluble than nitrogen would lead us to expect it in increased proportion in the dissolved gases of rain water. Experiment has confirmed this anticipation.   Some difficulty was at first experienced in collecting a sufficiency for the weighings in the large globe of nearly 2 litres capacity.   Attempts at extraction by means of a Töpler pump without heat were not very successful.   It was necessary to operate upon large quantities of water, and then the pressure of the liquid itself acted as an obstacle to the liberation of gas from all except the upper layers.   Tapping the vessel with a stick of wood promotes the liberation of gas in a remarkable manner, but to make this method effective, some means of circulating the water would have to be introduced.

Fig. 8.

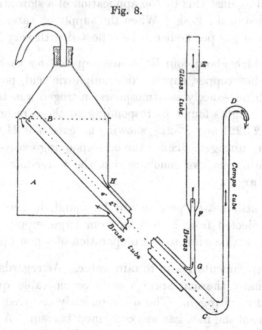

The extraction of the gases by heat proved to be more manageable. Although a large quantity of water has to be brought to or near 100° C., a prolonged boiling is not necessary, as it is not a question of collecting the whole of the gas contained in the water.   The apparatus employed, which worked very well after a little experience, will be understood from the accompanying figure.   The boiler *A* was constructed from an old oil-can, and was heated by an ordinary ring Bunsen burner.   For the supply and removal of water, two co-axial tubes of thin brass, and more than four feet in length, were applied upon the regenerative principle. The outgoing water flowed in the inner tube BC, continued from C to D by a prolongation of composition tubing. The inflowing water from a rain-water cistern was delivered into a glass tube at E, and passed through a brass connecting tube FG into the narrow annular space between the two principal tubes GH. The neck of the can was fitted with an india-rubber cork and delivery-tube, by means of which the gases were collected in the ordinary way. Any carbonic anhydride was removed by alkali before passage into the glass aspirating bottles used as gas-holders.

The convenient working of this apparatus depends very much upon the maintenance of a suitable relation between the heat and the supply of water. It is desirable that the water in the can should actually boil, but without a great development of steam; otherwise not only is there a waste of heat, and thus a smaller yield of gas, but the inverted flask used for the collection of the gas becomes inconveniently hot and charged with steam. It was found desirable to guard against this by the application of a slow stream of water to the external surface of the flask. When the supply of water is once adjusted, nearly half a litre of gas per hour can be collected with very little attention.

The gas, of which about four litres are required for each operation, was treated with red-hot copper, cupric oxide, sulphuric acid, potash, and finally phosphoric anhydride, exactly as atmospheric nitrogen was treated in former weighings. The weights found, corresponding to those recorded in § 1, were on two occasions 2·3221 and 2·3227, showing an excess of 24 milligrms. above the weight of true nitrogen. Since the corresponding excess for atmospheric nitrogen is 11 milligrms., we conclude that the water-nitrogen is relatively twice as rich in argon.

Unless some still better process can be found, it may be desirable to collect the gases ejected from boilers, or from large supply pipes which run over an elevation, with a view to the preparation of argon upon a large scale.

The above experiments relate to rain water. As regards spring water, it is known that many thermal springs emit considerable quantities of gas, hitherto regarded as nitrogen. The question early occurred to us as to what proportion, if any, of the new gas was contained therein. A notable example of a nitrogen spring is that at Bath, examined by Daubeny in 1833. With the permission of the authorities of Bath, Dr Arthur Richardson was kind enough to collect for us about 10 litres of the gases discharged from the King's Spring. A rough analysis on reception showed that it contained scarcely any oxygen and but little carbonic anhydride. Two determinations of density were made, the gas being treated in all respects as air, prepared by diffusion and unprepared, were treated for the isolation of atmospheric nitrogen. The results were :—

October 29 . . . . 2·30513
November 7 . . . . 2·30532

Mean . . . . 2·30522

The weight of the "nitrogen" from the Bath gas is thus about half-way between that of chemical and "atmospheric" nitrogen, suggesting that the proportion of argon is *less* than in air, instead of greater, as had been expected.

13. *Behaviour at Low Temperatures.*

A single experiment was made with an early sample of gas, of density 19·1, which certainly contained a considerable amount of nitrogen. On compressing it in a pressure apparatus to between 80 and 100 atmospheres pressure, and cooling to − 90° by means of boiling nitrous oxide, no appearance of liquefaction could be observed. As the critical pressure was not likely to be so high as the pressure to which it had been exposed, the non-liquefaction was ascribed to insufficient cooling.

VAPOUR-PRESSURES.

Temperature	Pressure	Temperature	Pressure	Temperature	Pressure
− 186·9	740·5 millims.	− 136·2	27·3 atms.	− 129·4	35·8 atms.
− 139·1	23·7 atms.	− 135·1	29·0 ,,	− 128·6	38·0 ,,
− 138·3	25·3 ,,	− 134·4	29·8 ,,	− 121·0	50·6 ,,

Gas	Critical temperature	Critical pressure	Boiling-point	Freezing-point	Freezing pressure	Density of gas	Density of liquid at boiling-point	Colour of liquid
		atms.	°	°	millims.			
Hydrogen, H_2	Below − 220·0°	20·0	?	?	?	1	?	Colourless
Nitrogen, N_2 .	− 146·0	35·0	− 194·4	− 214·0	60	14	0·885	,,
Carbon monoxide, CO...}	− 139·5	35·5	− 190·0	− 207·0	100	14	?	,,
Argon, A_1 ...	− 121·0	50·6	− 186·9	− 189·6	?	19·9	About 1·5	,,
Oxygen, O_2 ...	− 118·8	50·8	− 182·7	?	?	16	1·124	Bluish
Nitric oxide, NO ....}	− 93·5	71·2	− 153·6	− 167·0	138	15	?	Colourless
Methane, CH_4	− 81·8	54·9	− 164·0	− 185·8	80	8	0·415	,,

This supposition turned out to be correct. For, on sending a sample to Professor Olszewski, the author of most of the accurate measurements of the constants of gases at low temperatures, he was kind enough to submit it to examination. His results are published elsewhere; but, for convenience of reference, his tables, showing vapour-pressures, and giving a comparison between the constants of argon and those of other gases, are here reproduced.

14. *The ratio of the Specific Heats of Argon*.*

In order to decide regarding the elementary or compound nature of argon, experiments were made on the velocity of sound in it. It will be remembered that from the velocity of sound, the ratio of the specific heat at constant pressure to that at constant volume can be deduced by means of the equation

$$n\lambda = v = \sqrt{\left\{ \frac{e}{d}(1 + \alpha t)\frac{C_p}{C_v} \right\}},$$

where n is the frequency, λ is the wave-length of sound, v its velocity, e the isothermal elasticity, d the density, $(1 + \alpha t)$ the temperature-correction, C_p the specific heat at constant pressure, and C_v that at constant volume. In comparing two gases at the same temperature, each of which obeys Boyle's law with sufficient approximation and in using the same sound, many of these factors disappear, and the ratio of specific heats of one gas may be deduced from that of the other, if known, by the simple proportion

$$\lambda^2 d : \lambda'^2 d' :: 1\cdot408 : x,$$

where for example λ and d refer to air, of which the ratio is $1\cdot408$, according to the mean of observations by Röntgen ($1\cdot4053$), Wüllner ($1\cdot4053$), Kayser ($1\cdot4106$), and Jamin and Richard ($1\cdot41$).

The apparatus employed, although in principle the same as that usually employed, differed somewhat from the ordinary pattern, inasmuch as the tube was a narrow one, of 2 millims. bore, and the vibrator consisted of a glass rod, sealed into one end of the tube, so that about 15 centims. projected outside the tube, while 15 centims. was contained in the tube. By rubbing the projecting part longitudinally with a rag wet with alcohol, vibrations of exceedingly high pitch of the gas contained in the tube took place, causing waves which registered their nodes by the usual device of lycopodium powder. The temperature was that of the atmosphere and varied little from $17\cdot5°$; the pressure was also atmospheric, and varied only one millim. during the experiments. Much of the success of these experiments depends on so adjusting the length of the tube as to secure a good echo, else the wave-heaps are indistinct. But this is easily secured by attaching to its open end

* See Addendum, p. 185.

a piece of thick-walled india-rubber tubing, which can be closed by a clip at a spot which is found experimentally to produce good heaps at the nodes.

The accuracy of this instrument has frequently been tested; but fresh experiments were made with air, carbon dioxide, and hydrogen, so as to make certain that reasonably reliable results were obtainable. Of these an account is here given.

Gas in tube	Number of observations		Half-wave-length		Ratio $\frac{C_p}{C_v}$	
	I.	II.	I.	II.		
Air	3	2	19·60	19·59	1·408	Assumed
CO_2...	3	...	15·11	...	1·276	Found
H_2	3	...	73·0	...	1·376	Found

To compare these results with those of previous observers, the following numbers were obtained for carbon dioxide:—Cazin, 1·291; Röntgen, 1·305; De Lucchi, 1·292; Müller, 1·265; Wüllner, 1·311; Dulong, 1·339; Masson, 1·274; Regnault, 1·268; Amagat, 1·299; and Jamin and Richard, 1·29. It appears just to reject Dulong's number, which deviates so markedly from the rest; the mean of those remaining is 1·288, which is in sufficient agreement with that given above. For the ratio of the specific heats of hydrogen, we have:—Cazin, 1·410; Röntgen, 1·385; Dulong, 1·407; Masson, 1·401; Regnault, 1·400; and Jamin and Richard, 1·410. The mean of these numbers is 1·402. This number appears to differ considerably from the one given above. But it must be noted, first, that the wave-length which should have been found is 74·5, a number differing but little from that actually found; second, that the waves were long and that the nodes were somewhat difficult to place exactly; and third, that the atomic weight of hydrogen has been taken as unity, whereas it is more likely to be 1·01, if oxygen, as was done, be taken as 16. The atomic weight 1·01 raises the found value of the ratio to 1·399, a number differing but little from the mean value found by other observers.

Having thus established the trustworthiness of the method, we proceed to describe our experiments with argon.

Five series of measurements were made with the sample of gas of density 19·82. It will be remembered that a previous determination with the same gas gave as its density 19·90. The mean of these two numbers was therefore taken as correct, viz., 19·86.

The individual measurements are:

I.	II.	III.	IV.	V.	Mean
18·16	18·14	18·02	18·04	18·03	millims. 18·08

for the half-wave-length. Calculating the ratio of the specific heats, the number 1·644 is obtained.

The narrowness of the tube employed in these experiments might perhaps raise a doubt regarding the accuracy of the measurements, for it is conceivable that in so narrow a tube the viscosity of the gas might affect the results. We therefore repeated the experiments, using a tube of 8 millims. internal diameter.

The mean of eleven readings with air, at 18°, gave a half-wave-length of 34·62 millims. With argon in the same tube, and at the same temperature, the half-wave-length was, as a mean of six concordant readings, 31·64 millims. The density of this sample of argon, which had been transferred from a water gas-holder to a mercury gas-holder, was 19·82; and there is some reason to suspect the presence of a trace of air, for it had been standing for some time.

The result, however, substantially proves that the ratio previously found was correct. In the wide tube, $C_p : C_v :: 1·61 : 1$. Hence the conclusion must be accepted that the ratio of specific heats is practically 1·66 : 1.

It will be noticed that this is the theoretical ratio for a monatomic gas, that is, a gas in which all energy imparted to it at constant volume is expended in effecting translational motion. The only other gas of which the ratio of specific heats has been found to fulfil this condition is mercury at a high temperature*. The extreme importance of these observations will be discussed later.

15. *Attempts to induce Chemical Combination.*

A great number of attempts were made to induce chemical combination with the argon obtained by use of magnesium, but without any positive result. In such a case as this, however, it is necessary to chronicle negative results, if for no other reason but that of justifying its name, "argon." These will be detailed in order.

(*a*) *Oxygen in Presence of Caustic Alkali.*—This need not be further discussed here; the method of preparing argon is based on its inactivity under such conditions.

* Kundt and Warburg, *Pogg. Ann.* 157, p. 353, 1876.

(*b*) *Hydrogen.*—It has been mentioned that, in order to free argon from excess of oxygen, hydrogen was admitted, and sparks passed to cause combination of hydrogen and oxygen.   Here again caustic alkali was present, and argon appeared to be unaffected.

A separate experiment was, however, made in absence of water, though no special pains was taken to dry the mixture of gases.   The argon was admitted up to half an atmosphere pressure into a bulb, through whose sides passed platinum wires, carrying pointed poles of gas-carbon.   Hydrogen was then admitted until atmospheric pressure had been attained.   Sparks were then passed for four hours by means of a large induction coil, actuated by four storage cells.   The gas was confined in a bulb closed by two stop-cocks, and a small V-tube with bulbs was interposed, to act as a gauge, so that if expansion or contraction had taken place, the escape or entry of gas would be observable.   The apparatus, after the passage of sparks, was allowed to cool to the temperature of the atmosphere, and, on opening the stop-cock, the level of water in the V-tube remained unaltered.   It may therefore be concluded that, in all probability, no combination has occurred; or, that if it has, it was attended with no change of volume.

(*c*) *Chlorine.*—Exactly similar experiments were performed with dry, and afterwards with moist, chlorine.   The chlorine had been stored over strong sulphuric acid for the first experiment, and came in contact with dry argon.   Three hours sparking produced no change of volume.   A drop of water was admitted into the bulb.   After four hours sparking, the volume of the gas, after cooling, was diminished by about $\frac{1}{10}$ cub. centim., due probably to the solution of a little chlorine in the small quantity of water present.

(*d*) *Phosphorus.*—A piece of combustion-tubing, closed at one end, containing at the closed end a small piece of phosphorus, was sealed to the mercury reservoir containing argon; connected to the same reservoir was a mercury gauge and a Sprengel's pump.   After removing all air from the tubes, argon was admitted to a pressure of 600 millims.   The middle portion of the combustion-tube was then heated to bright redness, and the phosphorus was distilled slowly from back to front, so that its vapour should come into contact with argon at a red heat.   When the gas was hot, the level of the gauge altered; but, on cooling, it returned to its original level, showing that no contraction had taken place.   The experiment was repeated several times, the phosphorus being distilled through the red-hot tube from open to closed end, and *vice versâ*.   In each case, on cooling, no change of pressure was remarked.   Hence it may be concluded that phosphorus at a red heat is without action on argon.   It may be remarked parenthetically that no gaseous compound of phosphorus is known, which does not possess a volume different from the sum of those of its constituents.   That no solid compound was formed is sufficiently proved by the absence of contraction. The phosphorus was largely converted into the red modification during the experiment.

(*e*) *Sulphur.*—An exactly similar experiment was performed with sulphur, again with negative results. It may therefore be concluded that sulphur and argon are without action on each other at a red heat. And again, no gaseous compound of sulphur is known in which the volume of the compound is equal to the sum of those of its constituents.

(*f*) *Tellurium.*—As this element has a great tendency to unite with heavy metals, it was thought worth while to try its action. In this, and in the experiments to be described, a different form was given to the apparatus. The gas was circulated over the reagent employed, a tube containing it being placed in the circuit. The gas was dried by passage over soda-lime and phosphoric anhydride; it then passed over the tellurium or other reagent, then through drying tubes, and then back to the gas-holder. That combination did not occur was shown by the unchanged volume of gas in the gas-holder; and it was possible, by means of the graduated cylinder which admitted water to the gas-holder, to judge of as small an absorption as half a cubic centimetre. The tellurium distilled readily in the gas, giving the usual yellow vapours; and it condensed, quite unchanged, as a black sublimate. The volume of the gas, when all was cold, was unaltered.

(*g*) *Sodium.*—A piece of sodium, weighing about half a gramme, was heated in argon. It attacked the glass of the combustion tube, which it blackened, owing to liberation of silicon; but it distilled over in drops into the cold part of the tube. Again no change of volume occurred, nor was the surface of the distilled sodium tarnished; it was brilliant, as it is when sodium is distilled *in vacuo.* It may probably also be concluded from this experiment that silicon, even while being liberated, is without action on argon.

The action of compounds was then tried; those chosen were such as lead to oxides or sulphides. Inasmuch as the platinum-metals, which are among the most inert of elements, are attacked by fused caustic soda, its action was investigated.

(*h*) *Fused and Red-hot Caustic Soda.*—The soda was prepared from sodium, in an iron boat, by adding drops of water cautiously to a lump of the metal. When action had ceased, the soda was melted, and the boat introduced into a piece of combustion-tube placed in the circuit. After three hours circulation no contraction had occurred. Hence caustic soda has no action on argon.

(*i*) *Soda-lime at a red heat.*—Thinking that the want of porosity of fused caustic soda might have hindered absorption, a precisely similar experiment was carried out with soda-lime, a mixture which can be heated to bright redness without fusion. Again no result took place after three hours heating.

(*j*)   *Fused Potassium Nitrate* was tried under the impression that oxygen plus a base might act where oxygen alone failed.   The nitrate was fused, and kept at a bright red heat for two hours, but again without any diminution in volume of the argon.

(*k*)   *Sodium Peroxide.*—Yet another attempt was made to induce combination with oxygen and a base, by heating sodium peroxide to redness in a current of argon for over an hour, but also without effect.   It is to be noticed that metals of the platinum group would have entered into combination under such treatment.

(*l*)   *Persulphides of Sodium and Calcium.*—Soda-lime was heated to redness in an open crucible, and some sulphur was added to the red-hot mass, the lid of the crucible being then put on.   Combination ensued, with formation of polysulphides of sodium and calcium.   This product was heated to redness for three hours in a brisk current of argon, again with negative result. Again, metals of the platinum group would have combined under such treatment.

(*m*)   Some argon was shaken in a tube with nitro-hydrochloric acid. On addition of potash, so as to neutralise the acid, and to absorb the free chlorine and nitrosyl chloride, the volume of the gas was barely altered.   The slight alteration was evidently due to solubility in the aqueous liquid, and it may be concluded that no chemical action took place.

(*n*)   Bromine-water was also without effect.   The bromine vapour was removed with potash.

(*o*)   A mixture of potassium permanganate and hydrochloric acid, involving the presence of nascent chlorine, had no action, for on absorbing chlorine by means of potash, no alteration in volume had occurred.

(*p*)   Argon is not absorbed by platinum black.   A current was passed over a pure specimen of this substance; as usual, however, it contained occluded oxygen.   There was no absorption in the cold.   At 100° no action took place; and on heating to redness, by which the black was changed to sponge, still no evidence of absorption was noticed.   In all these experiments, absorption of half a cubic centimetre of argon could have at once been detected.

We do not claim to have exhausted the possible reagents.   But this much is certain, that the gas deserves the name " argon," for it is a most astonishingly indifferent body, inasmuch as it is unattacked by elements of very opposite character, ranging from sodium and magnesium on the one hand, to oxygen, chlorine, and sulphur on the other.   It will be interesting to see if fluorine also is without action, but for the present that experiment must be postponed, on account of difficulties of manipulation.

It will also be necessary to try whether the inability of argon to combine at ordinary or at high temperatures is due to the instability of its possible compounds, except when cold. Mercury vapour at 800° would present a similar instance of passive behaviour.

16. *General Conclusions.*

It remains, finally, to discuss the probable nature of the gas or gases which we have succeeded in separating from atmospheric air, and which has been provisionally named *argon*.

That argon is present in the atmosphere, and is not manufactured during the process of separation is amply proved by many lines of evidence. First, atmospheric nitrogen has a high density, while chemical nitrogen is lighter. That chemical nitrogen is a uniform substance is proved by the identity of properties of samples prepared by several different processes, and from several different compounds. It follows, therefore, that the cause of the high density of atmospheric nitrogen is due to the admixture with heavier gas. If that gas possesses the density of 20 compared with hydrogen as unity, atmospheric nitrogen should contain of it approximately 1 per cent. This is found to be the case, for on causing the nitrogen of the atmosphere to combine with oxygen in presence of alkali, the residue amounted to about 1 per cent.; and on removing nitrogen with magnesium the result is similar.

Second: This gas has been concentrated in the atmosphere by diffusion. It is true that it cannot be freed from oxygen and nitrogen by diffusion, but the process of diffusion increases relatively to nitrogen the amount of argon in that portion which does not pass through the porous walls. That this is the case is proved by the increase of density of that mixture of argon and nitrogen.

Third: On removing nitrogen from " atmospheric nitrogen " by means of magnesium, the density of the residue increases proportionately to the concentration of the heavier constituent.

Fourth: As the solubility of argon in water is relatively high, it is to be expected that the density of the mixture of argon and nitrogen, pumped out of water along with oxygen should, after removal of the oxygen, exceed that of " atmospheric nitrogen." Experiment has shown that the density is considerably increased.

Fifth: It is in the highest degree improbable that two processes, so different from each other, should each manufacture the same product. The explanation is simple if it be granted that these processes merely eliminate nitrogen from " atmospheric nitrogen."

Sixth: If the newly discovered gas were not in the atmosphere, the discrepancies in the density of "chemical" and "atmospheric" nitrogen would remain unexplained.

Seventh: It has been shown that pure nitrogen, prepared from its compounds, leaves a negligible residue when caused to enter into combination with oxygen or with magnesium.

There are other lines of argument which suggest themselves; but we think that it will be acknowledged that those given above are sufficient to establish the existence of argon in the atmosphere.

It is practically certain that the argon prepared by means of electric sparking with oxygen is identical with argon prepared by means of magnesium. The samples have in common:—

First: Spectra which have been found by Mr Crookes, Professor Schuster, and ourselves to be practically identical.

Second: They have approximately the same density. The density of argon, prepared by means of magnesium, was 19·9; that of argon, from sparking with oxygen, about 19·7; these numbers are practically identical.

Third: Their solubility in water is the same.

That argon is an element, or a mixture of elements, may be inferred from the observations of § 14. For Clausius has shown that if K be the energy of translatory motion of the molecules of a gas, and H their whole kinetic energy, then

$$\frac{K}{H} = \frac{3(C_p - C_v)}{2C_v},$$

C_p and C_v denoting as usual the specific heat at constant pressure and at constant volume respectively. Hence, if, as for mercury vapour and for argon (§ 14), the ratio of specific heats $C_p : C_v$ be $1\frac{2}{3}$, it follows that $K = H$, or that the whole kinetic energy of the gas is accounted for by the translatory motion of its molecules. In the case of mercury the absence of interatomic energy is regarded as proof of the monatomic character of the vapour, and the conclusion holds equally good for argon.

The only alternative is to suppose that if argon molecules are di- or polyatomic, the atoms acquire no relative motion, even of rotation, a conclusion improbable in itself and one postulating the sphericity of such complex groups of atoms.

Now a monatomic gas can be only an element, or a mixture of elements; and hence it follows that argon is not of a compound nature.

According to Avogadro, equal volumes of gases at the same temperature and pressure contain equal numbers of molecules. The molecule of hydrogen gas, the density of which is taken as unity, is supposed to consist of two atoms. Its molecular weight is therefore 2. Argon is approximately 20 times as heavy as hydrogen, that is, its molecular weight is 20 times as great as that of hydrogen, or 40. But its molecule is monatomic, hence its atomic weight, or, if it be a mixture, the mean of the atomic weights of the elements in that mixture, taken for the proportion in which they are present, must be 40.

This conclusion rests on the assumption that all the molecules of argon are monatomic. The result of the first experiment is, however, so nearly that required by theory, that there is room for only a small number of molecules of a different character. A study of the expansion of argon by heat is proposed, and would doubtless throw light upon this question.

There is evidence both for and against the hypothesis that argon is a mixture: for, owing to Mr Crookes' observations of the dual character of its spectrum; against, because of Professor Olszewski's statement that it has a definite melting-point, a definite boiling-point, and a definite critical temperature and pressure; and because on compressing the gas in presence of its liquid, pressure remains sensibly constant until all gas has condensed to liquid. The latter experiments are the well-known criteria of a pure substance; the former is not known with certainty to be characteristic of a mixture. The conclusions which follow are, however, so startling, that in our future experimental work we shall endeavour to decide the question by other means.

For the present, however, the balance of evidence seems to point to simplicity. We have, therefore, to discuss the relations to other elements of an element of atomic weight 40. We inclined for long to the view that argon was possibly one, or more than one, of the elements which might be expected to follow fluorine in the periodic classification of the elements—elements which should have an atomic weight between 19, that of fluorine, and 23, that of sodium. But this view is apparently put out of court by the discovery of the monatomic nature of its molecules.

The series of elements possessing atomic weights near 40 are:—

Chlorine	35·5
Potassium	39·1
Calcium	40·0
Scandium	44·0

There can be no doubt that potassium, calcium, and scandium follow legitimately their predecessors in the vertical columns, lithium, beryllium, and boron, and that they are in almost certain relation with rubidium, strontium, and (but not so certainly) yttrium. If argon be a single element, then there is reason to doubt whether the periodic classification of the elements is complete; whether, in fact, elements may not exist which cannot be fitted among those of which it is composed. On the other hand, if argon be a mixture of two elements, they might find place in the eighth group, one after chlorine and one after bromine. Assuming 37 (the approximate mean between the atomic weights of chlorine and potassium) to be the atomic weight of the lighter element, and 40 the mean atomic weight found, and supposing that the second element has an atomic weight between those of bromine, 80, and rubidium, 85·5, viz. 82, the mixture should consist of 93·3 per cent. of the lighter, and 6·7 per cent. of the heavier element. But it appears improbable that such a high percentage as 6·7 of a heavier element should have escaped detection during liquefaction.

If the atomic weight of the lighter element were 38, instead of 37, however, the proportion of heavier element would be considerably reduced. Still, it is difficult to account for its not having been detected, if present.

If it be supposed that argon belongs to the eighth group, then its properties would fit fairly well with what might be anticipated. For the series, which contains

$$\text{Si}_n{}^{\text{IV}}, \quad \text{P}_4{}^{\text{III and V}}, \quad \text{S}_{8 \text{ to } 2}{}^{\text{II to VI}}, \text{ and } \text{Cl}_2{}^{\text{I to VII}},$$

might be expected to end with an element of monatomic molecules, of no valency, i.e. incapable of forming a compound, or if forming one, being an octad; and it would form a possible transition to potassium, with its monovalence, on the other hand. Such conceptions are, however, of a speculative nature; yet they may be perhaps excused, if they in any way lead to experiments which tend to throw more light on the anomalies of this curious element.

In conclusion, it need excite no astonishment that argon is so indifferent to reagents. For mercury, although a monatomic element, forms compounds which are by no means stable at a high temperature in the gaseous state; and attempts to produce compounds of argon may be likened to attempts to cause combination between mercury gas at 800° and other elements. As for the physical condition of argon, that of a gas, we possess no knowledge why carbon, with its low atomic weight, should be a solid, while nitrogen is a gas, except in so far as we ascribe molecular complexity to the former and comparative molecular simplicity to the latter. Argon, with its comparatively low density and its molecular simplicity, might well be expected to rank among the gases. And its inertness, which has suggested its name, sufficiently explains why it has not previously been discovered as a constituent of compound bodies.

We would suggest for this element, assuming provisionally that it is not a mixture, the symbol A.

We have to record our thanks to Messrs Gordon, Kellas, and Matthews, and especially to Mr Percy Williams, for their assistance in the prosecution of this research.

ADDENDUM (by Professor W. Ramsay).
March 20, 1895.

Further determinations of the density of argon prepared by means of magnesium have been made. In each case the argon was circulated over magnesium for at least two hours after all absorption of nitrogen had stopped, as well as over red-hot copper, copper oxide, soda-lime, and phosphoric anhydride. The gas also passed out of the mercury gas-holder through phosphoric anhydride into the weighing globe. The results are in complete accordance with previous determinations of density; and for convenience of reference the former numbers are included in the table which follows.

DENSITY OF ARGON.

Date	Volume	Temperature	Pressure	Weight	Weight of 1 litre at 0° and 760 millims.	Density (O=16)
	cub. centims.	°	millims.	grm.		
(1) Nov. 26......	162·843	15·00	767·7	0·2773	1·7784	19·904
(2) „ 27......	162·843	16·00	769·0	0·2757	1·7717	19·823
(3) Dec. 22......	162·843	15·62	750·1	0·26915	1·7704	19·816
(4) Feb. 16......	162·843	13·45	771·1	0·2818	1·7834	19·959
(5) „ 19......	162·843	14·47	768·2	0·2789	1·7842	19·969
(6) „ 24......	162·843	17·85	764·4	0·2738	1·7810	19·932

The general mean is 19·900; or if Nos. (2) and (3) be rejected as suspiciously low, the mean of the remaining four determinations is 19·941. The molecular weight may therefore be taken as 39·9 without appreciable error.

The value of R in the gas-equation $R = pv/T$ has also been determined between $-89°$ and $+248°$. For this purpose, a gas-thermometer was filled with argon, and a direct comparison was made with a similar thermometer filled with hydrogen.

The method of using such a hydrogen-thermometer has already been described by Ramsay and Shields*. For the lowest temperature, the thermometer bulbs were immersed in boiling nitrous oxide; for atmospheric temperature, in running water; for temperatures near 100° in steam, and for the remaining temperatures, in the vapours of chlorobenzene, aniline, and quinolene.

* *Trans. Chem. Soc.* Vol. 63, pp. 835, 836. It is to be noticed that the value of R is not involved in using the hydrogen-thermometer; its constancy alone is postulated.

The results are collected in the following tables :—

HYDROGEN THERMOMETER.

Temperature	Pressure	Volume (corr.)	R
° C.	millims.		
13·04	763·6	1·00036	2·6705
99·84	992·6	1·00280	2·6697
130·62	1073·8	1·00364	2·6701
185·46	1218·5	1·00518	2·6716
248·66	1385·1	1·00703	2·6737
− 87·92	497·3	0·99756	2·6804

The value of R is thus practically constant, and this affords a proof that the four last temperatures have been estimated with considerable accuracy.

ARGON THERMOMETER.

	Temperature	Pressure	Volume (corr.)	R
	° C.	millims.		
Series I.......	14·15	701·7	1·000396	2·4446
	14·27	699·7	1·000401	2·4366
	14·40	702·6	1·000404	2·4462
	19·96	906·5	1·00280	2·4379
	100·06	904·8	1·00280	2·4322
	− 87·92	455·6	0·99756	2·4556

By mischance, air leaked into the bulb ; it was therefore refilled.

	Temperature	Pressure	Volume (corr.)	R
Series II. ...	130·58	1060·0	1·0037	2·6363
	185·46	1200·3	1·0052	2·6317

A bubble of argon leaked into the bulb, and the value of R increased.

	Temperature	Pressure	Volume (corr.)	R
Series III....	12·05	760·9	1·00034	2·6698
	12·61	761·3	1·00034	2·6728
	248·66	1384·0	1·0070	2·6717
	248·66	1376·9	1·0070	2·6580
	− 87·92	495·7	0·99756	2·6718

It may be concluded from these numbers, that argon undergoes no molecular change between − 88° and + 250°.

Further determinations of the wave-length of sound in argon have been made, the wider tube having been used. In every case the argon was as carefully purified as possible. In experiment (3) too much lycopodium dust was present in the tube; that is perhaps the cause of the low result. For completeness' sake, the original result in the narrow tube has also been given.

Date	Density	Half-wave-length		Temperature		Ratio
		In air	In argon	Air	Argon	
Dec. 6............	19·92	19·59	18·08	17·5°	17·5°	1·644
Feb. 15	19·96	33·73	31·00	6·7	6·5	1·641
„ 20	19·97	34·10	31·31	7·22	8·64	1·629
Mar. 19	19·94	34·23	31·68	11·20	11·49	1·659

The general mean of these numbers is 1·643; if (3) be rejected, it is 1·648. In the last experiment every precaution was taken. The half-wave-length in air is the mean of 11 readings, the highest of which was 34·67 and the lowest 34·00. They run:—

34·67 ; 34·06 ; 34·27 ; 34·39; 34·00; 34·00; 34·13; 34·20; 34·20; 34·33; 34·33.
11·25°; 11·00°; 10·80°; 10·8°; 10·0° ; 11·0° ; 11·3° ; 11·4° ; 11·4°; 11·6° ; 11·6°.

With argon the mean is also that of 11 readings, of which the highest is 31·83, and the lowest, 31·5. They are:—

31·5 ; 31·5 ; 31·66 ; 31·55 ; 31·83 ; 31·77 ; 31·81 ; 31·83 ; 31·83; 31·50 ; 31·66.
11·8° ; 11·8° ; 11·20°; 11·40°; 11·60°; 11·40°; 11·40°; 11·4° ; 11·5° ; 11·5° ; 11·4°.

If the atomic weight of argon is identical with its molecular weight, it must closely approximate to 39·9. But if there were some molecules of A_2 present, mixed with a much larger number of molecules of A_1, then the atomic weight would be correspondingly reduced. Taking an imaginary case, the question may be put:—What percentage of molecules of A_2 would raise the density of A_1 from 19·0 to 19·9? A density of 19·0 would imply an atomic weight of 38·0, and argon would fall into the gap between chlorine and potassium. Calculation shows that in 10,000 molecules, 474 molecules of A_2 would have this result, the remaining 9526 molecules being those of A_1.

Now if molecules of A_2 be present, it is reasonable to suppose that their number would be increased by lowering the temperature, and diminished by heating the gas. A larger change of density should ensue on lowering than on raising the temperature, however, as on the above supposition, there is not a large proportion of molecules of A_2 present.

But it must be acknowledged that the constancy of the found value of R is not favourable to this supposition.

A similar calculation is possible for the ratio of specific heats. Assuming the gas to contain 5 per cent. of molecules of A_2, and 95 per cent. of molecules of A_1 the value of γ, the ratio of specific heats, would be 1·648. All that can be said on this point is, that the found ratio approximates to this number; but whether the results are to be trusted to indicate a unit in the second decimal appears to me doubtful.

The question must therefore for the present remain open.

ADDENDUM.

April 9.

It appears worth while to chronicle an experiment of which an accident prevented the completion. It may be legitimately asked, Does magnesium not absorb any argon, or any part of what we term argon? To decide this question, about 500 grms. of magnesium nitride, mixed with metallic magnesium which had remained unacted on, during extraction of nitrogen from "air-nitrogen," was placed in a flask, to which a reservoir full of dilute hydrochloric acid was connected. The flask was coupled with a tube full of red-hot copper oxide, intended to oxidise the hydrogen which would be evolved by the action of the hydrochloric acid on the metallic magnesium. To the end of the copper oxide tube a gas-holder was attached, so as to collect any evolved gas; and the system was attached to a vacuum-pump, in order to exhaust the apparatus before commencing the experiment, as well as to collect all gas which should be evolved, and remain in the flask.

On admitting hydrochloric acid to the flask of magnesium nitride a violent reaction took place, and fumes of ammonium chloride passed into the tube of copper oxide. These gave, of course, free nitrogen. This had not been foreseen; it would have been well to retain these fumes by plugs of glass-wool. The result of the experiment was that about 200 cub. centims. of gas were collected. After sparking with oxygen in presence of caustic soda, the volume was reduced to 3 cub. centims. of a gas which appeared to be argon.

215.

ARGON.

[*Royal Institution Proceedings*, XIV. pp. 524—538, Ap. 1895.]

IT is some three or four years since I had the honour of lecturing here one Friday evening upon the densities of oxygen and hydrogen gases, and upon the conclusions that might be drawn from the results. It is not necessary, therefore, that I should trouble you to-night with any detail as to the method by which gases can be accurately weighed. I must take that as known, merely mentioning that it is substantially the same as is used by all investigators nowadays, and introduced more than fifty years ago by Regnault. It was not until after that lecture that I turned my attention to nitrogen; and in the first instance I employed a method of preparing the gas which originated with Mr Vernon Harcourt, of Oxford. In this method the oxygen of ordinary atmospheric air is got rid of with the aid of ammonia. Air is bubbled through liquid ammonia, and then passed through a red-hot tube. In its passage the oxygen of the air combines with the hydrogen of the ammonia, all the oxygen being in that way burnt up and converted into water. The excess of ammonia is subsequently absorbed with acid, and the water by ordinary desiccating agents. That method is very convenient; and, when I had obtained a few concordant results by means of it, I thought that the work was complete, and that the weight of nitrogen was satisfactorily determined. But then I reflected that it is always advisable to employ more than one method, and that the method that I had used—Mr Vernon Harcourt's method—was not that which had been used by any of those who had preceded me in weighing nitrogen. The usual method consists in absorbing the oxygen of air by means of red-hot copper; and I thought that I ought at least to give that method a trial, fully expecting to obtain forthwith a value in harmony with that already afforded by the ammonia method. The result, however, proved otherwise. The gas obtained by the copper method, as I may call it, proved to be one-thousandth part heavier than that obtained by the ammonia method; and, on repetition, that difference was only brought out more clearly. This was about three years ago. In order, if possible, to get further light upon a discrepancy which puzzled me very much, and which, at that time, I regarded only with disgust and impatience, I published a letter in *Nature** inviting criticisms from chemists who might be interested in such questions. I obtained various useful suggestions, but none going to the root of the matter. Several persons who wrote to me privately were inclined to think that the explanation was to be sought in a partial dissociation of the nitrogen derived from ammonia. For, before going further, I ought to explain that, in the nitrogen obtained by the ammonia method, some—about a seventh part—is derived from the ammonia, the larger part, however, being derived as usual from the atmosphere. If the chemically derived nitrogen were partly dissociated into its component atoms, then the lightness of the gas so prepared would be explained.

The next step in the enquiry was, if possible, to exaggerate the discrepancy. One's instinct at first is to try to get rid of a discrepancy, but I believe that experience shows such an endeavour to be a mistake. What one ought to do is to magnify a small discrepancy with a view to finding out the explanation; and, as it appeared in the present case that the root of the discrepancy lay in the fact that part of the nitrogen prepared by the ammonia method was nitrogen out of ammonia, although the greater part remained of common origin in both cases, the application of the principle suggested a trial of the weight of nitrogen obtained wholly from ammonia. This could easily be done by substituting pure oxygen for atmospheric air in the ammonia method, so that the whole, instead of only a part, of the nitrogen collected should be derived from the ammonia itself. The discrepancy was at once magnified some five times. The nitrogen so obtained from ammonia proved to be about one-half per cent. lighter than nitrogen obtained in the ordinary way from the atmosphere, and which I may call for brevity "atmospheric" nitrogen.

That result stood out pretty sharply from the first; but it was necessary to confirm it by comparison with nitrogen chemically derived in other ways. The table before you gives a summary of such results, the numbers being the weights in grams actually contained under standard conditions in the globe employed.

ATMOSPHERIC NITROGEN.

By hot copper (1892)	2·3103
By hot iron (1893)	2·3100
By ferrous hydrate (1894)	2·3102
Mean	2·3102

* [Vol. IV. p. 1.]

CHEMICAL NITROGEN.

From nitric oxide . . . .   2·3001

From nitrous oxide . . . .   2·2990

From ammonium nitrite purified at a red heat . . .   2·2987

From urea . . . .   2·2985

From ammonium nitrite purified in the cold . . . .   2·2987

                                                        ‾‾‾‾‾‾‾
                                            Mean 2·2990

The difference is about 11 milligrams, or about one-half per cent.; and it was sufficient to prove conclusively that the two kinds of nitrogen—the chemically derived nitrogen and the atmospheric nitrogen—differed in weight, and therefore, of course, in quality, for some reason hitherto unknown.

I need not spend time in explaining the various precautions that were necessary in order to establish surely that conclusion. One had to be on one's guard against impurities, especially against the presence of hydrogen, which might seriously lighten any gas in which it was contained. I believe, however, that the precautions taken were sufficient to exclude all questions of that sort, and the result, which I published about this time last year*, stood sharply out, that the nitrogen obtained from chemical sources was different from the nitrogen obtained from the air.

Well, that difference, admitting it to be established, was sufficient to show that some hitherto unknown gas is involved in the matter. It might be that the new gas was dissociated nitrogen, contained in that which was too light, the chemical nitrogen—and at first that was the explanation to which I leaned; but certain experiments went a long way to discourage such a supposition. In the first place, chemical evidence—and in this matter I am greatly dependent upon the kindness of chemical friends—tends to show that, even if ordinary nitrogen could be dissociated at all into its component atoms, such atoms would not be likely to enjoy any very long continued existence. Even ozone goes slowly back to the more normal state of oxygen; and it was thought that dissociated nitrogen would have even a greater tendency to revert to the normal condition. The experiment suggested by that remark was as follows—to keep chemical nitrogen—the too light nitrogen which might be supposed to contain dissociated molecules—for a good while, and to examine whether it changed in density. Of course it would be useless to shut up gas in a globe and weigh it, and then, after an interval, to weigh it again, for there would be no opportunity for any change of weight to occur, even although the gas within the globe had undergone some chemical alteration. It is necessary to re-establish the standard conditions of temperature and pressure which are always understood when we speak of filling a

* [Vol. IV. p. 104.]

globe with gas, for I need hardly say that filling a globe with gas is but a figure of speech.   Everything depends upon the temperature and pressure at which you work.   However, that obvious point being borne in mind, it was proved by experiment that the gas did not change in weight by standing for eight months—a result tending to show that the abnormal lightness was not the consequence of dissociation.

Further experiments were tried upon the action of the silent electric discharge—both upon the atmospheric nitrogen and upon the chemically derived nitrogen—but neither of them seemed to be sensibly affected by such treatment; so that, altogether, the balance of evidence seemed to incline against the hypothesis of abnormal lightness in the chemically derived nitrogen being due to dissociation, and to suggest strongly, as almost the only possible alternative, that there must be in atmospheric nitrogen some constituent heavier than true nitrogen.

At that point the question arose, What was the evidence that all the so-called nitrogen of the atmosphere was of one quality ? And I remember—I think it was about this time last year, or a little earlier—putting the question to my colleague, Professor Dewar.   His answer was that he doubted whether anything material had been done upon the matter since the time of Cavendish, and that I had better refer to Cavendish's original paper.   That advice I quickly followed, and I was rather surprised to find that Cavendish had himself put this question quite as sharply as I could put it.   Translated from the old-fashioned phraseology connected with the theory of phlogiston, his question was whether the inert ingredient of the air is really all of one kind; whether all the nitrogen of the air is really the same as the nitrogen of nitre. Cavendish not only asked himself this question, but he endeavoured to answer it by an appeal to experiment.

I should like to show you Cavendish's experiment in something like its original form.   He inverted a U-tube filled with mercury, the legs standing in two separate mercury cups.   He then passed up, so as to stand above the mercury, a mixture of nitrogen, or of air, and oxygen; and he caused an electric current from a frictional electrical machine like the one I have before me to pass from the mercury in the one leg to the mercury in the other, giving sparks across the intervening column of air.   I do not propose to use a frictional machine to-night, but I will substitute for it one giving electricity of the same quality of the construction introduced by Mr Wimshurst, of which we have a fine specimen in the Institution.   It stands just outside the door of the theatre, and will supply an electric current along insulated wires, leading to the mercury cups; and, if we are successful, we shall cause sparks to pass through the small length of air included above the columns of mercury. There they are; and after a little time you will notice that the mercury rises, indicating that the gas is sensibly absorbed under the influence of the sparks and of a piece of potash floating on the mercury. It was by that means that Cavendish established his great discovery of the nature of the inert ingredient in the atmosphere, which we now call nitrogen; and, as I have said, Cavendish himself proposed the question, as distinctly as we can do, Is this inert ingredient all of one kind? and he proceeded to test that question. He found, after days and weeks of protracted experiment, that, for the most part, the nitrogen of the atmosphere was absorbed in this manner, and converted into nitrous acid; but that there was a small residue remaining after prolonged treatment with sparks, and a final absorption of the residual oxygen. That residue amounted to about $\frac{1}{120}$ part of the nitrogen taken; and Cavendish draws the conclusion that, if there be more than one inert ingredient in the atmosphere, at any rate the second ingredient is not contained to a greater extent than $\frac{1}{120}$ part.

I must not wait too long over the experiment. Mr Gordon tells me that a certain amount of contraction has already occurred; and if we project the U upon the screen, we shall be able to verify the fact. It is only a question of time for the greater part of the gas to be taken up, as we have proved by preliminary experiments.

In what I have to say from this point onwards, I must be understood as speaking as much on behalf of Professor Ramsay as for myself. At the first, the work which we did was to a certain extent independent. Afterwards we worked in concert, and all that we have published in our joint names must be regarded as being equally the work of both of us. But, of course, Professor Ramsay must not be held responsible for any chemical blunder into which I may stumble to-night.

By his work and by mine the heavier ingredient in atmospheric nitrogen which was the origin of the discrepancy in the densities has been isolated, and we have given it the name of "argon." For this purpose we may use the original method of Cavendish, with the advantages of modern appliances. We can procure more powerful electric sparks than any which Cavendish could command by the use of the ordinary Ruhmkorff coil stimulated by a battery of Grove cells; and it is possible so to obtain evidence of the existence of argon. The oxidation of nitrogen by that method goes on pretty quickly. If you put some ordinary air, or, better still, a mixture of air and oxygen, in a tube in which electric sparks are made to pass for a certain time, then in looking through the tube, you observe the well-known reddish-orange fumes of the oxides of nitrogen. I will not take up time in going through the experiment, but will merely exhibit a tube already prepared (image on screen).

One can work more efficiently by employing the alternate currents from dynamo machines which are now at our command. In this Institution we have the advantage of a public supply; and if I pass alternate currents originating in Deptford through this Ruhmkorff coil, which acts as what is now called a "high potential transformer," and allow sparks from the secondary to pass in an inverted test tube between platinum points, we shall be able to show in a comparatively short time a pretty rapid absorption of the gases. The electric current is led into the working chamber through bent glass tubes containing mercury, and provided at their inner extremities with platinum points. In this arrangement we avoid the risk, which would otherwise be serious, of a fracture just when we least desired it. I now start the sparks by switching on the Ruhmkorff to the alternate current supply; and, if you will take note of the level of the liquid representing the quantity of mixed gases included, I think you will see after, perhaps, a quarter of an hour that the liquid has very appreciably risen, owing to the union of the nitrogen and the oxygen gases under the influence of the electrical discharge, and subsequent absorption of the resulting compound by the alkaline liquid with which the gas space is enclosed.

By means of this little apparatus, which is very convenient for operations upon a moderate scale, such as analyses of "nitrogen" for the amount of argon that it may contain, we are able to get an absorption of about 80 cubic centimetres per hour, or about 4 inches along this test tube, when all is going well. In order, however, to effect the isolation of argon on any considerable scale by means of the oxygen method, we must employ an apparatus still more enlarged. The isolation of argon requires the removal of nitrogen, and, indeed, of very large quantities of nitrogen, for, as it appears, the proportion of argon contained in atmospheric nitrogen is only about 1 per cent., so that for every litre of argon that you wish to get you must eat up some hundred litres of nitrogen. That, however, can be done upon an adequate scale by calling to our aid the powerful electric discharge now obtainable by means of the alternate current supply and high potential transformers.

In what I have done upon this subject I have had the advantage of the advice of Mr Crookes, who some years ago drew special attention to the electric discharge or flame, and showed that many of its properties depended upon the fact that it had the power of causing, upon a very considerable scale, a combination of the nitrogen and the oxygen of the air in which it was made.

I had first thought of showing in the lecture room the actual apparatus which I have employed for the concentration of argon; but the difficulty is that, as the apparatus has to be used, the working parts are almost invisible, and I came to the conclusion that it would really be more instructive as well as more convenient to show the parts isolated, a very little effort of imagination being then all that is required in order to reconstruct in the mind the actual arrangements employed.

First, as to the electric arc or flame itself. We have here a transformer made by Pike and Harris. It is not the one that I have used in practice; but it is convenient for certain purposes, and it can be connected by means of a switch with the alternate currents of 100 volts furnished by the Supply Company. The platinum terminals that you see here are modelled exactly upon the plan of those which have been employed in practice. I may say a word or two on the question of mounting. The terminals require to be very massive on account of the heat evolved. In this case they consist of platinum wire doubled upon itself six times. The platinums are continued by iron wires going through glass tubes, and attached at the ends to the copper leads. For better security, the tubes themselves are stopped at the lower ends with corks and charged with water, the advantage being that, when the whole arrangement is fitted by means of an indiarubber stopper into a closed vessel, you have a witness that, as long as the water remains in position, no leak can have occurred through the insulating tubes conveying the electrodes.

Now, if we switch on the current and approximate the points sufficiently, we get the electric flame. There you have it. It is, at present, showing a certain amount of soda. That in time would burn off. After the arc has once been struck, the platinums can be separated; and then you have two tongues of fire ascending almost independently of one another, but meeting above. Under the influence of such a flame, the oxygen and the nitrogen of the air combine at a reasonable rate, and in this way the nitrogen is got rid of. It is now only a question of boxing up the gas in a closed space, where the argon concentrated by the combustion of the nitrogen can be collected. But there are difficulties to be encountered here. One cannot well use anything but a glass vessel. There is hardly any metal available that will withstand the action of strong caustic alkali and of the nitrous fumes resulting from the flame. One is practically limited to glass. The glass vessel employed is a large flask with a single neck, about half full of caustic alkali. The electrodes are carried through the neck by means of an indiarubber bung provided also with tubes for leading in the gas. The electric flame is situated at a distance of only about half an inch above the caustic alkali. In that way an efficient circulation is established; the hot gases as they rise from the flame strike the top, and then as they come round again in the course of the circulation they pass sufficiently close to the caustic alkali to ensure an adequate removal of the nitrous fumes.

There is another point to be mentioned. It is necessary to keep the vessel cool; otherwise the heat would soon rise to such a point that there would be excessive generation of steam, and then the operation would come to a standstill. In order to meet this difficulty the upper part of the vessel is provided with a water-jacket, in which a circulation can be established. No doubt the glass is severely treated, but it seems to stand it in a fairly amiable manner.

By means of an arrangement of this kind, taking nearly three horse-power from the electric supply, it is possible to consume nitrogen at a reasonable rate. The transformers actually used are the "Hedgehog" transformers of Mr Swinburne, intended to transform from 100 volts to 2400 volts. By Mr Swinburne's advice I have used two such, the fine wires being in series so as to accumulate the electrical potential and the thick wires in parallel. The rate at which the mixed gases are absorbed is about seven litres per hour; and the apparatus, when once fairly started, works very well as a rule, going for many hours without attention. At times the arc has a trick of going out, and it then requires to be restarted by approximating the platinums. We have already worked 14 hours on end, and by the aid of one or two automatic appliances it would, I think, be possible to continue operations day and night.

The gases, air and oxygen in about equal proportions, are mixed in a large gas-holder, and are fed in automatically as required. The argon gradually accumulates; and when it is desired to stop operations the supply of nitrogen is cut off, and only pure oxygen allowed admittance. In this way the remaining nitrogen is consumed, so that, finally, the working vessel is charged with a mixture of argon and oxygen only, from which the oxygen is removed by ordinary well-known chemical methods. I may mention that at the close of the operation, when the nitrogen is all gone, the arc changes its appearance, and becomes of a brilliant blue colour.

I have said enough about this method, and I must now pass on to the alternative method which has been very successful in Professor Ramsay's hands—that of absorbing nitrogen by means of red-hot magnesium. By the kindness of Professor Ramsay and Mr Matthews, his assistant, we have here the full scale apparatus before us almost exactly as they use it. On the left there is a reservoir of nitrogen derived from air by the simple removal of oxygen. The gas is then dried. Here it is bubbled through sulphuric acid. It then passes through a long tube made of hard glass and charged with magnesium in the form of thin turnings. During the passage of the gas over the magnesium at a bright red heat, the nitrogen is absorbed in a great degree, and the gas which finally passes through is immensely richer in argon than that which first enters the hot tube. At the present time you see a tolerably rapid bubbling on the left, indicative of the flow of atmospheric nitrogen into the combustion furnace; whereas, on the right, the outflow is very much slower. Care must be taken to prevent the heat rising to such a point as to soften the glass. The concentrated argon is collected in a second gas-holder, and afterwards submitted to further treatment. The apparatus employed by Professor Ramsay in the subsequent treatment is exhibited in the diagram, and is very effective for its purpose; but I am afraid that the details of it would not readily be followed from any explanation that

I could give in the time at my disposal. The principle consists in the circulation of the mixture of nitrogen and argon over hot magnesium, the gas being made to pass round and round until the nitrogen is effectively removed from it. At the end that operation, as in the case of the oxygen method, proceeds somewhat slowly. When the greater part of the nitrogen is gone, the remainder seems to be unwilling to follow, and it requires somewhat protracted treatment in order to be sure that the nitrogen has wholly disappeared. When I say "wholly disappeared," that, perhaps, would be too much to say in any case. What we can say is that the spectrum test is adequate to show the presence, or at any rate to show the addition, of about one-and-a-half per cent. of nitrogen to argon as pure as we can get it; so that it is fair to argue that any nitrogen at that stage remaining in the argon is only a small fraction of one-and-a-half per cent.

I should have liked at this point to be able to give advice as to which of the two methods—the oxygen method or the magnesium method—is the easier and the more to be recommended; but I confess that I am quite at a loss to do so. One difficulty in the comparison arises from the fact that they have been in different hands. As far as I can estimate, the quantities of nitrogen eaten up in a given time are not very different. In that respect, perhaps, the magnesium method has some advantage; but, on the other hand, it may be said that the magnesium process requires a much closer supervision, so that, perhaps, fourteen hours of the oxygen method may not unfairly compare with eight hours or so of the magnesium method. In practice a great deal would depend upon whether in any particular laboratory alternate currents are available from a public supply. If the alternate currents are at hand, I think it may probably be the case that the oxygen method is the easier; but, otherwise, the magnesium method would, probably, be preferred, especially by chemists who are familiar with operations conducted in red-hot tubes.

I have here another experiment illustrative of the reaction between magnesium and nitrogen. Two rods of that metal are suitably mounted in an atmosphere of nitrogen, so arranged that we can bring them into contact and cause an electric arc to form between them. Under the action of the heat of the electric arc the nitrogen will combine with the magnesium; and if we had time to carry out the experiment we could demonstrate a rapid absorption of nitrogen by this method. When the experiment was first tried, I had hoped that it might be possible, by the aid of electricity, to start the action so effectively that the magnesium would continue to burn independently under its own developed heat in the atmosphere of nitrogen. Possibly, on a larger scale, something of this sort might succeed, but I bring it forward here only as an illustration. We turn on the electric current, and bring the magnesiums together. You see a brilliant green light, indicating the vaporisation of the magnesium. Under the influence of the heat the magnesium burns, and there is collected in the glass vessel a certain amount of brownish-looking powder which consists mainly of the nitride of magnesium. Of course, if there is any oxygen present it has the preference, and the ordinary white oxide of magnesium is formed.

The gas thus isolated is proved to be inert by the very fact of its isolation. It refuses to combine under circumstances in which nitrogen, itself always considered very inert, does combine—both in the case of the oxygen treatment and in the case of the magnesium treatment; and these facts are, perhaps, almost enough to justify the name which we have suggested for it. But, in addition to this, it has been proved to be inert under a considerable variety of other conditions such as might have been expected to tempt it into combination. I will not recapitulate all the experiments which have been tried, almost entirely by Professor Ramsay, to induce the gas to combine. Hitherto, in our hands, it has not done so, and I may mention that recently, since the publication of the abstract of our paper read before the Royal Society, argon has been submitted to the action of titanium at a red heat, titanium being a metal having a great affinity for nitrogen, and that argon has resisted the temptation to which nitrogen succumbs. We never have asserted, and we do not now assert, that argon can under no circumstances be got to combine. That would, indeed, be a rash assertion for any one to venture upon; and only within the last few weeks there has been a most interesting announcement by M. Berthelot, of Paris, that, under the action of the silent electric discharge, argon can be absorbed when treated in contact with the vapour of benzine. Such a statement, coming from so great an authority, commands our attention; and if we accept the conclusion, as I suppose we must do, it will follow that argon has, under those circumstances, combined.

Argon is rather freely soluble in water. That is a thing that troubled us at first in trying to isolate the gas; because, when one was dealing with very small quantities, it seemed to be always disappearing. In trying to accumulate it we made no progress. After a sufficient quantity had been prepared, special experiments were made on the solubility of argon in water. It has been found that argon, prepared both by the magnesium method and by the oxygen method, has about the same solubility in water as oxygen—some two-and-a-half times the solubility of nitrogen. This suggests, what has been verified by experiment, that the dissolved gases of water should contain a larger proportion of argon than does atmospheric nitrogen. I have here an apparatus of a somewhat rough description, which I have employed in experiments of this kind. The boiler employed consists of an old oil-can. The water is supplied to it and drawn from it by coaxial tubes of metal. The incoming cold water flows through the outer annulus between the two tubes. The outgoing hot water passes through the inner tube, which ends in the interior of the vessel at a higher level. By means of this arrangement the heat of the water which has done its work is passed on to the incoming water not yet in operation, and in that way a limited amount of heat is made to bring up to the boil a very much larger quantity of water than would otherwise be possible, the greater part of the dissolved gases being liberated at the same time. These are collected in the ordinary way. What you see in this flask is dissolved air collected out of water in the course of the last three or four hours. Such gas, when treated as if it were atmospheric nitrogen, that is to say after removal of the oxygen and minor impurities, is found to be decidedly heavier than atmospheric nitrogen to such an extent as to indicate that the proportion of argon contained is about double. It is obvious, therefore, that the dissolved gases of water form a convenient source of argon, by which some of the labour of separation from air is obviated. During the last few weeks I have been supplied from Manchester by Mr Macdougall, who has interested himself in this matter, with a quantity of dissolved gases obtained from the condensing water of his steam engine.

As to the spectrum, we have been indebted from the first to Mr Crookes, and he has been good enough to-night to bring some tubes which he will operate, and which will show you at all events the light of the electric discharge in argon. I cannot show you the spectrum of argon, for unfortunately the amount of light from a vacuum tube is not sufficient for the projection of its spectrum. Under some circumstances the light is red, and under other circumstances it is blue. Of course when these lights are examined with the spectroscope—and they have been examined by Mr Crookes with great care—the differences in the colour of the light translate themselves into different groups of spectrum lines. We have before us Mr Crookes' map, showing the two spectra upon a very large scale. The upper is the spectrum of the blue light; the lower is the spectrum of the red light; and it will be seen that they differ very greatly. Some lines are common to both; but a great many lines are seen only in the red, and others are seen only in the blue. It is astonishing to notice what trifling changes in the conditions of the discharge bring about such extensive alterations in the spectrum.

One question of great importance upon which the spectrum throws light is, Is the argon derived by the oxygen method really the same as the argon derived by the magnesium method? By Mr Crookes' kindness I have had an opportunity of examining the spectra of the two gases side by side, and such examination as I could make revealed no difference whatever in the two spectra, from which, I suppose, we may conclude either that the gases are absolutely the same, or, if they are not the same, that at any rate the ingredients by which they differ cannot be present in more than a small proportion in either of them.

My own observations upon the spectrum have been made principally at
atmospheric pressure. In the ordinary process of sparking, the pressure is
atmospheric; and, if we wish to look at the spectrum, we have nothing more
to do than to include a jar in the circuit, and to put a direct-vision prism to
the eye. At my request, Professor Schuster examined some tubes containing
argon at atmospheric pressure prepared by the oxygen method, and I have
here a diagram of a characteristic group. He also placed upon the sketch
some of the lines of zinc, which were very convenient as directing one exactly
where to look. See figure.

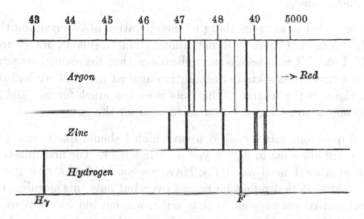

Within the last few days, Mr Crookes has charged a radiometer with
argon. When held in the light from the electric lamp, the vanes revolve
rapidly. Argon is anomalous in many respects, but not, you see, in this.

Next, as to the density of argon. Professor Ramsay has made numerous
and careful observations upon the density of the gas prepared by the magnesium method, and he finds a density of about 19·9 as compared with
hydrogen. Equally satisfactory observations upon the gas derived by the
oxygen method have not yet been made*, but there is no reason to suppose
that the density is different, such numbers as 19·7 having been obtained.

One of the most interesting matters in connection with argon, however, is
what is known as the ratio of the specific heats. I must not stay to elaborate
the questions involved, but it will be known to many who hear me that the
velocity of sound in a gas depends upon the ratio of two specific heats—the
specific heat of the gas measured at constant pressure, and the specific heat
measured at constant volume. If we know the density of a gas, and also the
velocity of sound in it, we are in a position to infer this ratio of specific heats;
and by means of this method, Professor Ramsay has determined the ratio in
the case of argon, arriving at the very remarkable result that the ratio of

* [See *Proc. Roy. Soc.* Vol. LIX. p. 198, 1896.]

specific heats is represented by the number 1·65, approaching very closely to
the theoretical limit, 1·67. The number 1·67 would indicate that the gas has
no energy except energy of translation of its molecules. If there is any other
energy than that, it would show itself by this number dropping below 1·67.
Ordinary gases, oxygen, nitrogen, hydrogen, &c., do drop below, giving the
number 1·4. Other gases drop lower still. If the ratio of specific heats is
1·65, practically 1·67, we may infer that the whole energy of motion is translational; and from that it would seem to follow by arguments which, however,
I must not stop to elaborate, that the gas must be of the kind called by
chemists monatomic.

I had intended to say something of the operation of determining the ratio
of specific heats, but time will not allow. The result is, no doubt, very
awkward. Indeed, I have seen some indications that the anomalous properties
of argon are brought as a kind of accusation against us. But we had the very
best intentions in the matter. The facts were too much for us; and all that
we can do now is to apologise for ourselves and for the gas.

Several questions may be asked, upon which I should like to say a word or
two, if you will allow me to detain you a little longer. The first question (I do
not know whether I need ask it) is, Have we got hold of a new gas at all?
I had thought that that might be passed over, but only this morning I read in
a technical journal the suggestion that argon was our old friend nitrous oxide.
Nitrous oxide has roughly the density of argon; but that, so far as I can see,
is the only point of resemblance between them.

Well, supposing that there is a new gas, which I will not stop to discuss,
because I think that the spectrum alone would be enough to prove it, the
next question that may be asked is, Is it in the atmosphere? This matter
naturally engaged our earnest attention at an early stage of the enquiry. I
will only indicate in a few words the arguments which seem to us to show
that the answer must be in the affirmative.

In the first place, if argon be not in the atmosphere, the original
discrepancy of densities which formed the starting-point of the investigation
remains unexplained, and the discovery of the new gas has been made upon a
false clue. Passing over that, we have the evidence from the blank experiments, in which nitrogen originally derived from chemical sources is treated
either with oxygen or with magnesium, exactly as atmospheric nitrogen is
treated. If we use atmospheric nitrogen, we get a certain proportion of argon,
about 1 per cent. If we treat chemical nitrogen in the same way we get, I
will not say absolutely nothing, but a mere fraction of what we should get had
atmospheric nitrogen been the subject. You may ask, Why do we get any
fraction at all from chemical nitrogen? It is not difficult to explain the small
residue, because in the manipulation of the gases large quantities of water are

used ; and, as I have already explained, water dissolves argon somewhat freely. In the processes of manipulation some of the argon will come out of solution, and it remains after all the nitrogen has been consumed.

Another wholly distinct argument is founded upon the method of diffusion introduced by Graham. Graham showed that if you pass gas along porous tubes you alter the composition, if the gas is a mixture. The lighter constituents go more readily through the pores than do the heavier ones. The experiment takes this form. A number of tobacco pipes—eight in the actual arrangement—are joined together in series with indiarubber junctions, and they are put in a space in which a vacuum can be made, so that the space outside the porous pipes is vacuous or approximately so. Through the pipes ordinary air is led. One end may be regarded as open to the atmosphere. The other end is connected with an aspirator so arranged that the gas collected is only some 2 per cent. of that which leaks through the porosities. The case is like that of an Australian river drying up almost to nothing in the course of its flow. Well, if we treat air in that way, collecting only the small residue which is less willing than the remainder to penetrate the porous walls, and then prepare " nitrogen " from it by removal of oxygen and moisture, we obtain a gas heavier than atmospheric nitrogen, a result which proves that the ordinary nitrogen of the atmosphere is not a simple body, but is capable of being divided into parts by so simple an agent as the tobacco pipe.

If it be admitted that the gas is in the atmosphere, the further question arises as to its nature.

At this point I would wish to say a word of explanation. Neither in our original announcement at Oxford, nor at any time since, until the 31st of January, did we utter a word suggesting that argon was an element; and it was only after the experiments upon the specific heats that we thought that we had sufficient to go upon in order to make any such suggestion in public. I will not insist that that observation is absolutely conclusive. It is certainly strong evidence. But the subject is difficult, and one that has given rise to some difference of opinion among physicists. At any rate this property distinguishes argon very sharply from all the ordinary gases.

One question which occurred to us at the earliest stage of the enquiry, as soon as we knew that the density was not very different from 21, was the question of whether, possibly, argon could be a more condensed form of nitrogen, denoted chemically by the symbol N_3. There seem to be several difficulties in the way of this supposition. Would such a constitution be consistent with the ratio of specific heats (1·65)? That seems extremely doubtful. Another question is, Can the density be really as high as 21, the number required on the supposition of N_3? As to this matter, Professor Ramsay has repeated his measurements of density, and he finds that he cannot get even so high as 20. To suppose that the density of argon is really 21, and that it appears to be 20 in consequence of nitrogen still mixed with it, would be to suppose a contamination with nitrogen out of all proportion to what is probable. It would mean some 14 per cent. of nitrogen, whereas it seems that from one-and-a-half to two per cent. is easily enough detected by the spectroscope. Another question that may be asked is, Would N_3 require so much cooling to condense it as argon requires?

There is one other matter on which I would like to say a word—the question as to what N_3 would be like if we had it. There seems to be a great discrepancy of opinions. Some high authorities, among whom must be included, I see, the celebrated Mendeleef, consider that N_3 would be an exceptionally stable body; but most of the chemists with whom I have consulted are of opinion that N_3 would be explosive, or, at any rate, absolutely unstable. That is a question which may be left for the future to decide. We must not attempt to put these matters too positively. The balance of evidence still seems to be against the supposition that argon is N_3, but for my part I do not wish to dogmatise.

A few weeks ago we had an eloquent lecture from Professor Rücker on the life and work of the illustrious Helmholtz. It will be known to many that during the last few months of his life Helmholtz lay prostrate in a semi-paralysed condition, forgetful of many things, but still retaining a keen interest in science. Some little while after his death we had a letter from his widow, in which she described how interested he had been in our preliminary announcement at Oxford upon this subject, and how he desired the account of it to be read to him over again. He added the remark, " I always thought that there must be something more in the atmosphere."

216.

ON THE STABILITY OR INSTABILITY OF CERTAIN FLUID MOTIONS. III.*

[*Proceedings of the London Mathematical Society,* XXVII. pp. 5–12, 1895.]

THE steady motions in question are those in which the velocity is parallel to a fixed line (x), and such that U is a function of y only. In the disturbed motion $U + u, v$, the infinitely small quantities u, v are supposed to be periodic functions of x, proportional to e^{ikx}, and, as dependent upon the time, to be proportional to e^{int}, where n is a constant, real or imaginary. Under these circumstances the equation determining v is

$$\left(\frac{n}{k} + U\right)\left(\frac{d^2v}{dy^2} - k^2v\right) - \frac{d^2U}{dy^2}v = 0. \quad\ldots\ldots\ldots\ldots(1)$$

The vorticity (Z) of the steady motion is $\frac{1}{2} dU/dy$. If throughout any layer Z be constant, d^2U/dy^2 vanishes, and, whenever $n + kU$ does not also vanish,

$$d^2v/dy^2 - k^2v = 0, \quad\ldots\ldots\ldots\ldots\ldots\ldots\ldots(2)$$

or

$$v = A e^{ky} + B e^{-ky}. \quad\ldots\ldots\ldots\ldots\ldots\ldots(3)$$

If there are several layers in each of which Z is constant, the various solutions of the form (3) are to be fitted together, the arbitrary constants being so chosen as to satisfy certain boundary conditions. The first of these conditions is evidently

$$\Delta v = 0. \quad\ldots\ldots\ldots\ldots\ldots\ldots\ldots\ldots(4)\dagger$$

The second may be obtained by integrating (1) across the boundary. Thus

$$\left(\frac{n}{k} + U\right) . \Delta\left(\frac{dv}{dy}\right) - \Delta\left(\frac{dU}{dy}\right) . v = 0. \quad\ldots\ldots\ldots\ldots(5)$$

---

* The two earlier papers upon this subject are to be found in *Proc. Lond. Math. Soc.* Vol. XI. p. 57, 1880 [Vol. I. p. 474]; Vol. XIX. p. 67, 1887 [Vol. III. p. 17]. The fluid is supposed to be destitute of viscosity.

† [Δ being the symbol of finite differences.]

At a fixed wall $\qquad v = 0.$

Equation (2) secures that the vorticity shall remain constant in each layer, and equation (3) that there shall be no slipping at the surface of transition. Equations (2) and (3) together may be regarded as expressing the continuity of the motion at the surface between the layers.

In the first of the papers above referred to, I have applied equation (1) to prove that, if d^2U/dy^2 be of one sign throughout the whole interval between two fixed walls, n can have no imaginary part. It is true that, if $n + kU$ vanishes anywhere, the expression for $d^2v/dy^2 - k^2v$ in (1) becomes infinite, unless indeed $v = 0$ at the place in question; and Lord Kelvin* considers that the "disturbing infinity" thus introduced vitiates the proof of stability. To this criticism it may be replied† that, "if n be complex, there is no disturbing infinity, and that, therefore, the argument does not fail, regarded as one for excluding complex values of n. What happens when n has a real value, such that $n + kU$ vanishes at an interior point, is a subject for further consideration."

In embarking upon this it will be convenient to take first the case of (2), (3), (4), (5), where the vorticity of the steady motion is uniform through layers of finite thickness. Any general conclusions arrived at in this way should at least throw light upon the extreme case where the number of the layers is infinitely great, and their thickness is infinitely small.

Starting from the first wall at $y = 0$, let the surfaces between the layers occur at $y = y_1$, $y = y_2$, &c., and let the values of U at these places be U_1, U_2, &c. In conformity with (4) and with the condition that $v = 0$, when $y = 0$, we may take in the first layer

$$v = v_1 = M_1 \sinh ky; \qquad \dots\dots\dots\dots(6)$$

in the second layer

$$v = v_2 = v_1 + M_2 \sinh k(y - y_1); \qquad \dots\dots\dots(7)$$

in the third layer

$$v = v_3 = v_2 + M_3 \sinh k(y - y_2); \qquad \dots\dots\dots(8)$$

and so on‡.

If the second fixed wall be in the rth layer at $y = y'$, then

$$M_1 \sinh ky' + M_2 \sinh k(y' - y_1) + \dots + M_r \sinh k(y' - y_{r-1}) = 0. \dots(9)$$

We have still to express the conditions (5) at the various surfaces of transition. At the first surface

$$v = M_1 \sinh ky_1, \qquad \Delta(dv/dy) = kM_2;$$

* Phil. Mag. Vol. xxiv. p. 275, 1887.
† Phil. Mag. Vol. xxxiv. p. 66, 1892. [Vol. iii. p. 580.]
‡ This is the process followed in the second of the papers cited, with a slight difference of notation.

at the second surface

$$v = M_1 \sinh ky_2 + M_2 \sinh k\,(y_2 - y_1), \qquad \Delta\,(dv/dy) = kM_3\,;$$

and so on. If we denote the values of $\Delta\,(dU/dy)$ at the various surfaces by Δ_1, Δ_2, &c., the conditions may be written

$$
\left.
\begin{aligned}
(n + kU_1)\,M_2 - \Delta_1\,.\,M_1 \sinh ky_1 &= 0 \\
(n + kU_2)\,M_3 - \Delta_2\,.\,\{M_1 \sinh ky_2 + M_2 \sinh k\,(y_2 - y_1)\} &= 0 \\
\cdots \qquad \cdots \qquad \cdots \qquad \cdots \qquad \cdots \qquad \cdots \qquad \cdots \;
\end{aligned}
\right\} \;\;\ldots(10)
$$

The $r-1$ equations (10) together with (9) suffice to determine n, and the $r-1$ ratios $M_1 : M_2 : M_3 : \ldots : M_r$. The determinantal equation in n is of degree $r-1$, the number of the surfaces of transition; and corresponding to each root there is an expression for v, definite except as regards a constant multiplier.

It is important to note that the disturbances thus expressed are such as leave the vorticity unaltered in the interior of every layer; that they relate, in fact, merely to waves upon the surfaces of transition. The additional vorticity due to the disturbance is proportional to $d^2v/dy^2 - k^2v$, and is equated to zero in (2). If we wish to consider the most general disturbance possible, we must provide for an arbitrary vorticity at every point.

The nature of the normal modes of disturbance not yet considered will be apparent from a comparison between (1) and (2). Even though $d^2U/dy^2 = 0$, the latter does not follow from the former, unless it be assumed that $n + kU$ is finite. Wherever $n + kU$ vanishes, that is, at the places where the wave velocity is equal to the stream velocity, (1) is satisfied, even though (2) be violated. Thus any value of $-kU$ to be found anywhere in the fluid is an admissible value of n, and the corresponding normal function (v) is obtained by allowing the arbitrary constants in (3) to be discontinuous at this place as well as at the surfaces of transition, subject of course to the condition that v itself shall be continuous. The new arbitrary constant thus disposable allows all the conditions to be satisfied with the value of n already prescribed.

The equations (9), (10) already found suffice for the present purpose if we introduce a fictitious surface of transition at the place in question. Suppose, for example, that $\Delta_3 = 0$ in the third of equations (10). It will follow either that $M_4 = 0$, or that $n + kU_3 = 0$. In the first alternative the constants A and B are continuous, and all local peculiarity disappears. The second alternative is the one with which we are now concerned. The equations suffice, as usual, to determine n (equal to $-kU_2$), as well as the ratios of the M's which give the form of the normal function. The mode of disturbance is such that a new vorticity is introduced at the place, or rather at the plane in question. In one sense this is the only new vorticity; but the waves upon the surfaces of transition involve changes of vorticity as regards given positions in space, though not as regards given portions of fluid.

We have now to consider what occurs at a second place in the fluid where the velocity happens to be the same as at the first place. The second place may be either within a layer of originally uniform vorticity or upon a surface of transition. In the first case nothing very special presents itself. If there be no new vorticity at the second place, the value of v is definite as usual, save as to an arbitrary multiplier. But, consistently with the given value of n, there may be new vorticity at the second as well as at the first place, and then the complete value of v for the given n may be regarded as composed of two parts, each proportional to one of the new vorticities, and each affected by an arbitrary multiplier.

If the second place lie upon a surface of transition, we have a state of things corresponding to the "disturbing infinity" in (1). In the above example, where $\Delta_3 = 0$, $n + kU_3 = 0$, we have now further to suppose that U_1, the velocity at the first surface of transition, coincides with U_3. From the first of equations (10), since $n + kU_1 = 0$, while Δ_1 and $\sinh ky_1$ are finite, we see that M_1 must vanish. Hence $v = 0$ throughout the entire layer from the wall $y = 0$ to $y = y_1$. The remainder of the motion from $y = y_1$ to $y = y'$ is to be determined as usual.

From the fact that $v = 0$, we might be tempted to infer that the surface in question behaves like a fixed wall. But a closer examination shows that the inference would be unwarranted. In order to understand this it may be well to investigate the relation between v and the displacement of the surface, supposed also to be proportional to $e^{int} \cdot e^{ikx}$. Thus, if the equation of the surface be

$$F = y - h e^{int + ikx} = 0, \quad \dots\dots\dots\dots\dots\dots\dots(11)$$

the condition to be satisfied is*

$$\frac{dF}{dt} + U_1 \frac{dF}{dx} + v \frac{dF}{dy} = 0, \quad \dots\dots\dots\dots\dots\dots(12)$$

so that

$$- ih(n + kU_1) + v = 0 \quad \dots\dots\dots\dots\dots\dots(13)$$

is the required relation. Using this, we see from the first of equations (10) that h does not vanish, but is given by

$$h = \frac{v}{i(n + kU_1)} = \frac{M_2}{i\Delta_1}. \quad \dots\dots\dots\dots\dots\dots(14)$$

The propagation of a wave at the same velocity as that at which the fluid moves does not entail the existence of a finite velocity v.

That v vanishes at a surface of transition where $n + kU = 0$ follows in general from (5), seeing that the value of $\Delta(dU/dy)$ is finite. That region of

* Lamb's *Hydrodynamics*, § 10.

the fluid, bounded by this surface and one of the fixed walls, which does not include the added vorticity, will in general remain undisturbed, but there may be exceptions when one of the values of n proper to this region (regarded as bounded by fixed walls) happens to coincide with that prescribed. It does not appear that the infinity which enters when $n + kU = 0$ disturbs any general conclusions as to the conditions of stability, or even seriously modifies the character of the solutions themselves.

When d^2U/dy^2 is finite, we must fall back upon equation (1). The character of the disturbing infinity at a place (say, $y = 0$) where $n + kU$ vanishes would be most satisfactorily investigated by means of the complete solution of some particular case. It is, however, sufficient to examine the form of solution in the neighbourhood of $y = 0$, and for this purpose the differential equation may be simplified. Thus, when y is small, $n + kU$ may be regarded as proportional to y, and d^2U/dy^2 as approximately constant. In comparison with the large term, k^2v may be neglected, and it suffices to consider

$$d^2v/dy^2 + y^{-1}v = 0, \quad\dots\dots(15)$$

a known constant multiplying y being omitted for brevity. This falls under the head of Riccati's equation

$$d^2v/dy^2 + y^\mu v = 0, \quad\dots\dots (16)$$

of which the solution* is in general (m fractional)

$$v = \sqrt{y}\,\{AJ_m(\xi) + BJ_{-m}(\xi)\}, \quad\dots\dots(17)$$

where

$$m = 1/(\mu + 2), \qquad \xi = 2my^{1/2m}. \quad\dots\dots(18)$$

When, as in the present case, m is integral, $J_{-m}(\xi)$ is to be replaced by the function of the second kind $Y_m(\xi)$. The general solution of (15) is accordingly

$$v = \sqrt{y}\,\{AJ_1(2\sqrt{y}) + BY_1(2\sqrt{y})\}. \quad\dots\dots(19)$$

In passing through zero y changes sign, and with it the character of the functions. If we regard (19) as applicable on the positive side, then on the negative side we may write

$$v = \sqrt{y}\,\{CJ_1(2\sqrt{y}) + DY_1(2\sqrt{y})\}, \quad\dots\dots(20)$$

the arguments of the functions in (20) being pure imaginaries.

The functions $J_1(z)$, $Y_1(z)$ are given by

$$J_1(z) = \frac{z}{2} - \frac{z^3}{2^2 \cdot 4} + \frac{z^5}{2^2 \cdot 4^2 \cdot 6} - \dots, \quad\dots\dots(21)$$

---

* Lommel, *Studien über die Bessel'schen Functionen*, § 31, Leipzig, 1868; Gray and Mathews' *Bessel s Functions*, p. 233, 1895.

$$Y_1(z) = \frac{1}{z}\left\{1 - \frac{z^2}{2^2} + \frac{z^4}{2^2 \cdot 4^2} - \frac{z^6}{2^2 \cdot 4^2 \cdot 6^2} + \dots\right\}$$

$$-\log z\left\{\frac{z}{2} - \frac{z^3}{2^2 \cdot 4} + \frac{z^5}{2^2 \cdot 4^2 \cdot 6} - \dots\right\}$$

$$+\frac{z}{2}S_1 - \frac{z^3}{2^2 \cdot 4}S_2 + \frac{z^5}{2^2 \cdot 4^2 \cdot 6}S_3 - \dots, \dots\dots\dots\dots(22)$$

where $$S_m = 1 + \tfrac{1}{2} + \tfrac{1}{3} + \dots + 1/m. \qquad\dots\dots\dots\dots(23)$$

When y is small, (19) gives

$$v = A\{y - \tfrac{1}{2}y^2\} + B\{\tfrac{1}{2}(1 - y + \tfrac{1}{4}y^2) - \log(2\sqrt{y})(y - \tfrac{1}{2}y^2) + yS_1 - \tfrac{1}{2}y^2 S_2\}; \dots(24)$$

so that ultimately

$$v = \tfrac{1}{2}B, \qquad dv/dy = A - \tfrac{1}{2}B\log y, \qquad d^2v/dy^2 = -A - \tfrac{1}{2}By^{-1}, \dots(25)$$

v remaining finite in any case.

We will now show that any value of $-kU$ is an admissible value of n in (1). The place where $n + kU = 0$ is taken as origin of y; and in the first instance we will suppose that $n + kU$ vanishes nowhere else. In the immediate neighbourhood of $y = 0$, the solutions applicable on the two sides are (19), (20), and they are subject to the condition that v shall be continuous. Hence, by (25), $B = D$, leaving three constants arbitrary. The manner in which the functions start from $y = 0$ being thus ascertained, their further progress is subject to the original equation (1), which completely defines them when the three arbitraries are known. In the present case two relations are given by the conditions to be satisfied at the fixed walls or other boundaries of the fluid, and thus is determined the entire form of v, save as to a constant multiplier. If, as must usually be the case, B and D are finite, there is infinite vorticity at the origin, but this is no more than occurs even when d^2U/dy^2 is zero throughout the region surrounding the origin.

Any other places at which $n + kU = 0$ may be treated in a similar manner, and the most general solution will contain as many arbitrary constants as there are places of infinite vorticity. But the vorticity need not be infinite merely because $n + kU = 0$; and, in fact, a particular solution may be obtained with only one infinite vorticity. At any other of the critical places, such, for example, as we may now suppose the origin to be, B and D may vanish, so that

$$v = 0, \qquad d^2v/dy^2 = A, \text{ or } C.$$

From this discussion it would seem that the infinities which present themselves when $n + kU = 0$ do not seriously interfere with the application of the general theory, so long as the square of the disturbance from steady motion is neglected. The value of conclusions relating only to infinitely small disturbances is another question.

When regard is paid to viscosity, the difficulties are of course much increased. In the particular case where the original vorticity is uniform, the problem of small disturbances has been solved by Lord Kelvin*, who shows that the motion is stable by the aid of a special solution not proportional to a simple exponential function of the time. If we retain the supposition of the present paper that the disturbance as a function of the time is proportional to e^{int}, we obtain an equation [(52) in Lord Kelvin's paper] which has been discussed by Stokes†. From his results it appears that it is not possible to find a solution applicable to an unlimited fluid which shall be periodic with respect to x, and remain finite when $y = \pm \infty$, and this whether n be real or complex. The cause of the failure would appear to lie in the fact, indicated by Lord Kelvin's solution, that the stability is ultimately of a higher order than can be expressed by any simple exponential function of the time.

[*Addendum, January*, 1896.—It may be well to emphasise more fully that the solutions of this paper only profess to apply in the limit, when the disturbances are infinitely small. The constant factor which represents the scale of the disturbance must be imagined to be so small that the actual disturbance *nowhere* rises to such a magnitude as to interfere with the approximations upon which (1) is founded. For example, in (25), although dv/dy is infinite at $y = 0$ relatively to its value at other places, it must still be regarded as infinitely small throughout in comparison with the quantities which define the steady motion.]

---

* *Phil. Mag.* Vol. xxiv. p. 191, 1887.
† *Camb. Phil. Trans.* Vol. x. p. 105, 1857.

217.

ON THE PROPAGATION OF WAVES UPON THE PLANE SURFACE SEPARATING TWO PORTIONS OF FLUID OF DIFFERENT VORTICITIES.

[*Proceedings of the London Mathematical Society*, XXVII. pp. 13—18, 1895.]

In former papers* I have considered the problem of the motion in two dimensions of inviscid incompressible fluid between two parallel walls. In the case where the steady motion is such that in each half of the layer included between the walls the vorticity is constant, it appeared that the motion is stable, small displacements of the surface separating the two vorticities being propagated as waves of constant amplitude. More particularly, if the velocity of the steady motion increase uniformly from zero at the walls to the value U in the middle stratum, a disturbance proportional to $e^{i(nt+kx)}$ requires that

$$n + kU = U/b \,.\, \tanh kb, \quad\dots\dots\dots\dots\dots\dots\dots(1)$$

where $2b$ is the distance between the walls. The wave-length is $2\pi/k$, and the fact that n is real indicates that the disturbance is stable.

Discussions upon the difficult question of the nature of the instability manifested by fluids in their flow through pipes of moderate bore seemed to make it desirable to push the investigation of the disturbance from some simple case of steady motion so far at least as to include the squares of the small quantities.

In the present paper the problem chosen for the purpose is that above referred to, simplified by excluding the fixed walls, or, what comes to the same thing, by supposing them removed to a distance very great in comparison with the wave-length of the disturbance. We suppose, then, that in the steady motion the surface of separation coincides with $y = 0$, that when y is positive the vorticity is $+\omega$, and that when y is negative the vorticity is $-\omega$.

---

* "On the Stability or Instability of certain Fluid Motions," *Proc. Lond. Math. Soc.* Vol. XI. p. 57, 1880 [Vol. I. p. 474]; Vol. XIX. p. 67, 1887 [Vol. III. p. 17].

In the disturbed motion the surface separating the two vorticities is displaced, so that its equation becomes $y = h \cos x$, k being put equal to unity for the sake of brevity.

In virtue of the incompressibility, the component velocities, denoted as usual by u and v, are connected with a stream-function ψ by the relations

$$u = d\psi/dy, \qquad v = - d\psi/dx. \qquad \qquad \ldots\ldots\ldots\ldots\ldots\ldots(2)$$

The vorticity is represented by $\frac{1}{2}\nabla^2\psi$, which is accordingly equal to $\pm \omega$. During the steady motion of the upper fluid, we have

$$\psi = \alpha + \beta y + \omega y^2. \qquad \qquad \ldots\ldots\ldots\ldots\ldots\ldots\ldots\ldots\ldots\ldots(3)$$

In consequence of the disturbance ψ deviates from the value given by (3); but, since, by a known theorem, the vorticity remains throughout equal to ω, the addition to ψ must satisfy $\nabla^2\psi = 0$. The additional terms must also satisfy the condition of being periodic in period 2π; and thus we obtain altogether as the expression for ψ during the disturbed motion

$$\psi = \alpha + \beta y + \omega y^2 + e^{-y}\,(A_1 \cos x + B_1 \sin x)$$
$$+ e^{-2y}\,(A_2 \cos 2x + B_2 \sin 2x) + \ldots, \qquad \ldots\ldots\ldots\ldots\ldots(4)$$

positive exponents being excluded by the condition to be satisfied when $y = +\infty$. Similarly in the lower fluid

$$\psi' = \alpha' + \beta'y - \omega y^2 + e^{y}\,(A_1' \cos x + B_1' \sin x)$$
$$+ e^{2y}\,(A_2' \cos 2x + B_2' \sin 2x) + \ldots. \qquad \ldots\ldots\ldots\ldots\ldots(5)$$

From these values of ψ, ψ' the velocities u, v at any point are deducible by (2).

We have still to satisfy the conditions at the surface of separation

$$y = h \cos x. \qquad \qquad \ldots\ldots\ldots\ldots\ldots\ldots\ldots\ldots\ldots\ldots\ldots\ldots(6)$$

It is necessary that u and v, as given by ψ and ψ', should there be continuous, any sliding of the one body of fluid upon the other being equivalent to a vortex-sheet, and therefore excluded by the conditions of the problem. Thus at the surface we must have

$$d\,(\psi - \psi')/dx = 0, \qquad d\,(\psi - \psi')/dy = 0. \qquad \ldots\ldots\ldots\ldots\ldots(7)$$

For the purposes of the first approximation, where only the first power of h is retained, y may be put equal to zero in the exponential terms so soon as the differentiations have been performed. Equations (7) give accordingly

$$- \sin x\,(A_1 - A_1') + \cos x\,(B_1 - B_1')$$
$$- 2 \sin 2x\,(A_2 - A_2') + 2 \cos 2x\,(B_2 - B_2') - \ldots\ldots\ldots\ldots = 0,$$

$$\beta - \beta' + 4\omega h \cos x - \cos x\,(A_1 + A_1') - \sin x\,(B_1 + B_1')$$
$$- 2 \cos 2x\,(A_2 + A_2') - 2 \sin 2x\,(B_2 + B_2') - \ldots\ldots\ldots\ldots = 0;$$

14—2

from which it appears that to this approximation all the coefficients with suffixes higher than unity must vanish. Also

$$B_1 = 0, \qquad B_1' = 0; \qquad \beta' = \beta, \qquad A_1' = A_1 = 2\omega h.$$

Thus
$$\psi = \alpha + \beta y + \omega y^2 + 2\omega h\, e^{-y} \cos x, \ldots\ldots\ldots\ldots\ldots(8)$$

$$\psi' = \alpha' + \beta y - \omega y^2 + 2\omega h\, e^{y} \cos x, \ldots\ldots\ldots\ldots\ldots(9)$$

are the values of ψ determined in accordance with (6) and the other prescribed conditions. From (8) or (9), we find as the values of u and v at the surface

$$u = \beta, \qquad v = 2\omega h \sin x, \ldots\ldots\ldots\ldots\ldots\ldots\ldots(10)$$

applicable when the form of the surface is that given by (6), at the moment, we may suppose, when $t = 0$.

By means of (10) it is possible to determine the form and position of the surface of separation at time dt, and thus to trace out its transformation. In the present case it will be simplest merely to verify that the propagation of the form (6) with a certain velocity (V) satisfies all the conditions. If

$$F(x,\, y,\, t) = y - h \cos (x - Vt) = 0 \ldots\ldots\ldots\ldots\ldots(11)$$

be the equation of the surface, the condition to be satisfied* is

$$DF/Dt = 0,$$

or
$$\frac{dF}{dt} + u \frac{dF}{dx} + v \frac{dF}{dy} = 0. \ldots\ldots\ldots\ldots\ldots(12)$$

Here, when $t = 0$,

$$\frac{dF}{dt} = - Vh \sin x, \qquad \frac{dF}{dx} = h \sin x, \qquad \frac{dF}{dy} = 1;$$

so that (12) becomes, with use of (10),

$$(- V + \beta + 2\omega)\, h \sin x = 0,$$

showing that (11) continues to represent the surface of separation at time dt, provided that

$$V = \beta + 2\omega. \ldots\ldots\ldots\ldots\ldots\ldots\ldots(13)$$

Accordingly, if (13) be satisfied, equation (11) suffices to represent the changes in the surface of separation for any length of time, or, in other words, the disturbance is propagated as a simple wave.

From (8) it appears that β represents the velocity in the steady motion when $y = 0$, and the result is in accordance with (1), where $\tanh kb = 1$. The disturbance may be supposed to be got rid of by the introduction of a flexible lamina at the surface of separation. If, by forces applied to it, the lamina be straightened out so as to coincide with $y = 0$, and be held there at rest, the steady motion is recovered.

* Lamb's *Hydrodynamics*, § 10.

In proceeding to further approximations, in which higher powers of h are retained, it appears either from the equations, or immediately from the symmetries involved, that all the B's vanish, so that cosines only occur in (4) and (5), that

$$A_1' = A_1, \quad A_3' = A_3, \quad A_5' = A_5, \text{ \&c.};$$

$$A_2' = -A_2, \quad A_4' = -A_4, \text{ \&c.};$$

and further that $\beta' = \beta$. Equations (4) and (5) may thus be written

$$\psi = \alpha + \beta y + \omega y^2 + A_1 e^{-y} \cos x + A_2 e^{-2y} \cos 2x + A_3 e^{-3y} \cos 3x + \dots, \quad \dots(14)$$

$$\psi' = \alpha' + \beta y - \omega y^2 + A_1 e^{y} \cos x - A_2 e^{2y} \cos 2x + A_3 e^{3y} \cos 3x - \dots. \quad \dots\dots(15)$$

A_1 is of order h, A_2 of order h^2, A_3 of order h^3, and so on. If we are content to neglect h^6, we may stop at A_5; and we find as the equations necessary in order to secure the continuity of u and v at the surface (6)

$$A_1 \left(2 + \frac{3h^2}{4} + \frac{h^4}{12}\right) = 4\omega h + 2A_2 \left(2h + \frac{4h^3}{3}\right) - 3A_3 \frac{9h^2}{4},$$

$$2A_2 (2 + 2h^2) = A_1 \left(h + \frac{h^3}{12}\right) + 3A_3 \cdot 3h,$$

$$3A_3 \left(2 + \frac{9h^2}{2}\right) = -A_1 \left(\frac{h^2}{4} + \frac{5h^4}{192}\right) + 2A_2 (2h + h^3) + 4A_4 \cdot 4h,$$

$$4A_4 \cdot 2 = A_1 \frac{h^3}{24} - 2A_2 \cdot h^2 + 3A_3 \cdot 3h,$$

$$5A_5 \cdot 2 = -A_1 \frac{h^4}{192} + 2A_2 \frac{h^3}{3} - 3A_3 \frac{9h^2}{4} + 4A_4 \cdot 4h.$$

From these equations the values of the constants may be determined by successive approximations. Thus, if we retain terms of the order h^2, A_3, A_4, \&c., vanish and

$$A_1 = 2\omega h, \qquad 2A_2 = \omega h^2.$$

This is the second approximation. The fifth approximation gives

$$A_1 = 2\omega h \left(1 + \frac{h^2}{8} - \frac{h^4}{96}\right), \qquad 2A_2 = \omega h^2 \left(1 + \frac{h^2}{3}\right), \quad \dots\dots\dots(16, 17)$$

$$3A_3 = \frac{3\omega h^3}{4}\left(1 + \frac{9h^2}{16}\right), \qquad 4A_4 = \frac{2\omega h^4}{3}, \qquad 5A_5 = \frac{125\omega h^5}{192}, \quad \dots\ \dots(18, 19, 20)$$

which values are to be substituted in (14), (15).

The next step is the investigation of the values of u, v at the surface (6). They are most conveniently expressed as

$$\tfrac{1}{2}d\,(\psi + \psi')/dy, \quad -\tfrac{1}{2}d\,(\psi + \psi')/dx.$$

We get, correct as far as h^5,

$$u = \beta + \omega h^2 - \tfrac{1}{4}\omega h^4 + \tfrac{1}{12}\omega h^4 \cos 2x, \quad \dots\dots\dots\dots(21)$$

$$v = 2\omega h \left(1 - \frac{h^2}{4} - \frac{5h^4}{192}\right) \sin x + \frac{271\omega h^5}{96} \sin 3x, \quad \dots\dots\dots(22)$$

the terms containing $\cos 4x$ in (21), and $\sin 5x$ in (22), vanishing to this order. If we substitute these values in (12), we obtain

$$h \sin x \{- V + \beta + 2\omega + \tfrac{1}{2}\omega h^2 - \tfrac{11}{32}\omega h^4\} + \tfrac{275}{96}\omega h^5 \sin 3x = 0. \quad \dots(23)$$

So far, then, as terms in h^4, the surface of separation (6) is propagated as a simple wave with velocity given by

$$V = \beta + 2\omega + \tfrac{1}{2}\omega h^2; \quad \dots\dots\dots\dots\dots(24)$$

but, if terms in h^5 are retained, a change of form manifests itself, corresponding to the term in $\omega h^5 \sin 3x$ outstanding in (23).

Hitherto the wave-length has been supposed to be 2π, but, if we now take it to be $2\pi/k$, (24) becomes

$$V = \beta + 2\omega/k . (1 + \tfrac{1}{4}k^2h^2), \quad \dots\dots\dots\dots (25)$$

as is evident by "dimensions." The velocity of propagation is that of the flow of the fluid in the steady motion at the place where

$$ky = 1 + \tfrac{1}{4}k^2h^2. \quad \dots\dots\dots\dots\dots(26)$$

So far as the present investigation can reach, there is no sign of the amplitude of a wave tending spontaneously to increase.

218.

ON SOME PHYSICAL PROPERTIES OF ARGON AND HELIUM*.

[*Proceedings of the Royal Society*, LIX. pp. 198—208, Jan. 1896.]

Density of Argon.

IN our original paper† are described determinations by Prof. Ramsay, of the density of argon prepared with the aid of magnesium. The volume actually weighed was 163 c.c., and the adopted mean result was 19·941, referred to $O_2 = 16$. At that time a satisfactory conclusion as to the density of argon prepared by the oxygen method of Cavendish had not been reached, although a preliminary result (19·7) obtained from a mixture of argon and oxygen‡ went far to show that the densities of the gases prepared by the two methods were the same. In order further to test the identity of the gases, it was thought desirable to pursue the question of density; and I determined, as the event proved, somewhat rashly, to attempt large scale weighings of pure argon with the globe of 1800 c.c. capacity employed in former weighings of gases‖ which could be obtained in quantity.

The accumulation of the 3 litres of argon, required for convenient working, involved the absorption of some 300 litres of nitrogen, or about 800 litres of the mixture with oxygen. This was effected at the Royal Institution with the apparatus already described§, and which is capable of absorbing the mixture at the rate of about 7 litres per hour. The operations extended themselves over nearly three weeks, after which the residual gases amounting to about 10 litres, still containing oxygen with a considerable quantity of

---

* [Some of the results here given were announced before the British Association at the Ipswich meeting. See *Report*, Sept. 13, 1895.]

† Rayleigh and Ramsay, *Phil. Trans.* A, Vol. CLXXXVI. pp. 221, 238, 1895. [Vol. IV. p. 130.]

‡ *Loc. cit.* p. 221. [Vol. IV. p. 165.]

‖ *Roy. Soc. Proc.* February, 1888 [Vol. III. p. 37]; February, 1892 [Vol. III. p. 534]; March, 1893 [Vol. IV. p. 39].

§ *Phil. Trans. loc. cit.* p. 219. [Vol. IV. p. 162.]

nitrogen, were removed to the country and transferred to a special apparatus where it could be prepared for weighing.

For this purpose the purifying vessel had to be arranged somewhat differently from that employed in the preliminary absorption of nitrogen. When the gas is withdrawn for weighing, the space left vacant must be filled up with liquid, and afterwards, when the gas is brought back for repurification, the liquid must be removed. In order to effect this, the working vessel (Fig. 7)* communicates by means of a siphon with a 10-litre "aspirating bottle," the ends of the siphon being situated in both cases near the bottom of the liquid. In this way the alkaline solution may be made to pass backwards and forwards, in correspondence with the desired displacements of gas.

There is, however, one objection to this arrangement which requires to be met. If the reserve alkali in the aspirating bottle were allowed to come into contact with air, it would inevitably dissolve nitrogen, and this nitrogen would be partially liberated again in the working vessel, and so render impossible a complete elimination of that gas from the mixture of argon and oxygen. By means of two more aspirating bottles an atmosphere of *oxygen* was maintained in the first bottle, and the outermost bottle, connected with the second by a rubber hose, gave the necessary control over the pressure.

Five glass tubes in all were carried through the large rubber cork by which the neck of the working vessel was closed. Two of these convey the electrodes: one is the siphon for the supply of alkali, while the fourth and fifth are for the withdrawal and introduction of the gas, the former being bent up internally, so as to allow almost the whole of the gaseous contents to be removed. The fifth tube, by which the gas is returned, communicates with the fall-tube of the Töpler pump, provision being made for the overflow of mercury. In this way the gas, after weighing, could be returned to the working vessel at the same time that the globe was exhausted. It would be tedious to describe in detail the minor arrangements. Advantage was frequently taken of the fact that *oxygen* could always be added with impunity, its presence in the working vessel being a necessity in any case.

When the nitrogen had been so far removed that it was thought desirable to execute a weighing, the gas on its way to the globe had to be freed from oxygen and moisture. The purifying tubes contained copper and copper oxide maintained at a red heat, caustic soda, and phosphoric anhydride. In all other respects the arrangements were as described in the memoir on the densities of the principal gases†, the weighing globe being filled at 0°, and at the pressure of the manometer gauge.

The process of purification with the means at my command proved to be

---

* *Phil. Trans. loc. cit.* p. 218. [Vol. IV. p. 163.]

† *Roy. Soc. Proc.* Vol. LIII. p. 134, 1893. [Vol. IV. p. 39.]

extremely slow. The gas contained more nitrogen than had been expected, and the contraction went on from day to day until I almost despaired of reaching a conclusion. But at last the visible contraction ceased, and soon afterwards the yellow line of nitrogen disappeared from the spectrum of the jar discharge *. After a little more sparking, a satisfactory weighing was obtained on May 22, 1895; but, in attempting to repeat, a breakage occurred, by which a litre of air entered, and the whole process of purification had to be re-commenced. The object in view was to effect, if possible, a *series* of weighings with intermediate sparkings, so as to obtain evidence that the purification had really reached a limit. The second attempt was scarcely more successful, another accident occurring when two weighings only had been completed. Ultimately a series of four weighings were successfully executed, from which a satisfactory conclusion can be arrived at.

May 22	3·2710	
June 4	3·2617	
June 7	3·2727	
June 13	3·2652	
June 18	3·2750	
June 25	3·2748	3·2746
July 2	3·2741	

The results here recorded are derived from the comparison of the weighings of the globe "full" with the mean of the preceding and following weighings "empty," and they are corrected for the errors of the weights and for the shrinkage of the globe when exhausted, as explained in former papers. In the last series, the experiment of June 13 gave a result already known to be too low. The gas was accordingly sparked for fourteen hours more. Between the weighings of June 18 and June 25 there was nine hours' sparking, and between those of June 25 and July 2 about eight hours' sparking. The mean of the last three, viz. 3·2746, is taken as the definitive result, and it is immediately comparable with 2·6276, the weight under similar circumstances of oxygen†. If we take $O_2 = 16$, we obtain for argon

19·940,

in very close agreement with Professor Ramsay's result.

The conclusion from the spectroscopic evidence that the gases isolated from the atmosphere by magnesium and by oxygen are essentially the same is thus confirmed.

---

* *Jan. 29.*—When the argon is nearly pure, the arc discharge (no jar connected) assumes a peculiar purplish colour, quite distinct from the greenish hue apparent while the oxidation of nitrogen is in progress and from the sky-blue observed when the residue consists mainly of oxygen.

† *Roy. Soc. Proc.* Vol. LIII. p. 144, 1893. [Vol. IV. p. 48.]

The Refractivity of Argon and Helium.

The refractivity of argon was next investigated, in the hope that it might throw some light upon the character of the gas. For this purpose absolute measurements were not required. It sufficed to compare the pressures necessary in two columns of air and argon of equal lengths, in order to balance the retardations undergone by light in traversing them.

The arrangement was a modification of one investigated by Fraunhofer, depending upon the interference of light transmitted through two parallel vertical slits placed in front of the object-glass of a telescope. If there be only one slit, and if the original source, either a distant point or a vertical line of light, be in focus, the field is of a certain width, due to "diffraction," and inversely as the width of the slit. If there be two equal parallel slits whose distance apart is a considerable multiple of the width of either, the field is traversed by bands of width inversely as the distance between the slits. If from any cause one of the portions of light be retarded relatively to the other, the bands are displaced in the usual manner, and can be brought back to the original position only by abolishing the relative retardation.

When the object is merely to see the interference bands in full perfection, the use of a telescope is not required. The function of the telescope is really to magnify the slit system*, and this is necessary when, as here, it is desired to operate separately upon the two portions of light. The apparatus is, however, extremely simple, the principal objection to it being the high magnifying power required, leading under ordinary arrangements to a great attenuation of light. I have found that this objection may be almost entirely overcome by the substitution of cylindrical lenses, magnifying in the horizontal direction only, for the spherical lenses of ordinary eye-pieces. For many purposes a single lens suffices, but it must be of high power. In the measurements about to be described most of the magnifying was done by a lens of home manufacture. It consisted simply of a round rod, about $\frac{1}{6}$ in. (4 mm.) in diameter, cut by Mr Gordon from a piece of plate glass†. This could be used alone; but as at first it was thought necessary to have a web, serving as a fixed mark to which the bands could be referred, the rod was treated as the object-glass of a compound cylindrical microscope, the eye-piece being a commercial cylindrical lens of $1\frac{1}{4}$ in. (31 mm.) focus. Both lenses were mounted on adjustable stands, so that the cylindrical axes could be made accurately vertical, or, rather, accurately parallel to the length of the original slit. The light from an ordinary paraffin lamp now sufficed, although the magnification was such as to allow the error of setting to be

* *Brit. Assoc. Report*, 1893, p. 703. [Vol. IV. p. 76.]

† Preliminary experiments had been made with ordinary glass cane and with tubes charged with water.

less than 1/20 of a band interval. It is to be remembered that with this arrangement the various parts of the length of a band correspond, not to the various parts of the original slit, but rather to the various parts of the object-glass. This departure from the operation of a spherical eye-piece is an advantage, inasmuch as optical defects show themselves by deformation of the bands instead of by a more injurious encroachment upon the distinction between the dark and bright parts.

Fig. 1.

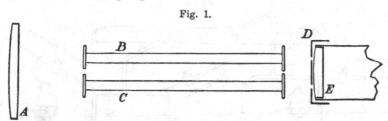

The collimating lens A (Fig. 1) is situated 23 ft. (7 metres) from the source of light. B, C are the tubes, one containing dry air, the other the gas to be experimented upon. They are 1 ft. (30·5 cm.) long, and of $\frac{1}{2}$ in. (1·3 cm.) bore, and they are closed at the ends with small plates of parallel glass cut from the same strip. E is the object-glass of the telescope, about 3 in. (7·6 cm.) in diameter. It is fitted with a cap, D, perforated by two parallel slits. Each slit is $\frac{1}{4}$ in. (6 mm.) wide, and the distance between the middle lines of the slits is $1\frac{1}{2}$ in. (38 mm.).

The arrangements for charging the tubes and varying the pressures of the gases are sketched in Fig. 2. A gas pipette, DE, communicates with the tube C, so that by motion of the reservoir E and consequent flow of mercury through the connecting hose, part of the gas may be transferred. The pressure was measured by a U-shaped manometer F, containing mercury. This was fitted below with a short length of stout rubber tubing G, to which was applied a squeezer H. The object of this attachment was to cause a rise of mercury in both limbs immediately before a reading, and thus to avoid the capillary errors that would otherwise have entered. A similar pipette and manometer were connected with the air-tube B. In order to be able, if desired, to follow with the eye a particular band during the changes of pressure (effected by small steps and alternately in the two tubes), diminutive windlasses were provided by which the motions of the reservoirs (E) could be made smooth and slow. In this way all doubt was obviated as to the identity of a band; but after a little experience the precaution was found to be unnecessary[*].

The manner of experimenting will now be evident. By adjustment of pressures the centre of the middle band was brought to a definite position,

---

[*] [For a description of a modified apparatus capable of working with an extremely small quantity of gas, see *Proc. Roy. Soc.* Vol. LXIV. p. 97, 1898.]

determined by the web or otherwise, and the pressures were measured. Both pressures were then altered and adjusted until the band was brought back precisely to its original position. The ratio of the changes of pressure is the inverse ratio of the refractivities $(\mu - 1)$ of the gases. The process may be repeated backwards and forwards any number of times, so as to eliminate in great degree errors of the settings and of the pressure readings.

Fig. 2.

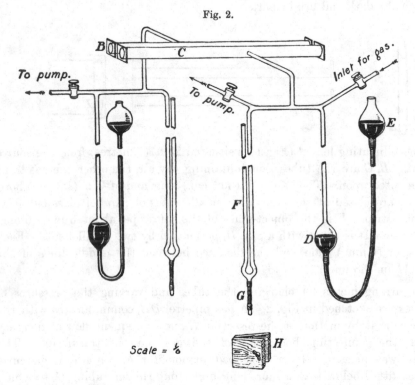

During these observations a curious effect was noticed, made possible by the independent action of the parts of the object-glass situated at various levels, as already referred to. When the bands were stationary, they appeared straight, or nearly so, but when in motion, owing to changes of pressure, they became curved, even in passing the fiducial position, and always in such a manner that the *ends* led. The explanation is readily seen to depend upon the temporary changes of temperature which accompany compression or rarefaction. The full effect of a compression, for example, would not be attained until the gas had cooled back to its normal temperature, and this recovery of temperature would occur more quickly at the top and bottom, where the gas is in proximity to the metal, than in the central part of the tube.

The success of the measures evidently requires that there should be no apparent movement of the bands apart from real retardations in the tubes.

As the apparatus was at first arranged, this condition was insufficiently satisfied. Although all the parts were carried upon the walls of the room, frequent and somewhat sudden displacements of the bands relatively to the web were seen to occur, probably in consequence of the use of wood in some of the supports. The observations could easily be arranged in such a manner that no systematic error could thence enter, but the agreement of individual measures was impaired. Subsequently a remedy was found in the use of a second system of bands, formed by light which passed just above the tubes, to which, instead of to the web, the moveable bands were referred. The coincidence of the two systems could be observed with accuracy, and was found to be maintained in spite of movements of both relatively to the web.

In the comparisons of argon and air (with nearly the same refractivities) the changes of pressure employed were about 8 in. (20 cm.), being deductions from the atmospheric pressure. In one observation of July 26, the numbers, representing suctions in inches of mercury, stood

Argon	Air
8·54	9·96
0·01	1·77
8·53	8·19

$$\text{Ratio} = 0·961,$$

signifying that 8·53 in. of argon balanced 8·19 in. of dry air. Four sets, during which the air and argon (from the globe as last filled for weighing) were changed, taken on July 17, 18, 19, 26, gave respectively for the final ratio 0·962, 0·961, 0·961, 0·960, or as the mean

$$\frac{\text{Refractivity of argon}}{\text{Refractivity of air}} = 0·961.$$

The evidence from the refractivities, as well as from the weights, is very unfavourable to the view that argon is an allotropic form of nitrogen such as would be denoted by N_3.

The above measurements, having been made with lamp-light, refer to the most luminous region of the spectrum, say in the neighbourhood of D. But since no change in the appearance of the bands at the two settings could be detected, the inference is that the dispersions of the two gases are approximately the same, so that the above ratio would not be much changed, even if another part of the spectrum were chosen. It may be remarked that the displacement actually compensated in the above experiments amounted to about forty bands, each band corresponding to about $\frac{1}{6}$ in. (5 mm.) pressure of mercury.

Similar comparisons have been made between air and helium. The latter gas, prepared by Professor Ramsay, was brought from London by

Mr W. Randall, who further gave valuable assistance in the manipulations. It appeared at once that the refractivity of helium was remarkably low, 13 in. pressure of the gas being balanced by less than 2 in. pressure of air. The ratios given by single comparisons on July 29 were 0·147, 0·146, 0·145, 0·146, mean 0·146; and on July 30 0·147, 0·147, 0·145, 0·145, mean 0·146. The observations were not made under ideal conditions, on account of the smallness of the changes of air pressure; but we may conclude that with considerable approximation

$$\frac{\text{Refractivity of helium}}{\text{Refractivity of air}} = 0\cdot146\,*.$$

The lowest refractivity previously known is that of hydrogen, nearly 0·5 of that of air.

Viscosity of Argon and Helium.

The viscosity was investigated by the method of passage through capillary tubes. The approximate formula has been investigated by O. Meyer†, on the basis of Stokes' theory for incompressible fluids. If the driving pressure $(p_1 - p_2)$ is not too great, the volume V_2 delivered in time t through a tube of radius R and length λ is given by

$$V_2 = \pi t \, \frac{p_1{}^2 - p_2{}^2}{2p_2} \, \frac{R^4}{8\eta\lambda},$$

the volume being measured at the lower pressure p_2, and η denoting the viscosity of the gas. In the comparison of different gases V_2, p_1, p_2, R, λ may be the same, and then η is proportional to t.

In the apparatus employed two gas pipettes and manometers, somewhat similar to those shown in Fig. 2, were connected by a capillary tube of very small bore and about 1 metre long. The volume V_2 was about 100 c.c., and was caused to pass by a pressure of a few centimetres of mercury, maintained as uniform as possible by means of the pipettes. There was a difficulty, almost inherent in the use of mercury, in securing the right pressures during the first few seconds of an experiment; but this was not of much importance as the whole time t amounted to several minutes. The apparatus was tested upon hydrogen, and was found to give the received numbers with sufficient accuracy. The results, referred to dry air, were for helium 0·96; and for argon 1·21, somewhat higher than for oxygen which at present stands at the head of the list of the principal gases‡.

---

* [1902. The sample must have contained impurity—probably *hydrogen*. Prof. Ramsay's latest result for the refractivity of helium referred to air is ·1238 (*Proc. Roy. Soc.* LXVII. p. 331, 1900).]

† *Pogg. Ann.* Vol. CXXVII. p. 270, 1866.

‡ [1902. Schultze (*Drude Ann.* VI. p. 310, 1901) finds for helium 1·086 in place of 0·96.]

Gas from the Bath Springs.

In the original memoir upon argon* results were given of weighings of the residue from the Bath gas after removal of oxygen, carbonic anhydride, and moisture, from which it appeared that the proportion of argon was only one-half of that contained in the residue, after similar treatment, from the atmosphere. After the discovery of helium by Professor Ramsay, the question presented itself as to whether this conclusion might not be disturbed by the presence in the Bath gas of helium, whose lightness would tend to compensate the extra density of argon.

An examination of the gas which had stood in my laboratory more than a year having shown that it still contained no oxygen, it was thought worth while to remove the nitrogen so as to determine the proportion that would refuse oxidation. For this purpose 200 c.c. were worked up with oxygen until the volume, free from nitrogen, was reduced to 8 c.c. On treatment with pyrogallol and alkali the residue measured 3·3 c.c., representing argon, and helium, if present. On sparking the residue at atmospheric pressure and examining the spectrum, it was seen to be mainly that of argon, but with an unmistakable exhibition of D_3. At atmospheric pressure this line appears very diffuse in a spectroscope of rather high power, but the place was correct.

From another sample of residue from the Bath gas, vacuum tubes were charged by my son, Mr R. J. Strutt, and some of them showed D_3 sharply defined and precisely coincident with the line of helium in a vacuum tube prepared by Professor Ramsay.

Although the presence of helium in the Bath gas is not doubtful, the quantity seems insufficient to explain the low density found in October, 1894. In order to reconcile that density with the proportion of residue $(3·3/200 = 0·016)$ found in the experiment just described, it would be necessary to suppose that the helium amounted to 25 per cent. of the whole residue of argon and helium. Experiment, however, proved that a mixture of argon and helium containing 10 per cent. of the latter gas showed D_3 more plainly than did the Bath residue. It is just possible that some of the helium was lost by diffusion during the long interval between the experiments whose results are combined in the above estimate.

Buxton Gas.

Gas from the Buxton springs, kindly collected for me by Mr A. McDougall, was found to contain no appreciable oxygen. The argon amounted to about

* Rayleigh and Ramsay, *Phil. Trans.* A, Vol. CLXXXVI. p. 227, 1895. [Vol. IV. p. 172.]

2 per cent. of the volume.   When its spectrum was examined, the presence of D_3 was suspected, but the appearance was too feeble to allow of a definite statement being made.   The proportion of helium is in any case very much lower than in the Bath gas.

Is Helium contained in the Atmosphere?

Apart from its independent interest, this question is important in connection with the density of atmospheric argon.   Since the spectrum of this gas does not show the line D_3, we may probably conclude that the proportion of helium is less than 3 per cent.; so that there would be less than 3×10^{-4} of helium in the atmosphere.   The experiment about to be described was an attempt to carry the matter further, and is founded upon the observation by Professor Ramsay, that the solubility of helium in water is only 0·007, less than one-fifth of that which we found for argon*.

It is evident that if a mixture of helium and argon be dissolved in water until there is only a small fraction remaining over, the proportion of helium will be much increased in the residue.   Two experiments have been made, of which that on October 6, 1895, was the more elaborate.   About 60 c.c. of argon were shaken for a long time with well-boiled water contained in a large flask.   When the absorption had ceased, the residue of 30 c.c. was sparked with a little oxygen until no nitrogen could be seen in the spectrum. It was then treated a second time with boiled water until its volume was reduced to $1\frac{1}{2}$ c.c.   With this vacuum tubes were charged by my son at two different pressures.   In none of them could D_3 be detected; nor was there any marked difference to be seen between the spectra of the washed and the unwashed argon.   If helium be present in the atmosphere, it must be in very small quantity, probably much less than a ten-thousandth part†.

* *Phil. Trans.* A, Vol. CLXXXVI. p. 225, 1895.   [Vol. IV. p. 170.]
† [1902.   The presence of traces of helium in the atmosphere is not doubtful.]

219.

ON THE AMOUNT OF ARGON AND HELIUM CONTAINED IN THE GAS FROM THE BATH SPRINGS*.

[*Proceedings of the Royal Society*, LX. pp. 56, 57, 1896.]

THE presence of helium in the residue after removal of nitrogen from this gas was proved in a former paper†, but there was some doubt as to the relative proportions of argon and helium. A fresh sample, kindly collected by Dr Richardson, has therefore been examined. Of this 2,500 c.c., submitted to electric sparks in presence of oxygen, gave a final residue of 37 c.c., after removal of all gases known until recently. The spectrum of the residue, observed at atmospheric pressure, showed argon, and the D_3 line of helium very plainly.

The easy visibility of D_3 suggested the presence of helium in some such proportion as 10 per cent., and this conjecture has been confirmed by a determination of the refractivity of the mixture. It may be remembered that while the refractivity of argon approaches closely that of air, the relative number being 0·961, the refractivity of helium (as supplied to me by Professor Ramsay) is very low, being only 0·146 on the same scale. If we assume that any sample of gas is a mixture of these two, its refractivity will determine the proportions in which the components are present.

The observations were made by an apparatus similar in character to that already described, but designed to work with smaller quantities of gas. The space to be filled is only about 12 c.c., and if the gas be at atmospheric pressure its refractivity may be fixed to about 1/1000 part. By working at pressures below atmosphere very fair results could be arrived at with quantities of gas ordinarily reckoned at only 3 or 4 c.c.

The refractivity found for the Bath residue after desiccation was 0·896 referred to air, so that the proportional amount of helium is 8 per cent. Referred to the original volume, the proportion of helium is 1·2 parts per thousand.

---

* I am reminded by Mr Whitaker that helium is appropriately associated with the Bath waters, which, according to some antiquaries, were called by the Romans *Aquæ Solis*.

† *Roy. Soc. Proc.* Vol. LIX. p. 206, 1896. [Vol. IV. p. 223.]

220.

THE REPRODUCTION OF DIFFRACTION GRATINGS.

[*Nature*, LIV. pp. 332, 333, 1896*.]

I HAVE first to apologise for the very informal character of the communication which I am about to make to the Club; I have not been able to put anything down upon paper, but I thought it might be interesting to some to hear an account of experiments that have now been carried on at intervals for a considerable series of years in the reproduction—mainly the photographic reproduction—of diffraction gratings. Probably most of you know that these consist of straight lines ruled very closely, very accurately, and parallel to one another, upon a piece of glass or speculum metal. Usually they are ruled with a diamond by the aid of a dividing machine; and in late years, particularly in the hands of Rutherfurd and Rowland, an extraordinary degree of perfection has been attained. It was many years ago—nearly 25 years, I am afraid—that I first began experiments upon the photographic reproduction of these divided gratings, each in itself the work of great time and trouble, and costing a good deal of money. At that time the only gratings available were made by Nobert, in Germany, of which I had two, each containing about a square inch of ruled surface, one of about 3,000 lines to the inch, and the other of about 6,000. It happened, by an accident, that the grating with 3,000 lines was the better of the two, in that it was more accurately ruled, and gave much finer definition upon the solar spectrum; the 6,000 line grating was brighter, but its definition was decidedly inferior, so that both had certain advantages according to the particular object in view.

If it comes to the question of how to make a grating by photography, probably the first idea to occur to one would be that it might be a comparatively simple matter to make a grating upon a large scale, and then

* [From a report of] an address delivered at the eighth annual conference of the Camera Club.

reduce it by photography, but if one goes into the figures the project is not found so promising. Take, for instance, a grating with 10,000 lines to the inch; if you magnified that, say 100 times, your lines would then be 100 to the inch; if you magnified it 1,000 times, they would still be 10 to the inch, and that would be a convenient size so far as interval between the lines was concerned; but think what would be the area required to hold a grating magnified to that extent. By the time you have magnified the inch by 100 or 1,000, you would want a wall of a house or of a cathedral to hold the grating. If the problem were proposed of ruling a grating with 6,000 lines, with a high degree of accuracy, it would be easier to do it on a microscopic scale than upon a large scale, leaving out of consideration the difficulty of reproducing it. And those difficulties would be insuperable, because, although with a good microscopic object-glass it would be easy to photograph lines which are much closer together than 3,000 or 6,000 to the inch, yet that could only be achieved over a very small area of surface—nothing like a square inch; and if it were required to cover a square inch with lines 6,000 to the inch, it would be beyond the power, not only, I believe, of any microscope, but of any lens that was ever made. So that that line of investigation does not fulfil the promise which at first it might appear to give; and, in fact, there is nothing simpler or better than to copy the original ruled by a dividing engine, by the simple process of contact printing.

For this purpose some precautions are required. You must use very flat glass, by preference it should be optically worked glass, although very good results may be obtained on selected pieces of ordinary plate. Of course, no one would think of making such a print by diffused daylight, but the sun itself, or a point of light from any suitable source, according to the nature of the photographic process which is adopted, permits quite well of the reproduction of any grating of a moderate degree of fineness. I have used almost all varieties of photographic processes in my time. In the days when I first worked, the various dry collodion processes were better understood than they are now; the old albumen process was extremely suitable for such work as this, on account of the almost complete absence of structure in the film, and the very convenient hardness of the surface, which made the result comparatively little liable to injury. I used with success the dry collodion processes, the tannin process among others, and also some of the direct printing methods, such as the collodio-chloride. The latter method, worked upon glass. gave excellent results, particularly if the finished print was treated with mercury in the way commonly used for intensification, except that, in the treatment of a grating with mercury, it is desirable to stop at the mercury and not to go on to the blackening process used in the intensification of negatives. From the visual point of view, the grating, after intensification—if one may use the term—with mercury, looks much less intense than before, but, nevertheless, the spectra seen when a point or slit of light is looked at through the grating becomes very much more brilliant.

I used another process at that time, more than twenty years ago, which gave excellent results, but had not the degree of certainty that I aimed at, namely, a bichromated gelatine process, similar to carbon printing, except that no pigment was employed. A glass plate was simply coated with bichromated gelatine of a suitable thickness—and a good deal depended upon hitting that off correctly; if the coating was too thin the grating showed a deficiency of brightness, whereas, if it was too thick, there might be a difficulty in getting it sufficiently uniform and smooth on the surface. However, I obtained excellent gratings by that process, most of them capable of showing the nickel line between the two well-known sodium or D lines in the solar spectrum, when suitably examined. The collodio-chloride process was comparatively slow, and bichromated gelatine required two or three minutes exposure to sunlight to produce a proper effect; but for the more sensitive developed negative processes a very much less powerful light or a reduced exposure was needed.

The performance of the copies was quite good, and, except where there was some obvious defect, I never could see that they were worse than the originals; in fact, in respect of brightness it not unfrequently happened that the copies were far superior to the originals, so that in many cases they would be more useful. I do not mean by that, however, that I would rather have a copy than an original if anyone wanted to make me a present. There seems to be some falling off in copies; so that they cannot well be copied again, and if you want to work upon spectra of an extremely high order, dispersed to a great extent laterally from the straight line, a copy would not be satisfactory. The reproduction of gratings on bichromated gelatine is easily and quickly accomplished; there is only the coating of the glass over-night, rapid drying to avoid crystallisation in the film, exposure, washing, and drying. In order to get the best effect it is usually desirable to treat the bichromated copies with hot water. It is a little difficult to understand what precisely happens. All photographers know that the action of light upon bichromated gelatine is to produce a comparative insolubility of the gelatine. In the carbon process, and many others in which gelatine is used, the gelatine which remains soluble, not having been sufficiently exposed to light, is fairly washed away in the subsequent treatment with warm water, but for that effect it is generally necessary to get at the back of the gelatine film, because on its face there is usually a layer which is so insoluble as not to allow of the washing away of any of the gelatine situated behind. But in the present case there is no question of transferring the film, which remains fixed to the glass, and therefore it is difficult to see how any gelatine could be dissolved out. However, under the action of water, the less exposed gelatine no doubt swells more than that which has received more exposure and has thus lost its affinity for water; and while the gelatine is wet it is reasonable that a rib-like structure should ensue, which is what would be required in order to make a grating, but when the gelatine dries, one would suppose that all would again become flat, and indeed that happens to a certain extent. The gratings lost a great deal of intensity in drying, but, if properly treated with warm water, the reduction does not go too far, and a considerable degree of intensity is left when the photograph is dry.

Although it belongs to another branch of the subject, a word may not be out of place as to the accuracy with which the gratings must be made. It seems a wonderful thing at first sight, to rule 6,000 lines to an inch at all, if you think of the smallest interval that you can readily see with the eye, perhaps one-hundredth of an inch, and remember that in these gratings there are sixty lines in the space of one-hundredth of an inch, and all disposed at rigorously equal intervals. Those familiar with optics will understand the importance of extreme accuracy if I give an illustration. Take the case of the two sodium lines in the spectrum, the D lines; they differ in wave-length by about a thousandth part; the dispersion—the extent to which the light is separated from the direct line—is in proportion to the wave-length of the light, and inversely as the interval between the consecutive lines on the grating; so that, if we had a grating in which the first half was ruled at the rate of 1,000 to the inch, and the second half at the rate of 1,001 to the inch, the one half would evidently do the same thing for one soda line as the other half of the grating was doing for the other soda line, and the two lines would be mixed together and confused. In order, therefore, to do anything like good work, it is necessary, not only to have a very great number of lines, but to have them spaced with most extraordinary precision; and it is wonderful what success has been reached by the beautiful dividing machines of Rutherfurd and Rowland. I have seen Rowland's machine at Baltimore, and have heard him speak of the great precautions required to get good results. The whole operation of the machine is automatic; the ruling goes on continuously day and night, and it is necessary to pay the most careful regard to uniformity of temperature, for the slightest expansion or contraction due to change of temperature of the different parts of the machine would bring utter confusion into the grating and its resulting spectrum.

The contact in printing has to be pretty close and the finer the grating the closer must the contact be. I experimented upon that point: one can get some kind of result, theoretically, by preparing a photographic film with a slightly convex surface and using that for the print; then, where the contact was closest, the original of course was very well impressed, and round that, one got different degrees of increasingly imperfect contact, and one could trace in the result what the effect of imperfect contact is. I found that, both with gratings of 3,000 and 6,000 lines to the inch, good enough contact was obtained with ordinary flat glass; but when you come to gratings of 17,000 or 20,000 lines to the inch the contact requires to be extremely close, and in order to get a good copy of a grating with 20,000 lines per inch it is necessary that there should nowhere be one ten-thousandth of an inch between the original and the printing surface—a degree of closeness not easily secured over the entire area. It is rather singular that though I published full accounts of this work a long time ago, and distributed a large number of copies, the process of reproducing gratings by photography did not become universally known, and was re-discovered in France, by Izarn, only two or three years since.

One reason why photographic reproduction is not practised to a very great extent, is, that the modern gratings—such as Rowland's—are ruled almost universally upon speculum metal. A grating upon speculum metal is very excellent for use, but does not well lend itself to the process of photographic copying, although I have succeeded to a certain extent in copying a grating ruled upon speculum metal. For this purpose the light had to pass first through the photographic film, then be reflected from the speculum metal, and so pass back again through the film. Gratings, such as could easily be made by copying from a glass original, are not readily produced from one on speculum metal, and I think that is the reason why the process has not come into more regular use. Glass is much more trying than speculum metal to the diamond, and that accounts for the latter being generally preferred for gratings; it is very hard, but has not ruinous effects upon the diamond; indeed the principal difficulty consists in getting a good diamond point, and maintaining it in a shape suitable for making the very fine cut which is required.

I may now allude to another method of photographic reproduction which I tried only last summer. It happened that I then went with Professor Meldola over Waterlow's large photo-mechanical printing establishment, and I was much interested, among other very interesting things, to see the use of the old bitumen process—the first photographic process known. It is used for the reproduction of cuts in black and white. A carefully cleansed zinc plate is coated with a varnish of bitumen dissolved in benzole, and exposed to sunlight for about two hours under a negative giving great contrast. Where the light penetrates the negative the bitumen becomes comparatively insoluble, and where it has been protected from the action of light it retains its original degree of solubility. When the exposed plate is treated with a solvent, turpentine or some milder solvent than benzole, the protected parts are dissolved away, leaving the bare metal; whereas the parts that have received the sunlight, being rendered insoluble, remain upon the metal and protect it in the subsequent etching process. I did not propose to etch metal, and, therefore, I simply used the bitumen varnish spread upon glass plates, and exposed the plates so prepared to sunshine for about two hours in contact with the grating. They were then developed, if one may use the phrase, with turpentine; and this is the part of the process which is the most difficult to manage. If you stop development early you get [without difficulty] a grating which gives fair spectra, but it may be deficient in intensity and brightness; if you push development the brightness increases up to a point at which the film disintegrates altogether. In this way one is tempted to pursue the process to the very last point, and, although one may succeed so far as to have a film which is quite intact so long as the turpentine is upon it, I have not succeeded in finding any method of getting rid of the turpentine without causing the disintegration of the film. In the commercial application of the process the bitumen is treated somewhat brutally—the turpentine is rinsed off with a jet of water; I have tried that, and many of my results have been very good. I have also tried to sling off the turpentine with the aid of a kind of centrifugal machine, but by either plan the [too tender] film is liable not to survive the treatment required for getting rid of the turpentine. If the solvent is allowed to remain we are in another difficulty, because then the developing action is continued and the result is lost. But if the process is properly managed, and development stopped at the right point, and if the film be of the right degree of thickness, you get an excellent copy. I have one here, 6,000 lines to the inch, which I think is about the very best copy I have ever made. The method gives results somewhat superior to the best that can be got with gelatine; but I would not recommend it in preference to the latter, because it is much more difficult to work unless some one can hit upon an improved manipulation.

I will not enlarge upon the importance of gratings; those acquainted with optics know how very important is the part played by diffraction gratings in optical research, and how the most delicate work upon spectra, requiring the highest degree of optical power, is made by means of gratings, ruled on speculum metal by Rowland. I suppose the reason why no professional photographer has taken up the production of photographic gratings, is the difficulty of getting the glass originals; they are very expensive, and indeed I do not know where they are now to be obtained. It seems a pity that photographic copies should not be more generally available. I have given a great many away myself; but educational establishments are increasing all over the country, and for the purpose of instructing students it is desirable that reasonably good gratings should be placed in their hands, to make them familiar with the measurements by which the wave-length of light is determined.

[1902. For earlier papers upon this subject see Vol. I. pp. 160, 199, 504.]

221.

THE ELECTRICAL RESISTANCE OF ALLOYS.

[*Nature*, LIV. pp. 154, 155, 1896.]

THE recent researches of Profs. Dewar and Fleming upon the electrical resistance of metals at low temperatures have brought into strong relief the difference between the behaviour of pure metals and of alloys. In the former case the resistance shows every sign of tending to disappear altogether as the absolute zero of temperature is approached, but in the case of alloys this condition of things is widely departed from, even when the admixture consists only of a slight impurity.

Some years ago it occurred to me that the apparent resistance of an alloy might be partly made up of thermo-electric effects, and as a rough illustration I calculated the case of a conductor composed of two metals arranged in alternate laminæ perpendicular to the direction of the current. Although a good many difficulties remain untouched, I think that the calculation may perhaps suggest something to those engaged upon the subject. At any rate it affords *à priori* ground for the supposition that an important distinction may exist between the resistances of pure and alloyed metals.

The general character of the effect is easily explained. According to the discovery of Peltier, when an electric current flows from one metal to another there is development or absorption of heat at the junction. The temperature disturbance thus arising increases until the conduction of heat through the laminæ balances the Peltier effects at the junctions, and it gives rise to a thermo-electromotive force opposing the passage of the current. Inasmuch as the difference of temperature at the alternate junctions is itself proportional to the current, so is also the reverse electromotive force thereby called into play. Now a reverse electromotive force proportional to current is indistinguishable experimentally from a *resistance*; so that the combination of laminated conductors exhibits a false resistance, having (so far as is known) nothing in common with the real resistance of the metals.

If e be the thermo-electric force of the couple for one degree difference of temperature of the junctions; t, t' the actual temperatures; then the electromotive force for one couple is $e(t-t')$. If we suppose that there are n similar couples per unit of length perpendicular to the lamination, the whole reverse electromotive force per unit of length is $ne(t-t')$. Again, if C be the current corresponding to unit of cross-section, the development of heat per second at each alternate junction is per unit of area $273 \times e \times C$, the actual temperature being in the neighbourhood of zero Cent. This is measured in ergs, and is to be equated to the heat conducted per second towards the cold junctions on the two sides. If k, k' be the conductivities for heat of the two metals, l and l' the corresponding thicknesses, the heat conducted per second is

$$(t-t')\{k/l + k'/l'\};$$

or if
$$l/(l+l') = p, \quad l'/(l+l') = q, \quad l+l' = 1/n,$$

the conducted heat is

$$n(t-t')\{k/p + k'/q\}.$$

In this expression $p+q=1$, the symbols p and q denoting the proportional amounts by volume in which the two metals are associated. Thus when a stationary state is reached

$$273 \times e \times C = n(t-t')\{k/p + k'/q\}.$$

This determines $(t-t')$ when C is given; and the whole back electromotive force per unit of thickness is rC, where

$$r = \frac{273 \times e^2}{k/p + k'/q}.$$

This is the expression for the false resistance per unit of thickness, which, it should specially be noted, is independent of n, the number of couples. The number of couples which co-operate is indeed increased by finer lamination, but the efficiency of each is decreased in the same proportion by the readier conduction of heat between the junctions. It is scarcely necessary to point out that the false resistance is called into play only by currents which flow *across* the laminæ.

In my original calculation the metals chosen for illustration were iron and copper. In this case (Everett's *C.G.S. System of Units*, p. 192) $c = 1600$. The conductivities are to be measured in ergs. For iron, $k = \cdot164 \times 4\cdot2 \times 10^7$;

for copper, $k' = 1\cdot11 \times 4\cdot2 \times 10^7$. Thus, if the metals are in equal volumes $(p = q = \frac{1}{2})$,

$$r = \frac{2 \times 273 \times 1600^2}{4\cdot2 \times 10^7 \times 1\cdot27} = 26\cdot2.$$

This is the thermo-electric addition to the true specific resistance, and is about $1\frac{1}{2}$ per cent. of that of copper. Such an addition may seem small; but it should be remembered that for the more distinctively thermo-electric metals e is much larger, and that it enters by its square. In any case it seems desirable that this complication should be borne in mind. The consequences which follow from recognised laws for laminated structures, however fine, must surely have some bearing upon the properties of alloys, although in this case the fineness is molecular.

222.

ON THE THEORY OF OPTICAL IMAGES, WITH SPECIAL REFERENCE TO THE MICROSCOPE.

[*Philosophical Magazine*, XLII. pp. 167—195, 1896.]

THE special subject of this paper has been treated from two distinct points of view. In the work of Helmholtz* the method followed is analogous to that which had long been used in the theory of the telescope. It consists in tracing the image representative of a mathematical point in the object, the point being regarded as self-luminous. The limit to definition depends upon the fact that owing to diffraction the image thrown even by a perfect lens is not confined to a point, but distends itself over a patch or disk of light of finite diameter. Two points in the object can appear fully separated only when the representative disks are nearly clear of one another. The application to the microscope was traced by means of a somewhat extended form of Lagrange's general optical theorem, and the conclusion was reached that the smallest resolvable distance ϵ is given by

$$\epsilon = \tfrac{1}{2}\lambda/\sin \alpha, \quad \dots\dots\dots\dots\dots\dots\dots\dots\dots\dots(1)$$

λ being the wave-length in the medium where the object is situated, and α the divergence-angle of the extreme ray (the semi-angular aperture) in the same medium. If λ_0 be the wave-length in vacuum,

$$\lambda = \lambda_0/\mu, \quad \dots\dots\dots\dots\dots\dots\dots\dots\dots\dots(2)$$

μ being the refractive index of the medium; and thus

$$\epsilon = \tfrac{1}{2}\lambda_0/\mu \sin \alpha. \quad \dots\dots\dots\dots\dots\dots\dots\dots(3)$$

The denominator $\mu \sin \alpha$ is the quantity now well known (after Abbe) as the "numerical aperture."

The extreme value possible for α is a right angle, so that for the microscopic limit we have

$$\epsilon = \tfrac{1}{2}\lambda_0/\mu. \quad \dots\dots\dots\dots\dots\dots\dots\dots\dots(4)$$

* *Pogg. Ann.* Jubelband, 1874.

The limit can be depressed only by a diminution in λ_0, such as photography makes possible, or by an increase in μ, the refractive index of the medium in which the object is situated.

This method, in which the object is considered point by point, seems the most straightforward, and to a great extent it solves the problem without more ado. When the representative disks are thoroughly clear of one another, the two points in which they originate are resolved, and on the other hand, when the disks overlap the points are not distinctly separated. Open questions can relate only to intermediate cases of partial overlapping and various degrees of resolution. In these cases (as has been insisted upon by Dr Stoney) we have to consider the relative phases of the overlapping lights before we can arrive at a complete conclusion.

If the various points of the object are self-luminous, there is no permanent phase-relation between the lights of the overlapping disks, and the resultant illumination is arrived at by simple addition of separate intensities. This is the situation of affairs in the ordinary use of a telescope, whether the object be a double star, the disk of the sun, the disk of the moon, or a terrestrial body. The distribution of light in the image of a double point, or of a double line, was especially considered in a former paper*, and we shall return to the subject later.

When, as sometimes happens in the use of the telescope, and more frequently in the use of the microscope, the overlapping lights have permanent phase-relations, these intermediate cases require a further treatment; and this is a matter of some importance as involving the behaviour of the instrument in respect to the finest detail which it is capable of rendering. We shall see that the image of a double point under various conditions can be delineated without difficulty.

In the earliest paper by Prof. Abbe†, which somewhat preceded that of Helmholtz, similar conclusions were reached; but the demonstrations were deferred, and, indeed, they do not appear ever to have been set forth in a systematic manner. Although some of the positions then taken up, as for example that the larger features and the finer structure of a microscopic object are delineated by different processes, have since had to be abandoned‡, the publication of this paper marks a great advance, and has contributed powerfully to the modern development of the microscope§. In

---

* "Investigations in Optics, with special reference to the Spectroscope," *Phil. Mag.* Vol. VIII. p. 266 (1879). [Vol. I. p. 415.]

† *Archiv f. Mikr. Anat.* Vol. IX. p. 413 (1873).

‡ Dallenger's edition of Carpenter's *Microscope*, p. 64, 1891.

§ It would seem that the present subject, like many others, has suffered from over-specialization, much that is familiar to the microscopist being almost unknown to physicists, and *vice versâ*. For myself I must confess that it is only recently, in consequence of a discussion between

Prof. Abbe's method of treating the matter the typical object is not a luminous *point*, but a *grating* illuminated by plane waves. Thence arise the well-known diffraction spectra, which are focused near the back of the object-glass in its principal focal plane. If the light be homogeneous the spectra are reduced to points, and the final image may be regarded as due to the simultaneous action of these points acting as secondary centres of light. It is argued that the complete representation of the object requires the co-operation of all the spectra. When only a few are present, the representation is imperfect; and when there is only one—for this purpose the central image counts as a spectrum—the representation wholly fails.

That this point of view offers great advantages, at least when the object under consideration is really a grating, is at once evident. More especially is this the case in respect of the question of the limit of resolution. It is certain that if one spectrum only be operative, the image must consist of a uniform field of light, and that no sign can appear of the real periodic structure of the object. From this consideration the resolving-power is readily deduced, and it may be convenient to recapitulate the argument for the case of perpendicular incidence. In Fig. 1 AB represents the axis,

Fig. 1.

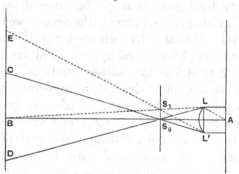

A being in the plane of the object (grating) and B in the plane of the image. The various diffraction spectra are focused by the lens LL' in the principal focal plane, S_0 representing the central image due to rays which issue normally from the grating. After passing S_0 the rays diverge in a cone corresponding to the aperture of the lens and illuminate a circle CD in the plane of the image, whose centre is B. The first lateral spectrum S_1 is formed by rays diffracted from the grating at a certain angle; and in the critical case the region of the image illuminated by the rays diverging from S_1 just includes B. The extreme ray S_1B evidently

Mr L. Wright and Dr G. J. Stoney in the *English Mechanic* (Sept., Oct., Nov., 1894; Nov. 8, Dec. 13, 1895; Jan. 17, 1896), that I have become acquainted with the distinguishing features of Prof. Abbe's work, and have learned that it was conducted upon different lines to that of Helmholtz. I am also indebted to Dr Stoney for a demonstration of some of Abbe's experiments.

proceeds from A, which is the image of B. The condition for the co-operation at B of the first lateral spectrum is thus that the angle of diffraction do not exceed the semi-angular aperture α. By elementary theory we know that the sine of the angle of diffraction is λ/ϵ, so that the action of the lateral spectrum requires that ϵ exceed $\lambda/\sin\alpha$. If we allow the incidence upon the grating to be oblique, the limit becomes $\frac{1}{2}\lambda/\sin\alpha$, as in (1).

We have seen that if one spectrum only illuminate B, the field shows no structure. If two spectra illuminate it with equal intensities, the field is occupied by ordinary interference bands, exactly as in the well-known experiments of Fresnel. And it is important to remark that the character of these bands is always the same, both as respects the graduation of light and shade, and in the fact that they have no focus. When more than two spectra co-operate, the resulting interference phenomena are more complicated, and there is opportunity for a completer representation of the special features of the original grating*.

While it is certain that the image ultimately formed may be considered to be due to the spectra focused at S_0, S_1..., the degree of conformity of the image to the original object is another question. From some of the expositions that have been given it might be inferred that if all the spectra emitted from the grating were utilized, the image would be a complete representation of the original. By considering the case of a very fine grating, which might afford no lateral spectra at all, it is easy to see that this conclusion is incorrect, but the matter stands in need of further elucidation. Again, it is not quite clear at what point the utilization of a spectrum really begins. All the spectra which the grating is competent to furnish are focused in the plane S_0S_1; and some of them might be supposed to operate partially even although the part of the image under examination is outside the geometrical cone defined by the aperture of the object-glass. For these and other reasons it will be seen that the

---

* These effects were strikingly illustrated in some observations upon gratings with 6,000 lines to the inch, set up vertically in a dark room and illuminated by sunlight from a distant vertical slit. The object-glass of the microscope was a quarter-inch. When the original grating, divided upon glass (by Nobert), was examined in this way, the lines were well seen if the instrument was in focus, but, as usual, a comparatively slight disturbance of focus caused all structure to disappear. When, however, a photographic copy of the same glass original, made with bitumen [p. 231], was substituted for it, very different effects ensued. The structure could be seen even although the object-glass were drawn back through $1\frac{1}{2}$ inch from its focused position; and the visible lines were twice as close, as if at the rate of 12,000 to the inch. The difference between the two cases is easily explained upon Abbe's theory. A soda flame viewed through the original showed a strong central image (spectrum of zero order) and comparatively faint spectra of the first and higher orders. A similar examination of the copy revealed very brilliant spectra of the first order on both sides, and a relatively feeble central image. The case is thus approximately the same as when in Abbe's experiment all spectra except the first (on the two sides) are blocked out.

spectrum theory*, valuable as it is, needs a good deal of supplementing, even when the representation of a grating under parallel light is in question.

When the object under examination is not a grating or a structure in which the pattern is repeated an indefinite number of times, but for example a double point, or when the incident light is not parallel, the spectrum theory, as hitherto developed, is inapplicable. As an extreme example of the latter case we may imagine the grating to be self-luminous. It is obvious that the problem thus presented must be within the scope of any complete theory, and equally so that here there are no spectra formed, as these require the radiations from the different elements of the grating to possess permanent phase-relations. It appears, therefore, to be a desideratum that the matter should be reconsidered from the older point of view, according to which the typical object is a point and not a grating. Such a treatment illustrates the important principle that the theory of resolving-power is essentially the same for all instruments. The peculiarities of the microscope arise from the fact that the divergence-angles are not limited to be small, and from the different character of the illumination usually employed; but, theoretically considered, these are differences of detail. The investigation can, without much difficulty, be extended to gratings, and the results so obtained confirm for the most part the conclusions of the spectrum theory.

It will be convenient to commence our discussion by a simple investigation of the resolving-power of an optical instrument for a self-luminous double point, such as will be applicable equally to the telescope and to the microscope. In Fig. 2 AB represents the axis, A being a point of the

Fig. 2.

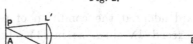

object and B a point of the image. By the operation of the object-glass LL' all the rays issuing from A arrive in the same phase at B. Thus if A be self-luminous, the illumination is a maximum at B, where all the secondary waves agree in phase. B is in fact the centre of the diffraction disk which constitutes the image of A. At neighbouring points the illumination is

---

* The special theory initiated by Prof. Abbe is usually called the "diffraction theory," a nomenclature against which it is necessary to protest. Whatever may be the view taken, any theory of resolving power of optical instruments must be a diffraction theory in a certain sense, so that the name is not distinctive. Diffraction is more naturally regarded as the obstacle to fine definition, and not, as with some exponents of Prof. Abbe's theory, the machinery by which good definition is brought about.

less, in consequence of the discrepancies of phase which there enter. In like manner, if we take a neighbouring point P in the plane of the object, the waves which issue from it will arrive at B with phases no longer absolutely accordant, and the discrepancy of phase will increase as the interval AP increases. When the interval is very small, the discrepancy of phase, though mathematically existent, produces no practical effect, and the illumination at B due to P is as important as that due to A, the intensities of the two luminous centres being supposed equal. Under these conditions it is clear that A and P are not separated in the image. The question is, to what amount must the distance AP be increased in order that the difference of situation may make itself felt in the image. This is necessarily a question of degree; but it does not require detailed calculations in order to show that the discrepancy first becomes conspicuous when the phases corresponding to the various secondary waves which travel from P to B range over about a complete period. The illumination at B due to P then becomes comparatively small, indeed for some forms of aperture evanescent. The extreme discrepancy is that between the waves which travel through the outermost parts of the object-glass at L and L'; so that, if we adopt the above standard of resolution, the question is, where must P be situated in order that the relative retardation of the rays PL and PL' may on their arrival at B amount to a wave-length (λ). In virtue of the general law that the reduced optical path is stationary in value, this retardation may be calculated without allowance for the different paths pursued on the further side of L, L', so that its value is simply $PL - PL'$. Now since AP is very small, $AL' - PL'$ is equal to $AP . \sin \alpha$, where α is the semi-angular aperture $L'AB$. In like manner $PL - AL$ has the same value, so that

$$PL - PL' = 2AP . \sin \alpha.$$

According to the standard adopted, the condition of resolution is therefore that AP, or ϵ, should exceed $\tfrac{1}{2}\lambda/\sin\alpha$, as in (1). If ϵ be less than this, the images overlap too much; while if ϵ greatly exceed the above value the images become unnecessarily separated.

In the above argument the whole space between the object and the lens is supposed to be occupied by matter of one refractive index, and λ represents the wave-length *in this medium* of the kind of light employed. If the restriction as to uniformity be violated, what we have ultimately to do with is the wave-length in the medium immediately surrounding the object.

The statement of the law of resolving-power has been made in a form appropriate to the microscope, but it admits also of immediate application to the telescope. If $2R$ be the diameter of the object-glass, and D the distance of the object, the angle subtended by AP is ϵ/D, and the angular resolving-power is given by

$$\frac{\lambda}{2D\sin\alpha} = \frac{\lambda}{2R}, \quad \dots\dots\dots\dots\dots\dots\dots\dots(5)$$

the well-known formula.

This method of derivation makes it obvious that there is no essential difference of principle between the two cases, although the results are conveniently stated in different forms. In the case of the telescope we have to do with a linear measure of aperture and an angular limit of resolution, whereas in the case of the microscope the limit of resolution is linear and it is expressed in terms of angular aperture.

In the above discussion it has been supposed for the sake of simplicity that the points to be discriminated are self-luminous, or at least behave as if they were such. It is of interest to enquire how far this condition can be satisfied when the object is seen by borrowed light. We may imagine that the object takes the form of an opaque screen, perforated at two points, and illuminated by distant sources situated behind.

If the source of light be reduced to a point, so that a single train of plane waves falls upon the screen, there is a permanent phase-relation between the waves incident at the two points, and therefore also between the waves scattered from them. In this case the two points are as far as possible from behaving as if they were self-luminous. If the incidence be perpendicular, the secondary waves issue in the same phase; but in the case of obliquity there is a permanent phase-difference. This difference, measured in wave-lengths, increases up to ϵ, the distance between the points, the limit being attained as the incidence becomes grazing.

When the light originates in distant independent sources, not limited to a point, there is no longer an absolutely definite phase-relationship between the secondary radiations from the two apertures; but this condition of things may be practically maintained, if the angular magnitude of the source be not too large. For example, if the source be limited to an angle θ round the normal to the screen, the maximum phase-difference measured in wave-lengths is $\epsilon\sin\theta$, so that if $\sin\theta$ be a small fraction of λ/ϵ, the finiteness of θ has but little effect. When, however, $\sin\theta$ is so great that $\epsilon\sin\theta$ becomes a considerable multiple of λ, the secondary radiations become approximately independent, and the apertures behave like self-luminous points. It is evident that even with a complete hemispherical illumination this condition can scarcely be attained when ϵ is less than λ.

The use of a condenser allows the widely-extended source to be dispensed with. By this means an image of a distant source composed of independently radiating parts, such as a lamp-flame, may be thrown upon the object, and it might at first sight be supposed that the problem under consideration was thus completely solved in all cases, inasmuch as the two apertures correspond to different parts of the flame. But we have to remember here and everywhere that optical images are not perfect, and that to a point of the flame corresponds in the image, not a point, but a disk of finite magnitude. When this consideration is taken into account, the same limitation as before is encountered.

For what is the smallest disk into which the condenser is capable of concentrating the light received from a distant point? Fig. 2 and the former argument apply almost without modification, and they show that the radius AP of the disk has the value $\frac{1}{2}\lambda/\sin\alpha$, where α is the semi-angular aperture of the condenser. Accordingly the diameter of the disk cannot be reduced below λ; and if ϵ be less than λ the radiations from the two apertures are only partially independent of one another.

It seems fair to conclude that the function of the condenser in micro-scopic practice is to cause the object to behave, at any rate in some degree, as if it were self-luminous, and thus to obviate the sharply-marked inter-ference-bands which arise when permanent and definite phase-relations are permitted to exist between the radiations which issue from various points of the object.

As we shall have occasion later to employ Lagrange's theorem, it may be well to point out how an instantaneous proof of it may be given upon the principles [especially that the optical distance measured along a ray is a minimum] already applied. As before, AB (Fig. 3) represents the

Fig. 3.

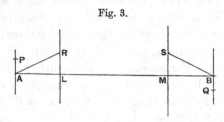

axis of the instrument, A and B being conjugate points. P is a point near A in the plane through A perpendicular to the axis, and Q is its image* in the perpendicular plane through B. Since A and B are conjugate, the optical distance between them is the same for all [ray-] paths, e.g. for

* [1902. In the original diagram Q was shown upon the wrong side of B. I owe the correction to a correspondence with Prof. Everett.]

ARSB and *ALMB*. [For the same reason the optical distance from P to Q is the same along the various rays, one of which lies infinitely near to *PRSQ* and another to *PLMQ*.] And, since AP, BQ are perpendicular to the axis, the optical distance from P to Q is the same (to the first order of small quantities [such as AP]) as from A to B. Consequently the optical distance *PRSQ* is the same as *ARSB*. Thus, if μ, μ' be the refractive indices in the neighbourhood of A and B respectively, α and β the divergence-angles *RAL*, *SBM* for a given ray, we have

$$\mu . AP . \sin \alpha = \mu' . BQ . \sin \beta, \quad\quad\quad\quad (6)$$

where AP, BQ denote the corresponding linear magnitudes of the two images. This is the theorem of Lagrange, extended by Helmholtz so as to apply to finite divergence-angles*.

We now pass on to the actual calculation of the images to be expected upon Fresnel's principles in the various cases that may arise. The origin of coordinates ($\xi = 0$, $\eta = 0$) in the focal plane is the geometrical image of the radiant point. If the vibration incident upon the lens be represented by $\cos(2\pi Vt/\lambda)$, where V is the velocity of light, the vibration at any point ξ, η in the focal plane is†

$$-\frac{1}{\lambda f}\iint \sin\frac{2\pi}{\lambda}\left\{Vt - f + \frac{x\xi + y\eta}{f}\right\} dx\,dy, \quad\quad (7)$$

in which f denotes the focal length, and the integration with respect to x and y is to be extended over the aperture of the lens. If for brevity we write

$$2\pi\xi/\lambda f = p, \quad\quad 2\pi\eta/\lambda f = q, \quad\quad\quad\quad (8)$$

(7) may be put into the form

$$-\frac{C}{\lambda f}\sin\frac{2\pi}{\lambda}(Vt - f) - \frac{S}{\lambda f}\cos\frac{2\pi}{\lambda}(Vt - f), \quad\quad (9)$$

where

$$S = \iint \sin(px + qy)\,dx\,dy, \quad\quad C = \iint \cos(px + qy)\,dx\,dy. \quad (10, 11)$$

It will suffice for our present purpose to limit ourselves to the case where the aperture is symmetrical with respect to x and y. We have then $S = 0$, and

$$C = \iint \cos px \cos qy\,dx\,dy, \quad\quad\quad\quad (12)$$

the phase of the vibration being the same at all points of the diffraction pattern.

---

* I learn from Czapski's excellent *Theorie der Optischen Instrumente* that a similar derivation of Lagrange's theorem from the principle of minimum path had already been given many years ago by Hockin (*Micros. Soc. Journ.* Vol. IV. p. 337, 1884).

† See for example *Enc. Brit.* "Wave Theory," p. 430 (1878). [Vol. III. p. 80.]

When the aperture is rectangular, of width a parallel to x, and of width b parallel to y, the limits of integration are from $-\frac{1}{2}a$ to $+\frac{1}{2}a$ for x, and from $-\frac{1}{2}b$ to $+\frac{1}{2}b$ for y. Thus

$$C = ab\, \frac{\sin(\pi\xi a/\lambda f)}{\pi\xi a/\lambda f}\, \frac{\sin(\pi\eta b/\lambda f)}{\pi\eta b/\lambda f}, \quad \dots\dots(13)$$

and by (9) the amplitude of vibration (irrespective of sign) is $C/\lambda f$. This expression gives the diffraction pattern due to a single point of the object whose geometrical image is at $\xi = 0$, $\eta = 0$. Sometimes, as in the application to a grating, we wish to consider the image due to a uniformly luminous *line*, parallel to η, and this can always be derived by integration from the expression applicable to a point. But there is a distinction to be observed according as the radiations from the various parts of the line are independent or are subject to a fixed phase-relation. In the former case we have to deal only with the *intensity*, represented by I^2 or $C^2/\lambda^2 f^2$; and we get

$$\int_{-\infty}^{+\infty} I^2 d\eta = \frac{a^2 b}{\lambda f}\, \frac{\sin^2(\pi\xi a/\lambda f)}{(\pi\xi a/\lambda f)^2} \quad \dots\dots(14)$$

by means of the known integral

$$\int_{-\infty}^{+\infty} \frac{\sin^2 x}{x^2}\, dx = \int_{-\infty}^{+\infty} \frac{\sin x}{x}\, dx = \pi. \quad \dots\dots(15)$$

This gives, as a function of ξ, the intensity due to a self-luminous line whose geometrical image coincides with $\xi = 0$.

Under the second head of a fixed phase-relation we need only consider the case where the radiations from the various parts of the line start in the *same* phase. We get, almost as before,

$$\frac{1}{\lambda f}\int_{-\infty}^{+\infty} C\, d\eta = a\, \frac{\sin(\pi\xi a/\lambda f)}{\pi\xi a/\lambda f} \quad \dots\dots(16)$$

for the expression of the resultant amplitude corresponding to ξ.

In order to make use of these results we require a table of the values of $\sin u/u$, and of $\sin^2 u/u^2$. The following will suffice for our purposes:—

TABLE I.

$\frac{4u}{\pi}$	$\frac{\sin u}{u}$	$\frac{\sin^2 u}{u^2}$	$\frac{4u}{\pi}$	$\frac{\sin u}{u}$	$\frac{\sin^2 u}{u^2}$	$\frac{4u}{\pi}$	$\frac{\sin u}{u}$	$\frac{\sin^2 u}{u^2}$
0	+1·0000	1·0000	6	−·2122	·0450	12	·0000	·0000
1	·9003	·8105	7	−·1286	·0165	13	−·0692	·0048
2	·6366	·4053	8	·0000	·0000	14	−·0909	·0083
3	·3001	·0901	9	+·1000	·0100	15	−·0600	·0036
4	·0000	·0000	10	·1273	·0162	16	·0000	·0000
5	−·1801	·0324	11	·0818	·0067			

When we have to deal with a single point or a single line only, this table gives directly the distribution of light in the image, u being equated to $\pi\xi a/\lambda f$. The illumination first vanishes when $u = \pi$, or $\xi/f = \lambda/a$.

On a former occasion[*] it has been shown that a self-luminous point or line at $u = -\pi$ is barely separated from one at $u = 0$. It will be of interest to consider this case under three different conditions as to phase-relationship: (i) when the phases are the same, as will happen when the illumination is by plane waves incident perpendicularly; (ii) when the phases are opposite; and (iii) when the phase-difference is a quarter period, which gives the same result for the intensity as if the apertures were self-luminous. The annexed table gives the numerical values required. In

TABLE II.

$\dfrac{4u}{\pi}$	$\dfrac{\sin u}{u} + \dfrac{\sin(u+\pi)}{u+\pi}$	$\dfrac{\sin u}{u} - \dfrac{\sin(u+\pi)}{u+\pi}$	$\sqrt{\left\{\dfrac{\sin^2 u}{u^2} + \dfrac{\sin^2(u+\pi)}{(u+\pi)^2}\right\}}$
-4...	$+1{\cdot}0000$	$-1{\cdot}0000$	$+1{\cdot}000$
-3...	$+1{\cdot}2004$	$-\ \cdot6002$	$+\ \cdot949$
-2...	$+1{\cdot}2732$	$\cdot0000$	$+\ \cdot900$
-1...	$+1{\cdot}2004$	$+\ \cdot6002$	$+\ \cdot949$
0...	$+1{\cdot}0000$	$+1{\cdot}0000$	$+1{\cdot}000$
1...	$+\ \cdot7202$	$+1{\cdot}0804$	$+\ \cdot918$
2...	$+\ \cdot4244$	$+\ \cdot8488$	$+\ \cdot671$
3...	$+\ \cdot1715$	$+\ \cdot4287$	$+\ \cdot326$
4...	$\cdot0000$	$\cdot0000$	$\cdot000$
5...	$-\ \cdot0800$	$-\ \cdot2801$	$-\ \cdot206$
6...	$-\ \cdot0849$	$-\ \cdot3395$	$-\ \cdot247$
7...	$-\ \cdot0468$	$-\ \cdot2105$	$-\ \cdot152$
8...	$\cdot0000$	$\cdot0000$	$\cdot000$
9...	$+\ \cdot0308$	$+\ \cdot1693$	$+\ \cdot122$
10...	$+\ \cdot0364$	$+\ \cdot2183$	$+\ \cdot156$
11...	$+\ \cdot0218$	$+\ \cdot1419$	$+\ \cdot101$
12...	$\cdot0000$	$\cdot0000$	$\cdot000$

cases (i) and (iii) the resultant amplitude is symmetrical with respect to the point $u = -\frac{1}{2}\pi$ midway between the two geometrical images; in case (ii) the sign is reversed, but this of course has no effect upon the intensity. Graphs of the three functions are given in Fig. 4, the geometrical images being at the points marked $-\pi$ and 0. It will be seen that while in case (iii), relating to self-luminous points or lines, there is an approach to separation,

* Phil. Mag. Vol. viii. p. 266, 1879. [Vol. i. p. 420.]

nothing but an accurate comparison with the curve due to a single source would reveal the duplicity in case (i). On the other hand, in case (ii), where there is a phase-difference of half a period between the radiations, the separation may be regarded as complete.

Fig. 4.

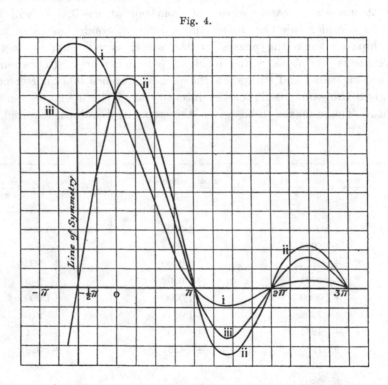

In a certain sense the last conclusion remains undisturbed even when the double point is still closer, and also when the aperture is of any other symmetrical form, *e.g.* circular. For at the point of symmetry in the image, midway between the two geometrical images of the radiant points, the component amplitudes are necessarily equal in numerical value and opposite in sign, so that the resultant amplitude or illumination vanishes. For example, suppose that the aperture is rectangular and that the points or lines are twice as close as before, the geometrical images being situated at $u = -\frac{1}{2}\pi$, $u = 0$. The resultant amplitude is represented by $f(u)$, where

$$f(u) = \frac{\sin u}{u} - \frac{\sin \left(u + \frac{1}{2}\pi\right)}{u + \frac{1}{2}\pi}. \quad \ldots\ldots\ldots\ldots\ldots(17)$$

The values of $f(u)$ are given in Table III. They show that the resultant vanishes at the place of symmetry $u = -\frac{1}{4}\pi$, and rises to a maximum at a point near $u = \frac{1}{2}\pi$, considerably beyond the geometrical image at $u = 0$. Moreover, the value of the maximum itself is much less than before, a feature which would become more and more pronounced as the points were taken closer.   At this stage the image becomes only a very incomplete representation of the object; but if the formation of a black line in the centre of the pattern be supposed to constitute resolution, then resolution occurs at all degrees of closeness*.   We shall see later, from calculations conducted by the same method, that a grating of an equal degree of closeness would show no structure at all but would present a uniformly illuminated field.

TABLE III.

$\frac{4u}{\pi}$	$f(u)$	$\frac{4u}{\pi}$	$f(u)$	$\frac{4u}{\pi}$	$f(u)$	$\frac{4u}{\pi}$	$f(u)$
-1......	$+\cdot00$	2......	$+\cdot64$	5......	$-\cdot05$	8......	$-\cdot13$
0......	$+\cdot36$	3......	$+\cdot48$	6......	$\cdot21$	9......	$+\cdot02$
1......	$+\cdot60$	4......	$+\cdot21$	7......	$-\cdot23$		

But before proceeding to such calculations we may deduce by Lagrange's theorem the interval ϵ in the original object corresponding to that between $u = 0$ and $u = \pi$ in the image, and thence effect a comparison with a grating by means of Abbe's theory.   The linear dimension (ξ) of the image corresponding to $u = \pi$ is given by $\xi = \lambda f / a$; and from Lagrange's theorem

$$\epsilon/\xi = \sin \beta / \sin \alpha, \dots\dots\dots\dots\dots\dots\dots(17\,a)$$

in which α is the "semi-angular aperture," and $\beta = a/2f$. Thus, corresponding to $u = \pi$,

$$\epsilon = \tfrac{1}{2}\lambda/\sin \alpha.$$

The case of a double point or line represented in Fig. 4 lies therefore at the extreme limit of resolution for a grating in which the period is the

---

* These results are easily illustrated experimentally. I have used two parallel slits, formed in films of tin-foil or of chemically deposited silver, of which one is conveniently made longer than the other. These slits are held vertically and are viewed through a small telescope, provided with a high-power eye-piece, whose horizontal aperture is restricted to a small width. The distance may first be so chosen that when backed by a neighbouring flame the double part of the slit just manifests its character by a faint shadow along the centre. If the flame is replaced by sunlight shining through a distant vertical slit, the effect depends upon the precise adjustment. When everything is in line the image is at its brightest, but there is now no sign of resolution of the double part of the slit. A very slight sideways displacement, in my case effected most conveniently by moving the telescope, brings in the half-period retardation, showing itself by a black bar down the centre. An increased displacement, leading to a relative retardation of three halves of a period, gives much the same result, complicated, however, by chromatic effects.

In conformity with theory the black bar down the image of the double slit may still be observed when the distance is increased much beyond that at which duplicity disappears under flame illumination.

For these experiments I chose the telescope, not only on account of the greater facility of manipulation which it allows, but also in order to make it clear that the theory is general, and that such effects are not limited, as is sometimes supposed, to the case of the microscope.

interval between the double points. And if the incidence of the light upon the grating were limited to be perpendicular, the period would have to be doubled before the grating could show any structure.

When the aperture is circular, of radius R, the diffraction pattern is symmetrical about the geometrical image ($p = 0$, $q = 0$), and it suffices to consider points situated upon the axis of ξ for which η (and q) vanish. Thus from (12)

$$C = \iint \cos px \, dx \, dy = 2 \int_{-R}^{+R} \cos px \, \sqrt{(R^2 - x^2)} \, dx. \quad \ldots\ldots\ldots\ldots(18)$$

This integral is the Bessel function of order unity, definable by

$$J_1(z) = \frac{z}{\pi} \int_0^\pi \cos(z \cos \phi) \sin^2 \phi \, d\phi. \quad \ldots\ldots\ldots\ldots\ldots(19)$$

Thus, if $x = R \cos \phi$,

$$C = \pi R^2 \frac{2 J_1(pR)}{pR}, \quad \ldots\ldots\ldots\ldots\ldots\ldots(20)$$

or, if we write $u = \pi \xi . 2R/\lambda f$,

$$C = \pi R^2 \frac{2 J_1(u)}{u}. \quad \ldots\ldots\ldots\ldots\ldots\ldots(21)*$$

This notation agrees with that employed for the rectangular aperture if we consider that $2R$ corresponds with a.

The illumination at various parts of the image of a double point may be investigated as before, especially if we limit ourselves to points which lie upon the line joining the two geometrical images. The only difference in the calculations is that represented by the substitution of $2J_1$ for sine. We shall not, however, occupy space by tables and drawings such as have been given for a rectangular aperture. It may suffice to consider the three principal points in the image due to a double source whose geometrical images are situated at $u = 0$ and $u = -\pi$, these being the points just mentioned, and that midway between them at $u = -\frac{1}{2}\pi$. The values of the functions required are

$$2J_1(0)/0 \quad = 1\cdot 0000 = \sqrt{\{1\cdot 0000\}}.$$
$$2J_1(\pi)/\pi \quad = \cdot 1812 = \sqrt{\{\cdot 03283\}}.$$
$$2J_1(\tfrac{1}{2}\pi)/\tfrac{1}{2}\pi = \cdot 7217 = \sqrt{\{\cdot 5209\}}.$$

In the case (corresponding to i. Fig. 4) where there is similarity of phase, we have at the geometrical images amplitudes $1\cdot 1812$ as against $1\cdot 4434$ at the point midway between. When there is opposition of phase, the first becomes $\pm \cdot 8188$, and the last zero†. When the phases differ by a quarter

---

* *Enc. Brit.* "Wave Theory," p. 432. [Vol. iii. p. 87.]

† The zero illumination extends to *all* points upon the line of symmetry.

period, or when the sources are self-luminous (iii. Fig. 4), the amplitudes at the geometrical images are $\sqrt{\{1\cdot0328\}}$ or $1\cdot0163$, and at the middle point $\sqrt{\{1\cdot0418\}}$ or $1\cdot0207$. The partial separation, indicated by the central depression in curve iii. Fig. 4, is thus lost when the rectangular aperture is exchanged for a circular one of equal width. It should be borne in mind that these results do not apply to a double *line*, which in the case of a circular aperture behaves differently from a double *point*.

There is one respect in which the theory is deficient, and the deficiency is the more important the larger the angular aperture. The formula (7) from which we start assumes that a radiant point radiates equally in all directions, or at least that the radiation from it after leaving the object-glass is equally dense over the whole area of the section. In the case of telescopes, and microscopes of moderate angular aperture, this assumption can lead to no appreciable error; but it may be otherwise when the angular aperture is very large. The radiation from an ideal centre of transverse vibrations is certainly not uniform in various directions, and indeed vanishes in that of primary vibration. If we suppose such an ideal source to be situated upon the axis of a wide-angled object-glass, we might expect the diffraction pattern to be less closely limited in that axial plane which includes the direction of primary vibration than in that which is perpendicular to it. The result for a double point illuminated by borrowed light would be a better degree of separation when the primary vibrations are perpendicular to the line of junction than when they are parallel to it.

Although it is true that complications and uncertainties under this head are not without influence upon the theory of the microscopic limit, it is not to be supposed that any considerable variation from that laid down by Abbe and Helmholtz is admissible. Indeed, in the case of a grating the theory of Abbe is still adequate, so far as the limit of resolution is concerned; for, as Dr Stoney has remarked, the irregularity of radiation in different directions tells only upon the relative brightness and not upon the angular position of the spectra. And it will remain true that there can be no resolution without the cooperation of two spectra at least.

In Table II. and Fig. 4 we have considered the image of a double point or line as formed by a lens of rectangular aperture. It is now proposed to extend the calculation to the case where the series of points or lines is infinite, constituting a *row* of points or a *grating*. The intervals are supposed to be strictly equal, and also the luminous intensities. When the aperture is rectangular, the calculation is the same whether we are dealing with a row of points or with a grating, but we have to distinguish according as the various centres radiate independently, viz., as if they were self-luminous, or are connected by phase-relations.   We will commence with the former case.

If the geometrical images of the various luminous points are situated at $u = 0$, $u = \pm v$, $u = \pm 2v$, &c., the expressions for the intensity at any point u of the field may be written as an infinite series,

$$I(u) = \frac{\sin^2 u}{u^2} + \frac{\sin^2 (u + v)}{(u + v)^2} + \frac{\sin^2 (u - v)}{(u - v)^2}$$

$$+ \frac{\sin^2 (u + 2v)}{(u + 2v)^2} + \frac{\sin^2 (u - 2v)}{(u - 2v)^2} + \dots . \quad \dots\dots\dots\dots(22)$$

Being an even function of u and periodic in period v, (22) may be expanded by Fourier's theorem in a series of cosines.   Thus

$$I(u) = I_0 + I_1 \cos \frac{2\pi u}{v} + \dots + I_r \cos \frac{2\pi ru}{v} + \dots\dots ; \quad \dots\dots(23)$$

and the character of the field of light will be determined when the values of the constants I_0, I_1, &c., are known.   For these we have as usual

$$I_0 = \frac{1}{v} \int_0^v I(u)\, du, \qquad I_r = \frac{2}{v} \int_0^v I(u) \cos \frac{2\pi ru}{v}\, du; \quad \dots\dots(24)$$

and it only remains to effect the integrations.   To this end we may observe that each term in the series (22) must in reality make an equal contribution to I_r.   It will come to the same thing whether, as indicated in (24), we integrate the sum of the series from 0 to v, or integrate a single term of it, *e.g.* the first, from $-\infty$ to $+\infty$.   We may therefore take

$$I_0 = \frac{1}{v} \int_{-\infty}^{+\infty} \frac{\sin^2 u}{u^2}\, du = \frac{\pi}{v}; \qquad I_r = \frac{2}{v} \int_{-\infty}^{+\infty} \frac{\sin^2 u}{u^2} \cos \frac{2\pi ru}{v}\, du. \quad \dots(25, 26)$$

To evaluate (26) we have

$$\int_{-\infty}^{+\infty} \frac{\sin^2 u \cos su}{u^2}\, du = \int_{-\infty}^{+\infty} \frac{1}{u} \frac{d}{du} (\sin^2 u \cos su)\, du,$$

and

$$\frac{d}{du} (\sin^2 u \cos su) = -\frac{s}{2} \sin su + \frac{2 + s}{4} \sin (2 + s) u + \frac{2 - s}{4} \sin (2 - s) u;$$

so that by (15) (s being positive)

$$\int_{-\infty}^{+\infty} \frac{\sin^2 u \cos su}{u^2}\, du = \pi \left\{ -\frac{s}{2} + \frac{2 + s}{4} \pm \frac{2 - s}{4} \right\},$$

the *minus* sign being taken when $2 - s$ is negative.

Hence

$$I_r = \frac{2\pi}{v} \left(1 - \frac{\pi r}{v} \right), \text{ or } 0, \quad \dots\dots\dots\dots\dots\dots(27)$$

according as v exceeds or falls short of $r\pi$.

We may now trace the effect of altering the value of v. When v is large, a considerable number of terms in the Fourier expansion (23) are of importance, and the discontinuous character of the luminous grating or row of points is fairly well represented in the image. As v diminishes, the higher terms drop out in succession, until when v falls below 2π only I_0 and I_1 remain. From this point onwards I_1 continues to diminish until it also finally disappears when v drops below π. The field is then uniformly illuminated, showing no trace of the original structure. The case $v = \pi$ is that of Fig. 4, and curve iii. shows that at a stage when an infinite series shows no structure, a *pair* of luminous points or lines of the same closeness are still in some degree separated. It will be remembered that $v = \pi$ corresponds to $\epsilon = \frac{1}{2}\lambda/\sin\alpha$, ϵ being the linear period of the original object and α the semi-angular aperture.

We will now pass on to consider the case of a grating or row of points perforated in an opaque screen and illuminated by plane waves of light. If the incidence be oblique, the phase of the radiation emitted varies by equal steps as we pass from one element to the next. But for the sake of simplicity we will commence with the case of perpendicular incidence, where the radiations from the various elements all start in the same phase. We have now to superpose amplitudes, and not as before intensities. If A be the resultant amplitude, we may write

$$A(u) = \frac{\sin u}{u} + \frac{\sin(u+v)}{u+v} + \frac{\sin(u-v)}{u-v} + \ldots\ldots$$

$$= A_0 + A_1 \cos\frac{2\pi u}{v} + \ldots + A_r \cos\frac{2\pi r u}{v} + \ldots \quad\ldots\ldots\ldots\ldots(28)$$

When v is very small, the infinite series identifies itself more and more nearly with the integral

$$\frac{1}{v}\int_{-\infty}^{+\infty} \frac{\sin u}{u}\, du, \text{ viz. } \frac{\pi}{v}.$$

In general we have, as in the last problem,

$$A_0 = \frac{1}{v}\int_{-\infty}^{+\infty} \frac{\sin u}{u}\, du; \quad A_r = \frac{2}{v}\int_{-\infty}^{+\infty} \frac{\sin u}{u}\cos\frac{2\pi r u}{v}\, du; \quad\ldots\ldots(29)$$

so that $A_0 = \pi/v$. As regards A_r, writing s for $2\pi r/v$, we have

$$A_r = \frac{1}{v}\int_{-\infty}^{+\infty} \frac{\sin(1+s)u + \sin(1-s)u}{u}\, du = \frac{\pi}{v}(1 \pm 1),$$

the lower sign applying when $(1 - s)$ is negative. Accordingly,

$$A(u) = \frac{\pi}{v}\left\{1 + 2\cos\frac{2\pi u}{v} + 2\cos\frac{4\pi u}{v} + \ldots\right\}, \quad\ldots\ldots\ldots\ldots(30)$$

the series being continued so long as $2\pi r < v$.

If the series (30) were continued *ad infinitum*, it would represent a discontinuous distribution, limited to the points (or lines) $u = 0$, $u = \pm v$, $u = \pm 2v$, &c., so that the image formed would accurately correspond to the original object. This condition of things is most nearly realised when v is very great, for then (30) includes a large number of terms. As v diminishes the higher terms drop out in succession, retaining however (in contrast with (27)) their full value up to the moment of disappearance. When v is less than 2π, the series is reduced to its constant term, so that the field becomes uniform. Under this kind of illumination, the resolving-power is only half as great as when the object is self-luminous.

These conclusions are in entire accordance with Abbe's theory. The first term of (30) represents the central image, the second term the *two* spectra of the first order, the third term the two spectra of the second order, and so on. Resolution fails at the moment when the spectra of the first order cease to cooperate, and we have already seen that this happens for the case of perpendicular incidence when $v = 2\pi$. The two spectra of any given order fail at the same moment.

If the series stops after the lateral spectra of the first order,

$$A\,(u) = \frac{\pi}{v}\left\{1 + 2\cos\frac{2\pi u}{v}\right\}, \quad\ldots\ldots\ldots\ldots\ldots\ldots(31)$$

showing a maximum intensity when $u = 0$, or $\frac{1}{2}v$, and zero intensity when $u = \frac{1}{3}v$, or $\frac{2}{3}v$. These bands are not the simplest kind of interference bands. The latter require the operation of two spectra only; whereas in the present case there are three—the central image and the two spectra of the first order.

We may now proceed to consider the case when the incident plane waves are inclined to the grating. The only difference is that we require now to introduce a change of phase between the image due to each element and its neighbour. The series representing the resultant amplitude at any point u may still be written

$$\frac{\sin u}{u} + \frac{\sin(u+v)}{u+v}e^{-imv} + \frac{\sin(u-v)}{u-v}e^{+imv} + \frac{\sin(u+2v)}{u+2v}e^{-2imv} + \ldots\ \ldots(32)$$

For perpendicular incidence $m = 0$. If γ be the obliquity, ϵ the grating-interval, λ the wave-length,

$$mv/2\pi = \epsilon\sin\gamma/\lambda. \quad\ldots\ldots\ldots\ldots\ldots\ldots(33)$$

The series (32), as it stands, is not periodic with respect to u in period v, but evidently it can differ from such a periodic series only by the factor e^{imu}.

The series

$$\frac{e^{-imu}\sin u}{u} + \frac{e^{-im(u+v)}\sin(u+v)}{u+v}$$

$$+ \frac{e^{-im(u-v)}\sin(u-v)}{u-v} + \frac{e^{-im(u+2v)}\sin(u+2v)}{u+2v} + \dots \dots(34)$$

is truly periodic, and may therefore be expanded by Fourier's theorem in periodic terms:

$$(34) = A_0 + iB_0 + (A_1 + iB_1)\cos(2\pi u/v) + (C_1 + iD_1)\sin(2\pi u/v) + \dots$$
$$+ (A_r + iB_r)\cos(2r\pi u/v) + (C_r + iD_r)\sin(2r\pi u/v) + \dots \quad \dots(35)$$

As before, if $s = 2r\pi/v$,

$$\tfrac{1}{2}v(A_r + iB_r) = \int_{-\infty}^{+\infty} \frac{e^{-imu}\sin u \cos su}{u}\,du;$$

so that $B_r = 0$, while

$$\tfrac{1}{2}v.A_r = \int_{-\infty}^{+\infty} \frac{\cos mu \sin u \cos su}{u}\,du. \quad\dots\dots(36)$$

In like manner $C_r = 0$, while

$$-\tfrac{1}{2}v.D_r = \int_{-\infty}^{+\infty} \frac{\sin mu \sin u \sin su}{u}\,du. \quad\dots\dots(37)$$

In the case of the zero suffix

$$B_0 = 0, \qquad vA_0 = \int_{-\infty}^{+\infty} \frac{\cos mu \sin u}{u}\,du. \quad\dots\dots(38)$$

When the products of sines and cosines which occur in (36) &c. are transformed in a well-known manner, the integration may be effected by (15). Thus

$$\cos mu \sin u \cos su = \tfrac{1}{4}\{\sin(1+m+s)u + \sin(1-m-s)u$$
$$+ \sin(1+m-s)u + \sin(1-m+s)u\};$$

so that

$$\tfrac{1}{2}v.A_r = \tfrac{1}{4}\pi\{[1+m+s]+[1-m-s]+[1+m-s]+[1-m+s]\} \dots(39)$$

where each symbol such as $[1+m+s]$ is to be replaced by ± 1, the sign being that of $(1+m+s)$. In like manner

$$-\tfrac{1}{2}v.D_r = \tfrac{1}{4}\pi\{[1+m-s]+[1-m+s]-[1+m+s]-[1-m-s]\}. \dots(40)$$

The rth terms of (35) are accordingly

$$\frac{\pi}{2v}\{e^{isu}([1+m+s]+[1-m-s]) + e^{-isu}([1+m-s]+[1-m+s])\};$$

or for the original series (32),

$$\frac{\pi}{2v}\{e^{i(m+s)u}([1+m+s]+[1-m-s])$$
$$+ e^{i(m-s)u}([1+m-s]+[1-m+s])\}. \dots(41)$$

For the term of zero order,

$$A_0 e^{imu} = \frac{\pi}{2v} e^{imu} ([1+m]+[1-m]). \quad \dots\dots\dots\dots(42)$$

From (41) we see that the term in $e^{i(m+s)u}$ vanishes unless $(m+s)$ lies between ± 1, and that then it is equal to $\pi/v \cdot e^{i(m+s)u}$; also that the term in $e^{i(m-s)u}$ vanishes unless $(m-s)$ lies between ± 1, and that it is then equal to $\pi/v \cdot e^{i(m-s)u}$. In like manner the term in e^{imu} vanishes unless m lies between ± 1, and when it does not vanish it is equal to $\pi/v \cdot e^{imu}$. This particular case is included in the general statement by putting $s=0$.

The image of the grating, or row of points, expressed by (32), is thus capable of representation by the sum of terms

$$\pi/v \cdot \{e^{imu} + e^{i(m+s_1)u} + e^{i(m-s_1)u} + e^{i(m+s_2)u} + \dots\} \dots\dots\dots\dots(43)$$

where $s_1 = 2\pi/v$, $s_2 = 4\pi/v$, &c., every term being included for which the coefficient of u lies between ± 1. Each of these terms corresponds to a spectrum of Abbe's theory, and represents plane progressive waves inclined at a certain angle to the plane of the image. Each spectrum when it occurs at all contributes equally, and it goes out of operation suddenly. If but one spectrum operates, the field is of uniform brightness. If two spectra operate, we have the ordinary interference bands due to two sets of plane waves crossing one another at a small angle of obliquity*.

Any consecutive pair of spectra give the same interference bands, so far as illumination is concerned. For

$$\frac{\pi}{v} \{e^{iu[m+2r\pi/v]} + e^{iu[m+2(r+1)\pi/v]}\} = \frac{2\pi}{v} \cos \frac{\pi u}{v} e^{iu[m+2(r+\frac{1}{2})\pi/v]},$$

of which the exponential factor influences only the phase.

In (43) the critical value of v for which the rth spectrum disappears is given by, when we introduce the value of m from (33),

$$\frac{v}{2\pi}\left(\frac{\epsilon \sin \gamma}{\lambda} \pm r\right) = \pm 1;$$

or, since (as we have seen)

$$\frac{2\pi}{v} = \frac{\epsilon \sin \alpha}{\lambda}, \quad \dots\dots\dots\dots\dots\dots(44)$$

$$\epsilon(\sin \gamma \mp \sin \alpha) = \mp r\lambda. \quad \dots\dots\dots\dots\dots(45)$$

This is the condition, according to elementary theory, in order that the rays forming the spectrum of the rth order should be inclined at the angle α, and so (Fig. 2) be adjusted to travel from A to B, through the edge of the lens L.

---

* *Enc. Brit.* "Wave Theory," p. 425. [Vol. iii. p. 59.]

The discussion of the theory of a rectangular aperture may here close. This case has the advantage that the calculation is the same whether the object be a row of points or a grating. A parallel treatment of other forms of aperture, e.g. the circular form, is not only limited to the first alternative, but applies there only to those points of the field which lie upon the line joining the geometrical images of the luminous points. Although the advantage lies with a more general method of investigation to be given presently, it may be well to consider the theory of a circular aperture as specially deduced from the formula (21) which gives the image of a single luminous centre.

If we limit ourselves to the case of parallel waves and perpendicular incidence, the infinite series to be discussed is

$$A(u) = \frac{J_1(u)}{u} + \frac{J_1(u+v)}{u+v} + \frac{J_1(u-v)}{u-v} + \frac{J_1(u+2v)}{u+2v} + \dots, \quad \dots(46)$$

where
$$u = \pi\xi \cdot 2R/\lambda f. \quad\dots(47)$$

Since A is necessarily periodic in period v, we may assume

$$A(u) = A_0 + A_1 \cos(2\pi u/v) + \dots + A_r \cos(2r\pi u/v) + \dots; \quad\dots(48)$$

and, as in the case of the rectangular aperture,

$$A_0 = \frac{1}{v}\int_{-\infty}^{+\infty} \frac{J_1(u)}{u}\, du, \quad A_r = \frac{2}{v}\int_{-\infty}^{+\infty} \frac{J_1(u)}{u} \cos\frac{2r\pi u}{v}\, du. \quad\dots(49)$$

These integrals may be evaluated. If a and b be real, and a be positive*,

$$\int_0^\infty e^{-ax} J_0(bx)\, dx = \frac{1}{\sqrt{(a^2+b^2)}}. \quad\dots(50)$$

Multiplying by $b\,db$ and integrating from 0 to b, we find

$$\int_0^\infty \frac{J_1(bx) e^{-ax}}{x}\, dx = \frac{\sqrt{(a^2+b^2)} - a}{b}. \quad\dots(51)$$

In this we write $b=1$, $a=is$, where s is real. Thus

$$\int_0^\infty \frac{J_1(x)\{\cos sx - i\sin sx\}}{x}\, dx = \sqrt{(1-s^2)} - is.$$

If $s^2 > 1$, we must write $i\sqrt{(s^2-1)}$ for $\sqrt{(1-s^2)}$. Hence, if $s<1$,

$$\int_0^\infty \frac{J_1(x)\cos sx}{x}\, dx = \sqrt{(1-s^2)}, \quad \int_0^\infty \frac{J_1(x)\sin sx}{x}\, dx = s; \quad\dots(52,53)$$

while, if $s>1$,

$$\int_0^\infty \frac{J_1(x)\cos sx}{x}\, dx = 0, \quad \int_0^\infty \frac{J_1(x)\sin sx}{x}\, dx = -\sqrt{(s^2-1)} + s. \quad\dots(54,55)$$

* Gray and Mathews, *Bessel's Functions*, 1895, p. 72.

We are here concerned only with (52),(54), and we conclude that $A_0 = 2/v$, and that

$$A_r = \frac{4\sqrt{(1-s^2)}}{v}, \quad \text{or} \quad 0, \quad \ldots\ldots\ldots\ldots\ldots\ldots\ldots(56)$$

according as s is less or greater than 1, viz. according as $2r\pi$ is less or greater than v.

If we compare this result with the corresponding one (30) for a rectangular aperture of equal width ($2R = a$), we see that the various terms representing the several spectra enter or disappear at the same time; but there is one important difference to be noted. In the case of the rectangular aperture the spectra enter suddenly and with their full effect, whereas in the present case there is no such discontinuity, the effect of a spectrum which has just entered being infinitely small. As will appear more clearly by another method of investigation, the discontinuity has its origin in the sudden rise of the ordinate of the rectangular aperture from zero to its full value.

In the method referred to the form of the aperture is supposed to remain symmetrical with respect to both axes, but otherwise is kept open, the integration with respect to x being postponed. Starting from (12) and considering only those points of the image for which η and q in equation (8) vanish, we have as applicable to the image of a single luminous source

$$C = \iint \cos px\, dx\, dy = 2 \int y \cos px\, dx \quad \ldots\ldots\ldots\ldots\ldots(57)$$

in which $2y$ denotes the whole height of the aperture at the point x. This gives the amplitude as a function of p. If there be a row of luminous points, from which start radiations in the same phase, we have an infinite series of terms, similar to (57) and derived from it by the addition to p of positive and negative integral multiples of a constant (p_1) representing the period. The sum of the series $A(p)$ is necessarily periodic, so that we may write

$$A(p) = A_0 + \ldots + A_r \cos(2r\pi p/p_1) + \ldots; \quad \ldots\ldots\ldots\ldots (58)$$

and, as in previous investigations, we may take

$$A_r = \int_{-\infty}^{+\infty} C \cos sp\, dp, \quad \ldots\ldots\ldots\ldots\ldots\ldots\ldots(59)$$

s (not quite the same as before) standing for $2r\pi/p_1$, and a constant factor being omitted. To ensure convergency we will treat this as the limit of

$$\int_{-\infty}^{+\infty} e^{\pm hp} C \cos sp\, dp \quad \ldots\ldots\ldots\ldots\ldots\ldots(60)$$

the sign of the exponent being taken negative, and h being ultimately made to vanish. Taking first the integration with respect to p, we have

$$\int_{-\infty}^{+\infty} e^{\pm hp} \cos xp \cos sp\, dp = \frac{h}{h^2+(x+s)^2} + \frac{h}{h^2+(x-s)^2};$$

and thus

$$A_r = \int \frac{hy\,dx}{h^2+(x+s)^2} + \int \frac{hy\,dx}{h^2+(x-s)^2},$$

in which h is to be made to vanish. In the limit the integrals receive sensible contributions only from the neighbourhoods of $x = \pm s$; and since

$$\int_{-\infty}^{+\infty} \frac{du}{1+u^2} = \pi, \quad\dots\dots\dots\dots\dots\dots(61)$$

we get
$$A_r = \pi\,(y_{x=-s} + y_{x=+s}) = 2\pi y_{x=s}. \quad\dots\dots\dots(62)$$

From (62) we see that the occurrence of the term in A_r, i.e. the appearance of the spectrum of the rth order, is associated with the value of a particular ordinate of the object-glass. If the ordinate be zero, i.e. if the abscissa exceed numerically the half-width of the object-glass, the term in question vanishes. The first appearance of it corresponds to

$$\tfrac{1}{2}a = 2r\pi/p_1 = r\lambda f/\xi_1,$$

in which a is the entire width of the object-glass and ξ_1 the linear period in the image. By (17a),

$$\frac{\lambda f}{\xi_1} = \frac{\lambda f \sin\beta}{\epsilon \sin\alpha} = \frac{\tfrac{1}{2}a\lambda}{\epsilon \sin\alpha};$$

so that the condition is, as before,

$$\epsilon \sin\alpha = r\lambda.$$

When A_r has appeared, its value is proportional to the ordinate at $x = s$. Thus in the case of a circular aperture ($a = 2R$) we have

$$y_{x=s} = R\,\sqrt{\{1 - r^2\lambda^2/\epsilon^2 \sin^2\alpha\}}. \quad\dots\dots\dots(63)$$

The above investigation relates to a row of luminous points emitting light of the same intensity and phase, and it is limited to those points of the image for which η (and q) vanish. If the object be a grating radiating under similar conditions, we have to retain $\cos qy$ in (12) and to make an integration with respect to q. Taking this first, and introducing a factor $e^{\pm kq}$, we have

$$\int_{-\infty}^{+\infty} e^{\pm kq} \cos qy\, dq = \frac{2k}{k^2+y^2}. \quad\dots\dots\dots(64)$$

This is now to be integrated with respect to y between the limits $-y$ and $+y$. If this range be finite, we have

$$\mathrm{Limit}_{k=0} \int_{-y}^{+y} \frac{2k\,dy}{k^2+y^2} = 2\pi, \quad\dots\dots\dots\dots(65)$$

independent of the length of the particular ordinate. Thus

$$C_1 = \int_{-\infty}^{+\infty} C\,dq = 2\pi \int \cos px\, dx, \dots\dots\dots\dots\dots(66)$$

the integration with respect to x extending over the range for which y is finite, that is, over the width of the object-glass. If this be $2R$, we have

$$\int_{-\infty}^{+\infty} C\,dq = 4\pi/p \,.\, \sin pR. \dots\dots\dots\dots(67)$$

From (67) we see that the image of a luminous line, all parts of which radiate in the same phase, is independent of the form of the aperture of the object-glass, being, for example, the same for a circular aperture as for a rectangular aperture of equal width. This case differs from that of a *self-luminous* line, the images of which thrown by circular and rectangular apertures are of different types*.

The comparison of (67) with (20), applicable to a circular aperture, leads to a theorem in Bessel's functions. For, when q is finite,

$$C = \pi R^2 \frac{2\,J_1\{\sqrt{(p^2+q^2)}\,R\}}{\sqrt{(p^2+q^2)}}; \dots\dots\dots\dots(68)$$

so that, setting $R=1$, we get

$$\int_0^\infty \frac{J_1\{\sqrt{(p^2+q^2)}\}}{\sqrt{(p^2+q^2)}}\,dq = \frac{\sin p}{p}. \dots\dots\dots\dots(69)\dagger$$

The application to a grating, of which all parts radiate in the same phase, proceeds as before. If, as in (58), we suppose

$$A(p) = A_0 + \dots + A_r \cos sp + \dots, \dots\dots\dots\dots(70)$$

we have
$$A_r = \int_{-\infty}^{+\infty} C_1 \cos sp\,dp; \dots\dots\dots\dots(71)$$

from which we find that A_r is $4\pi^2$ or 0, according as the ordinate is finite or not finite at $x=s$. The various spectra enter and disappear under the same conditions as prevailed when the object was a row of points; but now they enter discontinuously and retain constant values, instead of varying with the particular ordinate of the object-glass which corresponds to $x=s$.

We will now consider the corresponding problems when the illumination is such that each point of the row of points or of the grating radiates independently. The integration then relates to the intensity of the field as due to a single source.

* *Enc. Brit.* "Wave Theory," p. 434. [Vol. III. p. 92.]

† This may be verified by means of Neumann's formula (Gray and Mathews, *Bessel's Functions* (70), p. 27).

By (9), (10), (11), the intensity I^2 at the point (p, q) of the field, due to a single source whose geometrical image is situated at (0, 0) is given by

$$\lambda^2 f^2 I^2 = \{\iint \cos(px + qy)\,dx\,dy\}^2 + \{\iint \sin(px + qy)\,dx\,dy\}^2$$
$$= \iint \cos(px' + qy')\,dx'\,dy' \times \iint \cos(px + qy)\,dx\,dy$$
$$+ \iint \sin(px' + qy')\,dx'\,dy' \times \iint \sin(px + qy)\,dx\,dy$$
$$= \iiiint \cos\{p(x' - x) + q(y' - y)\}\,dx\,dy\,dx'\,dy', \quad \ldots\ldots\ldots\ldots(72)$$

the integrations with respect to x', y', as well as those with respect to x, y being over the area of the aperture.

In the present application to sources which are periodically repeated, the term in $\cos sp$ of the Fourier expansion representing the intensity at various points of the image has a coefficient found by multiplying (72) by $\cos sp$ and integrating with respect to p from $p = -\infty$ to $p = +\infty$. If the object be a row of points, we may take $q = 0$; if it be a grating, we have to integrate with respect also to q from $q = -\infty$ to $q = +\infty$.

Considering the latter case, and taking first the integrations with respect to p, q, we introduce the factors $e^{\mp hp \mp kq}$, the *plus* or *minus* being so chosen as to make the elements of the integral vanish at infinity. After the operations have been performed, h and k are to be supposed to vanish*. The integrations are performed as for (60), (64), and we get the sum of the two terms denoted by

$$\frac{2hk}{\{h^2 + (x' - x \pm s)^2\}\{k^2 + (y' - y)^2\}}. \quad \ldots\ldots\ldots\ldots(73)$$

We have still to integrate with respect to $dx\,dy\,dx'\,dy'$. As in (65), since the range for y' always includes y,

$$\text{Limit}_{k=0} \int \frac{2k\,dy'}{k^2 + (y' - y)^2} = 2\pi;$$

and we are left with

$$\iiint \frac{2\pi h\,dx\,dy\,dx'}{h^2 + (x' - x \pm s)^2}. \quad \ldots\ldots\ldots\ldots(74)$$

If s were zero, the integration with respect to x' would be precisely similar; but with s finite it will be only for certain values of x that $(x' - x \pm s)$ vanishes within the range of integration. Unless this evanescence takes place, the limit when h vanishes becomes zero. The effect of the integration with respect to x' is thus to limit the range of the subsequent integration with respect to x. The result may be written

$$2\pi^2 \iint dx\,dy \quad \ldots\ldots\ldots\ldots\ldots\ldots(75)$$

---

* The process is that employed by Stokes in his evaluation of the integral intensity, *Edin. Trans.* xx. p. 317 (1853). See also *Enc. Brit.* "Wave Theory," p. 431. [Vol. iii. p. 86.]

upon the understanding that, while the integration for y ranges over the whole vertical aperture, that for x is limited to such values of x as bring $x \mp s$ (as well as x itself) within the range of the horizontal aperture. The coefficient of the Fourier component of the intensity involving cos sp, or $\cos(2r\pi p/p_1)$, is thus proportional to a certain part of the area of the aperture. Other parts of the area are inefficient, and might be stopped off without influencing the result.

The limit to resolution, corresponding to $r = 1$, depends only on the width of the aperture, and is therefore for all forms of aperture the same as for the case of the rectangular aperture already fully investigated.

If the object be a row of points instead of a row of lines, $q = 0$, and there is no integration with respect to it. The process is nearly the same as above, and the result for the coefficient of the rth term in the Fourier expansion is proportional to $\int y^2 dx$, instead of $\int y\, dx$, the integration with respect to x being over the same parts of the aperture as when the object was a grating. The application to a circular aperture would lead to an evaluation of

$$\int_{-\infty}^{+\infty} \frac{J_1^2(u) \cos su}{u^2}\, du.$$

223.

THEORETICAL CONSIDERATIONS RESPECTING THE SEPARA-TION OF GASES BY DIFFUSION AND SIMILAR PROCESSES.

[*Philosophical Magazine*, XLII. pp. 493—498, 1896.]

THE larger part of the calculations which follow were made in connexion with experiments upon the concentration of argon from the atmosphere by the method of atmolysis *. When the supply of gas is limited, or when it is desired to concentrate the lighter ingredient, the conditions of the question are materially altered; but it will be convenient to take first the problem which then presented itself of the simple diffusion of a gaseous mixture into a vacuum, with special regard to the composition of the residue. The diffusion tends to alter this composition in the first instance only in the neighbourhood of the porous walls; but it will be assumed that the forces promoting mixture are powerful enough to allow of our considering the composition to be uniform throughout the whole volume of the residue, and variable only with time, on account of the unequal escape of the constituent gases.

Let x, y denote the quantities of the two constituents of the residue at any time, so that $- dx, - dy$ are the quantities diffused out in time dt. The values of dx/dt, dy/dt will depend upon the character of the porous partition and upon the actual pressure; but for our present purpose it will suffice to express dy/dx, and this clearly involves only the ratios of the constituents and of their diffusion rates. Calling the diffusion rates μ, ν, we have

$$\frac{dy}{dx} = \frac{\nu y}{\mu x}. \quad\dots\dots\dots\dots\dots\dots\dots\dots\dots\dots\dots\dots(1)$$

In this equation x, y may be measured on any consistent system that may be convenient. The simplest case would be that in which the residue is maintained at a constant volume, when x, y might be taken to represent

---

* Rayleigh and Ramsay, *Phil. Trans.* CLXXXVI. p. 206 (1895). [Vol. IV. p. 180.]

the partial pressures of the two gases. But the equation applies equally well when the volume changes, for example in such a way as to maintain the total pressure constant.

The integral of (1) is

$$y^{1/\nu} = Cx^{1/\mu}, \quad \dots\dots\dots\dots\dots\dots\dots(2)$$

where C is an arbitrary constant, or

$$y/x = Cx^{-1+\nu/\mu}. \quad \dots\dots\dots\dots\dots\dots(3)$$

If X, Y be simultaneous values of x, y, regarded as initial,

$$\frac{y/x}{Y/X} = \left(\frac{x}{X}\right)^{-1+\nu/\mu}, \quad \dots\dots\dots\dots\dots\dots(4)$$

so that

$$x = X\left(\frac{y/x}{Y/X}\right)^{\mu/(\nu-\mu)}. \quad \dots\dots\dots\dots\dots(5)$$

In like manner

$$y = Y\left(\frac{x/y}{X/Y}\right)^{\nu/(\mu-\nu)}. \quad \dots\dots\dots\dots\dots(6)$$

If we write

$$\frac{y/x}{Y/X} = r, \quad \dots\dots\dots\dots\dots\dots\dots(7)$$

r represents the enrichment of the residue as regards the second constituent, and we have from (5), (6),

$$\frac{x+y}{X+Y} = \frac{X}{X+Y}r^{\mu/(\nu-\mu)} + \frac{Y}{X+Y}r^{\nu/(\nu-\mu)}, \quad \dots\dots\dots(8)$$

an equation which exhibits the relation between the enrichment and the ratio of the initial and final total quantities of the mixture.

From (8), or more simply from (4), we see that as x diminishes with time the enrichment tends to zero or infinity, indicating that the residue becomes purer *without limit*, and this whatever may be the original proportions. Thus if the first gas (x) be the more diffusive $(\mu > \nu)$, the exponent on the right of (4) is negative; and this indicates that r becomes infinite, or that the first gas is ultimately eliminated from the residue. When the degree of enrichment required is specified, an easy calculation from (8) gives the degree to which the diffusion must be carried.

In Graham's atmolyser the gaseous mixture is caused to travel along a tobacco-pipe on the outside of which a vacuum is maintained. If the passage be sufficiently rapid to preclude sensible diffusion along the length of the pipe, the circumstances correspond to the above calculation; but the agreement with Graham's numbers is not good. Thus in one case given by him* of the atmolysis of a mixture containing equal volumes of oxygen and hydrogen, we have

$$Y/X = 1, \qquad y/x = 92\cdot78/7\cdot22,$$

* *Phil. Trans.* Vol. CLIII. p. 403 (1863).

so that $r = 13$ nearly. Thus, if in accordance with the view usually held $\mu/\nu = 4$, we should have from (8)

$$\frac{x+y}{X+Y} = \tfrac{1}{2} \times 13^{-\frac{3}{4}} + \tfrac{1}{2} \times 13^{-\frac{1}{4}} = \cdot229;$$

so that a reduction of the residue to ·229 of the initial quantity should have effected the observed enrichment. The initial and final volumes given by Graham are, however, 7·5 litres and ·45 litre, whose ratio is ·06. The inferior efficiency of the apparatus may have been due to imperfections in the walls or joints of the pipes. Such an explanation appears to be more probable than a failure of the law of independent diffusion of the component gases upon which the theoretical investigation is founded.

In the concentration of argon from a mixture of argon and nitrogen we have conditions much less favourable. In this case

$$\mu/\nu = \sqrt{20}/\sqrt{14} = \cdot077.$$

If an enrichment of $2:1$ is required and if the original mixture is derived from the atmosphere by removal of oxygen, the equation is

$$\frac{x+y}{X+Y} = \cdot99 \times 2^{-6\cdot13} + \cdot01 \times \cdot2^{-5\cdot13} = \cdot0142 + \cdot0029 = \cdot0171,$$

expressing the reduction needed. The results obtained experimentally (*loc. cit.*) were inferior in this case also.

When the object is the most effective separation of the components of a mixture, it is best, as supposed in the above theory, to maintain a vacuum on the further side of the porous wall. But we have sometimes to consider cases where the vacuum is replaced by an atmosphere of fixed composition, as in the well-known experiment of the diffusion of hydrogen into air through a porous plug. We will suppose that there are only two gases concerned and that the volume inside is given. The symbols x, y will then denote the partial pressures within the given volume, the constant partial pressures outside being α, β. Our equations may be written

$$dx = \mu(\alpha - x)\,dt, \qquad dy = \nu(\beta - y)\,dt, \dots\dots\dots\dots(9)$$

or on integration

$$x = \alpha + Ce^{-\mu t}, \qquad y = \beta + De^{-\nu t}, \dots\dots\dots\dots(10)$$

C, D being arbitrary constants.

After a sufficient time x, y reduce themselves respectively to α, β, as was to be expected.

The constants μ, ν are not known beforehand, depending as they do upon the specialities of the apparatus as well as upon the quality of the gases. If we eliminate t, we get

$$y - \beta = E (x - \alpha)^{\nu/\mu}, \dots\dots\dots\dots\dots\dots(11)$$

in which only the *ratio* ν/μ is involved.

As a particular case suppose that initially the inside volume is occupied by one pure gas and the outside by another, the initial pressures being unity. Then in (10)

$$\alpha = 0, \qquad \beta = 1, \qquad C = 1, \qquad D = -1;$$

we have

$$x = e^{-\mu t}, \qquad y = 1 - e^{-\nu t}, \dots\dots\dots\dots\dots(12)$$

and

$$x + y = 1 + e^{-\mu t} - e^{-\nu t} \dots\dots\dots\dots\dots(13)$$

gives the total internal pressure. When this is a maximum or minimum, $e^{(\mu - \nu)t} = \mu/\nu$, and the corresponding value is

$$x + y = 1 + \left(\frac{\mu}{\nu}\right)^{-\frac{\mu}{\mu - \nu}} \left\{1 - \frac{\mu}{\nu}\right\}. \dots\dots\dots\dots\dots(14)$$

Thus in the case of hydrogen escaping into oxygen, $\mu/\nu = 4$, and

$$x + y = 1 - 3 \times 4^{-\frac{4}{3}} = \cdot528,$$

the minimum being about half the initial pressure*.

Returning now to the separation of gases by diffusion into a vacuum, let us suppose that the difference between the gases is small, so that $(\nu - \mu)/\mu = \kappa$, a small quantity, and that at each operation one-half the total volume of the mixture is allowed to pass. In this case (8) becomes

$$\tfrac{1}{2} = \frac{X}{X + Y} r^{\frac{1}{\kappa}} + \frac{Y}{X + Y} r^{\frac{1+\kappa}{\kappa}} = r^{\frac{1}{\kappa}} \text{ nearly;}$$

so that

$$r = \frac{y/Y}{x/X} = (\tfrac{1}{2})^{\kappa}. \dots\dots\dots\dots\dots(15)$$

This gives the effect of the operation in question upon the composition of the residual gas. If s denote the corresponding symbol for the transmitted gas, we have

$$s = \frac{(Y - y)/Y}{(X - x)/X} = \frac{1 - y/Y}{1 - x/X} = \frac{1 - rx/X}{1 - x/X} = 1 + \frac{(1 - r) x/X}{1 - x/X} = 2 - r$$

approximately, since r is nearly equal to unity. Accordingly

$$\frac{1}{s} = \frac{1}{2 - r} = r \text{ nearly,}$$

so that approximately s and r are reciprocal operations. For example, if

---

* The most striking effects of this kind are when nitrous oxide, or dry ammonia gas, diffuse into the air through indiarubber. I have observed suctions amounting respectively to 53 and 64 centimetres of mercury.

starting with any proportions we collect the transmitted half, and submit it to another operation of the same sort, retaining the half not transmitted, the final composition corresponding to the operations sr is the same (approximately) as the composition with which we started, and the same also as would be obtained by operations taken in the reverse order, represented by rs. A complete scheme* on these lines is indicated in the diagram.

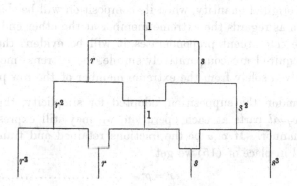

Representing the initial condition by unity, we may represent the result of the first operation by
$$\tfrac{1}{2}r + \tfrac{1}{2}s, \ \text{or} \ \tfrac{1}{2}(r + s),$$
in which the numerical coefficient gives the quantity of gas whose character is specified by the literal symbols. The second set of operations gives in the first instance
$$\tfrac{1}{4}r^2 + \tfrac{1}{4}sr + \tfrac{1}{4}rs + \tfrac{1}{4}s^2,$$
or, after admixture of the second and third terms (which are of the same quality),
$$\tfrac{1}{4}(r^2 + 2rs + s^2) = \left(\frac{r+s}{2}\right)^2.$$

In like manner the result of the third set of operations may be represented by $\left(\dfrac{r+s}{2}\right)^3$, and (as may be formally proved by "induction") of n sets of operations by
$$\left(\frac{r+s}{2}\right)^n. \quad\dots\dots\dots\dots\dots\dots\dots\dots\dots(16)$$

When we take account of the reciprocal character of r and s, this may be written
$$\frac{1}{2^n}\left\{ r^n + nr^{n-2} + \frac{n(n-1)}{1.2}r^{n-4} + \dots + nr^{-n+2} + r^{-n} \right\}, \quad\dots\dots\dots(17)$$
the number of parts into which the original quantity of gas is divided being

* It differs, however, from that followed by Prof. Ramsay in his recent researches (*Proc. Roy. Soc.* Vol. LX. p. 216, 1896).

$n + 1$. If n is even, the largest part, corresponding to the middle term, has the original composition*.

It is to be observed, however, that so far as the extreme concentration of the less diffusive constituent is concerned these complex operations are entirely unnecessary. The same result, represented by $(\frac{1}{2})^n r^n$ will be reached *at a single operation* by continuing the diffusion until the residue is reduced to $(\frac{1}{2})^n$ of the original quantity, when its composition will be that denoted by r^n. And even as regards the extreme member at the other end in which the more diffusive constituent preponderates, it will be evident that the operations really required are comparatively simple, the extreme member in each row being derived solely from the extreme member of the row preceding†.

If we abandon the supposition, adopted for simplicity, that the gas is divided into *equal* parts at each operation, we may still express the results in a similar manner. If ρ, σ be the fractions retained and transmitted, then $\rho + \sigma = 1$, and in place of (15) we get

$$r = \rho^k. \quad\dots\dots\dots\dots\dots\dots\dots\dots\dots(18)$$

The relation between r and s is

$$\rho r + \sigma s = 1; \quad\dots\dots\dots\dots\dots\dots\dots\dots(19)$$

and the various portions into which the gas is divided after n sets of operations are represented by the various terms of the expansion of

$$(\rho r + \sigma s)^n, \quad\dots\dots\dots\dots\dots\dots\dots\dots(20)$$

the Greek letters and the numerical coefficients giving the quantity of each portion, and the Roman letters giving the quality. But it must not be forgotten that this theory all along supposes the difference of diffusivities to be relatively small.

---

* There is here a formal analogy with the problem of determining the probability of a given combination of heads and tails in a set of n tosses of a coin; and the result of supposing n infinite may be traced as in the theory of errors.

† Possibly a better plan for the concentration of the lighter constituent would be diffusion along a column of easily absorbable gas, *e.g.* CO_2. The gas which arrives first at the remote end is infinitely rich in this constituent. [1902. See *Phil. Mag.* i. p. 105, 1901.]

224.

THE THEORY OF SOLUTIONS.

[*Nature*, LV. pp. 253, 254, 1897.]

As some recent *viva voce* remarks of mine have received an interpretation more wide than I intended, I shall be glad to be allowed to explain that when (now several years ago) I became acquainted with the work of van t' Hoff I was soon convinced of the great importance of the advances due to him and his followers. The subject has been prejudiced by a good deal of careless phraseology, and this is probably the reason why some distinguished physicists and chemists have refused their adhesion. It must be admitted, further, that the arguments of van t' Hoff are often insufficiently set out, and are accordingly difficult to follow. Perhaps this remark applies especially to his treatment of the central theorem, viz. the identification of the osmotic pressure of a dissolved gas with the pressure which would be exercised by the gas alone if it occupied the same total volume in the absence of the solvent. From this follows the formal extension of Avogadro's law to the osmotic pressure of dissolved gases, and thence by a natural hypothesis to the osmotic pressure of other dissolved substances, even although they may not be capable of existing in the gaseous condition. If I suggest a somewhat modified treatment, it is not that I see any unsoundness in van t' Hoff's argument, but because of the importance of regarding a matter of this kind from various points of view.

Let us suppose that we have to deal with an involatile liquid solvent, and that its volume, at the constant temperature of our operations, is unaltered by the dissolved gas—a question to which we shall return. We start with a volume v of gas under pressure p_0, and with a volume V of liquid just sufficient to dissolve the gas under the same pressure, and we propose to find what amount of work (positive or negative) must be done in order to bring the gas into solution reversibly. If we bring the gas at pressure p_0 into contact with the liquid, solution takes place irreversibly, but this difficulty may be overcome by a method which I employed for a similar purpose many years ago*. We begin by expanding the gas until its rarity is such that no sensible dissipation of energy occurs when contact with the liquid is established. The gas is then compressed and solution progresses under rising pressure until just as the gas disappears the pressure rises to p_0. The operations are to be conducted at constant temperature, and so slowly that the condition never deviates sensibly from that of equilibrium. The process is accordingly reversible.

In order to calculate the amount of work involved in accordance with the laws of Boyle and Henry, we may conveniently imagine the liquid and gas to be confined under a piston in a cylinder of unit cross-section. During the first stage contact is prevented by a partition inserted at the surface of the liquid. If the distance of the piston from this surface be x, we have initially $x = v$. At any stage of the expansion (x) the pressure p is given by $p = p_0 v / x$, and the work gained during the expansion is represented by

$$p_0 v \int_v^x \frac{dx}{x} = p_0 v \log \frac{x}{v},$$

x being a very large multiple of v. During the condensation, after the partition has been removed, the pressure upon the piston in a given position x is less than before. For the gas which was previously confined to the space x is now partly in solution. If s denote the solubility, the available volume is practically increased in the ratio $x : x + sV$, so that the pressure in position x is now given by

$$p = p_0 v / (x + sV),$$

and the work required to be done during the compression is

$$p_0 v \int_0^x \frac{dx}{x + sV} = p_0 v \log \frac{x + sV}{sV}.$$

On the whole the work lost during the double operation is

$$p_0 v \left\{ \log \frac{x + sV}{x} + \log \frac{v}{sV} \right\},$$

and of this the first part must be omitted, as x is indefinitely great. As regards the second part, we see that it is zero, since by supposition the quantity of liquid is such as to be just capable of dissolving the gas, so that $sV = v$. The conclusion then is that, upon the whole, there is no gain or loss of work in passing reversibly from the initial to the final state of things.

The remainder of the cycle, in which the gas is removed from solution and restored to its original state, may now be effected by the osmotic process

* "On the Work that may be gained during the mixing of Gases," *Phil. Mag.* Vol. XLIX. p. 311, 1875. [Vol. I. p. 242.]

of van t' Hoff*. For this purpose one "semi-permeable membrane," permeable to gas but not to liquid, is introduced just under the piston which rests at the surface of the liquid. A second, permeable to liquid but not to gas, is substituted as a piston for the bottom of the cylinder, and may be backed upon its lower side by pure solvent. By suitable proportional motions of the two pistons, the upper one being raised through the space v, and the lower through the space V, the gas may be expelled, the pressure of the gas retaining the constant value p_0, and the liquid (which has not yet been expelled) retaining a constant strength, and therefore a constant osmotic pressure P. When the expulsion is complete, the work done upon the lower piston is PV, and that recovered from the gas is p_0v, upon the whole $PV - p_0v$. Since this process, as well as the first, is reversible, and since the whole cycle has been conducted at constant temperature, it follows from the *second* law of thermo-dynamics that no work is lost or gained during the cycle, or that $PV = p_0v$. The osmotic pressure P is thus determined, and it is evident that its value is that of the pressure which the gas, as a gas, would exert in space V.

The objection may perhaps be taken that the assumption of unaltered volume of the liquid as the gas dissolves in it unduly limits the application of the argument. It is true that when finite pressures are in question, an expansion (or contraction) of the liquid would complicate the results; but we are concerned only, or at any rate primarily, with the osmotic pressure of *dilute* solutions. In this case the complications spoken of relate only to the second order of small quantities, and in our theory are accordingly to be dismissed.

* *Phil. Mag.* Vol. xxvi. p. 88, 1888.

225.

OBSERVATIONS ON THE OXIDATION OF NITROGEN GAS.

[*Chemical Society's Journal*, 71, pp. 181—186, 1897.]

THE observations here described were made in connexion with the isolation of argon by removal of the nitrogen from air, but they may, perhaps, possess a wider interest as throwing light upon the behaviour of nitrogen itself.

According to Davy*, the dissolved nitrogen of water is oxidised to nitrous (or nitric) acid when the liquid is submitted to electrolysis. "To make the experiment in as refined a form as possible, I procured two hollow cones of pure gold containing about 25 grains of water each, they were filled with distilled water connected together by a moistened piece of amianthus which had been used in the former experiments, and exposed to the action of a voltaic battery of 100 pairs. . . . In 10 minutes the water in the negative tube had gained the power of giving a slight blue tint to litmus paper: and the water in the positive tube rendered it red. The process was continued for 14 hours; the acid increased in quantity during the whole time, and the water became at last very sour to the taste. . . . The acid, as far as its properties were examined, agreed with pure nitrous acid having an excess of nitrous gas" (p. 6).

Further (p. 10), "I had never made any experiments, in which acid matter having the properties of nitrous acid was not produced, and the longer the operation the greater was the quantity which appeared. . . . It was natural to account for both these appearances, from the combination of nascent oxygene and hydrogene respectively; with the nitrogene of the common air dissolved in the water."

Davy was confirmed in his conclusion by experiments in which the

* *Phil. Trans.* 1807, p. 1.

electrolytic vessels were placed in a vacuum or in an atmosphere of hydrogen. There was then little or no reddening of the litmus, even after prolonged action of the battery.

If nitrogen could be oxidised in this way, the process would be a convenient one for the isolation of argon, for it could be worked on a large scale and be made self-acting. But it did not appear at all probable that nitrogen could take a direct part in the electrolysis. In that case, its oxidation would be a secondary action, due, perhaps, to the formation of peroxide of hydrogen. This consideration led me to try the effect of peroxide of sodium on dissolved nitrogen, but without success. The nitrogen dissolved in 1250 c.c. of tap water and liberated by boiling, was found to be 19·1 c.c., and it was not diminished by a previous addition of peroxide of sodium, with or without acid. Having failed in this direction, I endeavoured to repeat Davy's experiment nearly in its original form. The water was contained in two cavities bored in a block of paraffin, and connected by a wick of asbestos which had been previously ignited. By means of platinum terminals connected with a secondary battery, a potential difference of 100 volts was maintained between the cups. The whole was covered by a glass shade, to exclude any saline matter that might be introduced from the atmosphere. But, under these conditions, no difference in the behaviour of litmus when moistened with water from the two cups could be detected, even after 14 days' exposure to the 100 volts. When, however, the cover was removed, the litmus responded markedly after a day or two.

The failure of several attempts of this kind lead me to doubt the correctness of Davy's view, that the dissolved nitrogen of water is oxidised during electrolysis. At any rate, the action is so slow that the process holds out no promise of usefulness on a large scale.

In the oxidation of nitrogen by gaseous oxygen under the action of electric discharge, a question arises as to the influence of pressure. If the mass absorbed were proportional to pressure, or the volume independent of pressure, the electrical energy expended being the same, it might be desirable to work with highly condensed gases, in spite of the serious difficulties that must necessarily be encountered. That pressure would be favourable seems probable *a priori*, and is suggested by certain observations of Dr Frankland. My own early experiments pointed also in the same direction. A suitable mixture of nitrogen and oxygen, standing in an inverted test-tube over alkali, was sparked from a Ruhmkorff coil actuated by five Grove cells; when the total pressure was about three atmospheres, the mass absorbed was about three times that absorbed in the same time at the ordinary pressure.

This result made it necessary to proceed to operations upon a larger scale with the alternate current discharge. Experiments were first tried in

a small vessel (of 250 c.c.), which would be more easily capable of withstand-
ing internal pressure than a larger one.   In order to protect the glass, which
at the top was almost in contact with the electric flame, and to promote
absorption of the combined nitrogen, the alkali was used in the form of
a fountain, which struck the glass immediately over the flame, and washed
the whole of the internal surface*.   But, to my surprise, preliminary trials,
conducted at atmospheric pressure, showed that this apparatus was not
effective.   The rates of absorption were about 1600 c.c. per hour, the runs
themselves being for half-an-hour.   About double this rate had already been
obtained with the same electrical appliances and with stationary alkali.
Care having been taken that the quality of the mixture within the working
vessel was maintained throughout the run, the smaller efficiency could only
be connected with the confined space.

   As to the reason why a confined space should be unfavourable, it is
difficult to give a decided opinion.   Other things being the same, the surface
presented by the alkali will be diminished in a smaller vessel, and the absorption of the combined nitrogen may consequently be less rapid.   But it is
difficult to accept this explanation, in view of the favourable conditions
secured by the use of a fountain.   The gases, as they rise from the flame,
impinge directly upon the alkali, which is itself in rapid motion over the
whole internal surface.   It would almost seem as if the combined nitrogen,
as it leaves the flame, is not yet ready for absorption, and only becomes so
after the lapse of a certain time.   However this may be, the efficiency is in
practice improved by largely increasing the capacity of the working vessel.
A larger bottle, of 370 c.c. capacity, allowed a rate of 2000 c.c. per hour.
A flask of still greater capacity gave 3300 c.c. per hour, whilst with a larger
globe capable of holding $4\frac{1}{2}$ litres, a rate of 6800 c.c. per hour was obtained.
These experiments were all made at atmospheric pressure with a fountain of
alkali and with the electric flame in as nearly as possible a constant condition.
In the case of the smallest vessel, it was thought that the separation of the
platinum terminals may have been insufficient for the best effect, but the
loss due to this cause must have been relatively small.   Electrical instruments
connected with the primary circuit of the Ruhmkorff gave readings of 10
ampères and 41 volts.

   When the comparatively small vessel of 370 c.c. was used at a pressure
of about one additional atmosphere, the volume absorbed was about the same
as in the experiments with the same vessel at atmospheric pressure, thus
indicating a double efficiency.   This increased efficiency is, however, of no
practical importance, inasmuch as a higher efficiency still can be obtained at
atmospheric pressure by use of a larger vessel.   In order to clear up the
question, it was necessary to compare the efficiencies in a *large* vessel at

---

* Rayleigh and Ramsay, *Phil. Trans.* 1895, p. 217.   [Vol. IV. p. 162.]

different pressures, an operation involving considerable difficulty and even danger.

For this purpose, a glass globe, nearly spherical in form, and having a capacity of about 7 litres, was employed. The extra pressure was nearly an atmosphere and was obtained by gravity, the feed and return pipes for the alkaline fountain, as well as the pipe for the supply of water to the gas-holder, being carried to a higher level than that at which the rest of the apparatus stood. The rate of absorption (reduced to atmospheric pressure) was 6880 c.c. per hour. Experiments conducted at atmospheric pressure gave as a mean 6600 c.c.

In order to examine still further the influence of pressure, two experiments were tried under a total pressure of *half* an atmosphere. The reduced numbers were 5600, 5700 c.c. per hour. From these results, it would appear that the influence of pressure is slightly favourable. But, in comparing the results for one atmosphere and for half an atmosphere, it should be remembered that, in the latter case, aqueous vapour is responsible for a sensible part of the total pressure. At any rate, the results are much more nearly independent of pressure than proportional to pressure; so that the cases of large and small vessels are sharply distinguished, pressure appearing to be advantageous only where the space is too confined to admit of the best efficiency at a given pressure being reached.

Not sorry to be relieved from the obligation of designing a large scale apparatus to be worked at a high pressure, such as 20 or 100 atmospheres, I reverted to the ordinary pressure, and sought to obtain a high rate of absorption by employing a powerful electric flame contained in a large vessel whose walls were washed internally by an alkaline fountain. The electrical arrangements have been the subject of much consideration, and require to be different from what would naturally be expected. Since the voltage on the final platinums during discharge is only from 1600 to 2000, as measured by one of Lord Kelvin's instruments, it might be supposed that a commercial transformer, transforming from 100 volts to 2400 volts, would suffice for the purpose. When, however, the attempt is made, it is soon discovered that such an arrangement is quite unmanageable. When, after some difficulty, the arc is started, it is found that the electrical conditions are unstable. Things may go well for a time, but after perhaps some hours the current will rise and the platinums will become overheated and may melt. Even when two transformers were employed, so connected as to give on open secondary circuit nearly 4800 volts, the conditions were not steady enough for convenient practice. The transformer used in the experiments about to be described is by Messrs Swinburne, and is insulated with oil. On open secondary, the voltage is nearly 8000*, but it falls to 2000 or less when the

* Probably 6000 would have sufficed.

discharge is running. Even with this transformer, it was necessary to include in its primary (thick wire) circuit a self-induction coil, provided with a core consisting of a bundle of iron wires, and adjustable in position. As finally used, the adjustment was such that the electromotive force actually operative on the primary was only about 30 volts out of the 100 volts available at the mains of the public supply. This reduction of voltage does not, at any rate from a theoretical point of view, involve any loss of economy, and some such reduction seems to be essential to steadiness. Under these conditions, the current taken amounted to 40 ampères.

It is scarcely necessary to say that the watts actually delivered to the primary circuit of the transformer are less than the number (1200) derived by multiplication of volts and ampères. From some experiments made under similar conditions*, I have found that the factor of reduction—the cosine of the angle of lag—is about two-thirds, so that the watts taken in the above arrangement are about 800, representing a little more than a horse-power.

The working vessel, A, was of glass, spherical in form, and of 50 litres capacity. The neck was placed downwards, and was closed by a large rubber stopper, through which five tubes of glass penetrated. Two tubes of substantial construction carried the electrodes, B, C, arranged much as in a former apparatus†; two more, F and E, were required for the supply tube of the fountain and for the drain of liquid, whilst the fifth, D, was for the supply of gas. The external drowning of the vessel, formerly necessary, was now dispensed with; but a suitable cooling arrangement for the alkali (something like the worm of a condenser) had to be provided to obviate excessive accumulation of heat.

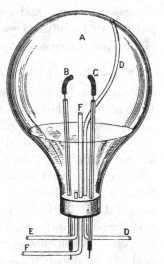

As the solution of alkali circulated entirely in the closed apparatus, it could lose none of its dissolved argon. It was maintained in circulation by a small centrifugal pump constructed of iron and driven from an electric motor.

The mixed gases (about 11 parts of oxygen to 9 parts of air) were supplied from a large gas-holder; but an auxiliary holder was also necessary in order to observe the rate of absorption. When the rate became unsatisfactory, the mixed gas in the working vessel was analysed and the necessary rectification effected.

In the earlier stages of the operation, the rate of absorption was about 21 litres per hour, and this, by proper attention, could be maintained without much loss until the accumulation of argon began to tell. If we take 20 litres as corresponding to 800 watts, we have 25 c.c. per watt-hour, an efficiency not very different from that found in operations on a much smaller scale.

The present apparatus works about three times as fast as the former one, in which the vessel was smaller and the alkali stationary. It is also more interesting to watch, as the electric flame is fully exposed to view. On the other hand, it is more complicated, owing to the use of a circulating pump, and probably requires closer attention. A failure of the fountain whilst the flame was established would doubtless soon lead to a disaster.

I have been efficiently aided throughout by Mr Gordon, who has not only fitted the apparatus, but has devised many of the contrivances necessary to meet the ever-recurring difficulties which must be expected in work of this character.

226.

ON THE PASSAGE OF ELECTRIC WAVES THROUGH TUBES, OR THE VIBRATIONS OF DIELECTRIC CYLINDERS.

[*Philosophical Magazine*, XLIII. pp. 125—132, 1897.]

General Analytical Investigation.

THE problem here proposed bears affinity to that of the vibrations of a cylindrical solid treated by Pochhammer* and others, but when the bounding conductor is regarded as perfect it is so much simpler in its conditions as to justify a separate treatment. Some particular cases of it have already been considered by Prof. J. J. Thomson†. The cylinder is supposed to be infinitely long and of arbitrary section; and the vibrations to be investigated are assumed to be periodic with regard both to the time (t) and to the coordinate (z) measured parallel to the axis of the cylinder, *i.e.*, to be proportional to $e^{i(mz+pt)}$.

By Maxwell's Theory, the components of electromotive intensity in the dielectric (P, Q, R) and those of magnetic induction (a, b, c) all satisfy equations such as,

$$\frac{d^2R}{dx^2} + \frac{d^2R}{dy^2} + \frac{d^2R}{dz^2} = \frac{1}{V^2}\frac{d^2R}{dt^2}, \quad \dots\dots\dots\dots\dots\dots(1)$$

V being the velocity of light; or since by supposition

$$d^2R/dz^2 = -m^2R, \qquad d^2R/dt^2 = -p^2R,$$

$$d^2R/dx^2 + d^2R/dy^2 + k^2R = 0, \quad \dots\dots\dots\dots\dots\dots(2)$$

where

$$k^2 = p^2/V^2 - m^2. \quad \dots\dots\dots\dots\dots\dots(3)\ddagger$$

---

* *Crelle*, Vol. XXXI. 1876.

† *Recent Researches in Electricity and Magnetism*, 1893, § 300.

‡ The k^2 of Prof. J. J. Thomson (*loc. cit.* § 262) is the negative of that here chosen for convenience.

The relations between P, Q, R and a, b, c are expressed as usual by

$$\frac{da}{dt} = \frac{dQ}{dz} - \frac{dR}{dy}, \quad \dots\dots\dots\dots\dots\dots\dots\dots(4)$$

and two similar equations; while

$$\frac{da}{dx} + \frac{db}{dy} + \frac{dc}{dz} = 0, \qquad \frac{d\Gamma}{dx} + \frac{dQ}{dy} + \frac{dR}{dz} = 0. \dots\dots\dots\dots(5, 6)$$

The conditions to be satisfied at the boundary are that the components of electromotive intensity parallel to the surface shall vanish. Accordingly

$$R = 0, \qquad P\frac{dx}{ds} + Q\frac{dy}{ds} = 0, \quad \dots\dots\dots\dots\dots(7, 8)$$

dx/ds, dy/ds being the cosines of the angles which the tangent (ds) at any point of the section makes with the axes of x and y.

Equations (2) and (7) are met with in various two-dimensional problems of mathematical physics. They are the equations which determine the free transverse vibrations of a stretched membrane whose fixed boundary coincides with that of the section of the cylinder. The quantity k^2 is limited to certain definite values, k_1^2, k_2^2, ..., and to each of these corresponds a certain normal function. In this way the possible forms of R are determined. A value of R which is zero throughout is also possible.

With respect to P and Q we may write

$$P = \frac{d\phi}{dx} + \frac{d\psi}{dy}, \qquad Q = \frac{d\phi}{dy} - \frac{d\psi}{dx}; \quad \dots\dots\dots\dots(9, 10)$$

where ϕ and ψ are certain functions, of which the former is given by

$$\nabla^2\phi = \frac{dP}{dx} + \frac{dQ}{dy} = -\frac{dR}{dz} = -imR. \quad \dots\dots\dots\dots(11)$$

There are thus two distinct classes of solutions; the first dependent upon ϕ, in which R has a finite value, while $\psi = 0$; the second dependent upon ψ, in which R and ϕ vanish.

For a vibration of the first class we have

$$P = d\phi/dz, \qquad Q = d\phi/dy, \quad \dots\dots\dots\dots\dots(12)$$

and

$$(\nabla^2 + k^2)\phi = 0. \dots\dots\dots\dots\dots\dots(13)$$

Accordingly by (11)

$$\phi = \frac{im}{k^2}R, \quad \dots\dots\dots\dots\dots\dots\dots(14)$$

and

$$P = \frac{im}{k^2}\frac{dR}{dx}, \qquad Q = \frac{im}{k^2}\frac{dR}{dy}, \quad \dots\dots\dots\dots\dots(15)$$

by which P and Q are expressed in terms of R supposed already known.

The boundary condition (7) is satisfied by the value ascribed to R, and the same value suffices also to secure the fulfilment of (8), inasmuch as

$$P\frac{dx}{ds} + Q\frac{dy}{ds} = \frac{im}{k^2}\frac{dR}{ds} = 0.$$

The functions P, Q, R being now known, we may express a, b, c. From (4)

$$\frac{da}{dt} = ipa = imQ - \frac{dR}{dy} = -\frac{m^2 + k^2}{k^2}\frac{dR}{dy};$$

so that

$$a = -\frac{m^2 + k^2}{ipk^2}\frac{dR}{dy}, \qquad b = \frac{m^2 + k^2}{ipk^2}\frac{dR}{dx}, \qquad c = 0. \quad \ldots\ldots(16)$$

In vibrations of the second class $R = 0$ throughout, so that (2) and (7) are satisfied, while k^2 is still at disposal. In this case

$$P = d\psi/dy, \qquad Q = -d\psi/dx, \quad \ldots\ldots\ldots\ldots\ldots\ldots(17)$$

and

$$(\nabla^2 + k^2)\,\psi = 0. \quad \ldots\ldots\ldots\ldots\ldots\ldots\ldots\ldots\ldots(18)$$

By the third of equations (4)

$$\frac{dc}{dt} = ipc = \frac{dP}{dy} - \frac{dQ}{dx} = \nabla^2\psi = -k^2\psi\,;$$

so that $\psi = -ipc/k^2$, and

$$P = -\frac{ip}{k^2}\frac{dc}{dy}, \qquad Q = \frac{ip}{k^2}\frac{dc}{dx}, \qquad R = 0. \quad \ldots\ldots\ldots(19)$$

Also by (4)

$$a = \frac{im}{k^2}\frac{dc}{dx}, \qquad b = \frac{im}{k^2}\frac{dc}{dy}. \quad \ldots\ldots\ldots\ldots\ldots\ldots\ldots(20)$$

Thus all the functions are expressed by means of c, which itself satisfies

$$(\nabla^2 + k^2)\,c = 0. \quad \ldots\ldots\ldots\ldots\ldots\ldots\ldots\ldots\ldots\ldots(21)$$

We have still to consider the second boundary condition (8). This takes the form

$$\frac{dc}{dy}\frac{dx}{ds} - \frac{dc}{dx}\frac{dy}{ds} = 0,$$

requiring that dc/dn, the variation of c along the normal to the boundary at any point, shall vanish. By (21) and the boundary condition

$$dc/dn = 0, \quad \ldots\ldots\ldots\ldots\ldots\ldots\ldots\ldots\ldots\ldots(22)$$

the form of c is determined, as well as the admissible values of k^2. The problem as regards c is thus the same as for the two-dimensional vibrations of gas within a cylinder which is bounded by rigid walls coincident with the conductor, or for the vibrations of a liquid under gravity in a vessel of the same form*.

All the values of k determined by (2) and (7), or by (21) and (22), are real,

* Phil. Mag. Vol. I. p. 272 (1876). [Vol. I. p. 265.]

but the reality of k still leaves it open whether m in (3) shall be real or imaginary. If we are dealing with free stationary vibrations m is given and real, from which it follows that p is also real. But if it be p that is given, m^2 may be either positive or negative. In the former case the motion is really periodic with respect to z; but in the latter z enters in the forms $e^{m'z}$, $e^{-m'z}$, and the motion becomes infinite when $z = +\infty$, or when $z = -\infty$, or in both cases. If the smallest of the possible values of k^2 exceeds p^2/V^2, m is necessarily imaginary, that is to say no periodic waves of the frequency in question can be propagated along the cylinder.

Rectangular Section.

The simplest case to which these formulæ can be applied is when the section of the cylinder is rectangular, bounded, we may suppose, by the lines $x - 0$, $x = \alpha$, $y = 0$, $y = \beta$.

As for the vibrations of stretched membranes*, the appropriate value of R applicable to solutions of the first class is

$$R - e^{i(mz+pt)} \sin(\mu\pi x/\alpha)\sin(\nu\pi y/\beta); \quad \ldots\ldots\ldots\ldots(23)$$

from which the remaining functions are deduced so easily by (15), (16) that it is hardly necessary to write down the expressions. In (23) μ and ν are integers, and by (13)

$$k^2 = \pi^2\left(\frac{\mu^2}{\alpha^2} + \frac{\nu^2}{\beta^2}\right), \quad \ldots\ldots\ldots\ldots(24)$$

whence
$$m^2 = p^2/V^2 - \pi^2\left(\frac{\mu^2}{\alpha^2} + \frac{\nu^2}{\beta^2}\right). \quad \ldots\ldots\ldots\ldots(25)$$

The lowest frequency which allows of the propagation of periodic waves along the cylinder is given by

$$\frac{p^2}{V^2} = \frac{\pi^2}{\alpha^2} + \frac{\pi^2}{\beta^2}. \quad \ldots\ldots\ldots\ldots(26)$$

If the actual frequency of a vibration having its origin at any part of the cylinder be much less than the above, the resulting disturbance is practically limited to a neighbouring finite length of the cylinder.

For vibrations of the second class we have

$$c = e^{i(mz+pt)}\cos(\mu\pi x/\alpha)\cos(\nu\pi y/\beta), \quad \ldots\ldots\ldots\ldots(27)$$

the remaining functions being at once deducible by means of (19), (20). The satisfaction of (22) requires that here again μ, ν be integers, and (21) gives

$$k^2 = \pi^2\left(\frac{\mu^2}{\alpha^2} + \frac{\nu^2}{\beta^2}\right), \quad \ldots\ldots\ldots\ldots(28)$$

identical with (24).

---

* Theory of Sound, § 195.

If $\alpha > \beta$, the smallest value of k corresponds to $\mu = 1$, $\nu = 0$. When $\nu = 0$, we have $k = \mu\pi/\alpha$, and if the factor $e^{i\,(mz+pt)}$ be omitted,

$$a = -\frac{im}{k}\sin kx, \qquad b = 0, \qquad c = \cos kx, \qquad \dots\dots\dots\dots(29)$$

$$P = 0, \qquad Q = -\frac{ip}{k}\sin kx, \qquad R = 0; \qquad \dots\dots\dots\dots(30)$$

a solution independent of the value of β. There is no solution derivable from $\mu = 0$, $\nu = 0$, $k = 0^{*}$.

Circular Section.

For the vibrations of the first class we have as the solution of (2) by means of Bessel's functions,

$$R = J_n(kr)\cos n\theta, \qquad\dots\dots\dots\dots\dots\dots(31)$$

n being an integer, and the factor $e^{i\,(mz+pt)}$ being dropped for the sake of brevity. In (31) an arbitrary multiplier and an arbitrary addition to θ are of course admissible. The value of k is limited to be one of those for which

$$J_n(kr') = 0 \dots\dots\dots\dots\dots\dots\dots\dots(32)$$

at the boundary where $r = r'$.

The expressions for P, Q, a, b, c in (15), (16) involve only dR/dx, dR/dy. For these we have

$$\frac{dR}{dx} = \frac{dR}{dr}\cos\theta - \frac{dR}{rd\theta}\sin\theta = kJ_n'(kr)\cos n\theta\cos\theta + \frac{n}{r}J_n(kr)\sin n\theta\sin\theta$$

$$= \tfrac{1}{2}k\cos(n-1)\theta\left\{J_n' + \frac{J_n}{kr}\right\} + \tfrac{1}{2}k\cos(n+1)\theta\left\{J_n' - \frac{J_n}{kr}\right\}$$

$$= \tfrac{1}{2}k\cos(n-1)\theta\, J_{n-1}(kr) - \tfrac{1}{2}k\cos(n+1)\theta\, J_{n+1}(kr), \qquad\dots\dots\dots(33)$$

according to known properties of these functions; and in like manner

$$\frac{dR}{dy} = \frac{dR}{dr}\sin\theta + \frac{dR}{rd\theta}\cos\theta$$

$$= -\tfrac{1}{2}k\sin(n-1)\theta\, J_{n-1}(kr) - \tfrac{1}{2}k\sin(n+1)\theta\, J_{n+1}(kr). \qquad\dots\dots(34)$$

These forms show directly that dR/dx, dR/dy satisfy the fundamental equation (2). They apply when n is equal to unity or any greater integer. When $n = 0$, we have

$$R = J_0(kr), \qquad\dots\dots\dots\dots\dots\dots(35)$$

$$\frac{dR}{dx} = -kJ_1(kr)\cos\theta, \qquad \frac{dR}{dy} = -kJ_1(kr)\sin\theta. \qquad\dots\dots\dots(36)$$

---

* For (18) would then become $\nabla^2\psi = 0$; and this, with the boundary condition $d\psi/dn = 0$, would require that P and Q, as well as R, vanish throughout.

The expressions for the electromotive intensity are somewhat simpler when the resolution is circumferential and radial:

circumf. component

$$= Q \cos \theta - P \sin \theta = \frac{im}{k^2} \frac{dR}{r d\theta} = -\frac{imn}{k^2 r} J_n(kr) \sin n\theta, \ldots \ldots \ldots (37)$$

radial component

$$= P \cos \theta + Q \sin \theta = \frac{im}{k^2} \frac{dR}{dr} = \frac{im}{k} J_n'(kr) \cos n\theta. \ldots \ldots \ldots (38)$$

If $n = 0$, the circumferential component vanishes.

Also for the magnetization

circ. comp. of magnetization

$$= b \cos \theta - a \sin \theta = \frac{m^2 + k^2}{ip k^2} \frac{dR}{dr} = \frac{m^2 + k^2}{ip k} J_n'(kr) \cos n\theta, \ldots \ldots (39)$$

rad. comp. of mag. $= a \cos \theta + b \sin \theta$

$$= -\frac{m^2 + k^2}{ip k^2} \frac{dR}{r d\theta} = \frac{n(m^2 + k^2)}{ip k^2 r} J_n(kr) \sin n\theta. \ldots \ldots (40)$$

The smallest value of k for vibrations of this class belongs to the series $n = 0$, and is such that $kr = 2\cdot 404$, r being the radius of the cylinder.

For the vibrations of the second class $R = 0$, and by (21),

$$c = J_n(kr) \cos n\theta, \ldots \ldots \ldots \ldots \ldots \ldots \ldots (41)$$

k being subject to the boundary condition

$$J_n'(kr') = 0. \ldots \ldots \ldots \ldots \ldots \ldots \ldots (42)$$

As in (33), (34),

$$\frac{dc}{dx} = \frac{dc}{dr} \cos \theta - \frac{dc}{r d\theta} \sin \theta$$

$$= \tfrac{1}{2} k \cos (n-1) \theta \, J_{n-1}(kr) - \tfrac{1}{2} k \cos (n+1) \theta \, J_{n+1}(kr), \ldots \ldots \ldots (43)$$

$$\frac{dc}{dy} = \frac{dc}{dr} \sin \theta + \frac{dc}{r d\theta} \cos \theta$$

$$= -\tfrac{1}{2} k \sin (n-1) \theta \, J_{n-1}(kr) - \tfrac{1}{2} k \sin (n+1) \theta \, J_{n+1}(kr), \ldots \ldots \ldots (44)$$

so that by (19), (20) all the functions are readily expressed.

When $n = 0$, we have

$$\frac{dc}{dx} = - k J_1(kr) \cos \theta, \qquad \frac{dc}{dy} = - k J_1(kr) \sin \theta. \ldots \ldots \ldots (45)$$

For the circumferential and radial components of magnetization we get circ. comp. of mag. $= b \cos \theta - a \sin \theta$

$$= \frac{im}{k^2} \frac{dc}{r d\theta} = - \frac{imn}{k^2 r} J_n (kr) \sin n\theta, \quad \dots\dots\dots\dots\dots(46)$$

rad. comp. of mag. $= a \cos \theta + b \sin \theta$

$$= \frac{im}{k^2} \frac{dc}{dr} = \frac{im}{k} J_n' (kr) \cos n\theta, \quad \dots\dots\dots\dots\dots(47)$$

corresponding to (37), (38) for vibrations of the first class.

In like manner equations analogous to (39), (40) now give the components of electromotive intensity. Thus

$$\text{circ. comp.} = Q \cos \theta - P \sin \theta = \frac{ip}{k^2} \frac{dc}{dr} = \frac{ip}{k} J_n' (kr) \cos n\theta, \quad \dots\dots\dots(48)$$

$$\text{rad. comp.} = P \cos \theta + Q \sin \theta = - \frac{ip}{k^2} \frac{dc}{r d\theta} = \frac{ipn}{k^2 r} J_n (kr) \sin n\theta. \quad \dots\dots(49)$$

The smallest value of k admissible for vibrations of the second class is of the series belonging to $n = 1$, and is such that $kr' = 1.841$, a smaller value than is admissible for any vibration of the first class. Accordingly no real wave of any kind can be propagated along the cylinder for which p/V is less than $1.841/r'$, where r' denotes the radius. The transition case is the two-dimensional vibration for which

$$c = e^{ipt} J_1 (1.841 r/r') \cos \theta, \qquad p = 1.841 V/r'. \quad \dots\dots\dots(50, 51)$$

227.

ON THE PASSAGE OF WAVES THROUGH APERTURES IN PLANE SCREENS, AND ALLIED PROBLEMS.

[*Philosophical Magazine*, XLIII. pp. 259—272, 1897.]

THE waves contemplated may be either aerial waves of condensation and rarefaction, or electrical waves propagated in a dielectric. Plane waves of simple type impinge upon a parallel screen. The screen is supposed to be infinitely thin, and to be perforated by some kind of aperture. Ultimately one or both dimensions of the aperture will be regarded as infinitely small in comparison with the wave-length (λ); and the method of investigation consists in adapting to the present purpose known solutions regarding the flow of incompressible fluids.

If ϕ be a velocity-potential satisfying

$$d^2\phi/dt^2 = V^2\nabla^2\phi, \quad\ldots\ldots\ldots\ldots\ldots\ldots\ldots\ldots\ldots(1)$$

where

$$\nabla^2 = d^2/dx^2 + d^2/dy^2 + d^2/dz^2,$$

the condition at the boundary may be (i) that $d\phi/dn = 0$, or (ii) that $\phi = 0$. The first applies directly to aerial vibrations impinging upon a fixed wall, and in this connexion has already been considered*.

If we assume that the vibration is everywhere proportional to e^{int}, (1) becomes

$$(\nabla^2 + k^2)\,\phi = 0, \quad\ldots\ldots\ldots\ldots\ldots\ldots\ldots\ldots(2)$$

where

$$k = n/V = 2\pi/\lambda. \quad\ldots\ldots\ldots\ldots\ldots\ldots\ldots\ldots(3)$$

It will conduce to brevity if we suppress the factor e^{int}. On this understanding the equation of waves travelling parallel to x in the positive direction, and accordingly incident upon the negative side of a screen situated at $x = 0$, is

$$\phi = e^{-ikx}. \quad\ldots\ldots\ldots\ldots\ldots\ldots\ldots\ldots\ldots(4)$$

* *Theory of Sound*, § 292.

When the solution is complete, the factor e^{int} is to be restored, and the imaginary part of the solution is to be rejected. The realized expression for the incident waves will therefore be

$$\phi = \cos(nt - kx). \quad \dots\dots\dots\dots\dots\dots\dots\dots(5)$$

Perforated Screen.—Boundary Condition $d\phi/dn = 0$.

If the screen be complete, the reflected waves under the above condition have the expression $\phi = e^{ikx}$.

Let us divide the actual solution into two parts χ and ψ, the first the solution which would obtain were the screen complete, the second the alteration required to take account of the aperture; and let us distinguish by the suffixes m and p the values applicable upon the negative (*minus*) and upon the positive side of the screen. In the present case we have

$$\chi_m = e^{-ikx} + e^{ikx}, \qquad \chi_p = 0. \quad \dots\dots\dots\dots\dots\dots(6)$$

This χ-solution makes $d\chi_m/dn = 0$, $d\chi_p/dn = 0$ over the whole plane $x = 0$, and over the same plane $\chi_m = 2$, $\chi_p = 0$.

For the supplementary solution, distinguished in like manner upon the two sides, we have

$$\psi_m = \iint \Psi_m \frac{e^{-ikr}}{r}\, dS, \qquad \psi_p = \iint \Psi_p \frac{e^{-ikr}}{r}\, dS. \quad \dots\dots\dots\dots(7)$$

where r denotes the distance of the point at which ψ is to be estimated from the element dS of the aperture, and the integration is extended over the whole of the area of aperture. Whatever functions of position Ψ_m, Ψ_p may be, these values on the two sides satisfy (2), and (as is evident from symmetry) they make $d\psi_m/dn$, $d\psi_p/dn$ vanish over the wall, viz. the unperforated part of the screen; so that the required condition over the wall for the complete solution $(\chi + \psi)$ is already satisfied. It remains to consider the further conditions that ϕ and $d\phi/dx$ shall be continuous across the aperture.

These conditions require that on the aperture

$$2 + \psi_m = \psi_p, \qquad d\psi_m/dx = d\psi_p/dx. \quad \dots\dots\dots\dots(8)^*$$

The second is satisfied if $\Psi_p = -\Psi_m$; so that

$$\psi_m = \iint \Psi_m \frac{e^{-ikr}}{r}\, dS, \qquad \psi_p = -\iint \Psi_m \frac{e^{-ikr}}{r}\, dS, \quad \dots\dots\dots\dots(9)$$

making the values of ψ_m and ψ_p equal and opposite at all corresponding points, viz. points which are images of one another in the plane $x = 0$. In

---

* The use of dx implies that the variation is in a fixed direction, while dn may be supposed to be drawn outwards from the screen in both cases.

order further to satisfy the first condition it suffices that over the area of aperture

$$\psi_m = -1, \qquad \psi_p = 1, \quad \ldots\ldots\ldots\ldots\ldots\ldots\ldots\ldots(10)$$

and the remainder of the problem consists in so determining Ψ_m that this shall be the case.

In this part of the problem we limit ourselves to the supposition that all the dimensions of the aperture are small in comparison with λ. For points at a distance from the aperture e^{-ikr}/r may then be removed from under the sign of integration, so that (9) becomes

$$\psi_m = \frac{e^{-ikr}}{r} \iint \Psi_m dS, \qquad \psi_p = -\frac{e^{-ikr}}{r} \iint \Psi_m dS. \quad \ldots\ldots\ldots\ldots(11)$$

The significance of $\iint \Psi_m dS$ is readily understood from an electrical interpretation. For in its application to a point, itself situated upon the area of aperture, e^{-ikr} in (9) may be identified with unity, so that ψ_m is the potential of a distribution of density Ψ_m on S. But by (10) this potential must have the constant value -1; so that $-\iint \Psi_m dS$, or $\iint \Psi_p dS$, represents the electrical *capacity* of a conducting disk having the size and shape of the aperture, and situated at a distance from all other electrified bodies. If we denote this by M, the solution applicable to points at a distance from the aperture may be written

$$\psi_m = -M \frac{e^{-ikr}}{r}, \qquad \psi_p = M \frac{e^{-ikr}}{r}. \quad \ldots\ldots\ldots\ldots\ldots(12)$$

To these are to be added the values of χ in (6). The realized solutions are accordingly

$$\phi_m = 2 \cos nt \cos kx - M \frac{\cos(nt - kr)}{r}, \quad \ldots\ldots\ldots\ldots(13)\cdot$$

$$\phi_p = M \frac{\cos(nt - kr)}{r}. \quad \ldots\ldots\ldots\ldots\ldots\ldots\ldots(14)$$

The value of M may be expressed* for an ellipse of semi-major axis a and eccentricity e. We have

$$M = \frac{a}{F(e)}, \quad \ldots\ldots\ldots\ldots\ldots\ldots\ldots\ldots(15)$$

F being the symbol of the complete elliptic function of the first kind. When $e = 0$, $F(e) = \frac{1}{2}\pi$; so that for a circle $M = 2a/\pi$.

It should be remarked that Ψ in (9) is closely connected with the normal velocity at dS. In general,

$$\frac{d\psi}{dx} = \iint \Psi \frac{d}{dx}\left(\frac{e^{-ikr}}{r}\right) dS. \quad \ldots\ldots\ldots\ldots(16)$$

---

* *Theory of Sound*, §§ 292, 306, where is given a discussion of the effect of ellipticity when area is given.

At a point (x) infinitely close to the surface, only the neighbouring elements contribute to the integral, and the factor e^{-ikr} may be omitted. Thus

$$\frac{d\psi}{dx} = -\iint \Psi \frac{x}{r^3} dS = -2\pi x \int_x^\infty \Psi \frac{rdr}{r^3} = -2\pi\Psi;$$

or

$$\Psi = -\frac{1}{2\pi} \frac{d\psi}{dn}, \quad\dots\dots\dots\dots\dots\dots\dots(17)$$

$d\psi/dn$ being the normal velocity at the point of the surface in question.

Boundary Condition $\phi = 0$.

We will now suppose that the condition to be satisfied on the walls is $\phi = 0$, although this case has no simple application to aerial vibrations. Using a similar notation to that previously employed, we have as the expression for the principal solution

$$\chi_m = e^{-ikx} - e^{ikx}, \qquad \chi_p = 0, \quad\dots\dots\dots\dots\dots\dots(18)$$

giving over the whole plane $(x = 0)$, $\chi_m = 0$, $\chi_p = 0$, $d\chi_m/dx = -2ik$, $d\chi_p/dx = 0$.

The supplementary solutions now take the form

$$\psi_m = \iint \frac{d}{dx}\left(\frac{e^{-ikr}}{r}\right)\Psi_m dS, \qquad \psi_p = \iint \frac{d}{dx}\left(\frac{e^{-ikr}}{r}\right)\Psi_p dS. \quad\dots\dots(19)$$

These give on the *walls* $\psi_m = \psi_p = 0$, and so do not disturb the condition of evanescence already satisfied by χ. It remains to satisfy over the *aperture*

$$\psi_m = \psi_p, \quad -2ik + d\psi_m/dx = d\psi_p/dx. \quad\dots\dots\dots\dots(20)$$

The first of these is satisfied if $\Psi_m = -\Psi_p$, so that ψ_m and ψ_p are equal at any pair of corresponding points upon the two sides. The values of $d\psi_m/dx$, $d\psi_p/dx$ are then opposite, and the remaining condition is also satisfied if

$$d\psi_m/dx = ik, \qquad d\psi_p/dx = -ik. \quad\dots\dots\dots\dots\dots(21)$$

Thus Ψ_m is to be such as to make $d\psi_m/dx = ik$; and, as in the proof of (17), it is easy to show that in (19)

$$\Psi_m = \psi_m/2\pi, \qquad \Psi_p = -\psi_p/2\pi, \quad\dots\dots\dots\dots\dots(22)$$

where ψ_m, ψ_p are the (equal) surface-values at dS.

When all the dimensions of S are small in comparison with the wavelength, (19) in its application to points at a sufficient distance from S assumes the form

$$\psi_p = \frac{ikx e^{-ikr}}{2\pi r^2} \iint \psi_p dS, \quad\dots\dots\dots\dots\dots\dots(23)$$

and it only remains to find what is the value of $\iint\psi_p dS$ which corresponds to $d\psi_p/dx = -ik$.

Now this correspondence is ultimately the same as if we were dealing with an absolutely incompressible fluid. If we imagine a rigid and infinitely thin plate (having the form of the aperture) to move normally through unlimited fluid with velocity u, the condition is satisfied that over the remainder of the plane the velocity-potential ψ vanishes. In this case the values of ψ at corresponding points upon the two sides are opposite; but if we limit our attention to the positive side, the conditions are the same as in the present problem. The kinetic energy of the motion is proportional to u^2, and we will suppose that twice the energy upon one side is hu^2. By Green's theorem this is equal to $-\iint \psi \cdot d\psi/dn \cdot dS$, or $-u \iint \psi \, dS$; so that $\iint \psi \, dS = -hu$. In the present application $u = -ik$, so that the corresponding value of $\iint \psi_p \, dS$ is ihk. Thus (23) becomes

$$\psi_p = -\frac{hk^2 x e^{-ikr}}{2\pi r^3} . \quad\quad\quad\quad\quad\quad (25)$$

The same algebraic expression gives ψ_m, if the *minus* sign be omitted; for as x itself changes sign in passing from one side to the other, the values of ψ_m and ψ_p at corresponding points are then equal.

The value of h can be determined in certain cases. For a circle* of radius c

$$h = \frac{4c^3}{3} ; \quad\quad\quad\quad\quad\quad\quad (26)$$

so that for a circular aperture the realized solution is

$$\phi_p = -\frac{8\pi c^3}{3\lambda^2} \frac{x}{r^3} \cos(nt - kr), \quad\quad\quad\quad (27)$$

$$\phi_m = 2 \sin nt \sin kx + \frac{8\pi c^3}{3\lambda^2} \frac{x}{r^3} \cos(nt - kr). \quad\quad (28)$$

It will be remarked that while in the first problem the wave (ψ) divergent from the aperture is proportional to the first power of the linear dimension, in the present case the amplitude is very much less, being proportional to the cube of that quantity.

The solution for an elliptic aperture is deducible from the general theory of the motion of an ellipsoid (a, b, c) through incompressible fluid†, by supposing $a = 0$, while b and c remain finite and unequal; but the general expression does not appear to have been worked out. When the eccentricity of the residual ellipse is small, I find that

$$h = \tfrac{4}{3}(bc)^{\frac{3}{2}}(1 - \tfrac{3}{64}e^4), \quad\quad\quad\quad\quad (29)$$

showing that the effect of moderate ellipticity is very small when the *area* is given.

* Lamb's *Hydrodynamics*, § 105.
† *Loc. cit.*, § 111.

From the solutions already obtained it is possible to derive others by differentiation. If, for example, we take the value of ϕ in the first problem and differentiate it with respect to x, we obtain a function which satisfies (2), which includes plane waves and their reflexion on the negative side, and which satisfies over the wall the condition of evanescence. It would seem at first sight as if this could be no other than the solution of the second problem, but the manner in which the linear dimension of the aperture enters suffices to show that it is not so. The fact is that although the proposed function vanishes over the plane part of the wall, it becomes infinite at the *edge*, and thus includes the action of *sources* there distributed. A similar remark applies to the solutions that might be obtained by differentiation of the second solution with respect to y or z, the coordinates measured parallel to the plane of the screen.

Reflecting Plate.—$d\phi/dn = 0$.

We now pass to the consideration of allied problems in which the transparent and opaque parts of the screen are interchanged. Under the above-written boundary condition the case is that of plane aerial waves incident upon a parallel infinitely thin plate, whose dimensions are ultimately supposed to be small in comparison with λ. The analytical process of solution may be illustrated by the following argument. Suppose a motion communicated to the plate identical with that which the air at that place would execute were the plate absent. It is evident that the propagation of the primary wave will then be undisturbed. The supplementary solution, representing the disturbance due to the plate, must then correspond to the reduction of the plate to rest, that is to a motion of the plate equal and opposite to that just imagined. The supplementary solution is accordingly analogous to that which occurs in the *second* of the problems already treated.

Using a similar notation, we have for the principal solution upon the two sides

$$\chi_m = \chi_p = e^{-ikx}, \dots\dots\dots\dots\dots\dots\dots\dots(30)$$

giving when $x = 0$

$$\chi_m = \chi_p = 1, \qquad d\chi_m/dx = d\chi_p/dx = -ik.$$

The supplementary solution is of the form (19), and gives upon the aperture, viz. the part of the plane $x = 0$ unoccupied by the plate, $\psi_m = \psi_p = 0$, and so does not disturb the continuity of ϕ. But in order that the continuity of $d\phi/dx$ may be maintained it is necessary that $\Psi_p = \Psi_m$; and then the values of ψ_m and ψ_p are *opposite* at any pair of corresponding points upon the two sides.

It remains to satisfy the necessary conditions at the plate itself. These are

$$\frac{d\chi_m}{dx} + \frac{d\psi_m}{dx} = 0, \qquad \frac{d\chi_p}{dx} + \frac{d\psi_p}{dx} = 0;$$

or, since $d\psi_m/dx$, $d\psi_p/dx$ are equal,

$$d\psi_m/dx = d\psi_p/dx = ik. \quad\ldots\ldots\ldots\ldots\ldots\ldots(31)$$

It follows that ψ_p has the opposite value to that expressed in (25); and the realized solution for a circular plate of radius c becomes

$$\phi_p = \cos(nt - kx) + \frac{8\pi c^3}{3\lambda^3}\, \frac{x}{r^2}\, \cos(nt - kr), \quad\ldots\ldots\ldots\ldots(32)$$

$$\phi_m = \cos(nt - kx) + \frac{8\pi c^3}{3\lambda^2}\, \frac{x}{r^2}\, \cos(nt - kr), \quad\ldots\ldots\ldots\ldots(33)$$

the analytical form being the same in the two cases.

It is important to notice that the reflexion from the plate is utterly different from the transmission by a corresponding aperture in an opaque screen, as given in (14), the former varying as the cube of the linear dimension, and the latter as the first power simply.

Reflecting Plate.—$\phi = 0$.

For the sake of completeness it may be well to indicate the solution of a fourth problem defined by the above heading. This has an affinity with the *first* problem, analogous to that of the third with the second. The form of χ is the same as in (30), and those for ψ_m, ψ_p the same as in (7). These make $d\psi_m/dx$, $d\psi_p/dx$ vanish on the aperture, and so do not disturb the continuity of $d\phi/dx$. But in order that the continuity of ϕ may also be maintained, we must have $\Psi_m = \Psi_p$, and not as in (9) $\Psi_m = -\Psi_p$. On the plate itself we must have

$$\psi_m = \psi_p = -1.$$

Accordingly ψ_m is the same as in (12), while ψ_p in (12) must have its sign reversed. The realized solution is

$$\phi_p = \phi_m = \cos(nt - kx) - M\frac{\cos(nt - kr)}{r}. \quad\ldots\ldots\ldots\ldots(34)$$

Two-dimensional Vibrations.

In the class of problems before us the velocity-potential of a point-source, viz. e^{-ikr}/r, is replaced by that of a linear source; and this in general is much more complicated. If we denote it by $D(kr)$, the expressions are*

$$D(kr) = -\left(\frac{\pi}{2ikr}\right)^{\frac{1}{2}} e^{-ikr} \left\{1 - \frac{1^2}{1 \cdot 8ikr} + \frac{1^2 \cdot 3^2}{1 \cdot 2 \cdot (8ikr)^2} - \cdots\right\}$$

$$= \left(\gamma + \log \frac{ikr}{2}\right) \left\{1 - \frac{k^2 r^2}{2^2} + \frac{k^4 r^4}{2^2 \cdot 4^2} - \cdots\right\}$$

$$+ \frac{k^2 r^2}{2^2} S_1 - \frac{k^4 r^4}{2^2 \cdot 4^2} S_2 + \frac{k^6 r^6}{2^2 \cdot 4^2 \cdot 6^2} S_3 - \cdots, \quad \dots\dots\dots(35)$$

where γ is Euler's constant ($\cdot 5772\ldots$), and

$$S_m = 1 + \tfrac{1}{2} + \tfrac{1}{3} + \cdots + 1/m.$$

Of these the first is "semiconvergent," and is applicable when kr is large; the second is fully convergent and gives the form of the function when kr is small.

Since the complete analytical theory is rather complicated, it may be convenient to give a comparatively simple derivation of the extreme forms, which includes all that is required for our present purpose, starting from the conception of a linear source as composed of distributed point-sources. If ρ be the distance of any element dx of the linear source from O, the point at which the potential is to be estimated, and r be the smallest value of ρ, so that $\rho^2 = r^2 + x^2$, we may take as the potential, constant factors being omitted,

$$\psi = -\int_0^\infty \frac{e^{-ik\rho}\, dx}{\rho} = -\int_r^\infty \frac{e^{-ik\rho}\, d\rho}{\sqrt{(\rho^2 - r^2)}}. \quad \dots\dots\dots\dots(36)$$

We have now to trace the form of (36) when kr is very great, and also when kr is very small. For the former case we replace ρ by $r+y$, thus obtaining

$$\psi = -\int_0^\infty \frac{e^{-ikr} e^{-iky}\, dy}{\sqrt{y} \cdot \sqrt{(2r + y)}}. \quad \dots\dots\dots\dots\dots(37)$$

When kr is very great, the approximate value of the integral in (37) may be obtained by neglecting the variation of $\sqrt{(2r + y)}$, since on account of the rapid fluctuation of sign caused by the factor e^{-iky} we need attend only to small values of y. Now, as is known,

$$\int_0^\infty \frac{\cos x\, dx}{\sqrt{x}} = \int_0^\infty \frac{\sin x\, dx}{\sqrt{x}} = \sqrt{\left(\frac{\pi}{2}\right)},$$

* See for example *Theory of Sound*, § 341.

so that in the limit

$$\psi = -(1-i)\sqrt{\left(\frac{\pi}{2kr}\right)}\,e^{-ikr} = -\sqrt{\left(\frac{\pi}{2ikr}\right)}\,e^{-ikr}, \quad\ldots\ldots\ldots(38)$$

in agreement with (35).

We have next to deduce the limiting form of (36) when kr is very small. For this purpose we may write it in the form

$$\psi = -\int_{r}^{\infty}\frac{e^{-ik\rho}\,d\rho}{\rho} - \int_{r}^{\infty}e^{-ik\rho}\left\{\frac{1}{\sqrt{(\rho^2-r^2)}} - \frac{1}{\rho}\right\}d\rho. \quad\ldots\ldots\ldots(39)$$

The first integral in (39) is well known. We have

$$-\int_{r}^{\infty}\frac{e^{-ik\rho}\,d\rho}{\rho} = \mathrm{Ci}\,(kr) - i\{\tfrac{1}{2}\pi + \mathrm{Si}\,(kr)\}$$

$$= \gamma + \log kr - \frac{k^2r^2}{2^2} + \frac{k^4r^4}{2\cdot3\cdot4^2} - \cdots$$

$$+ i\left\{\frac{\pi}{2} - kr + \frac{k^3r^3}{2\cdot3^3} - \cdots\right\}.$$

In the second integral of (39) the function to be integrated vanishes when ρ is great compared to r, and when ρ is not great in comparison with r, $k\rho$ is small and $e^{-ik\rho}$ may be identified with unity. Thus in the limit

$$\int_{r}^{\infty}e^{-ik\rho}\left\{\frac{1}{\sqrt{(\rho^2-r^2)}} - \frac{1}{\rho}\right\}d\rho = \left[\log\frac{\rho+\sqrt{(\rho^2-r^2)}}{\rho}\right]_{r}^{\infty} = \log 2;$$

and (39) becomes

$$\psi = \gamma + \log kr + \tfrac{1}{2}i\pi - \log 2 = \gamma + \log(\tfrac{1}{2}ikr), \quad\ldots\ldots\ldots(40)$$

in agreement with (35).

When kr is extremely small (40) may be considered for some purposes to reduce to $\log kr$; but the term $\tfrac{1}{2}i\pi$ is required in order to represent the equality of work done in the neighbourhood of the linear source and at a great distance from it.

We may now proceed to solve four problems relative to narrow slits and reflecting blades analogous to the four already considered in which the aperture or the reflecting plate was small in *both* its dimensions in comparison with the wave-length.

Narrow Slit.—Boundary Condition $d\phi/dn = 0$.

As in the former problem the principal solution is

$$\chi_m = e^{-ikx} + e^{ikx}, \qquad \chi_p = 0, \quad\ldots\ldots\ldots\ldots(41)$$

making $d\chi_m/dn$, $d\chi_p/dn$ vanish over the whole plane $x = 0$ and over the same plane $\chi_m = 2$, $\chi_p = 0$. The supplementary solution, which represents the effect of the slit, may be written

$$\psi_m = \int \Psi_m D\,(kr)\,dy, \qquad \psi_p = \int \Psi_p D\,(kr)\,dy, \quad \ldots\ldots\ldots(42)$$

Ψ_m, Ψ_p being certain functions of y to be determined, and the integration extending over the width of the slit from $y = -b$ to $y = +b$.

These additions do not disturb the condition to be satisfied over the wall. On the aperture continuity requires, as in (8), that

$$2 + \psi_m = \psi_p, \qquad d\psi_m/dx = d\psi_p/dx.$$

The second of these is satisfied by taking $\Psi_p = -\Psi_m$, so that at all corresponding pairs of points $\psi_m = -\psi_p$. It remains to determine Ψ_m so that on the aperture $\psi_m = -1$; and then by what has been said $\psi_p = +1$.

At a sufficient distance from the slit, supposed to be very narrow, $D\,(kr)$ may be removed from under the integral sign and also be replaced by its limiting form given in (35). Thus

$$\psi_m = -\left(\frac{\pi}{2ikr}\right)^{\frac{1}{2}} e^{-ikr} \int \Psi_m dy. \quad \ldots\ldots\ldots\ldots\ldots(43)$$

The condition by which Ψ_m is determined is that for all points upon the aperture

$$\int_{-b}^{+b} \Psi_m D\,(kr)\,dy = -1, \quad \ldots\ldots\ldots\ldots\ldots\ldots(44)$$

where, since kr is small throughout, the second limiting form given in (35) may be introduced.

From the known solution for the flow of incompressible fluid through a slit in an infinite plane we may infer that Ψ_m will be of the form $A\,(b^2 - y^2)^{-\frac{1}{2}}$, where A is some constant. Thus (44) becomes

$$A\left[(\gamma + \log \tfrac{1}{2}ik)\,\pi + \int_{-b}^{+b} \frac{\log\,(r)\,dy}{\sqrt{(b^2 - y^2)}}\right] = -1. \quad \ldots\ldots\ldots\ldots(45)$$

In this equation the first integral is obviously independent of the position of the point chosen, and if the form of Ψ_m has been rightly taken the second integral must also be independent of it. If its coordinate be η, lying between $\pm b$,

$$\int_{-b}^{+b} \frac{\log r\,dy}{\sqrt{(b^2 - y^2)}} = \int_{-b}^{\eta} \frac{\log\,(\eta - y)\,dy}{\sqrt{(b^2 - y^2)}} + \int_{\eta}^{b} \frac{\log\,(y - \eta)\,dy}{\sqrt{(b^2 - y^2)}},$$

and must be independent of η. This can be verified without much difficulty by assuming $\eta = b \sin \alpha$, $y = b \sin \theta$; but merely to determine A in (45) it suffices to consider the particular case of $\eta = 0$. Here

$$\int_{-b}^{+b} \frac{\log r \, dy}{\sqrt{(b^2 - y^2)}} = 2 \int_0^b \frac{\log y \, dy}{\sqrt{(b^2 - y^2)}} = 2 \int_0^{\frac12 \pi} \log (b \sin \theta) \, d\theta = \pi \log (\tfrac12 b).$$

Thus
$$A \, (\gamma \mid \log \tfrac14 ikb) \, \pi = -1,$$

and
$$\int_{-b}^{+b} \Psi_m \, dy = A \int_{-b}^{+b} \frac{dy}{\sqrt{(b^2 - y^2)}} = \pi A \, ;$$

so that (43) becomes
$$\psi_m = \frac{e^{-ikr}}{\gamma + \log (\tfrac14 ikb)} \left(\frac{\pi}{2ikr} \right)^{\frac12}. \quad \dots\dots\dots\dots\dots\dots(46)$$

From this ψ_p is derived by simply prefixing a negative sign.

The realized solution is obtained from (46) by omitting the imaginary part after introduction of the suppressed factor e^{int}. If the imaginary part of $\log (\tfrac14 ikb)$ be neglected, the result is

$$\psi_m = \left(\frac{\pi}{2kr} \right)^{\frac12} \cdot \frac{\cos (nt - kr - \tfrac14 \pi)}{\gamma + \log (\tfrac14 kb)}, \quad \dots\dots\dots\dots(47)$$

corresponding to
$$\chi_m = 2 \cos nt \cos kx. \quad \dots\dots\dots\dots\dots(48)$$

The solution (47) applies directly to aerial vibrations incident upon a perforated wall, and to an electrical problem which will be specified later. Perhaps the most remarkable feature of it is the very limited dependence of the transmitted vibration on the *width* ($2b$) of the aperture.

Narrow Slit.—Boundary Condition $\phi = 0$.

The principal solution is the same as in (18); and the conditions for the supplementary solution, to be satisfied over the aperture, are those expressed in (21). In place of (19)

$$\psi_m = - \int \frac{dD}{dx} \Psi_p \, dy, \qquad \psi_p = \int \frac{dD}{dx} \Psi_p \, dy \, ; \quad \dots\dots\dots(49)$$

the values of Ψ_m and Ψ_p being opposite, and those of ψ_m and ψ_p equal at corresponding points. At a distance we have

$$\psi_p = \frac{dD}{dx} \int_{-b}^{+b} \Psi_p \, dy, \quad \dots\dots\dots\dots\dots\dots (50)$$

in which
$$\frac{dD}{dx} = \frac{ikx}{r} \left(\frac{\pi}{2ikr} \right)^{\frac12} e^{-ikr}. \quad \dots\dots\dots\dots\dots(51)$$

There is a simple relation between the value of Ψ_p at any point of the aperture and that of ψ_p at the same point. For in the application of (49) to any point of the narrow aperture, $dD/dx = x/r^2$, showing that only those elements of the integral are sensible which lie infinitely near the point where ψ_p is to be estimated. The evaluation is effected by considering in the first instance a point for which x is finite, and afterwards passing to the limit. Thus

$$\psi_p = \int \frac{x}{x^2 + y^2}\, \Psi_p dy = \Psi_p \left[\tan^{-1} \frac{y}{x} \right]_{-b}^{+b} = \pi \Psi_p\,;$$

so that (50) becomes $\qquad \psi_p = \frac{1}{\pi} \frac{dD}{dx} \int_{-b}^{+b} \psi_p dy.$ ....(52)

It remains only to express the connexion between $\int \psi_p dy$ and the constant value of $d\psi_p/dx$ on the area of the aperture; and this is effected by the known solution for an incompressible fluid moving under similar conditions. The argument is the same as in the corresponding problem where the perforation is circular. In the motion (u) of a lamina of width $(2b)$ through infinite fluid, the whole kinetic energy per unit of length may be denoted by hu^2, and it appears from Green's theorem that $\int \psi_p dy = ihk$. The value of $h*$ is $\frac{1}{2}\pi b^2$; so that

$$\psi_p = - \frac{k^2 b^2 x}{2r} \left(\frac{\pi}{2ikr} \right)^{\frac{1}{2}} e^{-ikr}.$$ ....(53)

The same algebraical expression gives ψ_m, if the *minus* sign be omitted.

The realized solution from (53) is

$$\psi_p = - \frac{k^2 b^2 x}{2r} \left(\frac{\pi}{2kr} \right)^{\frac{1}{2}} \cos(nt - kr - \tfrac{1}{4}\pi),$$ ....(54)

corresponding to $\qquad\qquad \chi_m = 2 \sin nt \sin kx.$ ....(55)

Reflecting Blade.—Boundary Condition $d\phi/dn = 0$.

We have now to consider two problems which differ from the last in that the opaque and transparent parts of the screen are interchanged. As in the case of the circular aperture, we shall find that the correspondence lies between the reflecting blade under the condition $d\phi/dn = 0$ and the transmitting aperture under the condition $\phi = 0$, and reciprocally.

The principal solution remains as in (30). The supplementary solution must satisfy (31), where

$$\psi_m = \int \frac{dD}{dx} \Psi_p dy, \qquad \psi_p = \int \frac{dD}{dx} \Psi_p dy,$$ ....(56)

since Ψ_m and Ψ_p must be equal in order that the continuity of $d\phi/dx$ over

* Lamb's *Hydrodynamics*, § 71.

the aperture may be maintained. Thus ψ_m and ψ_p have opposite values at any pair of corresponding points.

If we compare these conditions with those by which (53) was determined, we see that ψ_m has the same value as in that case, but that the sign of ψ_p must be reversed. Thus in the present problem

$$\psi_m = \psi_p = \frac{k^2 b^2 x}{2r} \left(\frac{\pi}{2kr} \right)^{\frac{1}{2}} \cos{(nt - kr - \tfrac{1}{4}\pi)}, \quad \ldots\ldots\ldots\ldots(57)$$

corresponding to

$$\chi_m = \chi_p = \cos{(nt - kx)}. \quad \ldots\ldots\ldots\ldots\ldots\ldots\ldots(58)$$

Reflecting Blade.—Boundary Condition $\phi = 0$.

In this case χ still remains as in (30). The general forms for ψ_m, ψ_p are as in (42), which secure that $d\psi_m/dx$, $d\psi_p/dx$ shall vanish on the aperture (i.e. the part of the plane $x = 0$ unoccupied by the blade). But in order that the continuity of ϕ may also be maintained over that area we must have $\Psi_m = \Psi_p$. Thus ψ_m, ψ_p have equal values at corresponding points. On the blade itself $\psi_m = \psi_p = -1$.

A comparison of these conditions with those by which (46) was determined shows that in the present case

$$\psi_m = \psi_p = \frac{e^{-ikr}}{\gamma + \log{(\tfrac{1}{4}ikb)}} \left(\frac{\pi}{2ikr} \right)^{\frac{1}{2}}. \quad \ldots\ldots\ldots\ldots(59)$$

When $\log i$ in the denominator of (59) may be omitted, the realized form is that expressed by (47), and this corresponds to

$$\chi_m = \chi_p = \cos{(nt - kx)}. \quad \ldots\ldots\ldots\ldots\ldots\ldots(60)$$

Various Applications.

Of the eight problems, whose solutions have now been given, four have an immediate application to aerial vibrations, viz. those in which the condition on the walls is $d\phi/dn = 0$. The symbol ϕ then denotes the velocity-potential, and the condition expresses simply that the fluid does not penetrate the boundary. The four problems relating to two dimensions have also a direct application to electrical vibrations, if we suppose that the thin material constituting the screen (or the blade) is a perfect conductor. For if R denote the electromotive intensity parallel to z, the condition at the face of the conductor is $R = 0$; so that if R be written for ψ in (53), (59), we have the solutions for a narrow aperture in an infinite screen, and for a narrow reflecting blade respectively, corresponding to the incident wave

$R = e^{-ikx}$. A narrow aperture parallel to the electric vibrations transmits very much less than is reflected by a conductor elongated in the same direction.

The two other solutions relative to two dimensions find electrical application if we identify ϕ with c, the component of magnetic intensity parallel to z. For when the other components a and b are zero, the condition to be satisfied at the face of a conductor is $dc/dn = 0$. Thus (46), (57) apply to incident vibrations represented by $c = e^{-ikx}$. In this case the slit transmits much more than the blade reflects.

It may be remarked that in general problems of electrical vibration in *two dimensions* have simple acoustical analogues*. As an example we may refer to the reflexion of plane electric waves incident perpendicularly upon a *corrugated* surface, the acoustical analogue of which is treated in *Theory of Sound*, 2nd ed. § 272 *a*, and to the reflexion of electric waves from a conducting cylinder (§ 343).

* The comparison is not limited to the case of perfect conductors, but applies also when the obstacles, being non-conductors, differ from the surrounding medium in specific inductive capacity, or in magnetic permeability, or in both properties.

228.

THE LIMITS OF AUDITION.

[*Royal Institution Proceedings*, xv. pp. 417—418, 1897.]

IN order to be audible, sounds must be restricted to a certain range of pitch. Thus a sound from a hydrogen flame vibrating in a large resonator was inaudible, as being too low in pitch. On the other side, a bird-call, giving about 20,000 vibrations per second, was inaudible, although a sensitive flame readily gave evidence of the vibrations and permitted the wave-length to be measured. Near the limit of hearing the ear is very rapidly fatigued; a sound in the first instance loud enough to be disagreeable, disappearing after a few seconds. A momentary intermission, due, for example, to a rapid passage of the hand past the ear, again allows the sound to be heard.

The magnitude of vibration necessary for audition at a favourable pitch is an important subject for investigation. The earliest estimate is that of Boltzmann. An easy road to a superior limit is to find the amount of energy required to blow a whistle and the distance to which the sound can be heard (*e.g.* one-half a mile). Experiments upon this plan gave for the amplitude 8×10^{-8} cm., a distance which would need to be multiplied 100 times in order to make it visible in any possible microscope. Better results may be obtained by using a vibrating fork as a source of sound. The energy resident in the fork at any time may be deduced from the amplitude as observed under a microscope. From this the rate at which energy is emitted follows when we know the rate at which the vibrations of the fork die down (say to one-half). In this way the distance of audibility may be reduced to 30 metres, and the results are less liable to be disturbed by atmospheric irregularities. If s be the proportional condensation in the waves which are just capable of exciting audition, the results may be expressed :—

c'	frequency $= 256$	$s = 6\cdot0 \times 10^{-9}$
g'	„ $= 384$	$s = 4\cdot6 \times 10^{-9}$
c''	„ $= 512$	$s = 4\cdot6 \times 10^{-9}$

showing that the ear is capable of recognising vibrations which involve far less changes of pressure than the total pressure outstanding in our highest vacua.

In such experiments the whole energy emitted is very small, and contrasts strangely with the 60 horse-power thrown into the fog-signals of the Trinity House. If we calculate according to the law of inverse squares how far a sound absorbing 60 horse-power should be audible, the answer is 2700 kilometres! The conclusion plainly follows that there is some important source of loss beyond the mere diffusion over a larger surface. Many years ago Sir George Stokes calculated the effect of radiation upon the propagation of sound. His conclusion may be thus stated. The amplitude of sound propagated in plane waves would fall to half its value in six times the interval of time occupied by a mass of air heated above its surroundings in cooling through half the excess of temperature. There appear to be no data by which the latter interval can be fixed with any approach to precision; but if we take it at one minute, the conclusion is that sound would be propagated for six minutes, or travel over about seventy miles, without very serious loss from this cause.

The real reason for the falling off at great distances is doubtless to be found principally in atmospheric refraction due to variation of temperature, and of wind, with height. In a normal state of things the air is cooler overhead, sound is propagated more slowly, and a wave is tilted up so as to pass over the head of an observer at a distance. [Illustrated by a model.] The theory of these effects has been given by Stokes and Reynolds, and their application to the explanation of the vagaries of fog-signals by Henry. Progress would be promoted by a better knowledge of what is passing in the atmosphere over our heads.

The lecture concluded with an account of the observations of Preyer upon the delicacy of pitch perception, and of the results of Kohlrausch upon the estimation of pitch when the total number of vibrations is small. In illustration of the latter subject an experiment (after Lodge) was shown, in which the sound was due to the oscillating discharge of a Leyden battery through coils of insulated wire. Observation of the spark proved that the total number of (aerial) vibrations was four or five. The effect upon the pitch of moving one of the coils so as to vary the self-induction was very apparent.

229.

ON THE MEASUREMENT OF ALTERNATE CURRENTS BY MEANS OF AN OBLIQUELY SITUATED GALVANOMETER NEEDLE, WITH A METHOD OF DETERMINING THE ANGLE OF LAG.

[*Philosophical Magazine*, XLIII. pp. 343—349, 1897.]

IT is many years* since, as the result of some experiments upon induction, I proposed a soft iron needle for use with alternate currents in place of the permanently magnetized steel needle ordinarily employed in the galvanometer for the measurement of steady currents. An instrument of this kind designed for telephonic currents has since been constructed by Giltay; but, so far as I am aware, no application has been made of it to measurements upon a large scale, although the principle of alternately reversed magnetism is the foundation of several successful commercial instruments.

The theory of the behaviour of an elongated needle is sufficiently simple, so long as it can be assumed that the magnetism is made up of two parts, one of which is constant and the other proportional to the magnetizing force. If internal induced currents can be neglected, this assumption may be regarded as legitimate so long as the forces are small†. In the ordinary case of alternate currents, where upon the whole there is no transfer of electricity in either direction, the constant part of the magnetism has no effect; while the variable part gives rise to a deflecting couple proportional on the one hand to the mean value of the square of the magnetizing force or current, and upon the other to the sine of twice the angle between the direction of the force and the length of the needle. The deflecting couple is thus evanescent when the needle stands either parallel or perpendicular to the magnetizing force, and rises to a maximum at the angle of 45°. For practical

---

* *Brit. Assoc. Report*, 1868; *Phil. Mag.* Vol. III. p. 43 (1887). [Vol. I. p. 310.]

† *Phil. Mag.* Vol. XXIII. p. 225 (1887). [Vol. II. p. 579.]

purposes the law of proportionality to the mean square of current would seem to be trustworthy so long as no great change occurs in the frequency or type of current; otherwise eddy currents in the iron might lead to error, unless the metal were finely subdivided.

It is hardly to be supposed that for ordinary purposes a suspended iron needle would compete in convenience with the excellent instruments now generally available; but having found it suitable for a special purpose of my own, I think it may be worth while to draw to it the attention of those interested. In experiments upon the oxidation of nitrogen by the electric arc or flame it was desired to ascertain the relation between the electric power absorbed and the amount of nitrogen oxidized. A transformer with an unclosed magnetic circuit was employed to raise the potential from that of the supply to the 3000 volts or more needed at the platinum terminals. Commercial ampere-meters and volt-meters gave with all needed precision the current and potential at the primary of the transformer; but, as is well known, these data do not suffice for an estimate of power. The latter depends also upon the angle of lag, or retardation of current relatively to potential-difference. If this angle be θ, the power actually employed is to be found by multiplying the product of volts and amperes by $\cos \theta$, so that the actual power may be less to any extent than the apparent power represented by the simple product. Various watt-meters have been introduced for measuring the actual power directly, but I could not hear of one suitable for the large current of 40 amperes used at the Royal Institution. Working subsequently in the country I returned to the problem, and succeeded in determining the angle of lag very easily by means of the principle now to be explained.

The soft iron needle of 2 centim. in length, suspended by a fine torsion-fibre of glass and carrying a mirror in the usual way, is inclined at 45° to the direction of the magnetic force. This force is due to currents in *two* coils, the common axis of the coils being horizontal and passing through the centre of the needle. As in ordinary galvanometers, the mean plane of each coil may include the centre of the needle; but it was found better to dispose the

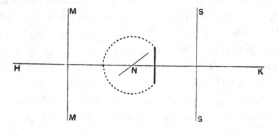

coils on opposite sides and at distances from the needle which could be varied. A plan of the arrangement is sketched diagrammatically in the woodcut, where MM, SS represent the two coils, the common axis HK passing through the centre of the needle N. If the currents in the coils are of the same frequency and of simple type, the magnetizing forces along HK may be denoted by $A \cos nt$, $B \cos (nt - \epsilon)$, ϵ being the phase-difference. If either force act alone, the deflecting couple is represented by A^2 or by B^2; but if the two forces cooperate the corresponding effect is

$$C^2 = A^2 + B^2 + 2AB \cos \epsilon, \quad \dots\dots\dots\dots\dots\dots\dots(1)$$

reducing itself to $(A + B)^2$ or $(A - B)^2$ only in the cases where ϵ is zero or two right angles. The method consists in measuring upon any common scale all the three quantities A^2, B^2, and C^2, from which ϵ can be deduced by trigono-metrical tables, or more simply in many cases by constructing the triangle whose sides are A, B, and C. The determination of the phase-difference between the currents is thus independent of any measurement of their absolute values.

The best method of estimating the deflecting couples may depend upon the circumstances of the particular case. The most accurate in principle is the restoration of the needle to the zero position by means of a torsion-head. But when the conditions are so arranged that the angular deflexions are moderate, it will usually suffice merely to read them, either objectively by a spot of light thrown upon a scale, or by means of a telescope. In any case where it may be desired to push the deflexions beyond the region where the law of proportionality can be relied upon, all risk of error may be avoided by comparison with another instrument of trustworthy calibration, one coil only of the soft iron apparatus being employed.

In certain cases the advantages which accompany the restoration of the zero position of the needle may be secured by causing the deflexions them-selves to assume a constant value, e.g. by making known changes of resistance in one or both of the circuits, or by motion of the coils altering their efficiencies in a known ratio.

In the particular experiments for which the apparatus was set up the coil MM (see woodcut) was reduced to a single turn of about 17 centim. diameter and conveyed the main current (about 10 amperes) which traversed the primary circuit of the transformer. This, it may be mentioned, was a home-made instrument, somewhat of the Ruhmkorff type, and was placed at a sufficient distance from the measuring apparatus. The shunt-coil SS was of somewhat less diameter, and contained 32 convolutions. The shunt-circuit included also two electric lamps, joined in series, and its terminals were connected with two points of the main circuit outside the apparatus, where the difference of potentials was about 40 volts. Provision was made for diverting the main current at pleasure from MM, and by means of a re-verser the direction of the current in SS could be altered, equivalent to a change of ϵ by 180°. The measurements to be made are the effects of MM and of SS acting separately, and of MM and SS acting together in one or both positions of the reverser.

The best arrangement of the details of observation will depend somewhat upon the particular value of ϵ to be dealt with. If this be 60°, or thereabouts, the method can be applied with peculiar advantage. For by preliminary adjustment of the coils, if movable, or by inclusion of (unknown) resistance in the shunt-circuit, the deflexions due to MM and SS may be made equal to one another; so that in the case supposed the same deflexion will ensue from the simultaneous action of the two currents in one of the ways in which they may be combined.

This condition of things was somewhat approached in the actual measures relating to the electric flame. Thus in one trial the coils were adjusted so as to make the deflexions, due to each of the currents acting singly, equal to one another. The value was 40 divisions of the scale. When both currents were turned on, the deflexion was $26\frac{1}{2}$ divisions. Thus

$$A^2 = B^2 = 40, \qquad A^2 + B^2 - 2AB \cos \epsilon = 26\tfrac{1}{2};$$

whence $\qquad\qquad\qquad \cos \epsilon = \cdot 67, \quad$ or $\; \epsilon = 48°.$

In a second experiment the deflexion due to both currents acting together was made equal to that of the main acting alone. Here

$$A^2 = 40, \qquad B^2 = 71, \qquad A^2 + B^2 - 2AB \cos \epsilon = 40;$$

whence $\qquad\qquad\qquad \cos \epsilon = \cdot 665.$

The accuracy was limited by the unsteadiness of the electric flame and of the primary currents (from a gas-driven De Méritens) rather than by want of delicacy in the measuring apparatus.

When the phase-difference is about a quarter of a period, $\cos \epsilon$ is small, and its value is best found by observing the effect of reversing the shunt-current while the main current continues running. The difference is $4AB \cos \epsilon$, from which, combined with a knowledge of A and B, the value of $\cos \epsilon$ is advantageously derived. If $\cos \epsilon$ is absolutely zero, the reversal does not alter the reading.

If the currents are in the same, or in opposite phases, it is possible to reduce the joint effect to zero by suitable adjustment of the coils or of the shunt resistance.

The application of principal interest is when the shunt-current may be assumed to have the same phase as the potential-difference at its terminals, for then $\cos \epsilon$ is the factor by which the true watts may be derived from the apparent watts. We will presently consider the question of the negligibility of the self-induction of the shunt-current, but before proceeding to this it may be well to show the application of the formulæ when the currents deviate from the sine type.

If a be the instantaneous current, and v the instantaneous potential-difference at the terminals, the work done is $\int av\,dt$. The readings of the soft iron galvanometer for either current alone may be represented by

$$A^2 = h^2 \int a^2 dt, \qquad B^2 = k^2 \int v^2 dt, \qquad \dots\dots\dots\dots\dots(2)$$

where h, k are constants depending upon the disposition of the apparatus. When both currents act, we have the readings

$$C_1^2 \text{ or } C_2^2 = \int (ha \pm kv)^2 dt. \qquad \dots\dots\dots\dots\dots\dots(3)$$

Taking the first alternative, we find

$$C_1^2 = h^2 \int a^2 dt + 2hk \int av\,dt + k^2 \int v^2 dt,$$

or

$$\frac{C_1^2 - A^2 - B^2}{2AB} = \frac{\int av\,dt}{\{\int a^2 dt \times \int v^2 dt\}^{\frac{1}{2}}}. \qquad \dots\dots\dots\dots\dots(4)$$

The fraction on the right of (4) is the ratio of true and apparent watts; and we see that, whether the currents follow the sine law or not, the ratio is given by $\cos \epsilon$, where, as before, ϵ is the angle of the triangle constructed with sides proportional to the square roots of the three readings.

Another formula for $\cos \epsilon$ is

$$\cos \epsilon = \frac{C_1^2 - C_2^2}{4AB}. \qquad \dots\dots\dots\dots\dots\dots\dots\dots\dots(5)$$

In the final formula (4) the factors of efficiency of the separate coils (h, k) do not enter. This result depends, however, upon the fulfilment of the condition of parallelism between the two coils. If the magnetic forces due to the coils be inclined at different angles χ, χ' to the length of the needle, we have in place of (3),

$$C^2 = \int (a \cos \chi + v \cos \chi')(a \sin \chi + v \sin \chi')\,dt$$

$$= \int [\tfrac{1}{2} a^2 \sin 2\chi + \tfrac{1}{2} v^2 \sin 2\chi' + av \sin (\chi + \chi')]\,dt; \qquad \dots\dots(6)$$

while

$$A^2 = \tfrac{1}{2} \sin 2\chi \int a^2 dt, \qquad B^2 = \tfrac{1}{2} \sin 2\chi' \int v^2 dt. \qquad \dots\dots\dots\dots(7)$$

Accordingly

$$\frac{\int av\,dt}{\{\int a^2 dt \times \int v^2 dt\}^{\frac{1}{2}}} = \frac{C^2 - A^2 - B^2}{2AB} \frac{\sqrt{\{\sin 2\chi . \sin 2\chi'\}}}{\sin (\chi + \chi')}, \qquad \dots\dots\dots(8)$$

in which the second fraction on the right represents the influence of the defect in parallelism. If χ and χ' are both nearly equal to 45°, then approximately

$$\frac{\sqrt{\{\sin 2\chi . \sin 2\chi'\}}}{\sin (\chi + \chi')} = 1 - \tfrac{1}{2}(\chi - \chi')^2. \qquad \dots\dots\dots\dots(9)$$

We have now to consider under what conditions the shunt-current may be assumed to be proportional to the instantaneous value of the potential-difference at its terminals. The obstacles are principally the self-induction of the shunt-coil itself, and the mutual induction between it and the coil which conveys the main current. As to the former, we know* that if the mean radius of a coil be a, and if the section be circular of radius c, and if n be the number of convolutions,

$$L = 4\pi n^2 a \left\{ \log \frac{8a}{c} - \frac{7}{4} \right\}. \quad \dots\dots\dots\dots\dots\dots\dots(10)$$

To take an example from the shunt-coil used in the experiments above referred to, where

$$a = 6 \text{ cm.,} \qquad c = 1 \text{ cm.,} \qquad n = 32,$$

L is of the order 10^5 cm. The time-constant of the shunt-circuit (τ) is equal to L/R, where R is the resistance in C.G.S. measure. If r be the resistance measured in ohms, $R = r \times 10^9$, so that

$$\tau = \frac{1}{r \times 10^4}.$$

What we are concerned with is the ratio of τ to the period of the currents; if the latter be $\frac{1}{100}$ second, the ratio is $1/100r$, so that if r be a good number of ohms—it must have exceeded 100 in the particular experiments—there is nothing to fear from self-induction. It would seem to follow generally that if the voltage be not too small, say not falling below 10 volts, there should be no difficulty in obtaining sufficient effect from a shunt-coil whose self-induction may be neglected. It may be remarked that since the efficiency of the coil varies as n, while L varies as n^2, it will be advantageous to keep n (and r) down so long as the self-induction of the whole shunt-circuit is mainly that of the coil.

If the main and the shunt-coils were wound closely together, the disturbance due to mutual induction would be of the same order of magnitude as that due to self-induction. If the coils are separated, as is otherwise convenient, the influence of mutual induction will be less, and may be neglected under the conditions above defined.

As to the effect of self-induction, if present, we know that the lag (ϕ) is given by

$$\tan \phi = Lp/R, \quad \dots\dots\dots\dots\dots\dots\dots\dots\dots(11)$$

where $p = 2\pi \times$ frequency. The angle of lag of the main current (θ), which it is the object of the measurements to determine, is then given by

$$\theta = \epsilon + \phi, \quad \dots\dots\dots\dots\dots\dots\dots\dots\dots(12)$$

ϵ being the phase-difference of the two currents as found directly from the observations.

---

* Maxwell's *Electricity*, § 706.

230.

ON THE INCIDENCE OF AERIAL AND ELECTRIC WAVES UPON SMALL OBSTACLES IN THE FORM OF ELLIPSOIDS OR ELLIPTIC CYLINDERS, AND ON THE PASSAGE OF ELECTRIC WAVES THROUGH A CIRCULAR APERTURE IN A CONDUCTING SCREEN.

[*Philosophical Magazine*, XLIV. pp. 28—52, 1897.]

THE present paper may be regarded as a development of previous researches by the author upon allied subjects. When the character of the obstacle differs only infinitesimally from that of the surrounding medium, a solution may be obtained independently of the size and the form which it presents. But when this limitation is disregarded, when, for example, in the case of aerial vibrations the obstacle is of arbitrary compressibility and density, or in the case of electric vibrations when the dielectric constant and the permeability are arbitrary, the solutions hitherto given are confined to the case of small spheres, or circular cylinders. In the present investigation extension is made to ellipsoids, including flat circular disks and thin blades.

The results arrived at are limiting values, strictly applicable only when the dimensions of the obstacles are infinitesimal, and at distances outwards which are infinitely great in comparison with the wave-length (λ). The method proceeds by considering in the first instance what occurs in an intermediate region, where the distance (r) is at once great in comparison with the dimensions of the obstacle and small in comparison with λ. Throughout this region and within it the calculation proceeds as if λ were infinite, and depends only upon the properties of the common potential. When this problem is solved, extension is made without much difficulty to the exterior region where r is great in comparison with λ, and where the common potential no longer avails.

At the close of the paper a problem of some importance is considered relative to the escape of electric waves through small circular apertures in metallic screens. The case of narrow elongated slits has already been treated*.

Obstacle in a Uniform Field.

The analytical problem with which we commence is the same whether the flow be thermal, electric, or magnetic, the obstacle differing from the surrounding medium in conductivity, specific inductive capacity, or permeability respectively. If ϕ denote its potential, the uniform field is defined by

$$\phi = ux + vy + wz ; \quad \dots\dots\dots\dots\dots(1)$$

u, v, w being the fluxes in the direction of fixed, arbitrarily chosen, rectangular axes. If ψ be the potential in the uniform medium due to the obstacle, so that the complete potential is $\phi + \psi$, ψ may be expanded in the series of spherical harmonics

$$\psi = \frac{S_0}{r} + \frac{S_1}{r^2} + \frac{S_2}{r^3} + \dots, \quad \dots\dots\dots\dots\dots(2)$$

the origin of r being within the obstacle. Since there is no source, S_0 vanishes. Further, at a great distance $S_2, S_3, \dots$ may be neglected, so that ψ there reduces to

$$\psi = \frac{S_1}{r^2} = \frac{A'x + B'y + C'z}{r^3}. \quad \dots\dots\dots\dots\dots(3)$$

The disturbance (3) corresponds to (1). If we express separately the parts corresponding to u, v, w, writing $A' = A_1 u + A_2 v + A_3 w$, &c., we have

$$r^3 \psi = u (A_1 x + B_1 y + C_1 z) + v (A_2 x + B_2 y + C_2 z) + w (A_3 x + B_3 y + C_3 z); \quad \dots\dots(4)$$

but the nine coefficients are not independent. By the law of reciprocity the coefficient of the x-part due to v must be the same as that of the y-part due to u, and so on†. Thus $B_1 = A_2$, &c., and we may write (4) in the form

$$r^3 \psi = u \frac{dF}{dx} + v \frac{dF}{dy} + w \frac{dF}{dz}, \quad \dots\dots\dots\dots\dots(5)$$

where $\quad F = \tfrac{1}{2}A_1 x^2 + \tfrac{1}{2}B_2 y^2 + \tfrac{1}{2}C_3 z^2 + B_1 xy + C_2 yz + C_1 zx. \quad \dots\dots(6)$

In the case of a body, like an ellipsoid, symmetrical with respect to three planes chosen as coordinate planes,

$$B_1 = C_2 = C_1 = 0,$$

---

* *Phil. Mag.* Vol. XLIII. p. 272. [Vol. IV. p. 295.]

† *Theory of Sound*, § 109. u and v may be supposed to be due to point-sources situated at a great distance R along the axes of x and y respectively.

and (4) reduces to

$$r^3\psi = A_1 ux + B_2 vy + C_3 wz. \quad\dots\dots\dots\dots\dots(7)$$

It will now be shown that by a suitable choice of coordinates this reduction may be effected in any case. Let u, v, w originate in a source at distance R, whose coordinates are x', y', z', so that $u = x'/R^3$, &c. Then (5) becomes

$$r^3 R^3 \psi = x'\frac{dF}{dx} + y'\frac{dF}{dy} + z'\frac{dF}{dz} = A_1 xx' + B_2 yy' + C_3 zz'$$

$$+ B_1(x'y + y'x) + C_2(y'z + z'y) + C_1(z'x + x'z)$$

$$= F(x + x',\ y + y',\ z + z') - F(x, y, z) - F(x', y', z').$$

Now by a suitable transformation of coordinates $F(x, y, z)$, and therefore $F(x', y', z')$ and $F(x + x', y + y', z + z')$, may be reduced to the form

$$A_1 x^2 + B_2 y^2 + C_3 z^2,\ \ \&c.$$

If this be done,

$$r^3 R^3 \psi = A_1 xx' + B_2 yy' + C_3 zz',$$

or reverting to u, v, w, reckoned parallel to the new axes,

$$r^3 \psi = A_1 ux + B_2 vy + C_3 wz, \quad\dots\dots\dots\dots\dots(8)$$

as in (7) for the ellipsoid. It should be observed that this reduction of the potential at a distance from the obstacle to the form (8) is independent of the question whether the material composing the obstacle is uniform.

For the case of the ellipsoid (a, b, c) of uniform quality the solution may be completely carried out. Thus [*], if T be the volume, so that

$$T = \tfrac{4}{3}\pi abc, \quad\dots\dots\dots\dots\dots\dots(9)$$

we have $$A_1 u = -AT, \quad B_2 v = -BT, \quad C_3 w = -CT, \quad\dots\dots\dots(10)$$

$$A = \frac{\kappa u}{1 + \kappa L}, \qquad B = \frac{\kappa v}{1 + \kappa M}, \qquad C = \frac{\kappa w}{1 + \kappa N}, \quad\dots\dots\dots(11)$$

where $$L = 2\pi abc \int_0^\infty \frac{d\lambda}{(a^2 + \lambda)^{\frac{3}{2}} (b^2 + \lambda)^{\frac{1}{2}} (c^2 + \lambda)^{\frac{1}{2}}}, \quad\dots\dots\dots(12)$$

with similar expressions for M and N.

In (11) κ denotes the susceptibility to magnetization. In terms of the permeability μ, analogous to conductivity in the allied problems, we have, if μ' relate to the ellipsoid and μ to the surrounding medium,

$$1 + 4\pi\kappa = \mu'/\mu, \quad\dots\dots\dots\dots\dots(13)$$

so that $$A = \frac{(\mu' - \mu) u}{4\pi\mu + (\mu' - \mu) L}, \quad\dots\dots\dots\dots(14)$$

with similar equations for B and C.

* The magnetic problem is considered in Maxwell's *Electricity and Magnetism*, 1873, § 437, and in Mascart's *Leçons*, 1896, §§ 52, 53, 276.

Two extreme cases are worthy of especial notice. If $\mu'/\mu = \infty$, the general equation for ψ becomes

$$-\frac{r^3 \psi}{T} = \frac{ux}{L} + \frac{vy}{M} + \frac{wz}{N} . \qquad \dots\dots\dots\dots\dots(15)$$

On the other hand, if $\mu'/\mu = 0$,

$$-\frac{r^3 \psi}{T} = \frac{ux}{L - 4\pi} + \frac{vy}{M - 4\pi} + \frac{wz}{N - 4\pi} . \qquad \dots\dots\dots\dots(16)$$

In the case of the sphere (a)

$$L = M = N = \tfrac{4}{3}\pi ; \qquad \dots\dots\dots\dots\dots\dots(17)$$

so that (15) becomes

$$\psi = -\frac{a^3}{r^3} (ux + vy + wz), \qquad \dots\dots\dots\dots\dots(18)$$

giving, when $r = a$, $\phi + \psi = 0$. This is the case of the perfect conductor.

In like manner for the non-conducting sphere (16) gives

$$\psi = \frac{a^3}{2r^3} (ux + vy + wz). \qquad \dots\dots\dots\dots\dots(19)$$

If the conductivity of the sphere be finite (μ'),

$$\psi = -\frac{a^3}{r^3} \frac{\mu' - \mu}{\mu' + 2\mu} (ux + vy + wz), \qquad \dots\dots\dots \dots\dots\dots(20)$$

which includes (18) and (19) as particular cases.

If the ellipsoid has two axes equal, and is of the planetary or flattened form,

$$b = c = \frac{a}{\sqrt{(1 - e^2)}}, \qquad T = \tfrac{4}{3}\pi c^3 \sqrt{(1 - e^2)} ; \qquad \dots\dots\dots\dots(21)$$

$$L = 4\pi \left\{ \frac{1}{e^2} - \frac{\sqrt{(1 - e^2)}}{e^3} \sin^{-1} e \right\}, \qquad \dots\dots\dots\dots\dots(22)$$

$$M = N = 2\pi \left\{ \frac{\sqrt{(1 - e^2)}}{e^3} \sin^{-1} e - \frac{1 - e^2}{e^2} \right\}. \qquad \dots\dots\dots\dots(23)$$

In the extreme case of a disk, when $e = 1$ nearly,

$$L = 4\pi - 2\pi^2 \sqrt{(1 - e^2)}, \qquad \dots\dots\dots\dots\dots\dots(24)$$

$$M = N = \pi^2 \sqrt{(1 - e^2)}. \qquad \dots\dots\dots\dots\dots\dots(25)$$

Thus in the limit from (14), (21) $TA = 0$, unless $\mu' = 0$; and when $\mu' = 0$,

$$TA = -\frac{2c^3 u}{3\pi} . \qquad \dots\dots\dots\dots\dots\dots\dots(26)$$

In like manner the limiting values of TB, TC are zero, unless $\mu' = \infty$, and then

$$TB = \frac{4c^3 v}{3\pi}, \qquad TC = \frac{4c^3 w}{3\pi} . \qquad \dots\dots\dots\dots\dots(27)$$

In all cases

$$\psi = -\frac{T(Ax + By + Cz)}{r^3} \quad \dots\dots\dots\dots\dots(28)$$

gives the disturbance due to the ellipsoid.

If the ellipsoid of revolution be of the ovary or elongated form,

$$a = b = c\sqrt{(1 - e^2)}; \quad \dots\dots\dots\dots\dots\dots(29)$$

$$L = M = 2\pi\left\{\frac{1}{e^2} - \frac{1 - e^2}{2e^3}\log\frac{1 + e}{1 - e}\right\}, \quad \dots\dots\dots\dots(30)$$

$$N = 4\pi\left\{\frac{1}{e^2} - 1\right\}\left\{\frac{1}{2e}\log\frac{1 + e}{1 - e} - 1\right\}. \quad \dots\dots\dots(31)^*$$

In the case of a very elongated ovoid L and M approximate to the value 2π, while N approximates to the form

$$N - 4\pi\frac{a^2}{c^2}\left(\log\frac{2c}{a} - 1\right), \quad \dots\dots\dots\dots\dots(32)$$

vanishing when $e = 1$.

In Two Dimensions.

The case of an elliptical cylinder in two dimensions may be deduced from (12) by making c infinite, when the integration is readily effected. We find

$$L = \frac{4\pi b}{a + b}, \qquad M = \frac{4\pi a}{a + b}. \quad \dots\dots\dots\dots(33)$$

A and B are then given by (14) as before, and finally

$$\psi = -\frac{ab(a + b)}{2r^2}\left\{\frac{(\mu' - \mu)ux}{\mu a + \mu' b} + \frac{(\mu' - \mu)vy}{\mu b + \mu' a}\right\}, \quad \dots\dots(34)$$

corresponding to

$$\phi = ux + vy. \quad \dots\dots\dots\dots\dots\dots(35)$$

In the case of circular section $L = M = 2\pi$, so that

$$\psi = -\frac{a^2}{r^2}\frac{\mu' - \mu}{\mu' + \mu}(ux + vy). \quad \dots\dots\dots\dots(36)$$

When $b = 0$, that is when the obstacle reduces itself to an infinitely thin blade, ψ vanishes unless $\mu' = 0$ or $\mu' = \infty$. In the first case

$$(\mu' = 0) \qquad \psi = \frac{a^2 vy}{2r^2}; \quad \dots\dots\dots\dots(37)$$

in the second

$$(\mu' = \infty) \qquad \psi = -\frac{a^2 ux}{2r^2}. \quad \dots\dots\dots\dots(38)$$

---

* There are slight errors in the values of L, M, N recorded for this case in both the works cited.

Aerial Waves.

We may now proceed to investigate the disturbance of plane aerial waves by obstacles whose largest diameter is small in comparison with the wavelength (λ). The volume occupied by the obstacle will be denoted by T; as to its shape we shall at first impose no restriction beyond the exclusion of very special cases, such as would involve resonance in spite of the small dimensions. The compressibilities and densities of the medium and of the obstacle are denoted by m, m'; σ, σ'; so that if V, V' be the velocities of propagation

$$V^2 = m/\sigma, \qquad V'^2 = m'/\sigma'. \qquad \qquad (39)$$

The velocity-potential of the undisturbed plane waves is represented by

$$\phi = e^{ikVt} \cdot e^{ikx}, \qquad \qquad (40)$$

in which $k = 2\pi/\lambda$. The time factor e^{ikVt}, which operates throughout, may be omitted for the sake of brevity.

The velocity-potential (ψ) of the disturbance propagated outwards from T may be expanded in spherical harmonic terms *

$$r\psi = e^{-ikr} \{S_0 + S_1 f_1(ikr) + S_2 f_2(ikr) + \ldots\}, \qquad \ldots\ldots(41)$$

where $\qquad f_n(ikr) = 1 + \dfrac{n(n+1)}{2 \cdot ikr} + \dfrac{(n-1)\ldots(n+2)}{2 \cdot 4 \cdot (ikr)^2}$

$$+ \ldots\ldots + \dfrac{1 \cdot 2 \cdot 3 \ldots 2n}{2 \cdot 4 \cdot 6 \ldots 2n \, (ikr)^n} \cdot \qquad \ldots\ldots\ldots(42)$$

At a great distance from the obstacle $f_n(ikr) = 1$; and the relative importance of the various harmonic terms decreases in going outwards with the order of the harmonic. For the present purpose we shall need to regard only the terms of order 0 and 1. Of these the term of order 0 depends upon the variation of compressibility, and that of order 1 upon the variation of density.

The relation between the variable part of the pressure δp, the condensation s, and ϕ is

$$V^2 s = - \frac{d\phi}{dt} = \frac{\delta p}{\sigma};$$

so that during the passage of the undisturbed primary waves the rate at which fluid enters the volume T (supposed for the moment to be of the same quality as the surrounding medium) is

$$T \frac{ds}{dt} = - \frac{T}{V^2} \frac{d^2\phi}{dt^2} = k^2 T. \qquad \qquad (43)$$

If the obstacle present an unyielding surface, its effect is to prevent the entrance of the fluid (43); that is, to superpose upon the plane waves such a

* *Theory of Sound*, §§ 323, 324.

disturbance as is caused by the *introduction* of (43) into the medium. Thus, if the potential of this disturbance be

$$\psi = S_0 \frac{e^{-ikr}}{r}, \quad \dots\dots\dots\dots\dots(44)$$

S_0 is to be determined by the condition that when $r = 0$

$$4\pi r^2 d\psi/dr = k^2 T,$$

so that $S_0 = - k^2 T/4\pi$, and

$$\psi = - \frac{k^2 T}{4\pi} \frac{e^{-ikr}}{r} = - \frac{\pi T}{\lambda^2} \frac{e^{-ikr}}{r}. \quad \dots\dots\dots\dots(45)$$

This result corresponds with $m' = \infty$ representing absolute incompressibility. The effect of finite compressibility, differing from that of the surrounding medium, is readily inferred by means of the pressure relation ($\delta p = ms$). The effect of the variation of compressibility at the obstacle is to increase the rate of introduction of fluid into T from what it would otherwise be in the ratio $m : m'$; and thus (45) now becomes

$$\psi = - \frac{\pi T}{\lambda^2} \frac{m' - m}{m'} \frac{e^{-ikr}}{r}; \quad \dots\dots\dots\dots(46)$$

or if we restore the factor e^{ikVt} and throw away the imaginary part of the solution,

$$\psi = - \frac{\pi T}{\lambda^2 r} \frac{m' - m}{m'} \cos k(Vt - r). \quad \dots\dots\dots(47)$$

This is superposed upon the primary waves

$$\phi = \cos k(Vt + x). \quad \dots\dots\dots\dots\dots(48)$$

When $m' = 0$, *i.e.*, when the material composing the obstacle offers no resistance to compression, (47) fails. In this case the condition to be satisfied at the surface of T is the evanescence of δp, or of the total potential ($\phi + \psi$). In the neighbourhood of the obstacle $\phi = 1$; and thus if M' denote the electrical "capacity" of a conducting body of form T situated in the open, $\psi = - M'/r$, r being supposed to be large in comparison with the linear dimension of T but small in comparison with λ. The latter restriction is removed by the insertion of the factor e^{-ikr}; and thus, in place of (46), we now have

$$\psi = - \frac{M'e^{-ikr}}{r}. \quad \dots\dots\dots\dots(49)$$

The value of M' may be expressed when T is in the form of an ellipsoid. For a sphere of radius R,

$$M' = R; \quad \dots\dots\dots\dots(50)$$

for a circular plate of radius R,

$$M' = 2R/\pi. \quad \dots\dots\dots\dots(51)$$

When the density of the obstacle (σ') is the same as that of the surrounding medium, (47) constitutes the complete solution. Otherwise the difference of densities causes an interference with the flow of fluid, giving rise to a disturbance of order 1 in spherical harmonics. This disturbance is independent of that already considered, and the flow in the neighbourhood of the obstacle may be calculated as if the fluid were incompressible. We thus fall back upon the problem considered in the earlier part of this paper, and the results will be applicable as soon as we have established the correspondence between density and conductivity.

In the present problem, if χ denote the whole velocity-potential, the conditions to be satisfied at any part of the surface of the obstacle are the continuity of $d\chi/dn$ and of $\sigma\chi$, the latter of which represents the pressure. Thus, if we regard $\sigma\chi$ as the variable, the conditions are the continuity of $(\sigma\chi)$ and of $\sigma^{-1}d(\sigma\chi)/dn$. In the conductivity problem the conditions to be satisfied by the potential (χ') are the continuity of χ' and of $\mu d\chi'/dn$.

In an expression relating only to the external region where σ is constant, it makes no difference whether we are dealing with $\sigma\chi$ or with χ; and accordingly there is correspondence between the two problems provided that we suppose the ratio of μ's in the one problem to be the reciprocal of the ratio of the σ's in the other.

We may now proceed to the calculation of the disturbance due to an obstacle, based upon the assumption that there is a region over which r is large compared with the linear dimension of T, but small in comparison with λ. Within this region ψ is given by (8) if the motion be referred to certain principal axes determined by the nature and form of the obstacle, the quantities u, v, w being the components of flow in the primary waves. By (41), (42), this is to be identified with

$$\psi = S_1 \frac{e^{-ikr}}{r}\left(1 + \frac{1}{ikr}\right), \qquad \ldots\ldots\ldots\ldots\ldots\ldots(52)$$

when r is small in comparison with λ; so that

$$S_1 = \frac{ik(A_1 ux + B_2 vy + C_3 wz)}{r}. \qquad \ldots\ldots\ldots\ldots\ldots(53)$$

At a great distance from T, (52) reduces to

$$\psi = \frac{ik(A_1 ux + B_2 vy + C_3 wz)e^{-ikr}}{r^2}, \qquad \ldots\ldots\ldots\ldots\ldots(54)$$

—a term of order 1, to be added to that of zero order given in (46).

In general, the axis of the harmonic in (54) is inclined to the direction of propagation of the primary waves; but there are certain cases of exception. For example, v and w vanish if the primary propagation be parallel to x (one of the principal axes). Again, as for a sphere or a cube, A_1, B_2, C_3 may be equal.

We will now limit ourselves to the case of the ellipsoid, and for brevity will further suppose that the primary waves move parallel to x, so that $v = w = 0$. The terms corresponding to u and v, if existent, are simply superposed. If, as hitherto, $\phi = e^{ikx}$, $u = ik$; so that by (14), σ being substituted for μ' and σ' for μ,

$$A = \frac{ik(\sigma - \sigma')}{4\pi\sigma' + (\sigma - \sigma')L} \cdot \quad\quad\quad\dots\dots\dots(55)$$

In the intermediate region by (28) $\psi = - TAx/r^3$, and thus at a great distance

$$\psi = - \frac{ikx\,TAe^{-ikr}}{r^2}; \quad\quad\dots\dots\dots\dots(56)$$

or on substitution of the values of A and k,

$$\psi = - \frac{\pi Tx e^{-ikr}}{\lambda^2 r^2} \cdot \frac{4\pi(\sigma' - \sigma)}{4\pi\sigma' + (\sigma - \sigma')L} \cdot \quad\dots\dots\dots(57)$$

Equations (46), (57) express the complete solution in the case supposed.

For an obstacle which is rigid and fixed, we may deduce the result by supposing in our equations $m' = \infty$, $\sigma' = \infty$. Thus

$$\psi = - \frac{\pi Te^{-ikr}}{\lambda^2 r} \left\{ 1 + \frac{x}{r} \frac{4\pi}{4\pi - L} \right\} \cdot \quad\dots\dots\dots(58)$$

Certain particular cases are worthy of notice. For the sphere $L = \frac{4}{3}\pi$, and

$$\psi = - \frac{\pi Te^{-ikr}}{\lambda^2 r} \left\{ 1 + \frac{3x}{2r} \right\} \cdot \quad\quad\dots\dots\dots(59)^*$$

If the ellipsoid reduce to an infinitely thin circular disk of radius c, $T = 0$ and the term of zero order vanishes. The term of the first order also vanishes if the plane of the disk be parallel to x. If the plane of the disk be perpendicular to x, $4\pi - L$ is infinitesimal. By (21), (24) we get in this case

$$\frac{4\pi T}{4\pi - L} = \frac{8c^3}{3};$$

so that

$$\psi = - \frac{8\pi c^3}{3\lambda^2} \frac{x}{r} \frac{e^{-ikr}}{r} \cdot \quad\quad\dots\dots\dots(60)$$

If the axis of the disk be inclined to that of x, ψ retains its symmetry with respect to the former axis, and is reduced in magnitude in the ratio of the cosine of the angle of inclination to unity.

In the case of the sphere the general solution is

$$\psi = - \frac{\pi Te^{-ikr}}{\lambda^2 r} \left\{ \frac{m' - m}{m'} + \frac{3x}{r} \frac{\sigma' - \sigma}{2\sigma' + \sigma} \right\} \cdot \quad\dots\dots(61)\dagger$$

---

* *Theory of Sound*, § 334.   † L c. cit. § 335.

Waves in Two Dimensions.

In the case of two dimensions (x, y) the waves diverging from a cylindrical obstacle have the expression, analogous to (41),

$$\psi = S_0 D_0(kr) + S_1 D_1(kr) + \dots , \qquad \dots\dots\dots\dots(62)^*$$

where $S_0, S_1 \dots$ are the plane circular functions of the various orders, and

$$D_0(kr) = -\left(\frac{\pi}{2ikr}\right)^{\frac{1}{2}} e^{-ikr}\left\{1 - \frac{1^2}{1 \cdot 8ikr} + \dots\right\}$$

$$= \left(\gamma + \log\frac{ikr}{2}\right)\left\{1 - \frac{k^2 r^2}{2^2} + \dots\right\} + \frac{k^2 r^2}{2^2} - \frac{3}{2}\frac{k^4 r^4}{2^2 \cdot 4^2} + \dots , \qquad \dots\dots(63)$$

$$D_1(kr) = \frac{dD_0(kr)}{d(kr)} = \left(\frac{\pi i}{2kr}\right)^{\frac{1}{2}} e^{-ikr}\left\{1 - \frac{-1 \cdot 3}{1 \cdot 8ikr} + \dots\right\}$$

$$= \frac{1}{kr}\left\{1 - \frac{k^2 r^2}{2^2} + \dots\right\} + \left(\gamma + \log\frac{ikr}{2}\right)\left\{\frac{kr}{2} - \frac{k^3 r^3}{2^2 \cdot 4} + \dots\right\}$$

$$+ \frac{kr}{2} - \frac{3}{2}\frac{k^3 r^3}{2^2 \cdot 4} + \dots . \qquad \dots\dots\dots\dots\dots\dots\dots\dots\dots(64)$$

As in the case of three dimensions already considered, the term of zero order in ψ depends upon the variation of compressibility. If we again begin with the case of an unyielding boundary, the constant S_0 is to be found from the condition that when $r = 0$

$$2\pi r \, d\psi / dr = k^2 T,$$

T denoting now the area of cross-section. When r is small,

$$\frac{dD_0(kr)}{dr} = \frac{1}{r};$$

and thus $S_0 = k^2 T / 2\pi$,

$$\psi = \frac{k^2 T}{2\pi} D_0(kr) = -\frac{k^2 T}{2\pi}\left(\frac{\pi}{2ikr}\right)^{\frac{1}{2}} e^{-ikr}, \dots \qquad \dots\dots\dots\dots(65)$$

when r is very great. This corresponds to (45).

In like manner, if the compressibility of the obstacle be finite,

$$\psi = -\frac{k^2 T}{\pi}\left(\frac{\pi}{2ikr}\right)^{\frac{1}{2}} \frac{m' - m}{2m'} e^{-ikr} \dots \qquad \dots\dots\dots\dots(66)$$

The factor $i^{-\frac{1}{2}} = e^{-\frac{1}{4}i\pi}$; and thus if we restore the time-factor e^{ikVt}, and reject the imaginary part of the solution, we have

$$\psi = -\frac{2\pi T}{r^{\frac{1}{2}}\lambda^{\frac{3}{2}}} \frac{m' - m}{2m'} \cos\frac{2\pi}{\lambda}(Vt - r - \tfrac{1}{8}\lambda), \qquad \dots\dots\dots\dots(67)$$

* See *Theory of Sound*, § 341; *Phil. Mag.* April, 1897, p. 266. [Vol. IV. p. 290.]

corresponding to the plane waves

$$\phi = \cos \frac{2\pi}{\lambda} (Vt + x). \quad \dots\dots\dots\dots\dots\dots(68)$$

In considering the term of the first order we will limit ourselves to the case of the cylinder of elliptic section, and suppose that one of the principal axes of the ellipse is parallel to the direction (x) of primary wave-propagation. Thus in (34), which gives the value of ψ at a distance from the cylinder which is great in comparison with a and b, but small in comparison with λ, we are to suppose $u = ik$, $v = 0$, at the same time substituting σ, σ' for μ', μ respectively. Thus for the region in question

$$\psi = \frac{ab \cdot ikx}{2r^2} \frac{(\sigma' - \sigma)(a + b)}{\sigma'a + \sigma b}; \quad \dots\dots\dots\dots\dots(69)$$

and this is to be identified with $S_1 D_1(kr)$ when kr is small, i.e. with S_1/kr. Accordingly

$$S_1 = \frac{x}{r} \frac{ik^2 ab}{2} \frac{(\sigma' - \sigma)(a + b)}{\sigma'a + \sigma b};$$

so that, at a distance r great in comparison with λ, ψ becomes

$$\psi = -\frac{k^2 T}{\pi} \left(\frac{\pi}{2ikr}\right)^{\frac{1}{2}} \frac{(\sigma' - \sigma)(a + b)}{2(\sigma'a + \sigma b)} \frac{x}{r} e^{-ikr}, \quad \dots\dots\dots\dots(70)$$

T being written for πab. The complete solution for a great distance is given by addition of (66) and (70), and corresponds to $\phi = e^{ikx}$.

In the case of circular section $(b = a)$ we have altogether *

$$\psi = -k^2 a^2 e^{-ikr} \left(\frac{\pi}{2ikr}\right)^{\frac{1}{2}} \left\{\frac{m' - m}{2m'} + \frac{\sigma' - \sigma}{\sigma' + \sigma} \frac{x}{r}\right\}, \quad \dots\dots\dots(71)$$

which may be realized as in (67). If the material be unyielding, the corresponding result is obtained by making $m' = \infty$, $\sigma' = \infty$ in (71). The realized value is then †

$$\psi = -\frac{2\pi \cdot \pi a^2}{r^{\frac{1}{2}}\lambda^{\frac{3}{2}}} \left(\frac{1}{2} + \frac{x}{r}\right) \cos \frac{2\pi}{\lambda} (Vt - r - \tfrac{1}{8}\lambda). \quad \dots\dots(72)$$

In general, if the material be unyielding, we get from (66), (70)

$$\psi = -k^2 ab e^{-ikr} \left(\frac{\pi}{2ikr}\right)^{\frac{1}{2}} \left(\frac{1}{2} + \frac{a + b}{2a} \frac{x}{r}\right). \quad \dots\dots\dots\dots(73)$$

The most interesting case of a difference between a and b is when one of them vanishes, so that the cylinder reduces to an infinitely thin blade. If

---

* *Theory of Sound*, § 343.
† *Loc. cit.* equation (17).

$b = 0$, ψ vanishes as to both its parts; but if $a = 0$, although the term of zero order vanishes, that of the first order remains finite, and we have

$$\psi = - \tfrac{1}{2}k^2 b^2 e^{-ikr} \left(\frac{\pi}{2ikr}\right)^{\tfrac{1}{2}} \frac{x}{r}, \quad \dots\dots\dots\dots(74)$$

in agreement with the value formerly obtained*.

It remains to consider the extreme case which arises when $m' = 0$. The term of zero order in circular harmonics, as given in (66), then becomes infinite, and that of the first order (70) is relatively negligible. The condition to be satisfied at the surface of the obstacle is now the evanescence of the total potential $(\phi + \psi)$, in which $\phi = 1$.

It will conduce to clearness to take first the case of the circular cylinder (a). By (62), (63) the surface condition is

$$S_0 \{\gamma + \log (\tfrac{1}{2}ika)\} + 1 = 0. \quad \dots\dots\dots\dots(75)$$

Thus at a distance r great in comparison with λ we have

$$\psi = \frac{e^{-ikr}}{\gamma + \log (\tfrac{1}{2}ika)} \left(\frac{\pi}{2ikr}\right)^{\tfrac{1}{2}}. \quad \dots\dots\dots\dots(76)$$

When the section of the obstacle is other than circular, a less direct process must be followed. Let us consider a circle of radius ρ concentric with the obstacle, where ρ is large in comparison with the dimensions of the obstacle but small in comparison with λ. Within this circle the flow may be identified with that of an incompressible fluid. On the circle we have

$$\phi + \psi = 1 + S_0 \{\gamma + \log (\tfrac{1}{2}ik\rho)\}, \quad \dots\dots\dots\dots(77)$$

$$2\pi d (\phi + \psi)/dr = 2\pi S_0, \quad \dots\dots\dots\dots(78)$$

of which the latter expresses the flow of fluid across the circumference. This flow in the region between the circle and the obstacle corresponds to the potential-difference (77). Thus, if R denote the electrical resistance between the two surfaces (reckoned of course for unit length parallel to z),

$$S_0 \{\gamma + \log (\tfrac{1}{2}ik\rho) - 2\pi R\} = 1, \quad \dots\dots\dots\dots(79)$$

and $\psi = S_0 D_0 (kr)$, as usual.

The value of S_0 in (79) is of course independent of the actual value of ρ, so long as it is large. If the obstacle be circular,

$$2\pi R = \log (\rho/a).$$

The problem of determining R for an elliptic section (a, b) can, as is well known, be solved by the method of conjugate functions. If we take

$$x = c \cosh \xi \cos \eta, \qquad y = c \sinh \xi \sin \eta, \quad \dots\dots\dots\dots(80)$$

* *Phil. Mag.* April 1897, p. 271. [Vol. IV. p. 295.] The primary waves are there supposed to travel in the direction of $+x$, but here in the direction of $-x$.

the confocal ellipses

$$\frac{x^2}{\cosh^2 \xi} + \frac{y^2}{\sinh^2 \xi} = c^2 \quad \dots\dots\dots\dots(81)$$

are the equipotential curves. One of these, for which ξ is large, can be identified with the circle of radius ρ, the relation between ρ and ξ being

$$\xi = \log(2\rho/c).$$

An inner one, for which $\xi = \xi_0$, is to be identified with the ellipse (a, b), so that

$$a = c \cosh \xi_0, \qquad b = c \sinh \xi_0,$$

whence

$$c^2 = a^2 - b^2, \qquad \tanh \xi_0 = b/a.$$

Thus

$$2\pi R = \xi - \xi_0 = \log \frac{2\rho}{a+b}; \quad \dots\dots\dots\dots (82)$$

and then (79) gives as applicable at a great distance

$$\psi = \frac{e^{-ikr}}{\gamma + \log\{\frac{1}{4}ik(a+b)\}} \left(\frac{\pi}{2ikr}\right)^{\frac{1}{2}}. \quad \dots\dots\dots\dots(83)$$

The result for an infinitely thin blade is obtained by merely putting $b = 0$ in (83).

For some purposes the imaginary part of the logarithmic term may be omitted. The realized solution is then

$$\psi = \left(\frac{\pi}{2kr}\right)^{\frac{1}{2}} \frac{\cos k(Vt - r - \frac{1}{8}\lambda)}{\gamma + \log\{\frac{1}{4}k(a+b)\}}, \quad \dots\dots \dots\dots\dots(84)$$

corresponding, as usual, to

$$\phi = \cos k(Vt + x). \quad \dots\dots\dots\dots\dots(85)$$

Electrical Applications.

The problems in two dimensions for aerial waves incident upon an obstructing cylinder of small transverse dimensions are analytically identical with certain electric problems which will now be specified. The general equation $(\bar{\nabla}^2 + k^2) = 0$ is satisfied in all cases. In the ordinary electrical notation $V^2 = 1/K\mu$, $V'^2 = 1/K'\mu'$; while in the acoustical problem $V^2 = m/\sigma$, $V'^2 = m'/\sigma'$. The boundary conditions are also of the same general form. Thus if the primary waves be denoted by $\gamma = e^{ikx}$, γ being the magnetic force parallel to z, the conditions to be satisfied at the surface of the cylinder are the continuity of γ and of $K^{-1} d\gamma/dn$. Comparing with the acoustical conditions we see that K replaces σ, and consequently (by the value of V^2) μ replaces $1/m$. These substitutions with that of γ, or c (the magnetic induction), for ψ and ϕ suffice to make (66), (70) applicable to the electrical problem. For example, in the case of the circular cylinder, we have for the dispersed wave

$$c = - k^2 a^2 e^{-ikr} \left(\frac{\pi}{2ikr}\right)^{\frac{1}{2}} \left\{\frac{\mu - \mu'}{2\mu} + \frac{K' - K}{K' + K}\frac{x}{r}\right\}, \qquad \dots\dots\dots(86)$$

corresponding to the primary waves

$$c = e^{ikx}. \qquad \dots\dots\dots\dots\dots\dots\dots(87)$$

An important particular case is obtained by making $K' = \infty$, $\mu' = 0$, in such a way that V' remains finite. This is equivalent to endowing the obstacle with the character of a perfect conductor, and we get

$$c = - k^2 a^2 e^{-ikr} \left(\frac{\pi}{2ikr}\right)^{\frac{1}{2}} \left\{\frac{1}{2} + \frac{x}{r}\right\}, \qquad \dots\dots\dots\dots(88)$$

which, when realized, coincides with (72).

The other two-dimensional electrical problem is that in which everything is expressed by means of R, the electromotive intensity parallel to z. The conditions at the surface are now the continuity of R and of $\mu^{-1}dR/dn$. Thus K and μ are simply interchanged, μ replacing σ and K replacing $1/m$ in (66), (70), ϕ and ψ also being replaced by R. In the case of the circular cylinder

$$R = - k^2 a^2 e^{-ikr} \left(\frac{\pi}{2ikr}\right)^{\frac{1}{2}} \left\{\frac{K - K'}{2K} + \frac{\mu' - \mu}{\mu' + \mu}\frac{x}{r}\right\}, \qquad \dots\dots\dots(89)$$

corresponding to the primary waves

$$R = e^{ikx}. \qquad \dots\dots\dots\dots\dots\dots\dots(90)$$

If in order to obtain the solution for a perfectly conducting obstacle we make $K' = \infty$, $\mu' = 0$, (89) becomes infinite, and must be replaced by the analogue of (83). Thus for the perfectly conducting circular obstacle

$$R = \frac{e^{-ikr}}{\gamma + \log\left(\frac{1}{2}ika\right)} \left(\frac{\pi}{2ikr}\right)^{\frac{1}{2}}, \qquad \dots\dots\dots\dots(91)$$

which may be realized as in (84).

The problem of a conducting cylinder is treated by Prof. J. J. Thomson in his valuable *Recent Researches in Electricity and Magnetism*, § 364; but his result differs from (84), not only in respect to the sign of $\frac{1}{8}\lambda$, but also in the value of the denominator*. The values here given are those which follow from the equations (9), (17) of § 343 *Theory of Sound*.

Electric Waves in Three Dimensions.

In the problems which arise under this head the simple acoustical analogue no longer suffices, and we must appeal to the general electrical

---

* It should be borne in mind that γ here is the same as Prof. Thomson's $\log \gamma$.

equations of Maxwell. The components of electric polarization (f, g, h) and of magnetic force (α, β, γ), being proportional to e^{ikVt}, all satisfy the fundamental equation

$$(\nabla^2 + k^2) = 0; \quad \dots\dots\dots\dots\dots\dots\dots\dots(92)$$

and they are connected together by such relations as

$$4\pi \frac{df}{dt} = \frac{d\gamma}{dy} - \frac{d\beta}{dz}, \quad \dots\dots\dots\dots\dots\dots\dots(93)$$

or

$$\frac{d\alpha}{dt} = 4\pi V^2 \left(\frac{dg}{dz} - \frac{dh}{dy}\right), \quad \dots\dots\dots\dots(94)$$

in which any differentiation with respect to t is equivalent to the introduction of the factor ikV. Further

$$\frac{df}{dx} + \frac{dg}{dy} + \frac{dh}{dz} = 0, \qquad \frac{d\alpha}{dx} + \frac{d\beta}{dy} + \frac{d\gamma}{dz} = 0. \quad \dots\dots\dots\dots(95)$$

The electromotive intensity (P, Q, R) and the magnetization (a, b, c) are connected with the quantities already defined by the relations

$$f, g, h = K(P, Q, R)/4\pi; \qquad a, b, c = \mu(\alpha, \beta, \gamma); \dots\dots\dots(96)$$

in which K denotes the specific inductive capacity and μ the permeability; so that $V^{-2} = K\mu$.

The problem before us is the investigation of the disturbance due to a small obstacle (K', μ') situated at the origin, upon which impinge primary waves denoted by

$$f_0 = 0, \qquad g_0 = 0, \qquad h_0 = e^{ikx}, \quad \dots\dots\dots\dots\dots(97)$$

or, as follows from (94),

$$\alpha_0 = 0, \qquad \beta_0 = 4\pi V e^{ikx}, \qquad \gamma_0 = 0. \quad \dots\dots\dots\dots(98)$$

The method of solution, analogous to that already several times employed, depends upon the principle that in the neighbourhood of the obstacle and up to a distance from it great in comparison with the dimensions of the obstacle but small in comparison with λ, the condition at any moment may be identified with a steady condition such as is determined by the solution of a problem in conduction. When this is known, the disturbance at a distance from the obstacle may afterwards be derived.

We will commence with the case of the *sphere*, and consider first the magnetic functions as disturbed by the change of permeability from μ to μ'. Since in the neighbourhood of the sphere the problem is one of steady distribution, α, β, γ are derivable from a potential. By (98), in which we may write $e^{ikx} = 1$, the primary potential is $4\pi Vy$; so that in (1) we are to take $u = 0$, $v = 4\pi V$, $w = 0$. Hence by (20) α, β, γ for the *disturbance* are given by

$$\alpha = d\psi/dx, \qquad \beta = d\psi/dy, \qquad \gamma = d\psi/dz,$$

where
$$\psi = -4\pi V \frac{\mu' - \mu}{\mu' + 2\mu} \frac{a^3 y}{r^3} . \qquad\dots\dots\dots\dots\dots\dots(99)$$

In like manner f, g, h are derivable from a potential χ. The primary potential is z simply, so that in (1), $u = 0, v = 0, w = 1$. Hence by (20)

$$\chi = -\frac{K' - K}{K' + 2K} \frac{a^3 z}{r^3} , \qquad\dots\dots\dots\dots\dots\dots(100)$$

from which f, g, h for the disturbance are derived by simple differentiations with respect to x, y, z respectively.

Since $f, g, h, \alpha, \beta, \gamma$ all satisfy (92), the values at a distance can be derived by means of (41). The terms resulting from (99), (100) are of the second order in spherical harmonics. When r is small,

$$r^{-1} e^{-ikr} f_2 (ikr) = -3/k^2 r^3,$$

and when r is great

$$r^{-1} e^{-ikr} f_2 (ikr) = r^{-1} e^{-ikr};$$

so that, as regards an harmonic of the second order, the value at a distance will be deduced from that in the neighbourhood of the origin by the introduction of the factor $-\frac{1}{3} k^2 r^2 e^{-ikr}$. Thus, for example, f in the neighbourhood of the origin is

$$f = \frac{d\chi}{dx} = \frac{K' - K}{K' + 2K} \frac{3a^3 xz}{r^5} ; \qquad\dots\dots\dots\dots\dots(101)$$

so that at a great distance we get

$$f = -\frac{K' - K}{K' + 2K} \frac{k^2 a^3 xz e^{-ikr}}{r^3} . \qquad\dots\dots\dots\dots\dots(102)$$

In this way the terms of the second order in spherical harmonics are at once obtained, but they do not constitute the complete solution of the problem. We have also to consider the possible occurrence of terms of other orders in spherical harmonics. Terms of order higher than the second are indeed excluded, because in the passage from r small to r great they suffer more than do the terms of the second order. But for a like reason it may happen that terms of order zero and 1 in spherical harmonics rise in relative importance so as to be comparable at a distance with the term of the second order, although relatively negligible in the neighbourhood of the obstacle. The factor, analogous to $-\frac{1}{3} k^2 r^2 e^{-ikr}$ for the second order, is for the first order $ikr e^{-ikr}$, and for zero order e^{-ikr}. Thus, although (101) gives the value of f with sufficient completeness for the neighbourhood of the obstacle, (102) may need to be supplemented by terms of the first and zero orders in spherical harmonics of the same importance as itself. The supplementary terms may be obtained without much difficulty from those already arrived at by means of the relations (93), (94), (95); but the process is rather cumbrous, and it seems better to avail ourselves of the forms deduced by Hertz * for electric vibrations radiated from a centre.

If we write $\Pi = Ae^{-ikr}/r$, the solution corresponding to an impressed electric force acting at the origin parallel to z is

$$f = \frac{d^2 \Pi}{dx\,dz}, \qquad g = -\frac{d^2 \Pi}{dy\,dz}, \qquad h = \frac{d^2 \Pi}{dx^2} + \frac{d^2 \Pi}{dy^2}; \quad \ldots\ldots(103)$$

$$\alpha = -4\pi \frac{d^2 \Pi}{dy\,dt}, \qquad \beta = 4\pi \frac{d^2 \Pi}{dx\,dt}, \qquad \gamma = 0. \quad \ldots\ldots\ldots(104)$$

These values evidently satisfy (92) since Π does so, and they harmonize with (93), (94), (95).

In the neighbourhood of the origin, where kr is small, e^{-ikr} may be identified with unity, so that $\Pi = A/r$. In this case (103) may be written

$$f = -\frac{d^2 \Pi}{dx\,dz}, \qquad g = -\frac{d^2 \Pi}{dy\,dz}, \qquad h = -\frac{d^2 \Pi}{dz^2},$$

and all that remains is to identify $-d\Pi/dz$ with χ in (100). Accordingly

$$A = -a^3 \frac{K' - K}{K' + 2K}. \quad \ldots\ldots\ldots\ldots\ldots\ldots(105)$$

The values of f, g, h in (103) are now determined. Those of α, β, γ are relatively negligible in the neighbourhood of the origin. At a great distance we have

$$f = -A \frac{d^2}{dx\,dz}\left(\frac{e^{-ikr}}{r}\right) = -\frac{A}{r}\frac{d^2 e^{-ikr}}{dx\,dz} = \frac{k^2 A e^{-ikr}}{r}\frac{xz}{r^2};$$

so that (103), (104) may be written

$$f, g, h = \frac{K' - K}{K' + 2K}\frac{k^2 a^3 e^{-ikr}}{r}\left(-\frac{xz}{r^2}, -\frac{yz}{r^2}, \frac{x^2 + y^2}{r^2}\right), \quad \ldots\ldots(106)$$

$$\frac{\alpha, \beta, \gamma}{4\pi V} = \frac{K' - K}{K' + 2K}\frac{k^2 a^3 e^{-ikr}}{r}\left(\frac{y}{r}, -\frac{x}{r}, 0\right). \quad \ldots\ldots\ldots(107)$$

These equations give the values of the functions for a disturbance radiating from a small spherical obstacle, so far as it depends upon $(K' - K)$. We have to add a similar solution dependent upon the change from μ to μ'. In this (103), (104) are replaced by

$$\frac{\alpha}{V^2} = -\frac{d^2 \Pi}{dx\,dy}, \qquad \frac{\beta}{V^2} = \frac{d^2 \Pi}{dx^2} + \frac{d^2 \Pi}{dz^2}, \qquad \frac{\gamma}{V^2} = -\frac{d^2 \Pi}{dz\,dy}; \quad \ldots\ldots(108)$$

$$4\pi f = -\frac{d^2 \Pi}{dz\,dt}, \qquad g = 0, \qquad 4\pi h = \frac{d^2 \Pi}{dx\,dt}, \quad \ldots\ldots\ldots\ldots(109)$$

---

* *Ausbreitung der electrischen Kraft*, Leipzig, 1892, p. 150. It may be observed that the solution for the analogous but more difficult problem relating to an elastic solid was given much earlier by Stokes (*Camb. Trans.*, Vol. IX. p. 1, 1849). Compare *Theory of Sound*, 2nd ed. § 378.

where $\Pi = Be^{-ikr}/r$, corresponding to an impressed magnetic force parallel to y. In the neighbourhood of the origin (108) becomes

$$\frac{\alpha}{V^2} = -\frac{d^2\Pi}{dx\,dy}, \quad \frac{\beta}{V^2} = -\frac{d^2\Pi}{dy^2}, \quad \frac{\gamma}{V^2} = -\frac{d^2\Pi}{dz\,dy},$$

so that ψ in (99) is to be identified with $-V^2 d\Pi/dy$. Thus

$$B = -\frac{4\pi a^3}{V}\frac{\mu'-\mu}{\mu'+2\mu}. \qquad\qquad\qquad\text{.........................(110)}$$

At a great distance we have

$$f, g, h = \frac{\mu'-\mu}{\mu'+2\mu}\frac{k^2 a^3 e^{-ikr}}{r}\left(\frac{z}{r},\ 0,\ -\frac{x}{r}\right); \quad\text{............(111)}$$

$$\frac{\alpha, \beta, \gamma}{4\pi V} = \frac{\mu'-\mu}{\mu'+2\mu}\frac{k^2 a^3 e^{-ikr}}{r}\left(-\frac{xy}{r^2},\ \frac{x^2+z^2}{r^2},\ -\frac{zy}{r^2}\right). \quad\text{......(112)}$$

By addition of (111) to (106) and of (112) to (107) we obtain the complete values of f, g, h, α, β, γ when both the dielectric constant and the permeability undergo variation. The disturbance corresponding to the primary waves $h = e^{ikx}$ is thus determined.

When the changes in the electric constants are small, (106), (111) may be written

$$f = \frac{\pi T}{\lambda^2 r}e^{-ikr}\left(-\frac{\Delta K}{K}\frac{xz}{r^2} + \frac{\Delta\mu}{\mu}\frac{z}{r}\right), \qquad\text{.................(113)}$$

$$g = \frac{\pi T}{\lambda^2 r}e^{-ikr}\left(-\frac{\Delta K}{K}\frac{yz}{r^2}\right), \qquad\qquad\text{.........................(114)}$$

$$h = \frac{\pi T}{\lambda^2 r}e^{-ikr}\left(\frac{\Delta K}{K}\frac{x^2+y^2}{r^2} - \frac{\Delta\mu}{\mu}\frac{x}{r}\right), \qquad\text{..............(115)}$$

where $T = \frac{4}{3}\pi a^3$, $k = 2\pi/\lambda$. These are the results given formerly * as applicable in this case to an obstacle of volume T and of arbitrary form. When the obstacle is spherical and $\Delta K/K$ is not small, it was further shown that $\Delta K/K$ should be replaced by $3(K'-K)/(K'+2K)$†, and similar reasoning would have applied to $\Delta\mu/\mu$.

The solution for the case of a spherical obstacle having the character of a perfect conductor may be derived from the general expressions by supposing that $K' = \infty$, and (in order that V' may remain finite) $\mu' = 0$. We get from (106), (111),

---

* "Electromagnetic Theory of Light," *Phil. Mag.* Vol. xii. p. 90 (1881). [Vol. i. p. 526.]

† [1902. The " 3 " was inadvertently omitted in the original of the present paper.]

$$f = -\frac{k^2 a^3 e^{-ikr}}{r} \left(\frac{xz}{r^2} + \frac{z}{2r} \right), \quad \text{......................(116)}$$

$$g = -\frac{k^2 a^3 e^{-ikr}}{r} \frac{yz}{r^2}, \quad \text{............................(117)}$$

$$h = +\frac{k^2 a^3 e^{-ikr}}{r} \left(\frac{x^2 + y^2}{r^2} + \frac{x}{2r} \right), \quad \text{..............(118)}$$

in agreement with the results of Prof. J. J. Thomson*. As was to be expected, in every case the vectors (f, g, h), (α, β, γ), (x, y, z) are mutually perpendicular.

Obstacle in the Form of an Ellipsoid.

The case of an ellipsoidal obstacle of volume T, whose principal axes are parallel to those of x, y, z, i.e. parallel to the directions of propagation and of vibration in the primary waves, is scarcely more complicated. The passage from the values of the disturbance in the neighbourhood of the obstacle to that at a great distance takes place exactly as in the case of the sphere. The primary magnetic potential in the neighbourhood of the obstacle is $4\pi Vy$, and thus, as before, $u = 0$, $v = 4\pi V$, $w = 0$ in (1). Accordingly, by (14), $A = 0$, $C = 0$; and (28) gives

$$\psi = -4\pi V \frac{\mu' - \mu}{4\pi\mu + (\mu' - \mu) M} \frac{Ty}{r^3}, \quad \text{.................(119)}$$

corresponding to (99) for the sphere. In like manner the electric potential is

$$\chi = -\frac{K' - K}{4\pi K + (K' - K) N} \frac{Tz}{r^3}. \quad \text{.................(120)}$$

These potentials give by differentiation the values of α, β, γ and f, g, h respectively in the neighbourhood of the ellipsoid. Thus at a great distance we obtain for the part dependent on $(K' - K)$, as generalizations of (106), (107),

$$f, g, h = \frac{K' - K}{4\pi K + (K' - K) N} \frac{k^2 T e^{-ikr}}{r} \left(-\frac{xz}{r^2}, \ -\frac{yz}{r^2}, \ \frac{x^2 + y^2}{r^2} \right); \ \text{...(121)}$$

$$\frac{\alpha, \beta, \gamma}{4\pi V} = \frac{K' - K}{4\pi K + (K' - K) N} \frac{k^2 T e^{-ikr}}{r} \left(\frac{y}{r}, \ -\frac{x}{r}, \ 0 \right). \ \text{......(122)}$$

To these are to be added corresponding terms dependent upon $(\mu' - \mu)$, viz.:—

$$f, g, h = \frac{\mu' - \mu}{4\pi\mu + (\mu' - \mu) M} \frac{k^2 T e^{-ikr}}{r} \left(\frac{z}{r}, \ 0, \ -\frac{x}{r} \right); \ \text{......(123)}$$

$$\frac{\alpha, \beta, \gamma}{4\pi V} = \frac{\mu' - \mu}{4\pi\mu + (\mu' - \mu) M} \frac{k^2 T e^{-ikr}}{r} \left(-\frac{xy}{r^2}, \ \frac{x^2 + z^2}{r^2}, \ -\frac{zy}{r^2} \right). \ \text{...(124)}$$

* Recent Researches, § 377, 1893.

The sum gives the disturbance at a distance due to the impact of the primary waves,

$$h_0 = e^{ikx}, \qquad \beta_0 = 4\pi V e^{ikx}, \quad \dots\dots\dots\dots\dots(125)$$

upon the ellipsoid T of dielectric capacity K' and of permeability μ'.

As in the case of the sphere, the result for an ellipsoid of perfect conductivity is obtained by making $K' = \infty$, $\mu' = 0$.   Thus

$$f = -\frac{k^2 e^{-ikr}}{r}\left(\frac{T}{N}\frac{xz}{r^2} + \frac{T}{4\pi - M}\frac{z}{r}\right), \quad \dots\dots\dots\dots(126)$$

$$g = -\frac{k^2 e^{-ikr}}{r}\frac{T}{N}\frac{yz}{r^2}, \quad \dots\dots\dots\dots\dots\dots\dots(127)$$

$$h = +\frac{k^2 e^{-ikr}}{r}\left(\frac{T}{N}\frac{x^2 + y^2}{r^2} + \frac{T}{4\pi - M}\frac{x}{r}\right). \quad \dots\dots\dots(128)$$

Next to the sphere the case of greatest interest is that of a flat circular disk (radius $= R$).   The volume of the obstacle then vanishes, but the effect remains finite in certain cases notwithstanding.   Thus, if the axis of the disk be parallel to x, that is to the direction of primary propagation, we have (21), (25),

$$\frac{T}{N} = \frac{4R^3}{3\pi}, \qquad \frac{T}{4\pi - M} = 0. \quad \dots\dots\dots\dots\dots(129)$$

In spite of its thinness, the plate being a perfect conductor disturbs the electric field in its neighbourhood; but the magnetic disturbance vanishes, the zero permeability having no effect upon the magnetic flow parallel to its face.   If the axis of the disk be parallel to y (see (24)),

$$\frac{T}{N} = \frac{4R^3}{3\pi}, \qquad \frac{T}{4\pi - M} = \frac{2R^3}{3\pi}; \quad \dots\dots\dots\dots(130)$$

and if the axis be parallel to z,

$$\frac{T}{N} = 0, \qquad \frac{T}{4\pi - M} = 0, \quad \dots\dots\dots\dots(131)$$

so that in this case the obstacle produces no effect at all.

Circular Aperture in Conducting Screen.

The problem proposed is the incidence of plane waves ($h_0 = e^{ikx}$) upon an infinitely thin screen at $x = 0$ endowed with perfect electric conductivity and perforated by a circular aperture.   In the absence of a perforation there would of course be no waves upon the negative side, and upon the positive side the effect of the screen would merely be to superpose the reflected waves denoted by $h_0 = -e^{-ikx}$.   We wish to calculate the influence of a small circular aperture of radius R.

In accordance with the general principle the condition of things is determined by what happens in the neighbourhood of the aperture, and this is substantially the same as if the wave-length were infinite. The problem is then expressible by means of a common potential. The magnetic force at a distance from the aperture on the positive side is altogether $8\pi V$, and on the negative side zero; while the condition to be satisfied upon the faces of the screen is that the force be entirely tangential. The general character of the flow is indicated in Fig. 1.

Fig. 1.   Fig. 2.   Fig. 3.

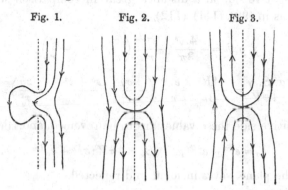

The problem here proposed is closely connected with those which we have already considered where no infinite screen was present, but a flat finite obstacle, which may be imagined to coincide with the proposed aperture. The primary magnetic field being $\beta = 4\pi V$, and the disk of radius R being of infinite permeability, the potential at a distance great compared with R (but small compared with λ) is by (27), (28)

$$\psi = -4\pi V \frac{4R^3}{3\pi} \frac{y}{r^3}. \qquad\qquad\qquad (132)$$

By the symmetry the part of the plane $x = 0$ external to the disk is not crossed by the lines of flow, and thus it will make no difference in the conditions if this area be filled up by a screen of zero permeability. On the other hand, the part of the plane $x = 0$ represented by the disk is met normally by the lines of flow. This state of things is indicated in Fig. 2.

The introduction of the lamina of zero permeability effects the isolation of the positive and negative sides. We may therefore now reverse the flow upon the negative side, giving the state of things indicated in Fig. 3. But the plate of infinite permeability then loses its influence and may be removed, so as to re-establish a communication between the positive and negative sides through an aperture. The passage from the present state of things to that of Fig. 1 is effected by superposition upon the whole field of $\beta = 4\pi V$, so as to destroy the field at a distance from the aperture upon the negative side and upon the positive side to double it.

As regards the solution of the proposed problem we have then on the positive side

$$\psi = 8\pi Vy - 4\pi V \frac{4R^3}{3\pi} \frac{y}{r^3}, \quad \text{...................}(133)$$

and on the negative side

$$\psi = 4\pi V \frac{4R^3}{3\pi} \frac{y}{r^3}. \quad \text{....................}(134)$$

Thus on the negative side at a distance great in comparison with the wave-length we get, as in (99), (111), (112),

$$f, g, h = - \frac{4R^3}{3\pi} \frac{k^2 e^{-ikr}}{r} \left(\frac{z}{r}, \quad 0, \quad -\frac{x}{r} \right), \quad \text{............}(135)$$

$$\frac{\alpha, \beta, \gamma}{4\pi V} = - \frac{4R^3}{3\pi} \frac{k^2 e^{-ikr}}{r} \left(-\frac{xy}{r^2}, \quad \frac{x^2 + z^2}{r^2}, \quad -\frac{zy}{r^2} \right). \quad \text{......}(136)$$

On the positive side these values are to be reversed, and addition made of

$$h_0 = e^{ikx} - e^{-ikx}, \qquad \beta_0 = 4\pi V (e^{ikx} + e^{-ikx}), \quad \text{.........}(137)$$

representing the plane waves incident and reflected.

The solution for h in (135) may be compared with that obtained (27), (28) in a former paper*, where, however, the primary waves were supposed to travel in the positive, instead of, as here, in the negative direction. It had at first been supposed that the solution for ϕ there given might be applied directly to h, which satisfies the condition (imposed upon ϕ) of vanishing upon the faces of the screen. If this were admitted, as also $g = 0$ throughout, the value of h would follow by (95). The argument was, however, felt to be insufficient on account of the discontinuities which occur at the *edge* of the aperture, and the value now obtained, though of the same form, is doubly as great.

* "On the Passage of Waves through Apertures in Plane Screens, and Allied Problems," *Phil. Mag.* Vol. XLIII. p. 264 (1897). [Vol. IV. p. 287.]

231.

ON THE PROPAGATION OF ELECTRIC WAVES ALONG CYLINDRICAL CONDUCTORS OF ANY SECTION.

[*Philosophical Magazine*, XLIV. pp. 199—204, 1897.]

THE problem of the propagation of waves along conductors has been considered by Mr Heaviside and Prof. J. J. Thomson, for the most part with limitation to the case of a wire of circular section with a coaxal sheath serving as a return. For practical applications it is essential to treat the conductivity of the wire as finite; but for some scientific purposes the conductivity may be supposed perfect without much loss of interest. Under this condition the problem is so much simplified that important extensions may be made in other directions. For example, the complete solution may be obtained for the case of parallel wires, even although the distance between them be not great in comparison with their diameters.

We may start from the general equations of Maxwell involving the electromotive intensity (P, Q, R) and the magnetic induction (a, b, c), introducing the supposition that all the functions are proportional to $e^{i\,(pt+mz)}$, and further that $m = p/V$, just as in the case of uninterrupted plane waves propagated parallel to z. Accordingly $d^2/dt^2 = V^2 d^2/dz^2$, and any equation such as

$$\frac{d^2P}{dx^2} + \frac{d^2P}{dy^2} + \frac{d^2P}{dz^2} = \frac{1}{V^2}\frac{d^2P}{dt^2} \quad \dots\dots\dots\dots\dots(1)$$

reduces to

$$\frac{d^2P}{dx^2} + \frac{d^2P}{dy^2} = 0. \quad \dots\dots\dots\dots\dots(2)$$

They may be summarized in the form

$$\left(\frac{d^2}{dx^2} + \frac{d^2}{dy^2}\right)(P, Q, R, a, b, c) = 0. \quad \dots\dots\dots(3)$$

The case to be here treated is characterized by the conditions $R = 0$, $c = 0$; but it would suffice to assume one of them, say the latter. Since in general throughout the dielectric

$$dc/dt = dP/dy - dQ/dx, \quad \dots\dots\dots\dots\dots\dots(4)$$

it follows that P and Q are derivatives of a function (ϕ), also proportional to $e^{i(pt+mz)}$, which as a function of x and y may be regarded as a potential since it satisfies the form (2). Thus $dP/dx + dQ/dy = 0$, from which it follows that dR/dz and R vanish. It will be convenient to express all the functions by means of ϕ. We have at once

$$P = d\phi/dx, \quad Q = d\phi/dy, \quad R = 0. \quad \dots\dots\dots\dots\dots(5)$$

Again, by the general equation analogous to (4), since $R = 0$, $ipa = imQ$; so that

$$a = V^{-1}d\phi/dy, \qquad b = -V^{-1}d\phi/dx, \qquad c = 0. \quad \dots\dots\dots\dots(6)$$

Thus the *same* function ϕ serves as a potential for P, Q and as a stream-function for a, b.

The problem is accordingly reduced to dependence upon a simple potential problem in two dimensions. Throughout the dielectric ϕ satisfies

$$d^2\phi/dx^2 + d^2\phi/dy^2 = 0. \quad \dots\dots\dots\dots\dots\dots(7)$$

At the boundary of a conductor, supposed to be perfect, the condition is that the electromotive intensity be entirely normal. So far as regards the component parallel to z this is satisfied already, since $R = 0$ throughout. The remaining condition is that ϕ be constant over the contour of any continuous conductor. This condition secures also that the magnetic induction shall be exclusively tangential.

It is to be observed that R is not equal to $d\phi/dz$. The former quantity vanishes throughout, while $d\phi/dz$ remains finite, since $\phi \propto e^{i(pt+mz)}$. Inasmuch as ϕ satisfies Laplace's equation in two dimensions, but not in three, it will be convenient to use language applicable to two dimensions, referring the conductors to their sections by the plane xy.

If a boundary of a conductor be in the form of a closed curve, the included dielectric is incapable of any vibration of the kind now under consideration. For a function satisfying (7) and retaining a constant value over a closed contour cannot deviate from that value in the interior. Thus the derivatives of ϕ vanish, and there is no disturbance. The question of dielectric vibrations within closed tubes, when m is not limited to equality with p/V, was considered in a former paper*.

* *Phil. Mag.* Vol. XLIII. p. 125 (1897). [Vol. IV. p. 276.]

For the case of a dielectric bounded by two planes perpendicular to x we may take

$$\phi = x e^{i(pt+mz)}, \quad \dots \dots \dots (8)$$

giving
$$P = e^{i(pt+mz)}, \quad Q = 0, \quad R = 0, \quad \dots \dots (9)$$
$$a = 0, \quad b = - V^{-1} e^{i(pt+mz)}, \quad c = 0, \quad \dots \dots (10)$$

in which, as usual, $m = p/V$. Since $Q = 0$, $R = 0$ throughout, the dielectric may be regarded as limited by conductors at any planes (perpendicular to x) that may be desired.

If the dielectric be bounded by conductors in the form of coaxal circular cylinders, we have the familiar wire with sheath return, first, I believe, considered on the basis of these equations by Mr Heaviside. We may take, with omission of a constant addition to $\log r$ which has here no significance,

$$\phi = \log r \cdot e^{i(pt+mz)}, \quad \dots \dots \dots (11)$$

giving
$$P, Q, R = e^{i(pt+mz)} \left(\frac{x}{r^2}, \frac{y}{r^2}, 0 \right), \quad \dots \dots (12)$$

$$V(a, b, c) = e^{i(pt+mz)} \left(\frac{y}{r^2}, -\frac{x}{r^2}, 0 \right). \quad \dots \dots (13)$$

And here again it makes no difference to these forms at what points (r_1, r_2) the dielectric is replaced by conductors.

For the moment these simple examples may suffice to illustrate the manner in which the propagation along z takes place, and to show that ϕ is determined by conditions completely independent of p and its associated m. In further discussions it will save much circumlocution to suppose that p and m are zero and thus to drop the exponential factor. The problem is then strictly reduced to two dimensions and relates to charges and steady currents upon cylindrical conductors, the currents being still entirely superficial. When ϕ is once determined for any case of this kind, the exponential factor may be restored at pleasure with an arbitrary value assigned to p and the corresponding value, viz. p/V, to m.

The usual expressions for electric and magnetic energies will then apply, everything being reckoned per unit length parallel to z. It suffices for practical purposes to limit ourselves to the case of a single outgoing and a single return conductor. We may then write

$$\text{Electric energy} = \frac{(\text{charge})^2}{2 \times \text{capacity}}, \quad \dots \dots \dots (14)$$

$$\text{Magnetic energy} = \tfrac{1}{2} \times \text{self-induction} \times (\text{current})^2; \quad \dots \dots (15)$$

and the value of the self-induction in the latter case is the reciprocal of that of the capacity in the former.

Thus, for a dielectric bounded by coaxal conductors at $r = r_1$ and $r = r_2$, we have $\phi = \log r$, and

$$\text{self-induction} = (\text{capacity})^{-1} = 2 \log \frac{r_2}{r_1}. \quad \ldots\ldots\ldots\ldots(16)$$

Among the cases for which the solution can be completely effected may be mentioned that of a dielectric bounded by confocal elliptical cylinders.

More important in practice is the case of parallel circular wires. In Lecher's arrangement, which has been employed by numerous experimenters, the wires are of equal diameter; and it is usually supposed to be necessary to maintain them at a distance apart which is very great in comparison with that diameter. The general theory above given shows that there is no need for any such restriction, the manner and velocity of propagation along the length being the same whatever may be the character of the cross-section of the system.

The form of ϕ, and the self-induction of the system, may be determined in this case, whatever may be the radii (a_1, a_2) of the wires and the distance (b) between their centres. If r_1, r_2 are the distances of any point P in the plane from fixed points O_1, O_2, the equipotential curves for which ϕ, equal to $\log(r_2/r_1)$, assumes constant values are a system of circles, two of which can be identified with the boundaries of the conductors. The details of the investigation, consisting mainly of the geometrical relations between the ultimate points O_1, O_2 and the circles of radii a_1, a_2, are here passed over. The result for the self-induction per unit length L, or for the capacity, may be written*

$$L = -2 \log \frac{b^2 - a_1^2 - a_2^2 - \sqrt{\{(b^2 - a_1^2 - a_2^2)^2 - 4a_1^2 a_2^2\}}}{2a_1 a_2}. \quad \ldots\ldots(17)$$

As was to be expected, L vanishes when $b = a_1 + a_2$, that is, when the conductors are just in contact.

When a_1, a_2 are small in comparison with b, the approximate value is

$$L = -2 \log \frac{a_1 a_2}{b^2} \left(1 + \frac{a_1^2 + a_2^2}{b^2}\right); \quad \ldots\ldots\ldots\ldots(18)$$

or, if $a_1 = a_2 = a$,

$$L = 4\left(\log \frac{b}{a} - \frac{a^2}{b^2}\right). \quad \ldots\ldots\ldots\ldots(19)$$

The first term of (19) is the value usually given. The same expression represents the reciprocal of the capacity of the system per unit length.

In the application of Lecher's arrangement to the investigation of refractive indices, we have to consider the effect of a variation of the dielectric

* Compare Macdonald, *Camb. Phil. Trans.* Vol. xv. p. 303 (1894).

occurring at planes for which z is constant. It will be seen that no new difficulty arises in the case of systems for which the appropriate function ϕ in two dimensions can be assigned.

Regarding ϕ as a given function, *e.g.* $\log r$ for the case of a coaxal wire and sheath (compare (11)), we may take as the solution for any length of uniform dielectric

$$P, Q, R = (A e^{i(pt+mz)} + B e^{i(pt-mz)})\left(\frac{d\phi}{dx}, \frac{d\phi}{dy}, 0\right), \quad \dots\dots(20)$$

$$V\mu\,(\alpha, \beta, \gamma) = (A e^{i(pt+mz)} - B e^{i(pt-mz)})\left(\frac{d\phi}{dy}, -\frac{d\phi}{dx}, 0\right), \quad \dots(21)$$

in which $\mu\alpha = a$, &c., and provision is made for waves travelling in both directions.

At a plane where the dielectric changes, the conditions to be satisfied are the continuity of P, Q and of α, β; and this is secured if

$$A e^{imz} + B e^{-imz}, \quad \dots\dots\dots\dots\dots\dots\dots\dots\dots(22)$$

$$\frac{1}{V\mu}(A e^{imz} - B e^{-imz}), \quad \dots\dots\dots\dots\dots\dots\dots\dots(23)$$

are continuous. It will be seen that the conditions are altogether independent of the section of the conductors, being the same in fact as if there were no conductors and we were dealing with infinite plane waves represented by $\phi = x$.

As a particular case we may suppose that waves travelling in the negative direction in the dielectric (V, μ) meet at $z = 0$ a dielectric of altered character (V', μ'). The expressions (20), (21) represent the incident (A) and reflected (B) waves. For the second medium it suffices to accent V and μ, writing also A' for A and 0 for B. Thus (22), (23) give

$$A + B = A', \qquad (A - B)/V\mu = A'/V'\mu', \quad \dots\dots\dots(24)$$

by which B and A' are determined. For the reflected wave

$$\frac{B}{A} = \frac{V'\mu' - V\mu}{V'\mu' + V\mu}; \quad \dots\dots\dots\dots\dots\dots\dots(25)$$

or if the difference between the dielectrics relate only to the dielectric constants (K, K'),

$$\frac{B}{A} = \frac{V' - V}{V' + V} = \frac{\sqrt{(K)} - \sqrt{(K')}}{\sqrt{(K)} + \sqrt{(K')}}, \quad \dots\dots\dots\dots(26)$$

in agreement with Young's well-known optical formula.

Whether the dielectric consist of uniform portions with discontinuous changes of character at the boundaries, or whether it be a continuous function of z, the solution of the problem is the same, whatever be the character of the cylindrical conductors. It is only the form of ϕ that is influenced by the latter consideration.

232.

THE ELECTRO-CHEMICAL EQUIVALENT OF SILVER.

[*Nature*, LVI. p. 292, 1897.]

In *Nature*, Vol. LVI. p. 259, Mr Griffiths points out that recent comparisons of the values of the mechanical equivalent of heat, obtained by mechanical and electrical methods, suggest that the adopted value of the equivalent of silver may be in error to the extent of $\frac{1}{1000}$. This adopted value rests, I believe, almost entirely upon experiments made by Kohlrausch, and by myself with Mrs Sidgwick in 1882; and the question has been frequently put to me as to the limits within which it is trustworthy. Such questions are more easily asked than answered, and experience shows that estimates of possible error given by experimenters themselves are usually framed in far too sanguine a spirit.

When our work was undertaken the generally accepted number was ·01136 obtained by Kohlrausch in 1873. Mascart had recently given ·01124, subsequently corrected to ·011156. The uncertainty, therefore, at that time amounted to at least 1 per cent. The experiments of Mrs Sidgwick and myself were very carefully conducted, and we certainly hoped to have attained an accuracy of $\frac{1}{2000}$. So far as errors that can be eliminated by repetition are concerned, this was doubtless the case, as is proved by an examination of our tabular results. But, as every experimenter knows, or ought to know, this class of errors is not really the most dangerous. Security is only to be obtained by coincidence of numbers derived by different methods and by different individuals. It was, therefore, a great satisfaction to find our number (*Phil. Trans.* 1884) (·011179) confirmed by that of Kohlrausch (·011183), resulting from experiments made at about the same time.

It would, however, in my opinion, be rash to exclude the question of an error of $\frac{1}{1000}$. Indeed, I have more than once publicly expressed surprise at the little attention given to this subject in comparison with that lavished upon the ohm. I do not know of any better method of measuring currents absolutely than that followed in 1882, but an ingenious critic would doubtless be able to suggest improvements in details. The only thing that has occurred to me is that perhaps sufficient attention was not given to the change in dimensions that must accompany the heating of the suspended coil when conveying the current of $\frac{1}{3}$ ampere. Recent experiments upon the coil (which exists intact) show that, as judged by resistance, the heating effect due to this current is $2\frac{1}{2}°$ C. But it does not appear possible that the expansion of mean radius thence arising could be comparable with $\frac{1}{1000}$. [See Vol. II. p. 278.]

233.

ON AN OPTICAL DEVICE FOR THE INTENSIFICATION OF PHOTOGRAPHIC PICTURES.

[*Philosophical Magazine*, XLIV. pp. 282—285, 1897.]

WHETHER from insufficient exposure or from other causes, it not unfrequently happens that a photographic negative is deficient in density, the ratio of light-transmissions for the transparent and opaque parts being too low for effective contrast. In many cases an adequate remedy is found in chemical processes of intensification, but modern gelatine plates do not always lend themselves well to this treatment.

The method now proposed may be described as one of using the negative twice over. Many years ago a pleasing style of portrait was current dependent upon a similar principle. A thin positive transparency is developed upon a collodion plate by acid pyrogallol. Viewed in the ordinary way by holding up to the light, the picture is altogether too faint; but when the film side is placed in contact with paper and the combination viewed by reflected light, the contrast is sufficient. Through the transparent parts the paper is seen with but little loss of brilliancy, while the opaque parts act, as it were, twice over, once before the light reaches the paper, and again after reflexion on its way to the eye. For this purpose it is necessary that the deposit, constituting the more opaque parts of the picture, be of such a nature as not itself to reflect light back to the eye in appreciable degree—a condition very far from being satisfied by ordinary gelatine negatives. But by a modification of the process the objection may be met without much difficulty.

To obtain an intensified copy (positive) of a feeble negative, a small source of illumination, *e.g.* a candle, is employed, and it is placed just alongside of the copying-lens. The white paper is replaced by a flat polished reflector, and the film side of the negative is brought into close contact with it. On the other side of the negative and pretty close to it is a field, or condensing, lens of such power that the light from the candle is made parallel by it. After reflexion the light again traverses the lens and forms an image of the candle centred upon the photographic copying-lens. The condenser must be large enough to include the picture and must be free from dirt and scratches; otherwise it does not need to be of good optical quality. If the positive is to preserve the original scale, the focal length of the condenser must be about twice that of the copying-lens.

In carrying this method into execution there are two points which require special attention. The first is the elimination of false light reflected from the optical surfaces employed. As regards the condensing-lens, the difficulty is easily met by giving it a moderate slope. But the light reflected from the glass face of the negative to be copied is less easily dealt with. If allowed to remain, it gives a uniform illumination over the whole field, which in many cases would go far to neutralize the advantages otherwise obtainable by the method. The difficulty arises from the parallelism of the two surfaces of the negative, and is obviated by using for the support of the film a glass whose faces are inclined. The false light can then be thrown to one side and rendered inoperative. In practice it suffices to bring into contact with the negative (taken as usual upon a parallel plate) a wedge-shaped glass of equal or greater area, the reflexion from the adjoining faces being almost destroyed by the interposition of a layer of turpentine. By these devices the false light is practically eliminated, and none reaches the sensitive film but what has twice traversed the original negative.

The other point requiring attention is to secure adequate superposition of the negative and its image in the associated reflector. On account of the slight lateral interval between the copying-lens and the source of light, the incidence of the rays upon the reflector is not accurately perpendicular, and thus any imperfection of contact between the negative film and the reflector leads to a displacement prejudicial to definition. The linear displacement is evidently $2t \sin \theta$, if t denote the interval between the surfaces and θ the angle of incidence, and it can be calculated in any particular case. It is the necessity for a small t that imposes the use of a speculum as a reflector. In practice 2θ can easily be reduced to $\frac{1}{12}$; so that if t were $\frac{1}{50}$ inch, the displacement would not exceed $\frac{1}{600}$ inch, and for most purposes might be disregarded*. The obliquity θ could be got rid of altogether by introducing the light with the aid of a parallel glass reflector placed at 45°; but this complication is hardly to be recommended.

The scale of the apparatus depends, of course, upon the size of the negatives to be copied. In my own experiments $\frac{1}{4}$-plates ($4\frac{1}{4}$ in. × $3\frac{1}{4}$ in.)

* If the glass of the negative were flat, its approximation to the reflector might be much closer than is here supposed.

were employed.   The condenser is of plate-glass 6 in. diameter and 36 in. focus.   The reflector is of silver deposited on glass*.   The wedge-shaped glass† attached to the negative with turpentine is 4 × 4 ins. and the angle between the faces is 2°.   The photographic lens is of 3 inch aperture and about 18 inch principal focus.   It stands at about 36 inches from the negative to be copied.   [Inch = 2·54 cm.]

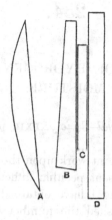

The accompanying sketch shows the disposition of some of the parts. It represents a section by a horizontal plane. A is the condensing-lens, B the wedge, C the negative temporarily cemented to B by fluid turpentine, D the speculum.

[1902.   An almost identical procedure had been described about three years earlier by Mach (Eder's *Jahrbuch für Photographie*).   The method of double transmission was employed in a former research (*Phil. Mag.* Oct. 1892; Vol. IV. of this collection, p. 10).]

---

* For a systematic use of the method a reflector of speculum metal would probably be preferable.

† It is one of those employed for a similar purpose in the projection of Newton's rings (*Proc. Roy. Inst.* March, 1893 ; *Nature*, Vol. XLVIII. p. 212 [Vol. IV. p. 54]).

234.

ON THE VISCOSITY OF HYDROGEN AS AFFECTED BY MOISTURE.

[Proceedings of the Royal Society, LXII. pp. 112—116, 1897.]

IN Sir W. Crookes's important work upon the viscosity of gases* the case of hydrogen was found to present peculiar difficulty. "With each improvement in purification and drying I have obtained a lower value for hydrogen, and have consequently diminished the number expressing the ratio of the viscosity of hydrogen to that of air. In 1876 I found the ratio to be 0·508. In 1877 I reduced this ratio to 0·462. Last year, with improved apparatus, I obtained the ratio 0·458, and I have now got it as low as 0·4439" (p. 425). The difficulty was attributed to moisture. Thus (p. 422): "After working at the subject for more than a year, it was discovered that the discrepancy arose from a trace of water obstinately held by the hydrogen—an impurity which behaved as I explain farther on in the case of air and water vapour."

When occupied in 1888 with the density of hydrogen, I thought that viscosity might serve as a useful test of purity, and I set up an apparatus somewhat on the lines of Sir W. Crookes. A light mirror, 18 mm. in diameter, was hung by a fine fibre (of quartz I believe) about 60 cm. long. A small attached magnet gave the means of starting the vibrations whose subsidence was to be observed. The viscosity chamber was of glass, and carried tubes sealed to it above and below. The window, through which the light passed to and fro, was of thick plate glass cemented to a ground face. This arrangement has great optical advantages, and though unsuitable for experiments involving high exhaustions, appeared to be satisfactory for the purpose in hand, viz., the comparison of various samples of hydrogen at atmospheric pressure. The Töpler pump, as well as the gas generating apparatus and purifying tubes, were connected by sealing. But I was not able to establish any sensible differences among the various samples of hydrogen experimented upon at that time.

* *Phil. Trans.* 1881, p. 387.

In view of the importance of the question, I have lately resumed these experiments. If hydrogen, carefully prepared and desiccated in the ordinary way, is liable to possess a viscosity of 10 per cent. in excess, a similar uncertainty in less degree may affect the density. I must confess that I was sceptical as to the large effect attributed to water vapour in gas which had passed over phosphoric anhydride. Sir W. Crookes himself described an experiment (p. 428) from which it appeared that a residue of water vapour in his apparatus indicated the viscosity due to hydrogen, and, without deciding between them, he offered two alternative explanations. Either the viscosity of water vapour is really the same as that of hydrogen, or under the action of the falling mercury in the Sprengel pump decomposition occurred with absorption of oxygen, so that the residual gas was actually hydrogen. It does not appear that the latter explanation can be accepted, at any rate as regards the earlier stages of the exhaustion, when a rapid current of aqueous vapour must set in the direction of the pump; but if we adopt the former, how comes it that small traces of water vapour have so much effect upon the viscosity of hydrogen?

It is a fact, as was found many years ago by Kundt and Warburg* (and as I have confirmed), that the viscosity of aqueous vapour is but little greater than that of hydrogen. The numbers (relatively to air) given by them are 0·5256 and 0·488. It is difficult to believe that small traces of a foreign gas having a six per cent. greater viscosity could produce an effect reaching to 10 per cent.

In the recent experiments the hydrogen was prepared from amalgamated zinc and sulphuric acid in a closed generator constituting in fact a Smee cell, and it could be liberated at any desired rate by closing the circuit externally through a wire resistance. The generating vessel was so arranged as to admit of exhaustion, and the materials did not need to be renewed during the whole course of the experiments. The gas entered the viscosity chamber from below, and could be made to pass out above through the upper tube (which served also to contain the fibre) into the pump head of the Töpler. By suitable taps the viscosity chamber could be isolated, when observations were to be commenced.

The vibrations were started by a kind of galvanometer coil in connexion (through a key) with a Leclanché cell. As a sample set of observations the following relating to hydrogen at atmospheric pressure and at 58° F., which had been purified by passage over fragments of sulphur and solid soda (without phosphoric anhydride), may be given:—

<hr>

* *Pogg. Ann.* 1875, Vol. CLV. p. 547.

OBSERVATIONS ON JUNE 7, 1897.

—	65·4	—	—	—
423·7	88·9	358·3	2·554	—
401·3	110·0	312·4	2·495	0·059
381·5	128·9	271·5	2·434	0·061
364·4	144·1	235·5	2·372	0·062
349·7	158·6	205·6	2·313	0·059
336·8	169·8	178·2	2·251	0·062
325·7	180·6	155·9	2·193	0·058
315·7	189·8	135·1	2·131	0·062
307·2	197·8	117·4	2·070	0·061
300·0	204·6	102·2	2·009	0·061
293·7	210·6	89·1	1·950	0·059
287·8	—	77·2	1·888	0·062

Mean log. dec. = 0·0604.

The two first columns contain the actually observed elongations upon the two sides. They require no correction, since the scale was bent to a circular arc centred at the mirror. The third column gives the actual arcs of vibration, the fourth their (common) logarithms, and the fifth the differences of these, which should be constant. The mean logarithmic decrement can be obtained from the first and last arcs only, but the intermediate values are useful as a check. The time of (complete) vibration was determined occasionally. It was constant, whether hydrogen or air occupied the chamber, at 26·2 seconds.

The observations extended themselves over two months, and it would be tedious to give the results in any detail. One of the points to which I attached importance was a comparison between hydrogen as it issued from the generator without any desiccation whatever and hydrogen carefully dried by passage through a long tube packed with phosphoric anhydride. The difference proved itself to be comparatively trifling. For the wet hydrogen there were obtained on May 10, 11, such log. decs. as 0·0594, 0·0590, 0·0591, or as a mean 0·0592. The dried hydrogen, on the other hand, gave 0·0588, 0·0586, 0·0584, 0·0590 on various repetitions with renewed supplies of gas, or as a mean 0·0587, about 1 per cent. smaller than for the wet hydrogen. It appeared that the dry hydrogen might stand for several days in the viscosity chamber without alteration of logarithmic decrement. It should be mentioned that the apparatus was set up underground, and that the changes of temperature were usually small enough to be disregarded.

In the next experiments the phosphoric tube was replaced by others containing sulphur (with the view of removing mercury vapour) and solid soda. Numbers were obtained on different days such as 0·0591, 0·0586, 0·0588, 0·0587, mean 0·0588, showing that the desiccation by soda was practically as efficient as that by phosphoric anhydride.

At this stage the apparatus was rearranged.   As shown by observations upon air (at 10 cm. residual pressure), the logarithmic decrements were increased, probably owing to a slight displacement of the mirror relatively to the containing walls of the chamber.   The sulphur and soda tubes were retained, but with the addition of one of hard glass containing turnings of magnesium.   Before the magnesium was heated the mean number for hydrogen (always at atmospheric pressure) was 0·0600.   The heating of the magnesium to redness, which it was supposed might remove residual water, had the effect of *increasing* the viscosity of the gas, especially at first*. After a few operations the logarithmic decrement from gas which had passed over the hot magnesium seemed to settle itself at 0·0606.   When the magnesium was allowed to remain cold, fresh fillings gave again 0·0602, 0·0601, 0·0598, mean 0·0600.   Dried air at 10 cm. residual pressure gave 0·01114, 0·01122, 0·01118, 0·01126, 0·01120, mean 0·01120.

In the next experiments a phosphoric tube was added about 60 cm. long and closely packed with fresh material.   The viscosity appeared to be slightly increased, but hardly more than would be accounted for by an accidental rise of temperature.   The mean uncorrected number may be taken as 0·0603.

The evidence from these experiments tends to show that residual moisture is without appreciable influence upon the viscosity of hydrogen; so much so that, were there no other evidence, this conclusion would appear to me to be sufficiently established.   It remains barely possible that the best desiccation to which I could attain was still inadequate, and that absolutely dry hydrogen would exhibit a less viscosity.   It must be admitted that an apparatus containing cemented joints and greased stop-cocks is in some respects at a disadvantage.   Moreover, it should be noticed that the ratio 0·0600 : 0·1120, viz. 0·536, for the viscosities of hydrogen and air is decidedly higher than that (0·500) deduced by Sir G. Stokes from Crookes's observations.   According to the theory of the former, a fair comparison may be made by taking, as above, the logarithmic decrements for hydrogen at atmospheric pressure, and for air at a pressure of 10 cm. of mercury.   I may mention that moderate rarefactions, down say to a residual pressure of 5 cm., had no influence on the logarithmic decrement observed with hydrogen.

I am not able to explain the discrepancy in the ratios thus exhibited. A viscous quality in the suspension, leading to a subsidence of vibrations independent of the gaseous atmosphere, would tend to diminish the apparent differences between various kinds of gas, but I can hardly regard this cause as operative in my experiments.   For actual comparisons of widely differing viscosities I should prefer an apparatus designed on Maxwell's principle, in which the gas subjected to shearing should form a comparatively thin layer bounded on one side by a moving plane and on the other by a fixed plane.

* The glass was somewhat attacked, and it is supposed that silicon compounds may have contaminated the hydrogen.

235.

ON THE PROPAGATION OF WAVES ALONG CONNECTED SYSTEMS OF SIMILAR BODIES.

[*Philosophical Magazine*, XLIV. pp. 356—362, 1897.]

FOR simplicity of conception the bodies are imagined to be similarly disposed at equal intervals (a) along a straight line. The position of each body, as displaced from equilibrium, is supposed to be given by *one* coordinate, which for the rth body is denoted by ψ_r. A wave propagated in one direction is represented by taking ψ_r proportional to $e^{i(nt+r\beta)}$. If we take an instantaneous view of the system, the disturbance is periodic when $r\beta$ increases by 2π, or when ra increases by $2\pi a/\beta$. This is the wave-length, commonly denoted by λ; so that, if $k = 2\pi/\lambda$, $k = \beta/a$. The velocity of propagation (V) is given by $V = n/k$; and the principal object of the investigation is to find the relation between n or V and λ.

The forces acting upon each body, which determine the vibration of the system about its configuration of equilibrium, are assumed to be due solely to the neighbours situated within a limited distance. The simplest case of all is that in which there is no mutual reaction between the bodies, the kinetic and potential energies of the system being then given by

$$T = \tfrac{1}{2} A_0 \Sigma \dot{\psi}_r^2, \qquad P = \tfrac{1}{2} C_0 \Sigma \psi_r^2, \qquad \dots\dots\dots\dots\dots(1)$$

similarity requiring that the coefficients A_0, C_0 be the same for all values of r. In this system each body vibrates independently, according to the equation

$$A_0 \ddot{\psi}_r + C_0 \psi_r = 0, \qquad \dots\dots\dots\dots\dots\dots\dots\dots(2)$$

and

$$n^2 = C_0/A_0. \qquad \dots\dots\dots\dots\dots\dots\dots\dots\dots(3)$$

The frequency is of course independent of the wave-length in which the phases may be arranged to repeat themselves, so that n is independent of k, while V equal to n/k varies inversely as k, or directly as λ. The propagation of waves along a system of this kind has been considered by Reynolds.

In the general problem the expression for P will include also products of ψ_r with the neighbouring coordinates ... ψ_{r-2}, ψ_{r-1}, ψ_{r+1}, ψ_{r+2} ..., and a similar statement holds good for T. Exhibiting only the terms which involve r, we may write

$$T = \ldots + \tfrac{1}{2}A_0\psi_r^2 - A_1\psi_r\psi_{r-1} - A_1\psi_r\psi_{r+1}$$
$$- A_2\psi_r\psi_{r-2} - A_2\psi_r\psi_{r+2} - \ldots, \quad \ldots\ldots(4)$$

$$P = \ldots + \tfrac{1}{2}C_0\psi_r^2 - C_1\psi_r\psi_{r-1} - C_1\psi_r\psi_{r+1}$$
$$- C_2\psi_r\psi_{r-2} - C_2\psi_r\psi_{r+2} - \ldots, \quad \ldots\ldots(5)$$

where A_1, A_2, ... C_1, C_2, ... are constants, finite for a certain number of terms and then vanishing. The equation for ψ_r is accordingly

$$A_0\ddot\psi_r - A_1\ddot\psi_{r-1} - A_1\ddot\psi_{r+1} - A_2\ddot\psi_{r-2} - A_2\ddot\psi_{r+2} - \ldots\ldots$$
$$+ C_0\psi_r - C_1\psi_{r-1} - C_1\psi_{r+1} - C_2\psi_{r-2} - C_2\psi_{r+2} - \ldots\ldots = 0. \quad \ldots\ldots(6)$$

In the other equations of the system r is changed, but without entailing any other alteration in (6). Since all the quantities ψ are proportional to e^{int}, the double differentiation is accounted for by the introduction of the factor $-n^2$. Making this substitution and remembering that ψ_r is also proportional to $e^{ir\beta}$, we get as the equivalent of any one of the equations (6)

$$n^2(A_0 - A_1e^{-i\beta} - A_1e^{i\beta} - A_2e^{-2i\beta} - A_2e^{2i\beta} - \ldots)$$
$$= C_0 - C_1e^{-i\beta} - C_1e^{i\beta} - C_2e^{-2i\beta} - C_2e^{2i\beta} - \ldots,$$

or

$$n^2 = \frac{C_0 - 2C_1\cos ka - 2C_2\cos 2ka - \ldots}{A_0 - 2A_1\cos ka - 2A_2\cos 2ka - \ldots}, \quad \ldots\ldots\ldots(7)$$

in which β is replaced by its equivalent ka. By (7) n is determined as a function of k and of the fundamental constants of the system.

In most of the examples which naturally suggest themselves A_1, A_2, ... vanish, so that T has the same simple form as in (1). If we suppose for brevity that A_0 is unity, (7) becomes

$$n^2 = C_0 - 2C_1\cos ka - 2C_2\cos 2ka - \ldots. \quad \ldots\ldots\ldots(8)$$

When the waves are very long, k approximates to zero. In the limit

$$n^2 = C_0 - 2C_1 - 2C_2 - \ldots. \quad \ldots\ldots\ldots\ldots(9)$$

If we call the limiting value C, we may write (8) in the form

$$n^2 = C + 4C_1\sin^2(\tfrac{1}{2}ka) + 4C_2\sin^2(ka) + \ldots. \quad \ldots\ldots\ldots(10)$$

In an important class of cases C vanishes, that is the frequency diminishes without limit as λ increases. If at the same time but one of the constants C_1, C_2, ... be finite, the equation simplifies. For example, if C_1 alone be finite,

$$n = 2C_1^{\frac{1}{2}}\sin(\tfrac{1}{2}ka). \quad \ldots\ldots\ldots\ldots(11)$$

In any case when n is known V follows immediately. Thus from (10) with C evanescent, we get

$$\frac{V^2}{a^2} = C_1 \frac{\sin^2 \frac{ka}{2}}{\left(\frac{ka}{2}\right)^2} + 4C_2 \frac{\sin^2 ka}{k^2 a^2} + 9C_3 \frac{\sin^2 \frac{3ka}{2}}{\left(\frac{3ka}{2}\right)^2} + \ldots \quad \ldots\ldots(12)$$

A simple case included under (11) is that of a stretched string, itself without mass, but carrying unit loads at equal intervals (a)*. The expression for the potential energy is

$$P = \ldots + \frac{T_1}{2a} (\psi_r - \psi_{r-1})^2 + \frac{T_1}{2a} (\psi_{r+1} - \psi_r)^2 + \ldots, \quad \ldots\ldots\ldots(13)$$

T_1 representing the tension. Thus by comparison with (5)

$$C_0 = 2T_1/a, \qquad C_1 = T_1/a, \qquad C_2 = 0, \quad \&c.;$$

so that by (8)

$$n^2 = \frac{2T_1}{a} - \frac{2T_1}{a} \cos ka,$$

$$n = \sqrt{\left(\frac{T_1}{a\mu}\right)} . 2 \sin (\tfrac{1}{2}ka), \quad \ldots\ldots\ldots\ldots\ldots(14)$$

μ being introduced to represent the mass of each load with greater generality. The value of V is obtained by division of (14) by k. In order more easily to compare with a known formula we may introduce the longitudinal density ρ, such that $\mu = a\rho$. Thus

$$V = \frac{n}{k} = \sqrt{\left(\frac{T_1}{\rho}\right)} . \frac{\sin (\tfrac{1}{2}ka)}{\tfrac{1}{2}ka}, \quad \ldots\ldots\ldots\ldots(15)$$

reducing to the well-known value of the constant velocity of propagation along a uniform string when a is made infinitesimal. Lord Kelvin's wave-model (*Popular Lectures and Addresses*, Vol. I. 2nd ed. p. 164) is also included under the class of systems for which P has the form (13).

Another example in which again $C_2, C_3 \ldots$ vanish is proposed by Fitzgerald†. It consists of a linear system of rotating magnets (Fig. 1) with their poles

Fig. 1.

close to one another and disturbed to an amount small compared with the distance apart of the poles. The force of restitution is here proportional to the sum of the angular displacements (ψ) of contiguous magnets, so that P is proportional to

$$\ldots + (\psi_r + \psi_{r-1})^2 + (\psi_r + \psi_{r+1})^2 + \ldots.$$

---

* See *Theory of Sound*, §§ 120, 148.
† *Brit. Assoc. Report*, 1893, p. 689.

Here $C_1 = -\tfrac{1}{2}C_0$, and (8) gives $n^2 = C_0(1 + \cos ka)$,

or
$$n = n_0 . \cos(\tfrac{1}{2}ka), \quad \dots\dots\dots\dots\dots\dots(16)$$

if n_0 represent the value of n appropriate to $k = 0$, *i.e.* to infinitely long waves. Here $n = 0$, when $\lambda = 2a$. In this case $\psi_{r+1} = -\psi_r$.

Fitzgerald considers, further, a more general linear system constructed by connecting a series of equidistant wheels by means of indiarubber bands. "By connecting the wheels each with its next neighbour we get the simplest system. If to this be superposed a system of connexion of each with its next neighbour but two, and so on, complex systems with very various relations between wave-length and velocity can be constructed depending on the relative strengths of the bands employed." If the bands may be crossed, the potential energy takes the form

$$P = \tfrac{1}{2}\gamma_1(\psi_r \pm \psi_{r-1})^2 + \tfrac{1}{2}\gamma_1(\psi_r \pm \psi_{r+1})^2$$
$$+ \tfrac{1}{2}\gamma_2(\psi_r \pm \psi_{r-2})^2 + \tfrac{1}{2}\gamma_2(\psi_r \pm \psi_{r+2})^2$$
$$+ \dots, \quad \dots\dots\dots\dots\dots\dots\dots\dots\dots\dots\dots\dots(17)$$

which is only less general than (5) by the limitation
$$\tfrac{1}{2}C_0 \pm C_1 \pm C_2 \pm \dots = 0. \quad \dots\dots\dots\dots\dots(18)$$

Prof. Fitzgerald appears to limit himself to the lower sign in the alternatives, so that C in (10) vanishes. This leads to (12), from which his result differs, but probably only by a slip of the pen.

If we take the upper sign throughout, (8) becomes
$$-\tfrac{1}{4}n^2 = C_1 \cos^2 \frac{ka}{2} + C_2 \cos^2 \frac{2ka}{2} + C_3 \cos^2 \frac{3ka}{2} + \dots. \quad \dots\dots(19)$$

It may be observed that Prof. Fitzgerald's system will have the most general potential energy possible (5), if in addition to the elastic connexions between the wheels there be introduced a force of restitution acting upon each wheel independently.

As an example in which C_2 is finite as well as C_1, let us imagine a system of masses of which each is connected to its immediate neighbours on the two sides by an elastic rod capable of bending but without inertia. Here

$$P = \dots + \tfrac{1}{2}c(2\psi_{r-1} - \psi_{r-2} - \psi_r)^2 + \tfrac{1}{2}c(2\psi_r - \psi_{r-1} - \psi_{r+1})^2$$
$$+ \tfrac{1}{2}c(2\psi_{r+1} - \psi_r - \psi_{r+2})^2 + \dots. \quad \dots\dots\dots(20)$$

A comparison with (5) gives
$$C_0 = 6c, \quad C_1 = 4c, \quad C_2 = -c,$$

so that
$$C = C_0 - 2C_1 - 2C_2 = 0.$$

Accordingly by (10),

$$n^2 = 16c \sin^2\left(\tfrac{1}{2}ka\right) - 4c \sin^2 ka = 16c \sin^4\left(\tfrac{1}{2}ka\right),$$

or
$$n = 4c^{\frac{1}{2}} \sin^2\left(\tfrac{1}{2}ka\right). \quad\dots\dots\dots\dots\dots\dots(21)$$

Thus far we have considered the propagation of waves along an unlimited series of bodies. If we suppose that the total number is m and that they form a *closed* chain, ψ must be such that

$$\psi_{r+m} = \psi_r, \quad\dots\dots\dots\dots\dots\dots(22)$$

from which it follows that

$$\beta = ka = 2s\pi/m, \quad\dots\dots\dots\dots\dots\dots(23)$$

s being an integer. Thus (8) becomes

$$n^2 = C_0 - 2C_1 \cos(2s\pi/m) - 2C_2 \cos(4s\pi/m) - \dots \quad\dots\dots(24)$$

When the chain, composed of a limited series of bodies, is open at the ends instead of closed, the general problem becomes more complicated. A simple example is that treated by Lagrange, of a stretched massless string, carrying a finite number of loads and fixed at its extremities*. The open chain of m magnets, for which

$$P = \tfrac{1}{2}(\psi_1 + \psi_2)^2 + \tfrac{1}{2}(\psi_2 + \psi_3)^2 + \dots + \tfrac{1}{2}(\psi_{m-1} + \psi_m)^2, \quad\dots\dots(25)$$

is considered by Fitzgerald. The equations are

$$\left.\begin{aligned}
\psi_1(1 - n^2) + \psi_2 &= 0,\\
\psi_1 + \psi_2(2 - n^2) + \psi_3 &= 0,\\
\dots\dots\dots\dots\dots\dots&\\
\psi_{r-1} + \psi_r(2 - n^2) + \psi_{r+1} &= 0,\\
\dots\dots\dots\dots\dots\dots&\\
\psi_{m-2} + \psi_{m-1}(2 - n^2) + \psi_m &= 0,\\
\psi_{m-1} + \psi_m(1 - n^2) &= 0,
\end{aligned}\right\} \quad\dots\dots\dots\dots(26)$$

of which the first and last may be brought under the same form as the others if we introduce ψ_0 and ψ_{m+1}, such that

$$\psi_0 + \psi_1 = 0, \qquad \psi_m + \psi_{m+1} = 0. \quad\dots\dots\dots\dots(27)$$

If we assume

$$\psi_r = \cos nt \, \sin(r\beta - \tfrac{1}{2}\beta), \quad\dots\dots\dots\dots(28)$$

the first of equations (27) is satisfied. The second is also satisfied provided that

$$\sin m\beta = 0, \quad \text{or} \quad \beta = s\pi/m. \quad\dots\dots\dots\dots(29)$$

* *Theory of Sound*, § 120.

The equations (26) are satisfied if

$$2\cos\beta + 2 - n^2 = 0,$$

that is, if

$$n = 2\cos(s\pi/2m). \quad\dots\dots\dots\dots\dots\dots(30)$$

In (29), (30) s may assume the m values 1 to m inclusive. In the last case $n = 0$, and $\beta = \pi$; and from (28),

$$\psi_r = -(-1)^r \cos nt.$$

The equal amplitudes and opposite phases of consecutive coordinates, *i.e.* angular displacements of the magnets, give rise to no potential energy, and therefore to a zero frequency of vibration. In the first case ($s = 1$) the angular deflexions are all in the same direction, and the frequency is the highest admissible. If at the same time m be very great, n reaches its maximum value, corresponding to parallel positions of all the magnets. If we call this value N, the generalized form of (30), applicable to all masses and degrees of magnetization, may be written

$$n = N\cos(s\pi/2m). \quad\dots\dots\dots\dots\dots\dots(31)$$

If m is great and s relatively small, (31) becomes approximately

$$n = N\left(1 - \frac{s^2\pi^2}{8m^2}\right); \quad\dots\dots\dots\dots\dots(32)$$

so that as s diminishes we have a series of frequencies approaching N as an upper limit, and are reminded (as Fitzgerald remarks) of certain groups of spectrum lines. A nearer approach to the remarkable laws of Balmer for hydrogen* and of Kayser and Runge for the alkalies is arrived at by supposing s constant while m varies. In this case, instead of supposing that the whole series of lines correspond to various modes of one highly compound system, we attribute each line to a different system vibrating in a given special mode. Apart from the better agreement of frequencies, this point of view seems the more advantageous as we are spared the necessity of selecting and justifying a special high value of m. If we were to take $s = 2$ in (31) and attribute to m integral values 3, 4, 5, ..., we should have a series of frequencies of the same general character as the hydrogen series, but still differing considerably in actual values.

There is one circumstance which suggests doubts whether the analogue of radiating bodies is to be sought at all in ordinary mechanical or acoustical systems vibrating about equilibrium. For the latter, even when gyratory terms are admitted, give rise to equations involving the *square* of the frequency; and it is only in certain exceptional cases, *e.g.* (31), that the frequency itself can be simply expressed. On the other hand, the formulæ

* Viz. $n = N(1 - 4m^{-2})$, with $m = 3, 4, 5$, &c.

and laws derived from observation of the spectrum appear to introduce more naturally the *first* power of the frequency. For example, this is the case with Balmer's formula. Again, when the spectrum of a body shows several doublets, the intervals between the components correspond closely to a constant difference of frequency, and could not be simply expressed in terms of squares of frequency. Further, the remarkable law, discovered independently by Rydberg and by Schuster, connecting the convergence frequencies of different series belonging to the same substance, points in the same direction.

What particular conclusion follows from this consideration, even if force be allowed to it, may be difficult to say. The occurrence of the first power of the frequency seems suggestive rather of kinematic relations* than of those of dynamics.

[1902. See further on the subject of the present paper, *Phil. Mag.* Dec. 1898, " On Iso-periodic Systems," Art. 242, below.]

* *E.g.* as in the phases of the moon.

236.

ON THE DENSITIES OF CARBONIC OXIDE, CARBONIC ANHYDRIDE, AND NITROUS OXIDE.

[*Proceedings of the Royal Society*, LXII. pp. 204—209, 1897.]

THE observations here recorded were carried out by the method and with the apparatus described in a former paper*, to which reference must be made for details. It must suffice to say that the globe containing the gas to be weighed was filled at 0° C., and to a pressure determined by a manometric gauge. This pressure, nearly atmospheric, is slightly variable with temperature on account of the expansion of the mercury and iron involved. The actually observed weights are corrected so as to correspond with a temperature of 15° C. of the gauge, as well as for the errors in the platinum and brass weights employed. In the present, as well as in the former, experiments I have been ably assisted by Mr George Gordon.

Carbonic Oxide.

This gas was prepared by three methods. In the first method a flask, sealed to the rest of the apparatus, was charged with 80 grams recrystallised ferrocyanide of potassium and 360 c.c. strong sulphuric acid. The generation of gas could be started by the application of heat, and with care it could be checked and finally stopped by the removal of the flame with subsequent application, if necessary, of wet cotton-wool to the exterior of the flask. In this way one charge could be utilised with great advantage for several fillings. On leaving the flask the gas was passed through a bubbler containing potash solution (convenient as allowing the rate of production to be more easily estimated) and thence through tubes charged with fragments of potash and phosphoric anhydride, all connected by sealing. When possible, the weight

---

* "On the Densities of the Principal Gases," *Roy. Soc. Proc.* Vol. LIII. p. 134, 1893. [Vol. IV. p. 39.]

of the globe *full* was compared with the mean of the preceding and following weights *empty*. Four experiments were made with results agreeing to within a few tenths of a milligram.

In the second set of experiments the flask was charged with 100 grams of oxalic acid and 500 c.c. strong sulphuric acid. To absorb the large quantity of CO_2 simultaneously evolved, a plentiful supply of alkali was required. A wash-bottle and a long nearly horizontal tube contained strong alkaline solution, and these were followed by the tubes containing solid potash and phosphoric anhydride as before. ·

For the experiments of the third set *oxalic* acid was replaced by *formic*, which is more convenient as not entailing the absorption of large volumes of CO_2. In this case the charge consisted of 50 grams formate of soda, 300 c.c. strong sulphuric acid, and 150 c.c. distilled water. The water is necessary in order to prevent action in the cold, and the amount requires to be somewhat carefully adjusted. As purifiers, the long horizontal bubbler was retained and the tubes charged with solid potash and phosphoric anhydride. In this set there were four concordant experiments. The immediate results stand thus :—

Carbonic Oxide.

From ferrocyanide . . . . . . 2·29843

„ oxalic acid . . . . 2·29852

„ formate of soda . . . . . 2·29854

Mean . . . . . . 2·29850

This corresponds to the number 2·62704 for oxygen*, and is subject to a correction (additive) of 0·00056 for the diminution of the external volume of the globe when exhausted.

The ratio of the densities of carbonic oxide and oxygen is thus

2·29906 : 2·62760 ;

so that if the density of oxygen be taken as 32, that of carbonic oxide will be 27·9989. If, as some preliminary experiments by Dr Scott† indicate, equal volumes may be taken as accurately representative of CO and of O_2, the atomic weight of carbon will be 11·9989 on the scale of oxygen = 16.

The very close agreement between the weights of carbonic oxide prepared in three different ways is some guarantee against the presence of an impurity of widely differing density. On the other hand, some careful experiments led Mr T. W. Richards‡ to the conclusion that carbonic oxide is liable to

* "On the Densities of the Principal Gases," *Roy. Soc. Proc.* Vol. LIII. p. 144, 1893. [Vol. IV. p. 39.]

† *Camb. Phil. Proc.* Vol. IX. p. 144, 1896.

‡ *Amer. Acad. Proc.* Vol. XVIII. p. 279, 1891.

contain considerable quantities of hydrogen or of hydrocarbons. From 5½ litres of carbonic oxide passed over hot cupric oxide he collected no less than 25 milligrams of water, and the evidence appeared to prove that the hydrogen was really derived from the carbonic oxide. Such a proportion of hydrogen would entail a deficiency in the weight of the globe of about 11 milligrams, and seems improbable in view of the good agreement of the numbers recorded. The presence of so much hydrogen in carbonic oxide is also difficult to reconcile with the well-known experiments of Professor Dixon, who found that prolonged treatment with phosphoric anhydride was required in order to render the mixture of carbonic oxide and oxygen inexplosive. In the presence of relatively large quantities of free hydrogen (or hydrocarbons) why should traces of water vapour be so important?

In an experiment by Dr Scott*, 4 litres of carbon monoxide gave only 1·3 milligrams to the drying tube after oxidation.

I have myself made several trials of the same sort with gas prepared from formate of soda exactly as for weighing. The results were not so concordant as I had hoped†, but the amount of water collected was even less than that given by Dr Scott. Indeed, I do not regard as proved the presence of hydrogen at all in the gas that I have employed‡.

Carbonic Anhydride.

This gas was prepared from hydrochloric acid and marble, and after passing a bubbler charged with a solution of carbonate of soda, was dried by phosphoric anhydride. Previous to use, the acid was caused to boil for some time by the passage of hydrochloric acid vapour from a flask containing another charge of the acid. In a second set of experiments the marble was replaced by a solution of carbonate of soda. There is no appreciable difference between the results obtained in the two ways; and the mean, corrected for the errors of weights and for the shrinkage of the globe when exhausted, is 3·6349, corresponding to 2·6276 for oxygen. The temperature at which the globe was charged was 0° C., and the actual pressure that of the manometric gauge at about 20°, reduction being made to 15° by the use of Boyle's law. From the former paper it appears that the actual height of the mercury column at 15° is 762·511 mm.

* *Chem. Soc. Trans.* 1897, p. 564.

† One obstacle was the difficulty of re-oxidising the copper reduced by carbonic oxide. I have never encountered this difficulty after reduction by hydrogen.

‡ In Mr Richards' work the gas in an imperfectly dried condition was treated with hot platinum black. Is it possible that the hydrogen was introduced at this stage?

Nitrous Oxide.

In preliminary experiments the gas was prepared in the laboratory, at as low a temperature as possible, from nitrate of ammonia, or was drawn from the iron bottles in which it is commercially supplied. The purification was by passage over potash and phosphoric anhydride. Unless special precautions are taken the gas so obtained is ten or more milligrams too light, presumably from admixture with nitrogen. In the case of the commercial supply, a better result is obtained by placing the bottles in an inverted position so as to draw from the *liquid* rather than from the *gaseous* portion.

Higher and more consistent results were arrived at from gas which had been specially treated. In consequence of the high relative solubility of nitrous oxide in water, the gas held in solution after prolonged agitation of the liquid with impure gas from any supply, will contain a much diminished proportion of nitrogen. To carry out this method on the scale required, a large (11-litre) flask was mounted on an apparatus in connexion with the lathe so that it could be vigorously shaken. After the dissolved air had been sufficiently expelled by preliminary passage of N_2O, the water was cooled to near 0° C. and violently shaken for a considerable time while the gas was passing in large excess. The nitrous oxide thus purified was expelled from solution by heat, and was used to fill the globe in the usual manner.

For comparison with the results so obtained, gas purified in another manner was also examined. A small iron bottle, fully charged with the commercial material, was cooled in salt and ice and allowed somewhat suddenly to blow off half its contents. The residue drawn from the bottle in one or other position was employed for the weighings.

Nitrous Oxide (1896).

Aug. 15	Expelled from water	. . . .		3·6359
„ 17	„ „	. . . .		3·6354
„ 19	From residue after blow off,	valve downwards	3·6364
„ 21	„ „	valve upwards .	3·6358
„ 22	„ „	valve downwards	3·6360
	Mean	. . . .		3·6359

The mean value may be taken to represent the corrected weight of the gas which fills the globe at 0° C. and at the pressure of the gauge (at 15°), corresponding to 2·6276 for oxygen.

One of the objects which I had in view in determining the density of nitrous oxide was to obtain, if it were possible, evidence as to the atomic weight of nitrogen. It may be remembered that observations upon the density of pure nitrogen, as distinguished from the atmospheric mixture containing argon which, until recently, had been confounded with pure nitrogen, led* to the conclusion that the densities of oxygen and nitrogen were as 16 : 14·003, thus suggesting that the atomic weight of nitrogen might really be 14 in place of 14·05, as generally received. The chemical evidence upon which the latter number rests is very indirect, and it appeared that a direct comparison of the weight of nitrous oxide and of its contained nitrogen might be of value. A suitable vessel would be filled, under known conditions, with the nitrous oxide, which would then be submitted to the action of a spiral of copper or iron wire rendered incandescent by an electric current. When all the oxygen was removed, the residual nitrogen would be measured, from which the ratio of equivalents could readily be deduced. The fact that the residual nitrogen would possess nearly the same volume as the nitrous oxide from which it was derived would present certain experimental advantages. If indeed the atomic weights were really as 14 : 16, the ratio (x) of volumes, after and before operations, would be given by

$$\frac{2\cdot 2996 \times x}{3\cdot 6359 - 2\cdot 2996 \times x} = \frac{14}{8},$$

whence

$$x = \frac{7 \times 3\cdot 6359}{11 \times 2\cdot 2996} = 1\cdot 0061,$$

3·6359 and 2·2996 being the relative weights of nitrous oxide and of nitrogen which (at 0° C. and at the pressure of the gauge) occupy the same volume. The integral numbers for the atomic weights would thus correspond to an expansion, after chemical reduction, of about one-half per cent.

But in practical operation the method lost most of its apparent simplicity. It was found that copper became unmanageable at a temperature sufficiently high for the purpose, and recourse was had to iron. Coils of iron suitably prepared and supported could be adequately heated by the current from a dynamo without twisting hopelessly out of shape; but the use of iron leads to fresh difficulties. The emission of carbonic oxide from the iron heated in vacuum continues for a very long time, and the attempt to get rid of this gas by preliminary treatment had to be abandoned. By final addition of a small quantity of oxygen (obtained by heating some permanganate of potash sealed up in one of the leading tubes) the CO could be oxidised to CO_2, and thus, along with any H_2O, be absorbed by a lump of potash placed beforehand in the working vessel. To get rid of superfluous oxygen, a coil of incandescent copper had then to be invoked, and thus the apparatus became rather complicated.

It is believed that the difficulties thus far mentioned were overcome, but nevertheless a satisfactory concordance in the final numbers was not attained.

* Rayleigh and Ramsay, *Phil. Trans.* Vol. CLXXXVI. p. 190, 1895. [Vol. IV. p. 133.]

In the present position of the question no results are of value which do not discriminate with certainty between 14·05 and 14·00. The obstacle appeared to lie in a tendency of the nitrogen to pass to higher degrees of oxidation. On more than one occasion mercury (which formed the movable boundary of an overflow chamber) was observed to be attacked. Under these circumstances I do not think it worth while to enter into further detail regarding the experiments in question.

The following summary gives the densities of the various gases relatively to air, all obtained by the same apparatus*. The last figure is of little significance.

Air free from H_2O and CO_2	1·00000
Oxygen	1·10535
Nitrogen and argon (atmospheric)	0·97209
Nitrogen	0·96737
Argon	1·37752
Carbonic oxide	0·96716
Carbonic anhydride	1·52909
Nitrous oxide	1·52951

The value obtained for hydrogen upon the same scale was 0·06960; but the researches of M. Leduc and of Professor Morley appear to show that this number is a little too high.

[1902. For the absolute densities of air and oxygen, see Vol. IV. p. 51.]

* Roy. Soc. Proc. Vol. LIII. p. 148, 1893; Vol. LV. p. 340, 1894; Phil. Trans. Vol. CLXXXVI. p. 189, 1895; Roy. Soc. Proc. Vol. LIX. p. 201, 1896. [Vol. IV. pp. 52, 104, 130, 215.]

237.

RÖNTGEN RAYS AND ORDINARY LIGHT.

[*Nature*, LVII. p. 607, 1898.]

ACCORDING to the theory of the Röntgen rays suggested by Sir G. Stokes*, and recently developed by Prof. J. J. Thomson†, their origin is to be sought in impacts of the charged atoms constituting the kathode-stream, whereby pulses of disturbance are generated in the ether. This theory has certainly much to recommend it; but I cannot see that it carries with it some of the consequences which have been deduced as to the distinction between Röntgen rays and ordinary luminous and non-luminous radiation. The conclusion of the authors above mentioned†, "that the Röntgen rays are not waves of very short wave-length, but impulses," surprises me. From the fact of their being highly condensed impulses, I should conclude on the contrary that they *are* waves of short wave-length. If short waves are inadmissible, longer waves are still more inadmissible. What then becomes of Fourier's theorem and its assertion that *any* disturbance may be analysed into regular waves?

Is it contended that previous to resolution (whether merely theoretical, or practically effected by the spectroscope) the vibrations of ordinary (*e.g.* white) light are regular, and thus distinguished from disturbances made up of impulses? This view was certainly supported in the past by high authorities, but it has been shown to be untenable by Gouy§, Schuster‖, and the present writer¶. A curve representative of white light, if it were drawn upon paper, would show no sequences of similar waves.

In the second of the papers referred to, I endeavoured to show in detail that white light might be supposed to have the very constitution now ascribed to the Röntgen radiation, except that of course the impulses would have to be less condensed. The peculiar behaviour of the Röntgen radiation with respect to diffraction and refraction would thus be attributable merely to the extreme shortness of the waves composing it.

[1902. In a reply to the above (*Nature*, LVIII. p. 8), Prof. Thomson expresses the opinion that "the difference between us is one of terminology."]

* *Manchester Memoirs*, Vol. XLI. No. 15, 1897.
† *Phil. Mag.* Vol. XLV. p. 172, 1898.
‡ See also Prof. S. P. Thompson's *Light Visible and Invisible* (London, 1897), p. 273.
§ *Journ. de Physique*, 1886, p. 354.
‖ *Phil. Mag.* Vol. XXXVII. p. 509, 1894.
¶ *Enc. Brit.* "Wave Theory," 1888. [Vol. III. p. 60.] *Phil. Mag.* Vol. XXVII. p. 461, 1889. [Vol. III. p. 270.]

238.

NOTE ON THE PRESSURE OF RADIATION, SHOWING AN APPARENT FAILURE OF THE USUAL ELECTROMAGNETIC EQUATIONS.

[*Philosophical Magazine*, XLV. pp. 522—525, 1898.]

FOLLOWING a suggestion of Bartoli, Boltzmann* and W. Wien† have arrived at the remarkable conclusion that that part of the energy of radiation from a black body at absolute temperature θ, which lies between wave-lengths λ and $\lambda + d\lambda$, has the expression

$$\theta^5 \phi (\theta \lambda) d\lambda, \quad \dots\dots\dots\dots\dots\dots\dots\dots\dots\dots\dots\dots(1)$$

where ϕ is an arbitrary function of the *single* variable $\theta \lambda$. The law of Stefan, according to which the total radiation is as θ^4, is therein included. The argument employed by these authors is very ingenious, and I think convincing when the postulates are once admitted. The most important of them relates to the *pressure* of radiation, supposed to be operative upon the walls within which the radiation is confined, and estimated at one-third of the *density* of the energy in the case when the radiation is alike in all directions. The argument by which Maxwell originally deduced the pressure of radiation not being clear to me, I was led to look into the question a little more closely, with the result that certain discrepancies have presented themselves which I desire to lay before those who have made a special study of the electric equations. The criticism which appears to be called for extends indeed much beyond the occasion which gave rise to it.

A straightforward calculation of the pressure exercised by plane electric waves incident perpendicularly upon a metallic reflector is given by Prof. J. J. Thomson‡. The face of the reflector coincides with $x = 0$, and in the vibrations under consideration the magnetic force reduces itself to the component (β) parallel to y, and the current to the component (w) parallel to z. The waves which penetrate the conducting mass die out more or less quickly according to the conductivity. If the conductivity is great, most of the energy is reflected, and such part as is propagated into the conductor is limited to a thin skin at $x = 0$. According to the usual equations the

---

* *Wied. Ann.* Vol. XXII. pp. 31, 291 (1884).

† *Berlin. Sitzungsber.* Feb. 1893.

‡ *Elements of Electricity and Magnetism*, Cambridge, 1895, § 241.

mechanical force exercised upon unit of area of the slice dx of the conductor is $-wb\,dx$, or altogether

$$\int_0^\infty wb\,dx. \qquad\qquad\qquad (2)$$

Here b denotes the magnetic induction, and is equal to $\mu\beta$, if μ be the permeability and β the magnetic force. Now

$$4\pi w = d\beta/dx,$$

so that the integral becomes

$$\frac{\mu}{8\pi}\{\beta_0^2 - \beta_\infty^2\}, \qquad\qquad\qquad (3)$$

where β_0 is the value of β within the conductor at $x=0$, and $\beta_\infty = 0$, if the conducting slab be sufficiently thick. Since there is no discontinuity of magnetic force at $x=0$, β_0 may be taken also to refer to the value at $x=0$ just *outside* the metallic surface.

The expression (3) gives the force at any moment; but we are concerned only with the mean value. Since the mean value of β_0^2 is one-half the maximum value, we have for the pressure

$$p = \frac{\mu}{16\pi}\beta^2_{\text{max.}} \qquad\qquad\qquad (4)$$

It only remains to compare with the density of the energy outside the metal, and we may limit ourselves to the case of complete reflexion. The constant energy of the stationary waves passes alternately between the electric and magnetic forms. If we estimate it at the moment of maximum magnetic force, we have

$$\text{energy} = \frac{1}{8\pi}\iiint \beta^2\,dx\,dy\,dz. \qquad\qquad (5)$$

In (5) β is variable with x. If $\beta_{\text{max.}}$ denote the maximum value which occurs at $x=0$, the mean of $\beta^2 = \tfrac12\beta^2_{\text{max.}}$ Thus

$$\text{density of energy} = \frac{\text{energy}}{\text{volume}} = \frac{1}{16\pi}\beta^2_{\text{max.}} \qquad\qquad (6)$$

Thus, if the permeability μ of the *metal* be unity, (4) and (6) coincide; and we conclude that in this case the pressure is equal to the density of the energy in the neighbourhood of the metal. This is Maxwell's result. When we consider radiation in all directions, the pressure is expressed as *one-third* of the density of energy.

The difficulty that I have to raise relates to the case where μ is not equal to unity. The conclusion in (4) that the pressure is proportional to μ would make havoc of the theory of Boltzmann and Wien and must, I think, be rejected. So long as the reflexion is complete—and it may be complete independently of μ—the radiation is similarly influenced, and (one would suppose) must exercise a similar force upon the reflector. But if the conclusion is impossible, where is the flaw in the process by which it is arrived at? Being unable to find any fault with the deduction above given (after Prof. J. J. Thomson), I was led to scrutinize more closely the fundamental equation itself; and I will now explain why it appears to me to be incorrect.

For this purpose let us apply it to the very simple case of a wire of circular section, parallel to z, moving in the direction of x across an originally uniform magnetic field (β). The uniformity of the field is disturbed in two ways: (i) by the operation of the current (w) flowing in the various filaments of the wire, and (ii) independently of a current, by the magnetic effect of the material composing the wire whose permeability (μ) is supposed to be great. In estimating as in (2) the mechanical force parallel to x operative upon the wire, we should have to integrate wb over the cross-section. In this w is supposed to be constant, and the local value is everywhere to be attribed to b. We may indeed, if we please, omit from b the part due to the currents in the wire, which will in the end contribute nothing to the result; but we are directed to use the actual value of b as disturbed by the presence of the magnetic material. In the particular case supposed, where μ is great, the value of b within the wire is uniform, and just twice as great as at a distance. It follows, when the integration is effected, that the force parallel to x acting upon the wire is greater (in the particular case doubly greater) than it would be if the value of μ were unity.

But this conclusion cannot be accepted. The force depends upon the number of lines of force to be crossed when the wire makes a movement parallel to x. And it is clear that the lines effectively crossed in such a movement are not the condensed lines due to the magnetic quality of the wire, but are to be reckoned from the intensity of the *undisturbed* field. The mechanical force cannot really depend upon μ, and the formula which leads to such a result must be erroneous.

As regards the problem of the pressure of radiation, I conclude that in this case also, and in spite of the formula, the permeability of the reflector is without effect, and that the consequences deduced by Boltzmann and Wien remain undisturbed.

Another investigation to which perhaps similar considerations will apply is that of the mechanical force between parallel slabs conveying rapidly alternating electric currents. Prof. J. J. Thomson's conclusion* is that the electromagnetic repulsion is μ times the electrostatic attraction, so that a balance will occur only when $\mu = 1$. It seems more probable that the factor μ should be omitted, and that balance between the two kinds of force is realized in every case.

[1902. See *Phil. Mag.* XLVI. p. 154, 1898, where Prof. J. J. Thomson returns to the consideration of the question above raised.]

* *Recent Researches in Electricity and Magnetism*, 1893, § 277.

239.

SOME EXPERIMENTS WITH THE TELEPHONE.

[Roy. Inst. Proc. XV. pp. 786—789, 1898; *Nature,* LVIII. pp. 429—430, 1898.]

EARLY estimates of the minimum current of suitable frequency audible in the telephone having led to results difficult of reconciliation with the theory of the instrument, experiments were undertaken to clear up the question. The currents were induced in a coil of known construction, either by a revolving magnet of known magnetic moment, or by a magnetised tuning-fork vibrating through a measured arc. The connexion with the telephone was completed through a resistance which was gradually increased until the residual current was but just easily audible. For a frequency of 512 the current was found to be 7×10^{-8} ampères[*]. This is a much less degree of sensitiveness than was claimed by the earlier observers, but it is more in harmony with what might be expected upon theoretical grounds.

In order to illustrate before an audience these and other experiments requiring the use of a telephone, a combination of that instrument with a sensitive flame was introduced. The gas, at a pressure less than that of the ordinary supply, issues from a pin-hole burner[†] into a cavity from which air is excluded (see figure). Above the cavity, and immediately over the burner, is mounted a brass tube, somewhat contracted at the top where ignition first occurs[‡]. In this arrangement the flame is in strictness only an indicator, the really sensitive organ being the jet of gas moving within the cavity and surrounded by a similar atmosphere. When the pressure is not too high, and the jet is protected from sound, the flame is rather tall and burns bluish. Under the influence of sound of suitable pitch the jet is dispersed. At first the flame falls, becoming for a moment almost invisible; afterwards it assumes a more smoky and luminous appearance, easily distinguishable from the unexcited flame.

When the sounds to be observed come through the air, they find access by a diaphragm of tissue paper with which the cavity is faced. This serves to admit vibration while sufficiently excluding air. To get the best results the gas pressure must be steady, and be carefully adjusted to the maximum (about 1 inch) at which the flame remains undisturbed. A hiss

---

[*] The details are given in *Phil. Mag.* Vol. XXXVIII. p. 285 (1894). [Vol. IV. p. 109.]

[†] The diameter of the pin-hole may be 0·03″. [inch = 2·54 cm.]

[‡] *Camb. Proc.* Vol. IV. p. 17, 1880. [Vol. I. p. 500.]

from the mouth then brings about the transformation, while a clap of the
hands or the sudden crackling of a
piece of paper often causes extinction,
especially soon after the flame has
been lighted.

When the vibrations to be indicated
are electrical, the telephone takes the
place of the disc of tissue paper, and it
is advantageous to lead a short tube
from the aperture of the telephone into
closer proximity with the burner. The
earlier trials of the combination were
comparative failures, from a cause that
could not at first be traced. As applied,
for instance, to a Hughes' induction
balance, the apparatus failed to indicate
with certainty the introduction of a
shilling into one of the cups, and the
performance, such as it was, seemed to
deteriorate after a few minutes' experimenting. At this stage an observation
was made which ultimately afforded a
clue to the anomalous behaviour. It
was found that the telephone became
dewed. At first it seemed incredible
that this could come from the water of
combustion, seeing that the lowest part
of the flame was many inches higher.
But desiccation of the gas on its way
to the nozzle was no remedy, and it
was soon afterwards observed that no
dewing ensued if the flame were all
the while under excitation, either from
excess of pressure or from the action
of sound. The dewing was thus connected with the *unexcited* condition.
Eventually it appeared that the flame
in this condition, though apparently
filling up the aperture from which it
issues, was nevertheless surrounded by
a descending current of air carrying
with it part of the moisture of combus-
tion. The deposition of dew upon the nozzle was thus presumably the source

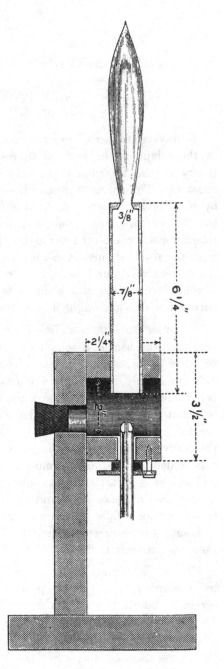

of the trouble, and a remedy was found in keeping the nozzle warm by means of a stout copper wire (not shown) conducting the heat downwards from the hot tube above.

The existence of the downward current could be made evident to private observation in various ways, perhaps most easily by projecting little scraps of tinder into the flame, whereupon bright sparks were seen to pass rapidly downwards. In this form the experiment could not be shown to an audience, but the matter was illustrated with the aid of a very delicate ether mano-meter devised by Professor Dewar. This was connected with the upper part of the brass tube by means of a small lateral perforation just below the root of the flame. The influence of sound and consequent passage of the flame from the unexcited to the excited condition was readily shown by the mano-meter, the pressure indicated being less in the former state of things.

The downward current is evidently closely associated with the change of appearance presented by the flame. In the excited state the gas issues at the large aperture above as from a reservoir at very low pressure. The unexcited flame rises higher, and must issue at a greater speed, carrying with it not only the material supplied from the nozzle, and constituting the original jet, but also some of the gaseous atmosphere in the cavity surround-ing it. The downward draught thus appears necessary in order to equalise the total issue from the upper aperture in the two cases.

Although the flame falls behind the ear in delicacy, the combination is sufficiently sensitive to allow of the exhibition of a great variety of in-teresting experiments. In the lecture the introduction of a threepenny piece into one of the cups of a Hughes' induction balance was made evident, the source of current being three Leclanché cells, and the interrupter being of the scraping contact type actuated by clockwork.

Among other experiments was shown one to prove that in certain cases the parts into which a rapidly alternating electric current is divided may be greater than the whole*. The divided circuit was formed from the three wires with which, side by side, a large flat coil is wound. One branch is formed by two of these wires connected in series, the other (in parallel with the first), by the third wire. Steady currents would traverse all three wires in the same direction. But the rapidly periodic currents from the interrupter distribute themselves so as to make the self-induction, and consequently the magnetic field, a minimum; and this is effected by the assumption of opposite values in the two branches, the ratio of currents being as $2:-1$. On the same scale the total or main current is $+1$. It was shown by means of the telephone and flame that the current in one branch was about the same (arithmetically) as in the main, and that the current in the other branch was much greater.

* See *Phil. Mag.* Vol. xxii. p. 496 (1886). [Vol. ii. p. 575.]

240.

LIQUID AIR AT ONE OPERATION.

[*Nature*, LVIII. p. 199, 1898.]

It is to be hoped that personal matters will not divert attention from the very interesting scientific questions involved. The liquefaction of air at one operation by Linde and Hampson is indeed a great feat, and a triumph for the principle of regeneration. But it must not be overlooked that to allow the air to expand without doing work, or rather to allow the work of expansion to appear as heat at the very place where the utmost cooling is desired, is very bad thermodynamics. The work of expansion should not be dissipated within, but be conducted to the exterior.

I understand that attempts to expand the air under a piston in a cylinder have led to practical difficulties connected with the low temperature. But surely a turbine of some sort might be made to work. This would occupy little space, and even if of low efficiency, would still allow a considerable fraction of the work of expansion to be conveyed away. The worst turbine would be better than none, and would probably allow the pressures to be reduced. It should be understood that the object is not so much to save the work, as to obviate the very prejudicial heating arising from its dissipation in the coldest part of the apparatus. It seems to me that the future may bring great developments in this direction, and that it may thus be possible to liquefy even hydrogen at one operation.

241.

ON THE CHARACTER OF THE IMPURITY FOUND IN NITROGEN GAS DERIVED FROM UREA [WITH AN APPENDIX CONTAINING DETAILS OF REFRACTOMETER].

[Proceedings of the Royal Society, LXIV. pp. 95—100, 1898.]

IT has already[*] been recorded that nitrogen, prepared from urea by the action of sodium hypobromite or hypochlorite, is contaminated with an impurity heavier than nitrogen. The weight of pure nitrogen in the globe employed being 2·299 grams, the gas obtained with hypochlorite was 36 milligrams, or about $1\frac{1}{2}$ per cent., heavier. "A test with alkaline pyrogallate appeared to prove the absence from this gas of free oxygen, and only a trace of carbon could be detected when a considerable quantity of the gas was passed over red-hot cupric oxide into solution of baryta." Most gases heavier than nitrogen are excluded from consideration by the thorough treatment with alkali to which the material in question is subjected. In view of the large amount of the impurity, and of the fact that it was removed by passage over red-hot iron, I inclined to identify it with nitrous oxide; but it appeared that there were strong chemical objections to this explanation, and so the matter was left open at that time. This summer I have returned to it; and although it is difficult to establish by direct evidence the presence of nitrous oxide, I think there can remain little doubt that this is the true explanation of the anomaly. I need scarcely say that there is here no question of argon beyond the minute traces that might be dissolved in the liquids employed.

In the present experiments hypochlorite has been employed, and the procedure has been the same as before. The generating bottle, previously exhausted, is first charged with the full quantity of hypochlorite solution, and the urea is subsequently fed in by degrees. The gas passes in succession over cold copper turnings, solid caustic soda, and phosphoric anhydride. In various experiments the excess of weight was found to be variable, from 23 to 36 milligrams. In order to identify the impurity it was desirable to have

---

[*] Rayleigh and Ramsay, *Phil. Trans.*, A (1895), p. 188. [Vol. IV. p. 131.]

as much of it as possible, and experiments were undertaken to find out the conditions of maximum weight. A change of procedure to one in which the urea was first introduced, so that the hypochlorite would always be on the point of exhaustion, led in the wrong direction, giving an excess of but 7 milligrams. Determinations of refractivity by the apparatus*, which uses only 12 c.c. of gas, allowed the substitution of a miniature generating vessel, and showed that the refractivity (and along with it the density) was increased by a previous heating of the hypochlorite to about 140° F. [60° C.]. Acting upon this information, arrangements were made for a preliminary heating of the large generating vessel and its charge, with the result that the excess of weight was raised to 55 milligrams, or about $2\frac{1}{2}$ per cent. of the whole. In any case heat is developed during the reaction, and the heavier weights of some of the earlier trials probably resulted from a more rapid generation of gas.

In seeking to obtain evidence as to the nature of the impurity, the most important question is as to the presence or the absence of carbon. The former experiment has been more than once repeated, with the result that the baryta showed a slight clouding. Parallel experiments, in which CO_2 was purposely introduced, indicated that the whole carbon in a charge of gas weighing 30 milligrams in excess was about 1 milligram. It is possible (though scarcely, I think, probable) that this carbon is not to be attributed to the gas at all, and in any case the amount appears to be too small to afford an explanation of the 30 milligrams excess of weight. If carbon be excluded, the range for conjecture is much narrowed. As to oxygen, only traces were found in most of the samples examined, whereas enormous quantities would be needed to explain the excessive weight. It should be noted, however, that the extra heavy sample, showing 55 milligrams excess, gave evidence of containing a more appreciable quantity of oxygen.

It seems difficult to suggest any other impurity than nitrous oxide which could account for the anomalous weight. Unfortunately there is no direct test for nitrous oxide, but so far as the examination has been carried, the behaviour of the gas is consistent with the view that this is the principal impurity. The gas as collected has no smell. The proportion of nitrous oxide indicated by the refractometer is nearly the same as that deduced from the weight. For example, the refractivity was observed of some of the gas which weighed 55 milligrams in excess. The proportion by volume (x) of N_2O in the whole required to explain the excess of weight is given by

$$ x \times \frac{22}{14} + 1 - x = \frac{2 \cdot 299 + 0 \cdot 055}{2 \cdot 299}, $$

whence $x = 0 \cdot 042.$

---

* Roy. Soc. Proc., Vol. LIX. p. 201, 1896 [Vol. IV. p. 218]; Vol. LX. p. 56, 1896 [Vol. IV. p. 225]. See also Appendix.

The refractivity (referred to *air* as unity) of the same gas was determined by two independent sets of observations as $1\cdot047$, $1\cdot048$; mean, $1\cdot0475$. If we assume that there are only nitrogen and nitrous oxide present, the proportion (x) of the latter can be deduced from the known refractivities ($\mu - 1$) of nitrous oxide, nitrogen, and air, which are respectively $0\cdot0005159$, $0\cdot0002977$, $0\cdot0002927$, the number for air being *less* than for nitrogen. Thus,

$$x \times 5159 + (1 - x) \times 2977 = 1\cdot0475 \times 2927,$$

giving $x = 0\cdot0408.$

The slight want of agreement can be explained by the presence of a little oxygen, the recognition of which would lead to a rise in the second value of x, and a fall in the first. Examination of the gas from the refractometer with alkaline pyrogallate proved that oxygen was actually present.

Evidence may also be obtained by exploding the gas with excess of hydrogen for which purpose oxy-hydrogen gas must be added. But when nitrous oxide is in question, operations over water are useless, while for the more exact procedure with mercury, experience and appliances were somewhat deficient. The contraction observed was rather in excess of the volume of nitrous oxide supposed to be present, but of this a good part is readily explained by a small proportion of free oxygen.

If the impurity is really nitrous oxide, it should admit of concentration by solution in water. To test this, about 1 litre of water (cooled with ice) was shaken with the contents of a globe (about 2 litres). The dissolved gases were then expelled by boiling, and were collected over water rendered alkaline, in order to guard against the introduction of CO_2. The quantity was, of course, too small for weighing, but it could readily be examined in the refractometer. Of one sample, after desiccation, the refractivity relatively to air was found to be as high as $1\cdot207$, although some air was known to have entered accidentally. The proportion of nitrous oxide in a mixture with nitrogen which would have this refractivity is $0\cdot255$. The impurity thus agrees with nitrous oxide in being very much more soluble in water than are the gases of the atmosphere.

In the analytical use of hypobromite for the determination of urea, it has been noticed[*] that the nitrogen collected is deficient by about 8 per cent., but the matter does not appear to have been further examined. The deficiency might be attributed to a part of the urea remaining undecomposed, but more probably to oxidation of nitrogen. In default of analysis any nitrogen collected as nitrous oxide would not appear anomalous, and the explanation suggested requires the formation in addition of higher oxides retained by the alkali.

* Russell and West, *Chem. Soc. Journ.*, Vol. XII. p. 749, 1874.

There is reason to suspect that nitrogen prepared by the action of chlorine upon ammonia is also contaminated with nitrous oxide, and this is a matter of interest, for the contamination in this case cannot well be referred to a carbon compound.   In two trials with distinct samples the refractivities were decidedly in excess of that of pure nitrogen.

APPENDIX.

Details of Refractometer.

Determinations of refractivity have proved so useful and can be made so readily and upon such small quantities of gas, that it may be desirable to give further details of the apparatus employed, referring for explanation of the principles involved to the former communication already cited.

The optical parts, other than the tubes containing the gases, are mounted independently of everything else upon a bar of T-iron 90 cm. in length over all.   The telescopes are cheap instruments, of about 3 cm. aperture and 30 cm. focus, from which the eye-pieces are removed.   At one end of the T-iron and in the focus of the collimating telescope the original slit is fixed. This requires to be rather narrow, and was made by scraping a fine line upon a piece of silvered glass.   At the further end the object-glass of the observing telescope carries two slits which give passage to the interfering pencils, and are situated opposite to the axes of the tubes holding the gases. The sole eye-piece is a short length of glass rod—the same as formerly described—of about 4 mm. diameter, which serves as horizontal magnifier. The gas tubes are of brass, about 20 cm. long and 6 mm. in bore.   These are soldered together side by side and are closed at the ends by plates of worked glass, so cemented as to obstruct as little as possible the passage of light immediately over the tubes.   There are two systems of bands, one formed by light which has traversed the gases within the tubes, the other by light which passes independently above; and an observation consists in so adjusting the pressures within the tubes that the two systems fit one another.   Unless some further provision be made, there is necessarily a dark interval between the two systems of bands corresponding to the thickness of the walls of the tubes and any projecting cement.   It is, perhaps, an improvement to bring the two sets of bands into closer juxtaposition.   The interval can be abolished with the aid of a bi-plate [see figure], formed of worked glass 4 or 5 mm. thick*. 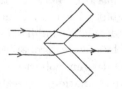 This is placed immediately in front of the object-glass of the observing telescope, the plane of junction of the two glasses being horizontal and at the level of the obstacles which are to be blotted out of the field of view.

* Compare Mascart, *Traité d'Optique*, Vol. i. p. 495, 1889.

The objects sought in the design of the remainder of the apparatus were (i) the use of a minimum of gas, and (ii) independence of other pumping appliances. To this end the glass tubes associated with each optical tube were arranged so as to serve both as manometer tubes and as a sort of Geissler pump. The two halves of the apparatus being independent and similar, it will suffice to speak of that which contained the gas to be investigated. The tubes in which the levels of mercury are observed are about 1 cm. in diameter. The fixed one, corresponding to the "pump-head" of a Geissler or Töpler, is 33 cm. in length, and is surmounted by a three-way tap, allowing it to be placed in communication either with the optical tube or with one of narrow bore ending in a U, drowned in a deep mercury trough. The bottom of the fixed tube, prolonged by 92 cm. of narrower bore, is connected through a hose of black rubber with the movable manometer tube. The latter is 70 cm. long and of one bore (1 cm.) throughout. It can either be held in the hand or placed in a groove (parallel to the fixed tube) along which it can slide. The four columns of mercury stand side by side, and the levels are referred by a cathetometer to a metre scale which occupies the central position. It is not proposed to describe the cathetometer in detail, but it may be mentioned that it is of home construction, and is mounted on centres attached to the floor and ceiling of the room. It sufficed to record the levels to tenths of millimetres. The whole apparatus was constructed by Mr Gordon.

If the glasses closing the optical tubes were perfect, there would be coincidence of bands corresponding to complete exhaustion of both optical tubes. A correction could be made for the residual error once for all determined, but it is safer to make two independent settings, one at pressures as nearly atmospheric as the case admits, and a second at minimum pressures. There are then in all eight readings to be combined. An example may be taken from a case already referred to:—

I.	II.	III.	IV.
9770	9371	9749	9790
7272	2165	2469	7445

Columns I, II refer to the anomalous nitrogen, III and IV to the dried air used as a standard of comparison. I and IV are the fixed manometer tubes in communication with the optical tubes. The reduction may be effected by subtraction of the rows:

| 2498 | 7206 | 7280 | 2345 |

Thus 4708, the difference between II and I, of the nitrogen balances 4935, the difference between III and IV, of air. The refractivity referred to air is accordingly $\frac{4935}{4708}$, or 1·048.

In this example the range of pressures for the air is 493·5 mm., or about two-thirds of an atmosphere.

Great care is sometimes required to ensure matching the same bands in the two settings. A mistake of one band in the above example would entail nearly 2 per cent. error in the final result, inasmuch as the whole number of bands concerned is about 96 per atmosphere of air, or about 62 over the range actually used. It is wise always to include a match with pressures about midway between the extremes. If the results harmonise, an error of a single band is excluded; and it is hardly possible to make a mistake of two bands.

As regards accuracy, independent final results usually agree to one-thousandth part.

242.

ON ISO-PERIODIC SYSTEMS.

[*Philosophical Magazine*, XLVI. pp. 567—569, 1898.]

IN general a system with m degrees of freedom vibrating about a configuration of equilibrium has m distinct periods, or frequencies, of vibration, but in particular cases two or more of these frequencies may be equal. The simple spherical pendulum is an obvious example of two degrees of freedom whose frequencies are equal. It is proposed to point out the properties of vibrating systems of such a character that all the frequencies are equal.

In the general case when a system is referred to its normal coordinates ϕ_1, ϕ_2, ... we have for the kinetic and potential energies*,

$$\left.\begin{array}{l} T = \tfrac{1}{2} a_1 \dot{\phi}_1{}^2 + \tfrac{1}{2} a_2 \dot{\phi}_2{}^2 + \dots \\ V = \tfrac{1}{2} c_1 \phi_1{}^2 + \tfrac{1}{2} c_2 \phi_2{}^2 + \dots \end{array}\right\}, \qquad \dots\dots\dots\dots\dots(1)$$

and for the vibrations

$$\phi_1 = A \cos(n_1 t - \alpha), \qquad \phi_2 = B \cos(n_2 t - \beta), \qquad \&c. \quad \dots\dots(2)$$

where $A, B, \dots \alpha, \beta \dots$ are arbitrary constants and

$$n_1{}^2 = c_1/a_1, \qquad n_2{}^2 = c_2/a_2, \qquad \&c. \dots\dots\dots\dots\dots(3)$$

If n_1, n_2, &c., are all equal, T and V are of the same form except as to a constant multiplier. By supposing α, $\beta \dots$ equal, we see that any prescribed ratios may be assigned to ϕ_1, $\phi_2 \dots$, so that vibrations of arbitrary type are normal and can be executed without constraint. In particular any parts of the system may remain at rest.

If x, y, z be the space coordinates (measured from the equilibrium position) of any point of the system, the most general values are given by

$$\left.\begin{array}{l} x = X_1 \cos nt + X_2 \sin nt \\ y = Y_1 \cos nt + Y_2 \sin nt \\ z = Z_1 \cos nt + Z_2 \sin nt \end{array}\right\}, \qquad \dots\dots\dots\dots\dots(4)$$

---

* See, for example, *Theory of Sound*, § 87.

where X_1, X_2, &c. are constants for each point. These equations indicate elliptic motion in the plane

$$x\,(Y_1 Z_2 - Z_1 Y_2) + y\,(Z_1 X_2 - X_1 Z_2) + z\,(X_1 Y_2 - Y_1 X_2) = 0.\ldots\ldots(5)$$

Thus every point of the system describes an elliptic orbit in the same periodic time.

An interesting case is afforded by a line of similar bodies of which each is similarly connected to its neighbours*. The general formula for n^2 is

$$n^2 = \frac{C_0 - 2C_1\cos ka - 2C_2\cos 2ka - \ldots}{A_0 - 2A_1\cos ka - 2A_2\cos 2ka - \ldots},\quad\ldots\ldots\ldots\ldots(6)$$

in which the constants C_0, C_1 ... refer to the potential, and A_1, A_2 ... to the kinetic energy. Here C_1, A_1 represent the influence of immediate neighbours distant a from one another, C_2, A_2 the influence of neighbours distant $2a$, and so on. Further, k denotes $2\pi/\lambda$, λ being the wave-length. If C_1, C_2 ..., A_1, A_2 ... vanish, each body is uninfluenced by its neighbours, and the case is one considered by Reynolds of a number of similar and disconnected pendulums hanging side by side at equal distances. It is obvious that a vibration of any type is normal and is executed in the same time. If we consider a progressive wave, its velocity is proportional to λ. A disturbance communicated to any region has no tendency to propagate itself; the " group velocity" is zero.

Although the line of disconnected pendulums is interesting and throws light upon the general theory of wave and group propagation, one can hardly avoid the feeling that it is only by compliment that it is regarded as a single system. It is therefore not without importance to notice that there are other cases for which n assumes a constant, and the group-velocity a zero, value. To this end it is only necessary that

$$C_0 : C_1 : C_2 : \ldots = A_0 : A_1 : A_2 : \ldots\quad\ldots\ldots\ldots\ldots\ldots(7)$$

If this condition be satisfied, the connexion of neighbouring bodies does not entail the propagation of disturbance. Any number of the bodies may remain at rest, and all vibrations have the same period.

We might consider particular systems for which C_2, C_3 ... A_2, A_3 ... vanish, while $C_1/C_0 = A_1/A_0$; but it is perhaps more interesting to draw an illustration from the case of continuous linear bodies. Consider a wire stretched with tension T_1, each element dx of which is urged to its position of equilibrium ($y=0$) by a force equal to $\mu y\,dx$. The potential energy† is given by

$$V = \tfrac{1}{2}\mu \int y^2 dx + \tfrac{1}{2}T_1 \int \left(\frac{dy}{dx}\right)^2 dx.\quad\ldots\ldots\ldots\ldots (8)$$

---

* *Phil. Mag.* Vol. XLIV. p. 356, 1897. [Vol. IV. p. 340.]
† See *Theory of Sound*, §§ 122, 162, 188.

If the "rotatory inertia" be included, the corresponding expression for the kinetic energy is

$$T = \tfrac{1}{2}\rho\omega \int \left(\frac{dy}{dt}\right)^2 dx + \tfrac{1}{2}\kappa^2\rho\omega \int \left(\frac{d^2y}{dt\,dx}\right)^2 dx, \quad \dots\dots\dots\dots(9)$$

in which ρ is the volume density, ω the area of cross section, and κ the radius of gyration of the cross section about an axis perpendicular to the plane of bending. In waves along an actual wire vibrating transversely the second term would be relatively unimportant, but there is no contradiction in the supposition that the rotatory term is predominant. The differential equation derived from (8) and (9) is

$$\frac{d^2y}{dt^2} - \kappa^2 \frac{d^4y}{dx^2\,dt^2} - a^2 \frac{d^2y}{dx^2} + c^2y = 0, \quad \dots\dots\dots\dots(10)$$

where
$$a^2 = T_1/\rho\omega, \qquad c^2 = \mu/\rho\omega. \quad \dots\dots\dots\dots(11)$$

If we suppose that there is no tension and no rotatory inertia, $a = 0$, $\kappa = 0$, and the solution of (10) may be written

$$y = \cos ct \,.\, y_1 + \sin ct \,.\, y_2, \quad \dots\dots\dots\dots(12)$$

y_1, y_2 being arbitrary functions of x. If $y_1 = \cos mx$, $y_2 = \sin mx$, (12) becomes

$$y = \cos (ct - mx), \quad \dots\dots\dots\dots(13)$$

and the velocity of propagation (c/m) is proportional to λ, equal to $2\pi/m$. This is the case of the disconnected pendulums.

On the other hand we may equally well suppose that c is zero and that the rotatory inertia is paramount, so that (10) reduces to

$$\kappa^2 \frac{d^4y}{dx^2\,dt^2} + a^2 \frac{d^2y}{dx^2} = 0.$$

The periodic part of the solution is again of the form (12), and has the same peculiar properties as before.

In the general case we have the solution for stationary vibrations

$$y = \sin mx \cos nt, \quad \dots\dots\dots\dots(14)$$

where $m = i\pi/l$, i being an integer, and

$$n^2 = \frac{c^2 + a^2m^2}{1 + \kappa^2m^2}. \quad \dots\dots\dots\dots(15)$$

This gives the frequencies for the various modes of vibration of a wire of length l fastened at the ends.

If $\kappa^2 = a^2/c^2$, n becomes independent of m as before.

If $\kappa^2 < a^2/c^2$, n^2 increases, as i and m increase, and approaches a finite upper limit a^2/κ^2. The series of frequencies is thus analogous to those met with in the spectra of certain bodies [*].

* Compare Schuster, *Nature*, Vol. LV. p. 200 (1890).

243.

ON JAMES BERNOULLI'S THEOREM IN PROBABILITIES.

[*Philosophical Magazine*, XLVII. pp. 246—251, 1899.]

IF p denote the probability of an event, then the probability that in μ trials the event will happen m times and fail n times is equal to a certain term in the expansion of $(p+q)^\mu$, namely,

$$\frac{\mu!}{m!\,n!}p^m q^n, \quad\dots\dots\dots\dots\dots\dots\dots\dots\dots\dots(1)$$

where $p+q=1$, $m+n=\mu$.

"Now it is known from Algebra that if m and n vary subject to the condition that $m+n$ is constant, the greatest value of the above term is when m/n is as nearly as possible equal to p/q, so that m and n are as nearly as possible equal to μp and μq respectively. We say *as nearly as possible*, because μp is not necessarily an integer, while m is. We may denote the value of m by $\mu p + z$, where z is some proper fraction, positive or negative; and then $n = \mu q - z$."

The rth term, counting onwards, in the expansion of $(p+q)^\mu$ after (1) is

$$\frac{\mu!}{m-r!\,n+r!}p^{m-r}q^{n+r}. \quad\dots\dots\dots\dots\dots\dots(2)$$

The approximate value of (2) when m and n are large numbers may be obtained with the aid of Stirling's theorem, viz.

$$\mu! = \mu^{\mu+\frac{1}{2}}e^{-\mu}\sqrt{(2\pi)}\left\{1+\frac{1}{12\mu}+\dots\right\}. \quad\dots\dots\dots\dots(3)$$

The process is given in detail after Laplace in Todhunter's *History of the Theory of Probability*, p. 549, from which the above paragraph is quoted. The expression for the rth term after the greatest is

$$\frac{e^{-\frac{\mu r^2}{2mn}}\sqrt{\mu}}{\sqrt{(2\pi mn)}}\left\{1+\frac{\mu rz}{mn}+\frac{r(n-m)}{2mn}-\frac{r^3}{6m^2}+\frac{r^3}{6n^2}\right\}; \quad\dots\dots\dots(4)$$

and that for the rth term before the greatest may be deduced by changing the sign of r in (4).

It is assumed that r^2 does not surpass μ in order of magnitude, and fractions of the order $1/\mu$ are neglected.

There is an important case in which the circumstances are simpler than in general. It arises when $p = q = \frac{1}{2}$, and μ is an even number, so that $m = n = \frac{1}{2}\mu$. Here z disappears *ab initio*, and (4) reduces to

$$\frac{2\,e^{-2r^2/\mu}}{\sqrt{(2\pi\mu)}}, \qquad \dots\dots\dots\dots\dots\dots\dots(5)$$

representing (2), which now becomes

$$\frac{\mu!}{2^\mu \cdot \frac{1}{2}\mu - r! \, \frac{1}{2}\mu + r!}. \qquad \dots\dots\dots\dots\dots(6)$$

An important application of (5) is to the theory of random vibrations. If μ vibrations are combined, each of the same phase but of amplitudes which are at random either $+1$ or -1, (5) represents the probability of $\frac{1}{2}\mu + r$ of them being positive vibrations, and accordingly $\frac{1}{2}\mu - r$ being negative. In this case, and in this case only, is the resultant $+2r$. Hence if x represent the resultant, the chance of x, which is necessarily an *even* integer, is

$$\frac{2\,e^{-x^2/2\mu}}{\sqrt{(2\pi\mu)}}.$$

The next greater resultant is $(x+2)$; so that when x is great the above expression may be supposed to correspond to a range for x equal to 2. If we represent the range by dx, the chance of a resultant lying between x and $x + dx$ is given by

$$\frac{e^{-x^2/2\mu}\,dx}{\sqrt{(2\pi\mu)}}. \qquad \dots\dots\dots\dots\dots\dots\dots(7)^*$$

Another view of this matter, leading to (5) or (7) without the aid of Stirling's theorem, or even of formula (1), is given (somewhat imperfectly) in *Theory of Sound*, 2nd ed. § 42 *a*. It depends upon a transition from an equation in finite differences to the well-known equation for the conduction of heat and the use of one of Fourier's solutions of the latter. Let $f(\mu, r)$ denote the chance that the number of events occurring (in the special application *positive* vibrations) is $\frac{1}{2}\mu + r$, so that the *excess* is r. Suppose that each random combination of μ receives *two* more random contributions—*two* in order that the whole number may remain even,—and inquire into the chance of a subsequent excess r, denoted by $f(\mu + 2, r)$. The excess after the addition can only be r if previously it were $r - 1$, r, or $r + 1$. In the first case the excess becomes r by the occurrence of both of the two new events,

* *Phil. Mag.* Vol. x. p. 75 (1880). [Vol. I. p. 491.]

of which the chance is $\frac{1}{4}$. In the second case the excess remains r in consequence of one event happening and the other failing, of which the chance is $\frac{1}{2}$; and in the third case the excess becomes r in consequence of the failure of both the new events, of which the chance is $\frac{1}{4}$. Thus

$$f(\mu+2, r) = \tfrac{1}{4} f(\mu, r-1) + \tfrac{1}{2} f(\mu, r) + \tfrac{1}{4} f(\mu, r+1). \quad \ldots\ldots(8)$$

According to the present method the limiting form of f is to be derived from (8). We know, however, that f has actually the value given in (6), by means of which (8) may be verified.

Writing (8) in the form

$$f(\mu+2, r) - f(\mu, r) = \tfrac{1}{4} f(\mu, r-1) - \tfrac{1}{2} f(\mu, r) + \tfrac{1}{4} f(\mu, r+1), \ldots(9)$$

we see that when μ and r are infinite the left-hand member becomes $2 df/d\mu$, and the right-hand member becomes $\tfrac{1}{4} d^2f/dr^2$, so that (9) passes into the differential equation

$$\frac{df}{d\mu} = \frac{1}{8} \frac{d^2f}{dr^2}. \quad \ldots\ldots\ldots\ldots\ldots(10)$$

In (9), (10) r is the excess of the actual occurrences over $\frac{1}{2}\mu$. If we take x to represent the difference between the number of occurrences and the number of failures, $x = 2r$ and (10) becomes

$$\frac{df}{d\mu} = \frac{1}{2} \frac{d^2f}{dx^2}. \quad \ldots\ldots\ldots\ldots\ldots(11)$$

In the application to vibrations $f(\mu, x)$ then denotes the chance of a resultant $+x$ from a combination of μ unit vibrations which are positive or negative at random.

In the formation of (10) we have supposed for simplicity that the addition to μ is 2, the lowest possible consistently with the total number remaining even. But if we please we may suppose the addition to be any even number μ'. The analogue of (8) is then

$$2^{\mu'} . f(\mu+\mu', r) = f(\mu, r - \tfrac{1}{2}\mu') + \mu' f(\mu, r - \tfrac{1}{2}\mu' + 1)$$
$$+ \frac{\mu'(\mu'-1)}{1.2} f(\mu, r - \tfrac{1}{2}\mu' + 2) + \ldots + f(\mu, r + \tfrac{1}{2}\mu');$$

and when μ is treated as very great the right-hand member becomes

$$f(\mu, r)\left\{1 + \mu' + \frac{\mu'(\mu'-1)}{1.2} + \ldots + \mu' + 1\right\}$$
$$+ \frac{1}{8}\frac{d^2f}{dr^2}\left\{1.\mu'^2 + \mu'(\mu'-2)^2 + \frac{\mu'(\mu'-1)}{1.2}(\mu'-4)^2\right.$$
$$\left. + \ldots\ldots + \mu'(\mu'-2)^2 + 1.\mu'^2\right\}.$$

The series which multiplies f is $(1+1)^{\mu'}$, or $2^{\mu'}$. The second series is equal to $\mu' \cdot 2^{\mu'}$, as may be seen by comparison of coefficients of x^2 in the equivalent forms

$$(e^x + e^{-x})^n = 2^n (1 + \tfrac{1}{2} x^2 + \dots)^n$$

$$= e^{nx} + n e^{(n-2)x} + \frac{n(n-1)}{1.2} e^{(n-4)x} + \dots .$$

The value of the left-hand member becomes simultaneously

$$2^{\mu'} \{f + \mu' df/d\mu\};$$

so that we arrive at the same differential equation (10) as before.

This is the well-known equation for the conduction of heat, and the solution developed by Fourier is at once applicable. The symbol μ corresponds to time and r to a linear coordinate. The special condition is that initially—that is when μ is relatively small—f must vanish for all values of r that are not small. We take therefore

$$f(\mu, r) = \frac{A}{\sqrt{\mu}} e^{-2r^2/\mu}, \quad \dots\dots\dots\dots\dots\dots(12)$$

which may be verified by differentiation.

The constant A may be determined by the understanding that $f(\mu, r) \, dr$ is to represent the chance of an excess lying between r and $r + dr$, and that accordingly

$$\int_{-\infty}^{+\infty} f(\mu, r) \, dr = 1. \quad \dots\dots\dots\dots\dots\dots(13)$$

Since $\int_{-\infty}^{+\infty} e^{-z^2} dz = \sqrt{\pi}$, we have

$$\frac{A}{\sqrt{\mu}} = \sqrt{\left(\frac{2}{\pi\mu}\right)}; \quad \dots\dots\dots\dots\dots\dots(14)$$

and, finally, as the chance that the excess lies between r and $r + dr$,

$$\sqrt{\left(\frac{2}{\pi\mu}\right)} e^{-2r^2/\mu} dr. \quad \dots\dots\dots\dots\dots\dots(15)$$

Another method by which A in (12) might be determined would be by comparison with (6) in the case of $r = 0$. In this way we find

$$\frac{A}{\sqrt{\mu}} = \frac{\mu!}{2^{\mu} \cdot \frac{1}{2}\mu! \, \frac{1}{2}\mu!} = \frac{1.3.5 \dots (\mu-1)}{2.4.6 \dots\dots\dots \mu}$$

$$= \sqrt{\left(\frac{2}{\pi\mu}\right)} \text{ by Wallis' theorem.}$$

If, as is natural in the problem of random vibrations, we replace r by x, denoting the difference between the number of occurrences and the number of failures, we have as the chance that x lies between x and $x + dx$

$$\frac{e^{-x^2/2\mu}\,dx}{\sqrt{(2\pi\mu)}}, \quad\dots\dots\dots\dots\dots\dots\dots(16)$$

identical with (7).

In the general case when p and q are not limited to the values $\frac{1}{2}$, it is more difficult to exhibit the argument in a satisfactory form, because the most probable numbers of occurrences and failures are no longer definite, or at any rate simple, fractions of μ. But the general idea is substantially the same. The excess of occurrences over the most probable number is still denoted by r, and its probability by $f(\mu, r)$. We regard r as continuous, and we then suppose that μ increases by unity. If the event occurs, of which the chance is p, the total number of occurrences is increased by unity. But since the most probable number of occurrences is increased by p, r undergoes only an increase measured by $1 - p$ or q. In like manner if the event fails, r undergoes a *decrease* measured by p. Accordingly

$$f(\mu + 1, r) = pf(\mu, r - q) + qf(\mu, r + p). \quad\dots\dots\dots\dots(17)$$

On the right of (17) we expand $f(\mu, r - q)$, $f(\mu, r + p)$ in powers of p and q. Thus

$$f(\mu, r + p) = f + \frac{df}{dr}p + \tfrac{1}{2}\frac{d^2f}{dr^2}p^2, \quad f(\mu, r - q) = f - \frac{df}{dr}q + \tfrac{1}{2}\frac{d^2f}{dr^2}q^2;$$

so that the right-hand member is

$$(p + q)f + \tfrac{1}{2}\frac{d^2f}{dr^2}(p^2q + pq^2), \quad \text{or} \quad f + \tfrac{1}{2}pq\frac{d^2f}{dr^2}.$$

The left-hand member may be represented by $f + df/d\mu$, so that ultimately

$$\frac{df}{d\mu} = \tfrac{1}{2}pq\frac{d^2f}{dr^2}. \quad\dots\dots\dots\dots\dots\dots(18)$$

Accordingly by the same argument as before the chance of an excess r lying between r and $r + dr$ is given by

$$\frac{1}{\sqrt{(2\pi pq\mu)}}e^{-r^2/2pq\mu}\,dr. \quad\dots\dots\dots\dots(19)$$

We have already considered the case of $p = q = \frac{1}{2}$. Another particular case of importance arises when p is very small, and accordingly q is nearly equal to unity. The whole number μ is supposed to be so large that $p\mu$, or m, representing the most probable number of occurrences, is also large. The general formula now reduces to

$$\frac{1}{\sqrt{(2\pi m)}} \, e^{-r^2/2m} dr, \dots\dots\dots\dots\dots\dots\dots(20)$$

which gives the probability that the number of occurrences shall lie between $m + r$ and $m + r + dr$. It is a function of m and r only.

The probability of the deviation from m lying between $\pm\, r$

$$= \frac{2}{\sqrt{(2\pi m)}} \int_0^r e^{-r^2/2m} dr = \frac{2}{\sqrt{\pi}} \int_0^\tau e^{-\tau^2} d\tau, \quad \dots\dots\dots\dots(21)$$

where $\tau = r/\sqrt{(2m)}$. This is equal to ·84 when $\tau = 1$·0, or $r = \sqrt{(2m)}$; so that the chance is comparatively small of a deviation from m exceeding $\pm \sqrt{(2m)}$. For example, if m is 50, there is a rather strong probability that the actual number of occurrences will lie between 40 and 60.

The formula (20) has a direct application to many kinds of statistics.

244.

ON THE COOLING OF AIR BY RADIATION AND CONDUCTION, AND ON THE PROPAGATION OF SOUND.

[*Philosophical Magazine*, XLVII. pp. 308—314, 1899.]

ACCORDING to Laplace's theory of the propagation of Sound the expansions (and contractions) of the air are supposed to take place without transfer of heat. Many years ago Sir G. Stokes* discussed the question of the influence of radiation from the heated air upon the propagation of sound. He showed that such small radiating power as is admissible would tell rather upon the intensity than upon the velocity. If x be measured in the direction of propagation, the factor expressing the diminution of amplitude is e^{-mx}, where

$$m = \frac{\gamma - 1}{\gamma} \frac{q}{2a} . \quad\dots\dots\dots\dots\dots\dots\dots\dots\dots(1)$$

In (1) γ represents the ratio of specific heats (1·41), a is the velocity of sound, and q is such that e^{-qt} represents the law of cooling by radiation of a small mass of air maintained at constant volume. If τ denote the time required to traverse the distance x, $\tau = x/a$, and (1) may be taken to assert that the amplitude falls to any fraction, *e.g.* one-half, of its original value in 7 times the interval of time required by a mass of air to cool to the same fraction of its original excess of temperature. "There appear to be no data by which the latter interval can be fixed with any approach to precision; but if we take it at one minute, the conclusion is that sound would be propagated for (seven) minutes, or travel over about (80) miles, without very serious loss from this cause†." We shall presently return to the consideration of the probable value of q.

Besides radiation there is also to be considered the influence of conductivity in causing transfer of heat, and further there are the effects of viscosity.

---

* *Phil. Mag.* [4] I. p. 305, 1851; *Theory of Sound*, § 247.
† *Proc. Roy. Inst.* April 9, 1897. [Vol. IV. p. 298.]

The problems thus suggested have been solved by Stokes and Kirchhoff*. If the law of propagation be

$$u = e^{-m'x} \cos (nt - x/a), \quad \dots\dots\dots\dots\dots\dots\dots(2)$$

then

$$m' = \frac{n^2}{2a^3} \left\{ \tfrac{4}{3}\mu' + \nu \frac{\gamma - 1}{\gamma} \right\}, \quad \dots\dots\dots\dots\dots\dots(3)$$

in which the frequency of vibration is $n/2\pi$, μ' is the kinematic viscosity, and ν the thermometric conductivity. In C.G.S. measure we may take $\mu' = \cdot 14$, $\nu = \cdot 26$, so that

$$\tfrac{4}{3}\mu' + \nu \frac{\gamma - 1}{\gamma} = \cdot 25.$$

To take a particular case, let the frequency be 256; then since $a = 33200$, we find for the time of propagation during which the amplitude diminishes in the ratio of $e : 1$,

$$(m'a)^{-1} = 3560 \text{ seconds.}$$

Accordingly it is only very high sounds whose propagation can be appreciably influenced by viscosity and conductivity.

If we combine the effects of radiation with those of viscosity and conduction, we have as the factor of attenuation

$$e^{-(m+m')x},$$

where

$$m + m' = \cdot 14 \, (q/a) + \cdot 12 \, (n^2/a^3). \quad \dots\dots\dots\dots\dots\dots(4)$$

In actual observations of sound we must expect the intensity to fall off in accordance with the law of inverse squares of distances. A very little experience of moderately distant sounds shows that in fact the intensity is in a high degree uncertain. These discrepancies are attributable to atmospheric refraction and reflexion, and they are sometimes very surprising. But the question remains whether in a uniform condition of the atmosphere the attenuation is sensibly more rapid than can be accounted for by the law of inverse squares. Some interesting experiments towards the elucidation of this matter have been published by Mr Wilmer Duff†, who compared the distances of audibility of sounds proceeding respectively from two and from eight similar whistles. On an average the eight whistles were audible only about one-fourth further than a pair of whistles; whereas, if the sphericity of the waves had been the only cause of attenuation, the distances would have been as 2 to 1. Mr Duff considers that in the circumstances of his experiments there was little opportunity for atmospheric irregularities, and he attributes the greater part of the falling off to radiation. Calculating from (1) he deduces a radiating power such that a mass of air at any given excess of temperature above its surroundings will (if its volume remain constant) fall by radiation to one-half of that excess in about one-twelfth of a second.

---

* *Pogg. Ann.* Vol. cxxxiv. p. 177, 1868; *Theory of Sound*, 2nd ed. § 348.

† *Phys. Review*, Vol. vi. p. 129, 1898.

In this paper I propose to discuss further the question of the radiating power of air, and I shall contend that on various grounds it is necessary to restrict it to a value hundreds of times smaller than that above mentioned. On this view Mr Duff's results remain unexplained. For myself I should still be disposed to attribute them to atmospheric refraction. If further experiment should establish a rate of attenuation of the order in question as applicable in uniform air, it will I think be necessary to look for a cause not hitherto taken into account. We might imagine a delay in the equaliza-tion of the different sorts of energy in a gas undergoing compression, not wholly insensible in comparison with the time of vibration of the sound. If in the dynamical theory we assimilate the molecules of a gas to hard smooth bodies which are nearly but not absolutely spherical, and trace the effect of a rapid compression, we see that at the first moment the increment of energy is wholly translational and thus produces a maximum effect in opposing the compression. A little later a due proportion of the excess of energy will have passed into rotational forms which do not influence the pressure, and this will accordingly fall off. Any effect of the kind must give rise to dissipation, and the amount of it will increase with the time required for the transformations, *i.e.* in the above mentioned illustration with the degree of approximation to the spherical form. In the case of absolute spheres no transformation of translatory into rotatory energy, or *vice versa*, would occur in a finite time. There appears to be nothing in the behaviour of gases, as revealed to us by experiment, which forbids the supposition of a delay capable of influencing the propagation of sound.

Returning now to the question of the radiating power of air, we may establish a sort of superior limit by an argument based upon the theory of exchanges, itself firmly established by the researches of B. Stewart. Consider a spherical mass of radius r, slightly and uniformly heated. Whatever may be the radiation proceeding from a unit of surface, it must be less than the radiation from an ideal black surface under the same conditions. Let us, however, suppose that the radiation is the same in both cases and inquire what would then be the rate of cooling. According to Bottomley[*] the emissivity of a blackened surface moderately heated is ·0001. This is the amount of heat reckoned in water-gram-degree units emitted in one second from a square centimetre of surface heated 1° C. If the excess of temperature be θ, the whole emission is

$$\theta \times 4\pi r^2 \times \cdot0001$$

On the other hand, the capacity for heat is

$$\tfrac{4}{3}\pi r^3 \times \cdot0013 \times \cdot24,$$

the first factor being the volume, the second the density, and the third the

---

[*] Everett, *C.G.S. Units*, 1891, p. 134.

specific heat of air referred, as usual, to water. Thus for the rate of cooling,

$$\frac{d\theta}{\theta dt} = -\frac{\cdot 0003}{\cdot 0013 \times \cdot 24 \times r} = -\frac{1}{r} \text{ very nearly,}$$

whence $\qquad\qquad\qquad\qquad \theta = \theta_0 e^{-t/r},$ ....(5)

θ_0 being the initial value of θ. The time in seconds of cooling in the ratio of $e:1$ is thus represented numerically by r expressed in centims.

When r is very great, the suppositions on which (5) is calculated will be approximately correct, and that equation will then represent the actual law of cooling of the sphere of air, supposed to be maintained uniform by mixing if necessary. But ordinary experience, and more especially the observations of Tyndall upon the diathermancy of air, would lead us to suppose that this condition of things would not be approached until r reached 1000 or perhaps 10,000 centims. For values of r comparable with the half wave-length of ordinary sounds, e.g. 30 centim., it would seem that the real time of cooling must be a large multiple of that given by (5). At this rate the time of cooling of a mass of air must exceed, and probably largely exceed, 60 seconds. To suppose that this time is one-twelfth of a second would require a sphere of air 2 millim. in diameter to radiate as much heat as if it were of blackened copper at the same temperature.

Although, if the above argument is correct, there seems little likelihood of the cooling of moderate masses of air being sensibly influenced by radiation, I thought it would be of interest to inquire whether the observed cooling (or heating) in an experiment on the lines of Clement and Desormes could be adequately explained by the conduction of heat from the walls of the vessel in accordance with the known conductivity of air. A nearly spherical vessel of glass of about 35 centim. diameter, well encased, was fitted, air-tight, with two tubes. One of these led to a manometer charged with water or sulphuric acid; the other was provided with a stopcock and connected with an air-pump. In making an experiment the stopcock was closed and a vacuum established in a limited volume upon the further side. A rapid opening and reclosing of the cock allowed a certain quantity of air to escape suddenly, and thus gave rise to a nearly uniform cooling of that remaining behind in the vessel. At the same moment the liquid rose in the manometer, and the observation consisted in noting the times (given by a metronome beating seconds) at which the liquid in its descent passed the divisions of a scale, as the air recovered the temperature of the containing vessel. The first record would usually be at the third or fourth second from the turning of the cock, and the last after perhaps 120 seconds. In this way data are obtained for a plot of the curve of pressure; and the part actually observed has to be supplemented by extrapolation, so as to go back to the zero of time (the moment of turning the tap) and to allow for the drop which might occur subsequent to the last observation. An estimate, which cannot be much in error, is thus obtained of the whole rise in pressure during the recovery of temperature, and for the time, reckoned from the commencement, at which the rise is equal to one-half of the total.

In some of the earlier experiments the whole rise of pressure (fall in the manometer) during the recovery of temperature was about 20 millim. of water, and the time of half recovery was 15 seconds. I was desirous of working with the minimum range, since only in this way could it be hoped to eliminate the effect of gravity, whereby the interior and still cool parts of the included air would be made to fall and so come into closer proximity to the walls, and thus accelerate the mean cooling. In order to diminish the disturbance due to capillarity, the bore of the manometer-tube, which stood in a large open cistern, was increased to about 18 millim.*, and suitable optical arrangements were introduced to render small movements easily visible. By degrees the range was diminished, with a prolongation of the time of half recovery to 18, 22, 24, and finally to about 26 seconds. The minimum range attained was represented by 3 or 4 millim. of water, and at this stage there did not appear to be much further prolongation of cooling in progress. There seemed to be no appreciable difference whether the air was artificially dried or not, but in no case was the moisture sufficient to develop fog under the very small expansions employed. The result of the experiments may be taken to be that when the influence of gravity was, as far as practicable, eliminated, the time of half recovery of temperature was about 26 seconds.

It may perhaps be well to give an example of an actual experiment. Thus in one trial on Nov. 1, the recorded times of passage across the divisions of the scale were 3, 6, 11, 18, 26, 35, 47, 67, 114 seconds. The divisions themselves were millimetres, but the actual movements of the meniscus were less in the proportion of about $2\frac{1}{2}:1$. A plot of these numbers shows that one division must be added to represent the movement between 0^s and 3^s, and about as much for the movement to be expected between 114^s and ∞. The whole range is thus 10 divisions (corresponding to 4 millim. at the meniscus), and the mid-point occurs at 26^s. On each occasion 3 or 4 sets of readings were taken under given conditions with fairly accordant results.

It now remains to compare with the time of heating derived from theory. The calculation is complicated by the consideration that when during the process any part becomes heated, it expands and compresses all the other parts, thereby developing heat in them. From the investigation which

---

* It must not be forgotten that too large a diameter is objectionable, as leading to an augmentation of volume during an experiment, as the liquid falls.

follows *, we see that the time of half recovery t is given by the formula

$$t = \frac{\cdot 184\gamma a^2}{\pi^2 \nu}, \quad \dots\dots\dots\dots\dots\dots\dots\dots\dots(6)$$

in which a is the radius of the sphere, γ the ratio of specific heats (1·41), and ν is the thermometric conductivity, found by dividing the ordinary or calorimetric conductivity by the thermal capacity of unit volume. This thermal capacity is to be taken with volume constant, and it will be less than the thermal capacity with pressure constant in the ratio of $\gamma : 1$. Accordingly ν/γ in (6) represents the latter thermal capacity, of which the experimental value is ·00128 × ·239, the first factor representing the density of air referred to water. Thus, if we take the calorimetric conductivity at ·000056, we have in C.G.S. measure

$$\nu = \cdot 258, \qquad \nu/\gamma = \cdot 183 \, ;$$

and thence

$$t = \cdot 102 a^2.$$

In the present apparatus a, determined by the contents, is 16·4 centim., whence

$$t = 27 \cdot 4 \text{ seconds.}$$

The agreement of the observed and calculated values is quite as close as could have been expected, and confirms the view that the transfer of heat is due to conduction, and that the part played by radiation is insensible. From a comparison of the experimental and calculated curves, however, it seems probable that the effect of gravity was not wholly eliminated, and that the later stages of the phenomenon, at any rate, may still have been a little influenced by a downward movement of the central parts.

* See next paper.

245.

ON THE CONDUCTION OF HEAT IN A SPHERICAL MASS OF AIR CONFINED BY WALLS AT A CONSTANT TEMPERATURE.

[*Philosophical Magazine*, XLVII. pp. 314—325, 1899.]

It is proposed to investigate the subsidence to thermal equilibrium of a gas slightly disturbed therefrom and included in a solid vessel whose walls retain a constant temperature. The problem differs from those considered by Fourier in consequence of the mobility of the gas, which may give rise to two kinds of complication. In the first place gravity, taking advantage of the different densities prevailing in various parts, tends to produce circulation. In many cases the subsidence to equilibrium must be greatly modified thereby. But this effect diminishes with the amount of the temperature disturbance, and for infinitesimal disturbances the influence of gravity disappears. On the other hand, the second complication remains, even though we limit ourselves to infinitesimal disturbances. When one part of the gas expands in consequence of reception of heat by radiation or conduction, it compresses the remaining parts, and these in their turn become heated in accordance with the laws of gases. To take account of this effect a special investigation is necessary.

But although the fixity of the boundary does not suffice to prevent local expansions and contractions and consequent motions of the gas, we may nevertheless neglect the inertia of these motions since they are very slow in comparison with the free oscillations of the mass regarded as a resonator. Accordingly the pressure, although variable with time, may be treated as uniform at any one moment throughout the mass.

In the usual notation*, if s be the condensation and θ the excess of temperature, the pressure p is given by

$$p = k\rho \, (1 + s + \alpha\theta). \quad \dotfill (1)$$

* *Theory of Sound*, § 247.

The effect of a small sudden condensation s is to produce an elevation of temperature, which may be denoted by βs. Let dQ be the quantity of heat entering the element of volume in the time dt, measured by the rise of temperature which it would produce, if there were no "condensation." Then

$$\frac{d\theta}{dt} = \beta \frac{ds}{dt} + \frac{dQ}{dt} ; \quad\quad\quad\quad\quad (2)$$

and, if the passage of dQ be the result of radiation and conduction, we have

$$\frac{dQ}{dt} = \nu \nabla^2 \theta - q\theta. \quad\quad\quad\quad\quad (3)$$

In (3) ν represents the "thermometric conductivity" found by dividing the conductivity by the thermal capacity of the gas (per unit volume), at constant volume. Its value for air at $0°$ and atmospheric pressure may be taken to be $·26$ cm²./sec. Also q represents the radiation, supposed to depend only upon the excess of temperature of the gas over that of the enclosure.

If $dQ = 0$, $\theta = \beta s$, and in (1)

$$p = k\rho \{1 + (1 + \alpha\beta) s\} ;$$

so that

$$1 + \alpha\beta = \gamma, \quad\quad\quad\quad\quad (4)$$

where γ is the well-known ratio of specific heats, whose value for air and several other gases is very nearly $1·41$.

In general from (2) and (3)

$$\frac{d\theta}{dt} = \beta \frac{ds}{dt} + \nu \nabla^2 \theta - q\theta. \quad\quad\quad\quad\quad (5)$$

In order to find the normal modes into which the most general subsidence may be analysed, we are to assume that s and θ are functions of the time solely through the factor e^{-ht}. Since p is uniform, $s + \alpha\theta$ must by (1) be of the form He^{-ht}, where H is some constant; so that if for brevity the factor e^{-ht} be dropped,

$$s + \alpha\theta = H ; \quad\quad\quad\quad\quad (6)$$

while from (5)

$$\nu \nabla^2 \theta + (h - q) \theta = h\beta s. \quad\quad\quad\quad\quad (7)$$

Eliminating s between (5) and (7), we get

$$\nabla^2 \theta + m^2 (\theta - C) = 0, \quad\quad\quad\quad\quad (8)$$

where

$$m^2 = \frac{h\gamma - q}{\nu}, \quad\quad C = \frac{h\beta H}{h\gamma - q}. \quad\quad\quad\quad\quad (9)$$

These equations are applicable in the general case, but when radiation and conduction are both operative the equation by which m is determined becomes rather complicated. If there be no conduction, $\nu = 0$. The solution is then very simple, and may be worth a moment's attention.

Equations (6) and (7) give

$$\theta = \frac{h\beta H}{h\gamma - q}, \qquad s = \frac{(h - q) H}{h\gamma - q}. \qquad \dots\dots\dots\dots(10)$$

Now the mean value of s throughout the mass, which does not change with the time, must be zero; so that from (10) we obtain the alternatives

(i) $h = q$,   (ii) $H = 0$.

Corresponding to (i) we have with restoration of the time-factor

$$\theta = (H/\alpha)\,e^{-qt}, \qquad s = 0. \qquad \dots\dots\dots\dots\dots(11)$$

In this solution the temperature is uniform and the condensation zero throughout the mass. By means of it any initial *mean* temperature may be provided for, so that in the remaining solutions the mean temperature may be considered to be zero.

In the second alternative $H = 0$, so that $s = -\alpha\theta$. Using this in (7) with ν evanescent, we get

$$(h\gamma - q)\,\theta = 0. \qquad \dots\dots\dots\dots\dots\dots(12)$$

The second solution is accordingly

$$\theta = \phi\,(x,\,y,\,z)\,e^{-qt/\gamma}, \qquad s = -\alpha\phi\,(x,\,y,\,z)\,e^{-qt/\gamma}, \qquad \dots\dots\dots(13)$$

where ϕ denotes a function arbitrary throughout the mass, except for the restriction that its mean value must be zero.

Thus if Θ denote the initial value of θ as a function of x, y, z, and Θ_0 its mean value, the complete solution may be written

$$\left.\begin{array}{l}\theta = \Theta_0 e^{-qt} + (\Theta - \Theta_0)\,e^{-qt/\gamma} \\[2mm] s = \qquad -\alpha(\Theta - \Theta_0)\,e^{-qt/\gamma}\end{array}\right\}, \qquad \dots\dots\dots\dots(14)$$

giving

$$s + \alpha\theta = \alpha\Theta_0 e^{-qt}. \qquad \dots\dots\dots\dots\dots(15)$$

It is on (15) that the variable part of the pressure depends.

When the conductivity ν is finite, the solutions are less simple and involve the form of the vessel in which the gas is contained. As a first example we may take the case of gas bounded by two parallel planes perpendicular to x, the temperature and condensation being even functions of x measured from the mid-plane. In this case $\nabla^2 = d^2/dx^2$, and we get

$$\theta = C + A\cos mx, \qquad -s/\alpha = D + A\cos mx, \qquad \dots\dots\dots(16)$$

$$s + \alpha\theta = \alpha C - \alpha D = H. \qquad \dots\dots\dots\dots(17)$$

By (9), (17)

$$C = \frac{h\beta H}{h\gamma - q}, \qquad D = \frac{(q-h)H}{\alpha(h\gamma - q)}. \qquad (18)$$

There remain two conditions to be satisfied. The first is simply that $\theta = 0$ when $x = \pm a$, $2a$ being the distance between the walls. This gives

$$C + A \cos ma = 0. \qquad (19)$$

The remaining condition is given by the consideration that the mean value of s, proportional to $\int s\,dx$, must vanish. Accordingly

$$ma \cdot D + \sin ma \cdot A = 0. \qquad (20)$$

From (18), (19), (20) we have as the equation for the admissible values of m,

$$\frac{\tan ma}{ma} = \frac{\alpha\beta q - vm^2}{\alpha\beta(q + vm^2)}, \qquad (21)$$

reducing for the case of evanescent q to

$$\frac{\tan ma}{ma} = -\frac{1}{\alpha\beta}. \qquad (22)$$

The general solution may be expressed in the series

$$\left.\begin{array}{l} \theta = A_1 e^{-h_1 t}\theta_1 + A_2 e^{-h_2 t}\theta_2 + \dots \\ s = A_1 e^{-h_1 t}s_1 + A_2 e^{-h_2 t}s_2 + \dots \end{array}\right\}, \qquad (23)$$

where $h_1, h_2, \dots$ are the values of h corresponding according to (9) with the various values of m, and $\theta_1, \theta_2 \dots$ are of the form

$$\left.\begin{array}{l} \theta_1 = \cos m_1 x - \cos m_1 a \\ s_1 = -\alpha(\cos m_1 x - \sin m_1 a / m_1 a) \end{array}\right\}. \qquad (24)$$

It only remains to determine the arbitrary constants $A_1, A_2, \dots$ to suit prescribed initial conditions. We will limit ourselves to the simpler case of $q = 0$, so that the values of m are given by (22). With use of this relation and putting for brevity $a = 1$, we find from (24)

$$\int_0^1 \theta_1 \theta_2\,dx = \frac{\alpha\beta + 1}{\alpha\beta}\cos m_1 \cos m_2,$$

$$\int_0^1 s_1 s_2\,dx = -\frac{\alpha\beta + 1}{\beta^2}\cos m_1 \cos m_2;$$

so that

$$\int_0^1 \theta_1 \theta_2\,dx + \beta/\alpha \cdot \int_0^1 s_1 s_2\,dx = 0, \qquad (25)$$

θ_1, θ_2 being any (different) functions of the form (24). Also

$$\int_0^1 \theta_1^2\,dx + \beta/\alpha \cdot \int_0^1 s_1^2\,dx = \frac{1+\alpha\beta}{2}\left\{1 + \frac{\cos^2 m_1}{\alpha\beta}\right\}. \qquad (26)$$

There is now no difficulty in finding A_1, A_2, ... to suit arbitrary initial values of θ and its associated s, *i.e.* so that

$$\left.\begin{array}{l} \Theta = A_1 \theta_1 + A_2 \theta_2 + \ldots \\ S = A_1 s_1 + A_2 s_2 + \ldots \end{array}\right\} \quad \ldots\ldots\ldots\ldots\ldots\ldots\ldots(27)$$

Thus to determine A_1,

$$\int_0^1 (\Theta\theta_1 + \beta/\alpha . Ss_1)\, dx = A_1 \int_0^1 (\theta_1^2 + \beta/\alpha . s_1^2)\, dx$$

$$+ A_2 \int_0^1 (\theta_1\theta_2 + \beta/\alpha . s_1 s_2)\, dx + \ldots\ldots$$

in which the coefficients of A_2, A_3 ... vanish by (25); so that by (26)

$$A_1 \left\{1 + \frac{\cos^2 m_1}{\alpha\beta}\right\} = \frac{2}{1 + \alpha\beta} \int_0^1 (\Theta\theta_1 + \beta/\alpha . Ss_1)\, dx. \quad \ldots\ldots\ldots(28)$$

An important particular case is that in which Θ is constant, and accordingly $S = 0$. Since

$$\int_0^1 \theta_1\, dx = \frac{\sin m_1}{m_1} - \cos m_1 = -\frac{1 + \alpha\beta}{\alpha\beta} \cos m_1,$$

we have

$$A_1 = -\frac{2\Theta \cos m_1}{\alpha\beta + \cos^2 m_1}. \quad \ldots\ldots\ldots\ldots\ldots\ldots (29)$$

For the pressure we have

$$\theta + s/\alpha = A_1 e^{-h_1 t}\left(-\cos m_1 + \frac{\sin m_1}{m_1}\right) + \ldots\ldots$$

$$= -\frac{\alpha\beta + 1}{\alpha\beta} \cos m_1 . A_1 e^{-h_1 t} + \ldots\ldots ,$$

or in the particular case of (29),

$$\theta + s/\alpha = 2\Theta \frac{1 + \alpha\beta}{\alpha\beta} \frac{\cos^2 m_1 e^{-h_1 t}}{\alpha\beta + \cos^2 m_1} + \ldots \quad \ldots\ldots\ldots(30)$$

If $\beta = 0$, we fall back upon a problem of the Fourier type. By (22) in that case

$$ma = \tfrac{1}{2}\pi\,(1, 3, 5, \ldots) \quad \text{and} \quad \cos^2 ma = \alpha^2\beta^2/m^2 a^2,$$

so that (30) becomes

$$2\Theta \left(\frac{e^{-h_1 t}}{m_1^2 a^2} + \frac{e^{-h_2 t}}{m_2^2 a^2} + \ldots\right), \quad \ldots\ldots\ldots\ldots\ldots(31)$$

or initially

$$\frac{8\Theta}{\pi^2}\left(\frac{1}{1^2} + \frac{1}{3^2} + \frac{1}{5^2} + \ldots\right), \quad i.e.\ \Theta.$$

The values of h are given by

$$h = \frac{\nu\pi^2}{4\gamma a^2}\,(1^2, 3^2, 5^2, ..). \quad \ldots\ldots\ldots\ldots\ldots(32)$$

We will now pass on to the more important practical case of a spherical envelope of radius a. The equation (8) for $(\theta - C)$ is identical with that which determines the vibrations of air* in a spherical case, and the solution may be expanded in Laplace's series. The typical term is

$$\theta - C = (mr)^{-\frac{1}{2}} J_{n+\frac{1}{2}}(mr) \cdot Y_n, \quad \text{......................(33)}$$

Y_n being the surface spherical harmonic of order n where $n = 0, 1, 2, 3 \ldots$, and J the symbol of Bessel's functions. In virtue of (6) we may as before equate $- s/\alpha - D$, where D is another constant, to the right-hand member of (33). The two conditions yet to be satisfied are that $\theta = 0$ when $r = a$, and that the mean value of s throughout the sphere shall vanish.

When the value of n is greater than zero, the first of these conditions gives $C = 0$ and the second $D = 0$; so that

$$\theta = - s/\alpha = (mr)^{-\frac{1}{2}} J_{n+\frac{1}{2}}(mr) \cdot Y_n, \quad \text{..................(34)}$$

and $s + \alpha\theta = 0$. Accordingly these terms contribute nothing to the pressure. It is further required that

$$J_{n+\frac{1}{2}}(ma) = 0, \quad \text{..............................(35)}$$

by which the admissible values of m are determined. The roots of (35) are discussed in *Theory of Sound*, § 206... ; but it is not necessary to go further into the matter here, as interest centres rather upon the case $n = 0$.

If we assume symmetry with respect to the centre of the sphere, we may replace ∇^2 in (8) by $\dfrac{1}{r} \dfrac{d^2}{dr^2} r$, thus obtaining

$$\frac{d^2 r (\theta - C)}{dr^2} + m^2 r (\theta - C) = 0, \quad \text{....................(36)}$$

of which the general solution is

$$\theta = C + A \frac{\cos mr}{mr} + B \frac{\sin mr}{mr}.$$

But for the present purpose the term $r^{-1} \cos mr$ is excluded, so that we may write

$$\theta = C + B \frac{\sin mr}{mr}, \quad - s/\alpha = D + B \frac{\sin mr}{mr}, \quad \text{........(37)}$$

giving

$$s + \alpha\theta = \alpha (C - D) = H. \quad \text{....................(37 bis)}$$

The first special condition gives

$$maC + B \sin ma = 0. \quad \text{......................(38)}$$

The second, that the mean value of s shall vanish, gives on integration

$$\tfrac{1}{3} m^3 a^3 D + B (\sin ma - ma \cos ma) = 0. \quad \text{..............(39)}$$

* *Theory of Sound*, Vol. II. ch. xvii.

Equations (18), derived from (9) and (37 *bis*), giving C and D in terms of H, hold good as before. Thus

$$\frac{D}{C} = \frac{q-h}{h\alpha\beta} = \frac{\alpha\beta q - vm^2}{\alpha\beta(q + vm^2)}. \qquad\qquad\dots\dots(40)$$

Equating this ratio to that derived from (38), (39), we find

$$\frac{3}{m^2 a^2} \frac{ma \cos ma - \sin ma}{\sin ma} = \frac{vm^2 - \alpha\beta q}{\alpha\beta(vm^2 + q)}. \qquad\dots\dots(41)$$

This is the equation from which m is to be found, after which h is given by (9).

In the further discussion we will limit ourselves to the case of $q = 0$, when (41) reduces to

$$m^2 = 3\alpha\beta(m \cot m - 1), \qquad\qquad\dots\dots\dots(42)$$

in which a has been put equal to unity. Here by (40)

$$D = - C/\alpha\beta.$$

Thus we may set, as in (23),

$$\left. \begin{aligned} \theta &= B_1 e^{-h_1 t}\theta_1 + B_2 e^{-h_2 t}\theta_2 + \dots\dots \\ s &= B_1 e^{-h_1 t}s_1 + B_2 e^{-h_2 t}s_2 + \dots\dots \end{aligned} \right\}, \qquad\dots\dots\dots(43)$$

in which

$$\theta_1 = \frac{\sin m_1 r}{m_1 r} - \frac{\sin m_1 a}{m_1 a}, \qquad s_1 = -\alpha \frac{\sin m_1 r}{m_1 r} - \frac{1}{\beta} \frac{\sin m_1 a}{m_1 a}, \quad\dots(44)$$

and by (9) $\quad h_1 = vm_1^2/\gamma$. Also

$$s_1/\alpha + \theta_1 = -\frac{1 + \alpha\beta}{\alpha\beta} \frac{\sin m_1 a}{m_1 a}. \qquad\qquad\dots\dots\dots(45)$$

The process for determining B_1, B_2, ... follows the same lines as before. By direct integration from (44) we find

$$\frac{2m_1 m_2}{1 + \alpha\beta} \int_0^1 (\theta_1 \theta_2 + \beta/\alpha . s_1 s_2) r^2 dr$$

$$= \frac{\sin(m_1 - m_2)}{m_1 - m_2} - \frac{\sin(m_1 + m_2)}{m_1 + m_2} + \frac{2 \sin m_1 \sin m_2}{3\alpha\beta},$$

a being put equal to unity. By means of equation (42) satisfied by m_1 and m_2 we may show that the quantity on the right in the above equation vanishes. For the sum of the first two fractions is

$$\frac{2m_2 \sin m_1 \cos m_2 - 2m_1 \sin m_2 \cos m_1}{m_1^2 - m_2^2},$$

of which the denominator by (42) is equal to

$$3\alpha\beta(m_1 \cot m_1 - m_2 \cot m_2).$$

Accordingly $\quad\quad\quad \int_0^1 (\theta_1\theta_2 + \beta/\alpha \cdot s_1 s_2)\, r^2 dr = 0. \quad ....(46)$

Also

$$\frac{2m_1^2}{1+\alpha\beta} \int_0^1 (\theta_1^2 + \beta/\alpha \cdot s_1^2)\, r^2 dr = 1 - \frac{\sin 2m_1}{2m_1} + \frac{2\sin^2 m_1}{3\alpha\beta}. \quad ......(47)$$

To determine the arbitrary constants $B_1 \ldots$ from the given initial values of θ and s, say Θ and S, we proceed as usual. We limit ourselves to the term of zero order in spherical harmonics, *i.e.* to the supposition that θ, s are functions of r only. The terms of higher order in spherical harmonics, if present, are treated more easily, exactly as in the ordinary theory of the conduction of heat. By (43)

$$\left. \begin{aligned} \Theta &= B_1\theta_1 + B_2\theta_2 + \ldots\ldots \\ S &= B_1 s_1 + B_2 s_2 + \ldots\ldots \end{aligned} \right\} ; \quad ....(48)$$

and thus $\quad \int_0^1 (\Theta\theta_1 + \beta/\alpha \cdot S s_1)\, r^2 dr = B_1 \int_0^1 (\theta_1^2 + \beta/\alpha \cdot s_1^2)\, r^2 dr$

$$+ B_2 \int_0^1 (\theta_1\theta_2 + \beta/\alpha \cdot s_1 s_2)\, r^2 dr + \ldots\ldots ,$$

in which the coefficients of B_2, B_3, ... vanish by (46). The coefficient of B_1 is given by (47). Thus

$$B_1 \left\{ 1 - \frac{\sin 2m_1}{2m_1} + \frac{2\sin^2 m_1}{3\alpha\beta} \right\} = \frac{2m_1^2}{1+\alpha\beta} \int_0^1 (\Theta\theta_1 + \beta/\alpha \cdot S s_1)\, r^2 dr, \quad ......(49)$$

by which B_1 is determined.

An important particular case is that where Θ is constant and accordingly S vanishes. Now with use of (42)

$$\int_0^1 \theta_1 r^2 dr = \frac{\sin m_1 - m_1 \cos m_1}{m_1^3} - \frac{\sin m_1}{3m_1} = -\frac{(1+\alpha\beta)\sin m_1}{3\alpha\beta m_1} ;$$

so that

$$B_1 \left\{ 1 - \frac{\sin 2m_1}{2m_1} + \frac{2\sin^2 m_1}{3\alpha\beta} \right\} = -\frac{2m_1 \sin m_1 \cdot \Theta}{3\alpha\beta}. \quad ....(50)$$

B_1, B_2, ... being thus known, θ and s are given as functions of the time and of the space coordinates by (43), (44).

To determine the pressure in this case we have from (45)

$$\frac{\theta + s/\alpha}{\Theta} = \frac{1+\alpha\beta}{\alpha\beta} \sum \frac{\sin^2 m \cdot e^{-ht}}{\sin^2 m + \frac{3\alpha\beta}{2}\left(1 - \frac{\sin 2m}{2m}\right)}, \quad ......(51)$$

the summation extending to all the values of m in (42). Since (for each term) the mean value of s is zero, the right-hand member of (51) represents also $\bar{\theta}/\Theta$, where $\bar{\theta}$ is the mean value of θ.

If in (51) we suppose $\beta = 0$, we fall back upon a known Fourier solution, relative to the mean temperature of a spherical solid which, having been initially at uniform temperature Θ throughout, is afterwards maintained at zero all over the surface. From (42) we see that in this case $\sin m$ is small and of order β. Approximately

$$\sin m = 3\alpha\beta/m ;$$

and (51) reduces to

$$\frac{\bar{\theta}}{\Theta} = \frac{6}{\pi^2}\left(\frac{e^{-h_1 t}}{1^2} + \frac{e^{-h_2 t}}{2^2} + \frac{e^{-h_3 t}}{3^2} + \dots\right), \quad \dots\dots\dots\dots(52)$$

of which by a known formula the right-hand member identifies itself with unity when $t = 0$. By (9) with restoration of a,

$$h = (1^2, 3^2, 5^2, \dots)\, \nu\pi^2/a^2. \quad \dots\dots\dots\dots(53)$$

In the general case we may obtain from (42) an approximate value applicable when m is moderately large. The first approximation is $m = i\pi$, i denoting an integer. Successive operations give

$$m = i\pi + \frac{3\alpha\beta}{i\pi} - \frac{18\alpha^2\beta^2 + 9\alpha^3\beta^3}{i^3\pi^3}. \quad \dots\dots\dots\dots(54)$$

In like manner we find approximately in (51)

$$\frac{\sin^2 m\,(1 + \alpha\beta)/\alpha\beta}{\sin^2 m + \dfrac{3\alpha\beta}{2}\left(1 - \dfrac{\sin 2m}{2m}\right)} = \frac{6\,(1 + \alpha\beta)}{i^2\pi^2}\left\{1 - \frac{15\alpha\beta + 9\alpha^2\beta^2}{i^2\pi^2}\right\}, \quad \dots(55)$$

showing that the coefficients of the terms of *high order* in (51) differ from the corresponding terms in (52) only by the factor $(1 + \alpha\beta)$ or γ.

In the numerical computation we take $\gamma = 1\cdot41$, $\alpha\beta = \cdot41$. The series (54) suffices for finding m when i is greater than 2. The first two terms are found by trial and error with trigonometrical tables from (42). In like manner the approximate value of the left-hand member of (51) therein given suffices when i is greater than 3. The results as far as $i = 12$ are recorded in the annexed table.

i	m/π	Left-hand member of (55)	i	m/π	Left-hand member of (55)
1......	1·0994	·4942	7......	7·0177	·0175
2......	2·0581	·1799	8......	8·0156	·0134
3......	3·0401	·0871	9......	9·0138	·0106
4......	4·0305	·0510	10......	10·0125	·0086
5......	5·0246	·0332	11......	11·0113	·0071
6......	6·0206	·0233	12......	12·0104	·0060

Thus the solution (51) of our problem is represented by

$$\bar{\theta}/\Theta = \cdot4942\,e^{-(1\cdot0994)^2 t'} + \cdot1799\,e^{-(2\cdot0581)^2 t'} + \dots, \quad \dots\dots\dots(56)$$

where by (9), with omission of q and restoration of a,

$$t'/t = \pi^2 v / \gamma a^2. \qquad \qquad \ldots\ldots\ldots\ldots\ldots\ldots\ldots\ldots\ldots\ldots(57)$$

The numbers entered in the third column of the above table would add up to unity if continued far enough. The verification is best made by a comparison with the simpler series (52). If with t zero we call this series Σ' and the present series Σ, both Σ and Σ' have unity for their sum, and accordingly $\gamma\Sigma' - \Sigma = \gamma - 1$, or

$$\frac{6\gamma}{\pi^2}\left(\frac{1}{1^2} + \frac{1}{2^2} + \frac{1}{3^2} + \ldots\right) - \Sigma = \gamma - 1 = \cdot 41.$$

Here $6\gamma/\pi^2 = \cdot 8573$, and the difference between this and the first term of Σ, $i.e.$ $\cdot 4942$, is $\cdot 3631$. The differences of the second, third, &c. terms are $\cdot 0344$, $\cdot 0082$, $\cdot 0026$, $\cdot 0011$, $\cdot 0005$, $\cdot 0000$, &c., making a total of $\cdot 4099$.

We are now in a position to compute the right-hand member of (56) as a function of t'. The annexed table contains sufficient to give an idea

t'	(56)	t'	(56)	t'	(56)
·00......	1·0000	·40......	·3401	·90......	·1705
·05......	·7037	·50......	·2926	1·00......	·1502
·10......	·6037	·60......	·2538	1·50......	·0809
·20......	·4811	·70......	·2215	2·00......	·0441
·30......	·4002	·80......	·1940		

of the course of the function. It is plotted in the figure. The second entry ($t' = \cdot 05$) requires the inclusion of 9 terms of the series. After $t' = \cdot 7$ two terms suffice; and after $t' = 2\cdot 0$ the first term represents the series to four places of decimals.

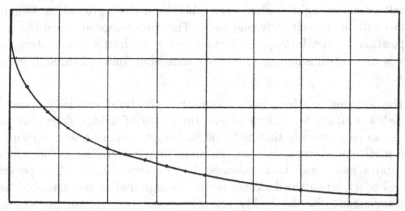

By interpolation we find that the series attains the value $\cdot 5$ when

$$t' = \cdot 184. \qquad \qquad \ldots\ldots\ldots\ldots\ldots\ldots\ldots\ldots\ldots\ldots(58)$$

246.

TRANSPARENCY AND OPACITY.

[*Proc. Roy. Inst.* XVI. pp. 116—119, 1899; *Nature*, LX. pp. 64, 65, 1899.]

ONE kind of opacity is due to absorption; but the lecture dealt rather with that deficiency of transparency which depends upon irregular reflections and refractions. One of the best examples is that met with in Christiansen's experiment. Powdered glass, all from one piece and free from dirt, is placed in a bottle with parallel flat sides. In this state it is quite opaque; but if the interstices between the fragments are filled up with a liquid mixture of bisulphide of carbon and benzole, carefully adjusted so as to be of equal refractivity with the glass, the mass becomes optically homogeneous, and therefore transparent. In consequence, however, of the different dispersive powers of the two substances, the adjustment is good for one part only of the spectrum, other parts being scattered in transmission much as if no liquid were employed, though, of course, in a less degree. The consequence is that a small source of light, backed preferably by a dark ground, is seen in its natural outlines but strongly coloured. The colour depends upon the precise composition of the liquid, and further varies with the temperature, a few degrees of warmth sufficing to cause a transition from red through yellow to green.

The lecturer had long been aware that the light regularly transmitted through a stratum from 15 to 20 mm. thick was of a high degree of purity, but it was only recently that he found to his astonishment, as the result of a more particular observation, that the range of refrangibility included was but two and a half times that embraced by the two D-lines. The poverty of general effect, when the darkness of the background is not attended to, was thus explained; for the highly monochromatic and accordingly attenuated light from the special source is then overlaid by diffused light of other colours.

More precise determinations of the range of light transmitted were subsequently effected with thinner strata of glass powder contained in cells formed of parallel glass. The cell may be placed between the prisms of the spectroscope and the object-glass of the collimator. With the above mentioned liquids a stratum 5 mm. thick transmitted, without appreciable disturbance, a range of the spectrum measured by 11·3 times the interval of the D's. In another cell of the same thickness an effort was made to reduce the difference of dispersive powers. To this end the powder was of plate glass and the liquid oil of cedar-wood adjusted with a little bisulphide of carbon. The general transparency of this cell was the highest yet observed. When it was tested upon the spectrum, the range of refrangibility transmitted was estimated at 34 times the interval of the D's.

As regards the substitution of other transparent solid material for glass, the choice is restricted by the presumed necessity of avoiding appreciable double refraction. Common salt is singly refracting, but attempts to use it were not successful. Opaque patches always interfered. With the idea that these might be due to included mother-liquor, the salt was heated to incipient redness, but with little advantage. Transparent rock-salt artificially broken may, however, be used with good effect, but there is some difficulty in preventing the approximately rectangular fragments from arranging themselves too closely.

The principle of evanescent refraction may also be applied to the spectroscope. Some twenty years ago, an instrument had been constructed upon this plan*. Twelve 90° prisms of Chance's "dense flint" were cemented in a row upon a strip of glass (Fig. 1), and the whole was immersed in a liquid mixture of bisulphide of carbon with a little benzole. The dispersive power of the liquid exceeds that of the solid, and the difference amounts to about three-quarters of the dispersive power of Chance's "extra dense flint." The

Fig. 1.

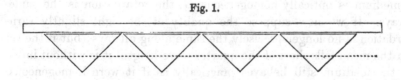

resolving power of the latter glass is measured by the number of centimetres of available thickness, if we take the power required to resolve the D-lines as unity. The compound spectroscope had an available thickness of 12 inches or 30 cm., so that its theoretical resolving power (in the yellow region of the spectrum) would be about 22. With the aid of a reflector the prism could be used twice over, and then the resolving power is doubled.

* [Vol. I. p. 456.]

One of the objections to a spectroscope depending upon bisulphide of carbon is the sensitiveness to temperature. In the ordinary arrangement of prisms the refracting edges are vertical. If, as often happens, the upper part of a fluid prism is warmer than the lower, the definition is ruined, one degree (Centigrade) of temperature making nine times as great a difference of refraction as a passage from D_1 to D_2. The objection is to a great extent obviated by so mounting the compound prism that the refracting edges are *horizontal*, which of course entails a horizontal slit. The disturbance due to a stratified temperature is then largely compensated by a change of focus.

In the instrument above described the dispersive power is great—the D-lines are seen widely separated with the naked eye—but the aperture is inconveniently small ($\frac{1}{2}$-inch). In the new instrument exhibited the prisms (supplied by Messrs Watson) are larger, so that a line of ten prisms occupies 20 inches. Thus, while the resolving power is much greater, the dispersion is less than before*.

In the course of the lecture the instrument was applied to show the duplicity of the reversed soda lines. The interval on the screen between the centres of the dark lines was about half an inch.

It is instructive to compare the action of the glass powder with that of the spectroscope. In the latter the disposition of the prisms is regular, and in passing from one edge of the beam to the other there is complete substitution of liquid for glass over the whole length. For one kind of light there is no relative retardation; and the resolving power depends upon the question of what change of wave-length is required in order that its relative retardation may be altered from zero to the quarter wave-length. All kinds of light for which the relative retardation is less than this remain mixed. In the case of the powder we have similar questions to consider. For one kind of light the medium is optically homogeneous, *i.e.* the retardation is the same along all rays. If we now suppose the quality of the light slightly varied, the retardation is no longer precisely the same along all rays; but if the variation from the mean falls short of the quarter wave-length, it is without importance, and the medium still behaves practically as if it were homogeneous. The difference between the action of the powder and that of the regular prisms in the spectroscope depends upon this, that in the latter there is complete substitution of glass for liquid along the extreme rays, while in the former the paths of all the rays lie partly through glass and partly through liquid in nearly the same proportions. The difference of retardations along various rays is thus a question of a deviation from an average.

* [1902. When carefully used this instrument gives about as good definition in the green as a first-rate Rowland grating.]

It is true that we may imagine a relative distribution of glass and liquid that would more nearly assimilate the two cases. If, for example, the glass consisted of equal spheres resting against one another in cubic order, some rays might pass entirely through glass and others entirely through liquid, and then the quarter wave-length of relative retardation would enter at the same total thickness in both cases. But such an arrangement would be highly unstable; and, if the spheres be packed in close order, the extreme relative retardation would be much less. The latter arrangement, for which exact results could readily be calculated, represents the glass powder more nearly than does the cubic order.

A simplified problem, in which the element of chance is retained, may be constructed by supposing the particles of glass replaced by thin parallel discs which are distributed entirely at random over a certain stratum. We may go further and imagine the discs limited to a particular plane. Each disc is supposed to exercise a minute retarding influence on the light which traverses it, and they are supposed to be so numerous that it is improbable that a ray can pass the plane without encountering a large number. A certain number (m) of encounters is more probable than any other, but if every ray encountered the same number of discs, the retardation would be uniform and lead to no disturbance.

It is a question of Probabilities to determine the chance of a prescribed number of encounters, or of a prescribed deviation from the mean. In the notation of the integral calculus the chance of the deviation from m lying between $\pm r$ is*

$$\frac{2}{\sqrt{\pi}} \int_0^\tau e^{-\tau^2} d\tau,$$

where $\tau = r/\sqrt{(2m)}$. This is equal to ·84 when $\tau = 1·0$, or $r = \sqrt{(2m)}$; so that the chance is comparatively small of a deviation from m exceeding $\pm \sqrt{(2m)}$.

To represent the glass powder occupying a stratum of 2 cm. thick, we may perhaps suppose that $m = 72$. There would thus be a moderate chance of a difference of retardations equal to, say, one-fifth of the extreme difference corresponding to a substitution of glass for liquid throughout the whole thickness. The range of wave-lengths in the light regularly transmitted by the powder would thus be about five times the range of wave-lengths still unseparated in a spectroscope of equal (2 cm.) thickness. Of course, no calculation of this kind can give more than a rough idea of the action of the powder, whose disposition, though partly a matter of chance, is also influenced by mechanical considerations; but it appears, at any rate, that the character

---

* See *Phil. Mag.* 1899, Vol. xlvii. p. 251. [Vol. iv. p. 375.]

of the light regularly transmitted by the powder is such as may reasonably be explained.

As regards the size of the grains of glass, it will be seen that as great or a greater degree of purity may be obtained in a given thickness from coarse grains as from fine ones, but the light not regularly transmitted is dispersed through smaller angles. Here again the comparison with the regularly disposed prisms of an actual spectroscope is useful.

At the close of the lecture the failure of transparency which arises from the presence of particles small compared to the wave-length of light was discussed. The tints of the setting sun were illustrated by passing the light from the electric lamp through a liquid in which a precipitate of sulphur was slowly forming*. The lecturer gave reasons for his opinion that the blue of the sky is not wholly, or even principally, due to particles of foreign matter. The molecules of air themselves are competent to disperse a light not greatly inferior in brightness to that which we receive from the sky.

* *Op. cit.* 1881, Vol. xii. p. 96.   [Vol. i. p. 531.]

247.

ON THE TRANSMISSION OF LIGHT THROUGH AN ATMOSPHERE CONTAINING SMALL PARTICLES IN SUSPENSION, AND ON THE ORIGIN OF THE BLUE OF THE SKY.

[*Philosophical Magazine*, XLVII. pp. 375—384, 1899.]

THIS subject has been treated in papers published many years ago[*]. I resume it in order to examine more closely than hitherto the attenuation undergone by the primary light on its passage through a medium containing small particles, as dependent upon the number and size of the particles. Closely connected with this is the interesting question whether the light from the sky can be explained by diffraction from the molecules of air themselves, or whether it is necessary to appeal to suspended particles composed of foreign matter, solid or liquid. It will appear, I think, that even in the absence of foreign particles we should still have a blue sky[†].

The calculations of the present paper are not needed in order to explain the general character of the effects produced. In the earliest of those above

---

[*] *Phil. Mag.* XLI. pp. 107, 274, 447 (1871); XII. p. 81 (1881). [Vol. I. pp. 87, 104, 518.]

[†] My attention was specially directed to this question a long while ago by Maxwell in a letter which I may be pardoned for reproducing here. Under date Aug. 28, 1873, he wrote :—

"I have left your papers on the light of the sky, &c. at Cambridge, and it would take me, even if I had them, some time to get them assimilated sufficiently to answer the following question, which I think will involve less expense to the energy of the race if you stick the data into your formula and send me the result....

"Suppose that there are N spheres of density ρ and diameter s in unit of volume of the medium. Find the index of refraction of the compound medium and the coefficient of extinction of light passing through it.

"The object of the enquiry is, of course, to obtain data about the size of the molecules of air. Perhaps it may lead also to data involving the density of the æther. The following quantities are known, being combinations of the three unknowns,

M = mass of molecule of hydrogen ;

N = number of molecules of any gas in a cubic centimetre at 0° C. and 760 B.

s = diameter of molecule in any gas :—

referred to I illustrated by curves the gradual reddening of the transmitted light by which we see the sun a little before sunset. The same reasoning proved, of course, that the spectrum of even a vertical sun is modified by the atmosphere in the direction of favouring the waves of greater length.

For such a purpose as the present it makes little difference whether we speak in terms of the electromagnetic theory or of the elastic solid theory of light; but to facilitate comparison with former papers on the light from the sky, it will be convenient to follow the latter course. The small particle of volume T is supposed to be small in all its dimensions in comparison with the wave-length (λ), and to be of optical density D' differing from that (D) of the surrounding medium. Then, if the incident vibration be taken as unity, the expression for the vibration scattered from the particle in a direction making an angle θ with that of *primary vibration* is

$$\frac{D'-D}{D}\frac{\pi T}{r\lambda^2}\sin\theta\cos\frac{2\pi}{\lambda}(bt-r)^*, \quad \dots\dots\dots\dots(1)$$

r being the distance from T of any point along the secondary ray.

In order to find the whole emission of energy from T we have to integrate the square of (1) over the surface of a sphere of radius r. The element of area being $2\pi r^2\sin\theta d\theta$, we have

$$\int_0^\pi \frac{\sin^2\theta}{r^2}2\pi r^2\sin\theta d\theta = 4\pi\int_0^{\frac{1}{2}\pi}\sin^3\theta d\theta = \frac{8\pi}{3};$$

so that the energy emitted from T is represented by

$$\frac{8\pi^3}{3}\frac{(D'-D)^2}{D^2}\frac{T^2}{\lambda^4}, \quad \dots\dots\dots\dots\dots(2)$$

Known Combinations.

$MN=$ density.

Ms^2 from diffusion or viscosity.

Conjectural Combination.

$\frac{6M}{\pi s^3}=$ density of molecule.

"If you can give us (i) the quantity of light scattered in a given direction by a stratum of a certain density and thickness; (ii) the quantity cut out of the direct ray; and (iii) the effect of the molecules on the index of refraction, which I think ought to come out easily, we might get a little more information about these little bodies.

"You will see by *Nature*, Aug. 14, 1873, that I make the diameter of molecules about $\frac{1}{1000}$ of a wave-length.

"The enquiry into scattering must begin by accounting for the great observed transparency of air. I suppose we have no numerical data about its absorption.

"But the index of refraction can be numerically determined, though the observation is of a delicate kind, and a comparison of the result with the dynamical theory may lead to some new information."

Subsequently he wrote, "Your letter of Nov. 17 quite accounts for the observed transparency of any gas." So far as I remember, my argument was of a general character only.

* The factor π was inadvertently omitted in the original memoir.

on such a scale that the energy of the primary wave is unity per unit of wave-front area.

The above relates to a single particle. If there be n similar particles per unit volume, the energy emitted from a stratum of thickness dx and of unit area is found from (2) by introduction of the factor $n\,dx$. Since there is no waste of energy on the whole, this represents the loss of energy in the primary wave. Accordingly, if E be the energy of the primary wave,

$$\frac{1}{E}\frac{dE}{dx} = -\frac{8\pi^3 n}{3}\frac{(D'-D)^2}{D^2}\frac{T^2}{\lambda^4};\quad\dots\dots(3)$$

whence

$$E = E_0 e^{-hx},\quad\dots\dots(4)$$

where

$$h = \frac{8\pi^3 n}{3}\frac{(D'-D)^2}{D^2}\frac{T^2}{\lambda^4}.\quad\dots\dots(5)$$

If we had a sufficiently complete expression for the scattered light, we might investigate (5) somewhat more directly by considering the resultant of the primary vibration and of the secondary vibrations which travel in the same direction. If, however, we apply this process to (1), we find that it fails to lead us to (5), though it furnishes another result of interest. The combination of the secondary waves which travel in the direction in question has this peculiarity, that the phases are no more distributed at random. The intensity of the secondary light is no longer to be arrived at by addition of individual intensities, but must be calculated with consideration of the particular phases involved. If we consider a number of particles which all lie upon a primary ray, we see that the phases of the secondary vibrations which issue along this line are all the same.

The actual calculation follows a similar course to that by which Huygens' conception of the resolution of a wave into components corresponding to the various parts of the wave-front is usually verified. [See for example Vol. III. p. 74.] Consider the particles which occupy a thin stratum dx perpendicular to the primary ray x. Let AP (Fig. 1) be this stratum and O the point where the vibration is to be estimated. If $AP=\rho$, the element of volume is $dx\,.\,2\pi\rho\,d\rho$, and the number of particles to be found in it is deduced by introduction of the factor n. Moreover, if $OP=r$, $AO=x$, $r^2=x^2+\rho^2$, and $\rho\,d\rho=r\,dr$. The resultant at O of all the secondary vibrations which issue from the stratum dx is by (1), with $\sin\theta$ equal to unity,

Fig. 1.

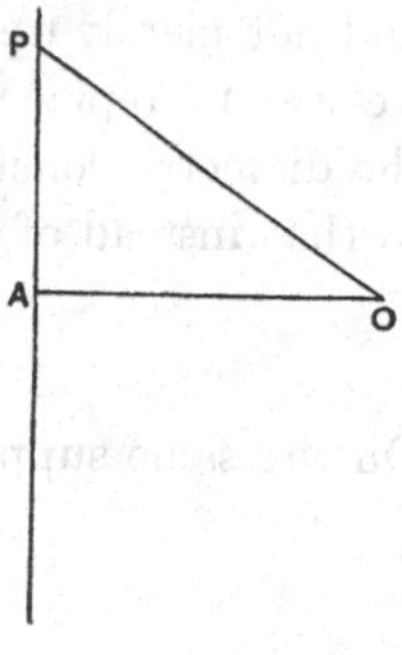

$$n\,dx\,.\int_x^\infty \frac{D'-D}{D}\frac{\pi T}{r\lambda^2}\cos\frac{2\pi}{\lambda}(bt-r)\,2\pi r\,dr,$$

or

$$n\,dx\,.\frac{D'-D}{D}\frac{\pi T}{\lambda}\sin\frac{2\pi}{\lambda}(bt-x).\quad\dots\dots(6)$$

To this is to be added the expression for the primary wave itself, supposed to advance undisturbed, viz., $\cos \dfrac{2\pi}{\lambda} (bt - x)$, and the resultant will then represent the whole actual disturbance at O as modified by the particles in the stratum dx.

It appears, therefore, that to the order of approximation afforded by (1) the effect of the particles in dx is to modify the phase, *but not the intensity*, of the light which passes them. If this be represented by

$$\cos \frac{2\pi}{\lambda} (bt - x - \delta), \quad \dots\dots\dots\dots\dots\dots\dots(7)$$

δ is the *retardation* due to the particles, and we have

$$\delta = nT dx \,(D' - D)/2D. \quad \dots\dots\dots\dots\dots\dots(8)$$

If μ be the refractive index of the medium as modified by the particles, that of the original medium being taken as unity, $\delta = (\mu - 1)\,dx$, and

$$\mu - 1 = nT (D' - D)/2D. \quad \dots\dots\dots\dots\dots\dots(9)$$

If μ' denote the refractive index of the material composing the particles regarded as continuous, $D'/D = \mu'^2$, and

$$\mu - 1 = \tfrac{1}{2} nT (\mu'^2 - 1), \quad \dots\dots\dots\dots\dots\dots(10)$$

reducing to

$$\mu - 1 = nT (\mu' - 1) \quad \dots\dots\dots\dots\dots\dots\dots(11)$$

in the case where $\mu' - 1$ can be regarded as small.

It is only in the latter case that the formulæ of the elastic-solid theory are applicable to light. In the electric theory, to be preferred on every ground except that of easy intelligibility, the results are more complicated in that when $(\mu' - 1)$ is not small, the scattered ray depends upon the shape and not merely upon the volume of the small obstacle. In the case of *spheres* we are to replace $(D' - D)/D$ by $3 (K' - K)/(K' + 2K)$, where K, K' are the dielectric constants proper to the medium and to the obstacle respectively[*]; so that instead of (10)

$$\mu - 1 = \frac{3nT}{2}\,\frac{\mu'^2 - 1}{\mu'^2 + 2}. \quad \dots\dots\dots\dots\dots\dots(12)$$

On the same suppositions (5) is replaced by

$$h = 24\pi^3 n\,\frac{(\mu'^2 - 1)^2}{(\mu'^2 + 2)^2}\,\frac{T^2}{\lambda^4}. \quad \dots\dots\dots\dots\dots(13)$$

On either theory

$$h = \frac{32\pi^3 (\mu - 1)^2}{3n\lambda^4}, \quad \dots\dots\dots\dots\dots\dots(14)$$

* *Phil. Mag.* XII. p. 98 (1881). [Vol. I. p. 533.] For the corresponding theory in the case of an *ellipsoidal* obstacle, see *Phil. Mag.* Vol. XLIV. p. 48 (1897). [Vol. IV. p. 305.]

a formula giving the coefficient of transmission in terms of the refraction, and of the *number of particles per unit volume.*

We have seen that when we attempt to find directly from (1) the effect of the particles upon the transmitted primary wave, we succeed only so far as regards the retardation. In order to determine the attenuation by this process it would be necessary to supplement (1) by a term involving

$$\sin 2\pi \, (bt - r)/\lambda;$$

but this is of higher order of smallness. We could, however, reverse the process and determine the small term in question *à posteriori* by means of the value of the attenuation obtained indirectly from (1), at least as far as concerns the secondary light emitted in the direction of the primary ray.

The theory of these effects may be illustrated by a completely worked out case, such as that of a small rigid and fixed spherical obstacle (radius c) upon which plane waves of sound impinge*. It would take too much space to give full details here, but a few indications may be of use to a reader desirous of pursuing the matter further.

The expressions for the terms of orders 0 and 1 in spherical harmonics of the velocity-potential of the secondary disturbance are given in equations (16), (17), § 334. With introduction of approximate values of γ_0 and γ_1, viz.

$$\gamma_0 + kc = \tfrac{1}{3}k^3c^3, \qquad \gamma_1 + kc = \tfrac{1}{2}\pi + \tfrac{1}{8}k^3c^3,$$

we get

$$[\psi_0] + [\psi_1] = -\frac{k^2c^3}{3r}\left(1 + \frac{3\mu}{2}\right)\cos k\,(at - r) + \frac{k^5c^6}{9r}\left(1 - \frac{3\mu}{4}\right)\sin k\,(at - r), \ldots (15)\dagger$$

in which c is the radius of the sphere, and $k = 2\pi/\lambda$. This corresponds to the primary wave

$$[\phi] = \cos k\,(at + x), \quad\ldots\ldots\ldots\ldots\ldots\ldots\ldots(16)$$

and includes the most important terms from all sources in the multipliers of $\cos k\,(at - r)$, $\sin k\,(at - r)$. Along the course of the primary ray ($\mu = -1$) it reduces to

$$[\psi_0] + [\psi_1] = \frac{k^2c^3}{6r}\cos k\,(at - r) + \frac{7k^5c^6}{36r}\sin k\,(at - r). \quad\ldots\ldots(17)$$

We have now to calculate by the method of Fresnel's zones the effect of a distribution of n spheres per unit volume. We find, corresponding to (6), for the effect of a layer of thickness dx,

$$2\pi n\,dx\,\{\tfrac{1}{6}kc^3 \sin k\,(at + x) - \tfrac{7}{36}k^4c^6 \cos k\,(at + x)\}. \quad\ldots\ldots\ldots(18)$$

---

* *Theory of Sound*, 2nd ed. § 334.
† [1902. μ here denotes the sine of the latitude.]

To this is to be added the expression (16) for the primary wave. The coefficient of $\cos k\,(at+x)$ is thus altered by the particles in the layer dx from unity to $(1-\frac{7}{18}k^4c^6\pi n\,dx)$, and the coefficient of $\sin k\,(at+x)$ from 0 to $\frac{1}{3}kc^3\pi n\,dx$. Thus, if E be the energy of the primary wave,

$$dE/E = -\tfrac{7}{9}k^4c^6\pi n\,dx\,;$$

so that if, as in (4), $E = E_0 e^{-hx}$,

$$h = \tfrac{7}{9}\pi n k^4 c^6. \quad\dots\dots\dots\dots\dots\dots\dots\dots(19)$$

The same result may be obtained indirectly from the *first* term of (15). For the whole energy emitted from one sphere may be reckoned as

$$\frac{k^4c^6}{9r^2}\int_{-1}^{+1} 2\pi r^2\,(1+\tfrac{3}{2}\mu)^2\,d\mu = \frac{7\pi k^4 c^6}{9}, \quad\dots\dots\dots\dots(20)$$

unity representing the energy of the primary wave per unit area of wavefront. From (20) we deduce the same value of h as in (19).

The first term of (18) gives the refractivity of the medium. If δ be the retardation due to the spheres of the stratum dx,

$$\sin k\delta = \tfrac{1}{3}kc^3\pi n\,dx,$$

or

$$\delta = \tfrac{1}{3}\pi n c^3\,dx. \quad\dots\dots\dots\dots\dots\dots\dots(21)$$

Thus, if μ be the refractive index as modified by the spheres, that of the original medium being unity,

$$\mu - 1 = \tfrac{1}{3}\pi n c^3 = \tfrac{1}{4}p, \quad\dots\dots\dots\dots\dots\dots(22)$$

where p denotes the (small) ratio of the volume occupied by the spheres to the whole volume. This result agrees with equations formerly obtained for the refractivity of a medium containing spherical obstacles disposed in cubic order*.

Let us now inquire what degree of transparency of air is admitted by its molecular constitution, *i.e.*, in the absence of all foreign matter. We may take $\lambda = 6 \times 10^{-5}$ centim., $\mu - 1 = {\cdot}0003$; whence from (14) we obtain as the distance x, equal to $1/h$, which light must travel in order to undergo attenuation in the ratio $e:1$,

$$x = 4{\cdot}4 \times 10^{-13} \times n. \quad\dots\dots\dots\dots\dots\dots(23)$$

The completion of the calculation requires the value of n. Unfortunately this number—according to Avogadro's law the same for all gases—can hardly be regarded as known. Maxwell† estimates the number of molecules under standard conditions as 19×10^{18} per cub. centim. If we use this value of n, we find

$$x = 8{\cdot}3 \times 10^6 \text{ cm.} = 83 \text{ kilometres,}$$

* *Phil. Mag.* Vol. xxxiv. p. 499 (1892). [Vol. iv. p. 35.] Suppose $m=\infty$, $\sigma=\infty$.

† "Molecules," *Nature*, viii. p. 440 (1873).

as the distance through which light must pass in air at atmospheric pressure before its intensity is reduced in the ratio of $2·7 : 1$.

Although Mount Everest appears fairly bright at 100 miles distance as seen from the neighbourhood of Darjeeling, we cannot suppose that the atmosphere is as transparent as is implied in the above numbers; and of course this is not to be expected, since there is certainly suspended matter to be reckoned with. Perhaps the best data for a comparison are those afforded by the varying brightness of stars at various altitudes. Bouguer and others estimate about ·8 for the transmission of light through the entire atmosphere from a star in the zenith. This corresponds to 8·3 kilometres of air at standard pressure. At this rate the transmission through 83 kilometres would be $(·8)^{10}$, or ·11, instead of $1/e$ or ·37. It appears then that the actual transmission through 83 kilometres is only about 3 times less than that calculated (with the above value of n) from molecular diffraction without any allowance for foreign matter at all. And we may conclude that the light scattered from the molecules would suffice to give us a blue sky, not so very greatly darker than that actually enjoyed.

If n be regarded as altogether unknown, we may reverse our argument, and we then arrive at the conclusion that n cannot be greatly less than was estimated by Maxwell. A lower limit for n, say 7×10^{18} per cubic centimetre, is somewhat sharply indicated. For a still smaller value, or rather the increased individual efficacy which according to the observed refraction would be its accompaniment, must lead to a less degree of transparency than is actually found. When we take into account the known presence of foreign matter, we shall probably see no ground for any reduction of Maxwell's number.

The results which we have obtained are based upon (14), and are as true as the theories from which that equation was derived. In the electromagnetic theory we have treated the molecules as spherical continuous bodies differing from the rest of the medium merely in the value of their dielectric constant. If we abandon the restriction as to sphericity, the results will be modified in a manner that cannot be precisely defined until the shape is specified. On the whole, however, it does not appear probable that this consideration would greatly affect the calculation as to transparency, since the particles must be supposed to be oriented in all directions indifferently. But the theoretical conclusion that the light diffracted in a direction perpendicular to the primary rays should be *completely* polarized may well be seriously disturbed. If the view, suggested in the present paper, that a large part of the light from the sky is diffracted from the molecules themselves, be correct, the observed incomplete polarization at 90° from the Sun may be partly due to the molecules behaving rather as elongated bodies with indifferent orientation than as spheres of homogeneous material.

Again, the suppositions upon which we have proceeded give no account of *dispersion*. That the refraction of gases increases as the wave-length diminishes is an observed fact; and it is probable that the relation between refraction and transparency expressed in (14) holds good for each wave-length. If so, the falling off of transparency at the blue end of the spectrum will be even more marked than according to the inverse fourth power of the wave-length.

An interesting question arises as to whether (14) can be applied to highly compressed gases and to liquids or solids. Since approximately $(\mu - 1)$ is proportional to n, so also is h according to (14). We have no reason to suppose that the purest water is any more transparent than (14) would indicate; but it is more than doubtful whether the calculations are applicable to such a case, where the fundamental supposition, that the phases are entirely at random, is violated. When the volume occupied by the molecules is no longer very small compared with the whole volume, the fact that two molecules cannot occupy the same space detracts from the random character of the distribution. And when, as in liquids and solids, there is some approach to a regular spacing, the scattered light must be much less than upon a theory of random distribution.

Hitherto we have considered the case of obstacles small compared to the wave-length. In conclusion it may not be inappropriate to make a few remarks upon the opposite extreme case and to consider briefly the obstruction presented, for example, by a shower of rain, where the diameters of the drops are large multiples of the wave-length of light.

The full solution of the problem presented by spherical drops of water would include the theory of the rainbow, and if practicable at all would be a very complicated matter. But so far as the direct light is concerned, it would seem to make little difference whether we have to do with a spherical refracting drop, or with an opaque disk of the same diameter. Let us suppose then that a large number of small disks are distributed at random over a plane parallel to a wave-front, and let us consider their effect upon the direct light at a great distance behind. The plane of the disks may be divided into a system of Fresnel's zones, each of which will by hypothesis include a large number of disks. If α be the area of each disk, and ν the number distributed per unit of area of the plane, the efficiency of each zone is diminished in the ratio $1 : 1 - \nu\alpha$, and, so far as the direct wave is concerned, this is the only effect. The *amplitude* of the direct wave is accordingly reduced in the ratio $1 : 1 - \nu\alpha$, or, if we denote the relative opaque area by m, in the ratio $1 : 1 - m$*. A second operation of the same kind will reduce the

---

* The *intensity* of the direct wave is $1 - 2m$, and that of the scattered light m, making altogether $1 - m$.

amplitude to $(1-m)^2$, and so on. After x passages the amplitude is $(1-m)^x$, which if m be very small may be equated to e^{-mx}. Here mx denotes the whole opaque area passed, reckoned per unit area of wave-front; and it would seem that the result is applicable to any sufficiently sparse random distribution of obstacles.

It may be of interest to give a numerical example. If the unit of length be the centimetre and x the distance travelled, m will denote the projected area of the drops situated in one cubic centimetre. Suppose now that a is the radius of a drop, and n the number of drops per cubic centimetre, then $m = n\pi a^2$. The distance required to reduce the *amplitude* in the ratio $e : 1$ is given by

$$x = 1/n\pi a^2.$$

Suppose that $a = \frac{1}{20}$ centim., then the above-named reduction will occur in a distance of one kilometre ($x = 10^5$) when n is about 10^{-3}, *i.e.* when there is about one drop of one millimetre diameter per litre.

It should be noticed that according to this theory a distant point of light seen through a shower of rain ultimately becomes invisible, not by failure of definition, but by loss of intensity either absolutely or relatively to the scattered light.

248.

THE INTERFEROMETER.

[*Nature*, LIX. p. 533, 1899.]

THE questions raised by Mr Preston (*Nature*, March 23) can only be fully answered by Prof. Michelson himself; but as one of the few who have used the interferometer in observations involving high interference, I should like to make a remark or two. My opportunity was due to the kindness of Prof. Michelson, who some years ago left in my hands a small instrument of his model.

I do not understand in what way the working is supposed to be prejudiced by "diffraction." My experience certainly suggested nothing of the sort, and I do not see why it is to be expected upon theoretical grounds.

The estimation of the "visibility" of the bands, and the deduction of the structure of the spectrum line from the visibility curve, are no doubt rather delicate matters. I have remarked upon a former occasion (*Phil. Mag.* November, 1892)* that, strictly speaking, the structure cannot be deduced from the visibility curve without an auxiliary assumption. But in the application to radiation in a magnetic field the assumption of symmetry would appear to be justified.

My observations were made with a modification of the original apparatus, which it may be worth while briefly to describe. In order to increase the retardation it is necessary to move backwards, parallel to itself, one of the perpendicularly reflecting mirrors. Unless the ways upon which the sliding piece travels are extremely true, this involves a troublesome readjustment of the mirror after each change of distance. The difficulty is avoided by the use of a fluid surface as reflector, which after each movement automatically sets itself rigorously horizontal. If mercury be contained in a glass dish, the depth must be considerable, and then the surface is inconveniently mobile. A better plan is to use a thin layer standing on a piece of copper plate carefully amalgamated. A screw movement for raising and lowering the mercury reflector is still desirable, though not absolutely necessary.

* [Vol. IV. p. 15.]

249.

ON THE CALCULATION OF THE FREQUENCY OF VIBRATION OF A SYSTEM IN ITS GRAVEST MODE, WITH AN EXAMPLE FROM HYDRODYNAMICS.

[Philosophical Magazine, XLVII. pp. 566—572, 1899.]

WHEN the expressions for the kinetic (T) and potential (V) energy of a system moving about a configuration of stable equilibrium are given, the possible frequencies of vibration are determined by an algebraic equation of degree (in the square of the frequency) equal to the number of independent motions of which the system is capable. Thus in the case of a system whose position is defined by *two* coordinates q_1 and q_2, we have

$$\left. \begin{aligned} T &= \tfrac{1}{2} L \dot{q}_1{}^2 + M \dot{q}_1 \dot{q}_2 + \tfrac{1}{2} N \dot{q}_2{}^2, \\ V &= \tfrac{1}{2} A q_1{}^2 + B q_1 q_2 + \tfrac{1}{2} C q_2{}^2; \end{aligned} \right\} \quad \dots\dots\dots\dots\dots(1)$$

and if in a free vibration the coordinates are proportional to $\cos pt$, the determinantal equation is

$$\begin{vmatrix} A - p^2 L, & B - p^2 M \\ B - p^2 M, & C - p^2 N \end{vmatrix} = 0, \quad \dots\dots\dots\dots(2)$$

viz.:

$$p^4 (LN - M^2) + p^2 (2MB - LC - NA) + AC - B^2 = 0. \quad \dots\dots\dots(3)$$

And whatever be the number of coordinates, the possible frequencies are given by a determinantal equation analogous to (2).

When the determinantal equation is fully expressed, the smallest root, or indeed any other root, can be found by the ordinary processes of successive approximation. In many of the most interesting cases, however, the number of coordinates is infinite, and the inclusion of even a moderate number of them in the expressions for T and V would lead to laborious calculations. We may then avail ourselves of the following method of approximating to the value of the smallest root.

The method is founded upon the principle* that the introduction of a constraint can never lower, and must in general raise, the frequency of any mode of a vibrating system. The first constraint that we impose is the evanescence of one coordinate, say the last. The lowest frequency of the system thus constrained is higher than the lowest frequency of the unconstrained system. Next impose as an *additional* constraint the evanescence of the last coordinate but one. The lowest frequency is again raised. If we continue this process until only one coordinate is left free to vary, we obtain a series of continually increasing quantities as the lowest frequencies of the various systems. Or, if we contemplate the operations in the reverse order, we obtain a series of decreasing quantities ending in the precise quantity sought. The first of the series, resulting from the sole variation of the first coordinate, is given by an equation of the first degree, viz. $A - p^2 L = 0$. The second is the lower root of the determinant (2) of the second order. The third is the lowest root of a determinant of the third order formed by the addition of one row and one column to (2), and so on. This series of quantities may accordingly be regarded as successive approximations to the value required. Each is nearer than its predecessor to the truth, and all (except of course the last itself) are too high.

The practical success of the method must depend upon the choice of coordinates and of the order in which they are employed. The object is so to arrange matters that the variation of the first two or three coordinates shall allow a good approximation to the actual mode of vibration.

The example by which I propose to illustrate the method is one already considered by Prof. Lamb. It is that of the transverse vibration of a liquid mass, contained in a horizontal cylindrical vessel, and of such quantity that the free surface contains the axis of the cylinder ($r = 0$). If we measure θ vertically downwards, the fluid is limited by $r = 0$, $r = c$, and by $\theta = -\tfrac{1}{2}\pi$, $\theta = +\tfrac{1}{2}\pi$. Between the above limits of θ and when $r = c$ the motion must be exclusively tangential.

In the gravest mode of vibration the fluid swings from one side to the other in such a manner that the horizontal motions are equal and the vertical motions opposite at any two points which are images of one another in the line $\theta = 0$. This relation, which holds also at the two halves of the free surface, implies a stream-function ψ which is symmetrical with respect to $\theta = 0$.

Let η, denoting the elevation of the surface at a distance r from the centre on the side for which $\theta = \tfrac{1}{2}\pi$, be expressed by

$$\eta = - 2q_2 (r/c) + 4q_4 (r/c)^3 - 6q_6 (r/c)^5 + \dots ; \quad \dots\dots\dots(4)$$

---

* *Theory of Sound*, §§ 88, 89. [See Vol. I. p. 170.]

then the potential energy for the whole mass (supposed to be of unit density) is given by

$$V = 2\int_0^c \tfrac{1}{2}g\eta^2 dr = 4gc\left(\tfrac{1}{3}q_2^2 - \tfrac{4}{5}q_2q_4 + \tfrac{4}{7}q_4^2 + \ldots\right). \quad \ldots\ldots\ldots\ldots(5)$$

The more difficult part of the problem lies in determining the motion and in the calculation of the kinetic energy. It may be solved by the method of Sir G. Stokes, who treated a particular case, corresponding in fact to our first approximation in which (4) reduces to its first term. It is required to find the motion of an incompressible fluid in two dimensions within the semicylinder, the normal velocity being zero over the whole of the curved boundary ($r = c$, $\tfrac{1}{2}\pi > \theta > -\tfrac{1}{2}\pi$) and over the flat boundary having values prescribed by (4). If ψ be the stream-function, satisfying

$$d^2\psi/dx^2 + d^2\psi/dy^2 = 0,$$

the conditions are that ψ shall be symmetrical with respect to $\theta = 0$, that it be constant when $r = c$ from $\theta = 0$ to $\theta = \tfrac{1}{2}\pi$, and that when $\theta = \tfrac{1}{2}\pi$,

$$d\psi/dr = d\eta/dt = -2\dot{q}_2(r/c) + 4\dot{q}_4(r/c)^3 - \ldots,$$

or

$$\psi/c = -\dot{q}_2(r/c)^2 + \dot{q}_4(r/c)^4 - \dot{q}_6(r/c)^6 + \ldots. \quad \ldots\ldots\ldots(6)$$

At the edge, where $r = c$,

$$\psi/c = -\dot{q}_2 + \dot{q}_4 - \dot{q}_6 - \ldots, \quad \ldots\ldots\ldots\ldots\ldots(7)$$

and this value must obtain also over the curved boundary.

The conditions may be satisfied* by assuming

$$\psi/c = \dot{q}_2(r/c)^2\cos 2\theta + \dot{q}_4(r/c)^4\cos 4\theta + \ldots$$
$$+ \Sigma A_{2n+1}(r/c)^{2n+1}\cos(2n+1)\theta, \quad \ldots\ldots\ldots\ldots(8)$$

in which $n = 0$, 1, 2, &c. This form satisfies Laplace's equation and the condition of symmetry since cosines of θ alone occur. When $\theta = \tfrac{1}{2}\pi$, it reduces to (6). It remains only to secure the reduction to (7) when $r = c$, and this can be effected by Fourier's method. It is required that from $\theta = 0$ to $\theta = \tfrac{1}{2}\pi$

$$\Sigma A_{2n+1}\cos(2n+1)\theta = -\dot{q}_2(1+\cos 2\theta) + \dot{q}_4(1-\cos 4\theta) - \ldots \ldots(9)$$

It will be convenient to write

$$A_{2n+1} = \dot{q}_2 A_{2n+1}^{(2)} + \dot{q}_4 A_{2n+1}^{(4)} + \ldots, \quad \ldots\ldots\ldots\ldots(10)$$

so that

$$\Sigma A_{2n+1}^{(2s)}\cos(2n+1)\theta = (-1)^s - \cos 2s\theta. \quad \ldots\ldots\ldots(11)$$

In (11) s may have the values 1, 2, 3, &c.

* Lamb's *Hydrodynamics*, § 72.

The values of the constants in (11) are to be found as usual. Since

$$2 \int_0^{\frac{1}{2}\pi} \cos(2n+1)\,\theta \, . \, \cos(2m+1)\,\theta \, d\theta$$

vanishes when m and n are different, and when m and n coincide has the value $\frac{1}{2}\pi$, and since

$$2 \int_0^{\frac{1}{2}\pi} \{(-1)^s - \cos 2s\theta\} \cos(2n+1)\,\theta \, d\theta$$

$$= (-1)^{s+n} \left\{ -\frac{1}{2n+2s+1} + \frac{2}{2n+1} - \frac{1}{2n-2s+1} \right\},$$

we get

$$A_{2n+1}^{(2s)} = (-1)^{s+n}\frac{2}{\pi} \left\{ -\frac{1}{2n+2s+1} + \frac{2}{2n+1} - \frac{1}{2n-2s+1} \right\}, \quad \ldots(12)$$

in which $s = 1, 2, 3, \&c., \; n = 0, 1, 2, \&c.$

The value of ψ in (8) is now completely determined when $\dot{q}_2$, &c. are known. The velocity-potential ϕ is deducible by merely writing sines, in place of cosines, of the multiples of θ.

We have now to calculate the kinetic energy T of the motion thus expressed, supposing for brevity that the density is unity. We have in general

$$2T = \int \phi \, \frac{d\phi}{dn} \, ds, \quad \ldots \ldots \ldots \ldots \ldots \ldots \ldots \ldots (13)$$

where dn is drawn normally outwards and the integration extends over the whole contour. In the present case, however, $d\phi/dn$ vanishes over the circular boundary, so that the integration may be limited to the plane part. Of this the two halves contribute equally. Now when $\theta = \frac{1}{2}\pi$,

$$\phi/c = \Sigma (-1)^n A_{2n+1}(r/c)^{2n+1}, \quad \ldots \ldots \ldots \ldots \ldots \ldots (14)$$

$$d\phi/dn = d\phi/r d\theta = -2\dot{q}_2 (r/c) + 4\dot{q}_4 (r/c)^3 - \ldots \quad \ldots \ldots \ldots (15)$$

Thus

$$T = \Sigma (-1)^n c^2 A_{2n+1} \left\{ -\frac{2\dot{q}_2}{2n+3} + \frac{4\dot{q}_4}{2n+5} - \ldots \right\}, \quad \ldots \ldots \ldots (16)$$

where A_{2n+1} is given by (10) and (12); it is of course a quadratic function of $\dot{q}_2$, $\dot{q}_4$, &c.

The summation with respect to n is easily effected in particular cases by decomposition into partial fractions according to the general formula

$$\frac{1}{(2n+2s+1)(2n+2s'+1)} = \frac{1}{2(s-s')} \left\{ \frac{1}{2n+2s'+1} - \frac{1}{2n+2s+1} \right\}. \ldots(17)$$

If $s' = -s$, we have

$$\Sigma \frac{1}{(2n+2s+1)(2n-2s+1)} = \frac{1}{4s}\left(\frac{1}{2n-2s+1} - \frac{1}{2n+2s+1}\right)$$

$$= \frac{1}{4s}\left\{\left(-\frac{1}{2s-1} - \frac{1}{2s-3} - \cdots - 1 + 1 + \frac{1}{3} + \cdots + \frac{1}{2s-1}\right.\right.$$

$$\left.+ \frac{1}{2s+1} + \cdots\right) - \left(\frac{1}{2s+1} + \frac{1}{2s+3} + \cdots\right)\right\} = 0. \quad \ldots\ldots\ldots(18)$$

If $s' = s$, (17) fails, but we have by a known formula

$$\Sigma \frac{1}{(2n+2s+1)^2} = \frac{\pi^2}{8} - 1 - \frac{1}{3^2} - \frac{1}{5^2} - \cdots - \frac{1}{(2s-1)^2}. \quad \ldots\ldots(19)$$

Thus for the term in $\dot{q}_2{}^2$, we have in (16)

$$\frac{4c^2\dot{q}_2{}^2}{\pi} \Sigma \frac{1}{2n+3}\left\{-\frac{1}{2n+3} + \frac{2}{2n+1} - \frac{1}{2n-1}\right\}, \quad \ldots\ldots\ldots(20)$$

in which by (18) $\Sigma (2n+3)^{-1}(2n-1)^{-1} = 0,$

by (17) $\Sigma (2n+3)^{-1}(2n+1)^{-1} = \frac{1}{2}\Sigma (2n+1)^{-1} - \frac{1}{2}\Sigma (2n+3)^{-1}$

$$= \frac{1}{2}\left(1 + \frac{1}{3^2} + \frac{1}{5^2} + \cdots\right) - \frac{1}{2}\left(\frac{1}{3^2} + \frac{1}{5^2} + \cdots\right) = \frac{1}{2},$$

and by (19) $\Sigma (2n+3)^{-2} = \frac{1}{8}\pi^2 - 1.$

The complete term (20) in $\dot{q}_2{}^2$ is accordingly

$$\frac{4c^2\dot{q}_2{}^2}{\pi}(2 - \frac{1}{8}\pi^2). \quad \ldots\ldots\ldots\ldots\ldots\ldots\ldots\ldots(21)$$

The first approximation to p^2 is therefore from (5), (21)

$$p^2 = \frac{\pi}{3(2-\frac{1}{8}\pi^2)}\frac{g}{c}, \quad \ldots\ldots\ldots\ldots\ldots\ldots\ldots\ldots(22)$$

or $p = 1\cdot1690\,(g/c)^{\frac{1}{2}}, \quad \ldots\ldots\ldots\ldots\ldots\ldots\ldots(23)$

which is Prof. Lamb's result [*].

For the second approximation we require also the terms in (16) which involve $\dot{q}_4{}^2$ and $\dot{q}_2\dot{q}_4$, and they are calculated as before. The term in $\dot{q}_4{}^2$ is

$$\frac{8c^2\dot{q}_4{}^2}{\pi} \Sigma \frac{1}{2n+5}\left(-\frac{1}{2n+5} + \frac{2}{2n+1} - \frac{1}{2n-3}\right) = \frac{8c^2\dot{q}_4{}^2}{\pi}\left(\frac{16}{9} - \frac{\pi^2}{8}\right). \quad \ldots(24)$$

The term in $\dot{q}_2\,\dot{q}_4$ is made up of two parts. Its complete value is

$$-\frac{64c^2}{9\pi}\dot{q}_2\dot{q}_4. \quad \ldots\ldots\ldots\ldots\ldots\ldots\ldots\ldots(25)$$

* Hydrodynamics, § 238.

Thus

$$T = \frac{4c^2}{\pi}\left(2 - \frac{\pi^2}{8}\right)\dot{q}_2{}^2 - \frac{64c^2}{9\pi}\dot{q}_2\dot{q}_4 + \frac{8c^2}{\pi}\left(\frac{16}{9} - \frac{\pi^2}{8}\right)\dot{q}_4{}^2 + \dots, \quad \dots\dots(26)$$

which with (5) gives materials for the second approximation. In proceeding to this we may drop the symbols c and g, which can at any moment be restored by consideration of dimensions. Also the factor 8 may be omitted from the expressions for T and V. On this understanding we have by comparison with (1),

$$A = \frac{1}{3}, \qquad B = -\frac{2}{5}, \qquad C = \frac{4}{7};$$

$$L = \frac{2}{\pi} - \frac{\pi}{8}, \qquad M = -\frac{8}{9\pi}, \qquad N = \frac{32}{9\pi} - \frac{\pi}{4},$$

or on introduction of the value of π,

$$L = \cdot2439204, \qquad M = -\cdot2829420, \qquad N = \cdot3463696.$$

The coefficients of the quadratic (3) are thence found to be

$$LN - M^2 = \cdot00443040, \qquad AC - B^2 = \cdot0304762,$$

$$2MB - LC - NA = -\cdot0284860;$$

whence on restoration of the factor $(g/c)^{\frac{1}{2}}$,

$$p_1 = 1\cdot1644\,(g/c)^{\frac{1}{2}}, \qquad p_2 = 2\cdot2525\,(g/c)^{\frac{1}{2}}, \quad \dots\dots\dots(27)$$

the first of which constitutes the *second* approximation to the value of p in cos pt, corresponding to the gravest mode of vibration. The small difference between (23) and (27) shows the success of the method and indicates that (27) is but very little in excess of the truth.

If the result were of special importance it would be quite practicable to take another step in the approximation, determining p^2 as the lowest root of a *cubic* equation.

A question naturally suggests itself as to the significance of the value of p_2 in (27). The general theory of constraints* shows that it may be regarded as a first, but probably a rather rough, approximation to the frequency of the second lowest mode of the complete system. Just as for the gravest mode of all, the second lowest roots of the series of determinants (of the 2nd, 3rd, and following orders) form successive approximations to the true value, each value being lower and truer than its predecessor. The second approximation would be the middle root of the cubic above mentioned. But for this purpose it is doubtful whether the method is practical.

* *Theory of Sound*, 2nd ed. § 92 a.

250.

THE THEORY OF ANOMALOUS DISPERSION.

[*Philosophical Magazine*, XLVIII. pp. 151, 152, 1899.]

I HAVE lately discovered that Maxwell, earlier than Sellmeier or any other writer, had considered this question. His results are given in the Mathematical Tripos Examination for 1869 (see *Cambridge Calendar* for that year). In the paper for Jan. 21, $1\frac{1}{2}^{h}$—4^{h}, Question IX. is:—

" Show from dynamical principles that if the elasticity of a medium be such that a tangential displacement η (in the direction of y) of one surface of a stratum of thickness a calls into action a force of restitution equal to $E\eta/a$ per unit of area, then the equation of propagation of such displacements is

$$\rho \frac{d^2\eta}{dt^2} = E \frac{d^2\eta}{dx^2}.$$

"Suppose that every part of this medium is connected with an atom of other matter by an attractive force varying as distance, and that there is also a force of resistance between the medium and the atoms varying as their relative velocity, the atoms being independent of each other; show that the equations of propagation of waves in this compound medium are

$$\rho \frac{d^2\eta}{dt^2} - E \frac{d^2\eta}{dx^2} = \sigma \left(p^2 \zeta + R \frac{d\zeta}{dt} \right) = -\sigma \left(\frac{d^2\eta}{dt^2} + \frac{d^2\zeta}{dt^2} \right),$$

where ρ and σ are the quantities of the medium and of the atoms respectively in unit of volume, η is the displacement of the medium, and $\eta + \zeta$ that of the atoms, $\sigma p^2 \zeta$ is the attraction, and $\sigma R d\zeta/dt$ is the resistance to the relative motion per unit of volume.

"If one term of the value of η be $Ce^{-x/l}\cos n\,(t-x/v)$, show that

$$\frac{1}{v^2}+\frac{1}{l^2n^2}=\frac{\rho+\sigma}{E}+\frac{\sigma n^2}{E}\,\frac{p^2-n^2}{(p^2-n^2)^2+R^2n^2},$$

$$\frac{2}{vln}=\frac{\sigma n^2}{E}\,\frac{Rn}{(p^2-n^2)^2+R^2n^2}.$$

"If σ be very small, one of the values of v^2 will be less than E/ρ, and if R be very small v will diminish as n increases, except when n is nearly equal to p, and in the last case l will have its lowest values. Assuming these results, interpret them in the language of the undulatory theory of light."

If we suppose that $R=0$,

$$\frac{1}{v^2}=\frac{\rho}{E}+\frac{\sigma}{E}\,\frac{p^2}{p^2-n^2},$$

and

$$\mu^2=\frac{v_0^2}{v^2}=1+\frac{\sigma}{\rho}\,\frac{p^2}{p^2-n^2},$$

if v_0 be the velocity corresponding to $\sigma=0$.

251.

INVESTIGATIONS IN CAPILLARITY:—THE SIZE OF DROPS.— THE LIBERATION OF GAS FROM SUPERSATURATED SOLUTIONS. — COLLIDING JETS. — THE TENSION OF CONTAMINATED WATER-SURFACES.—A CURIOUS OBSERVATION.

[*Philosophical Magazine*, XLVIII. pp. 321—337, 1899.]

The Size of Drops.

THE relation between the diameter of a tube and the weight of the drop which it delivers appears to have been first investigated by Tate*, whose experiments led him to the conclusion that "other things being the same, the weight of a drop of liquid is proportional to the diameter of the tube in which it is formed." Sufficient time must of course be allowed for the formation of the drops; otherwise no simple results can be expected. In Tate's experiments the period was never less than 40 seconds.

The magnitude of a drop delivered from a tube, even when the formation up to the phase of instability is infinitely slow, cannot be calculated *à priori*. The weight is sometimes equated to the product of the capillary tension (T) and the circumference of the tube ($2\pi a$), but with little justification. Even if the tension at the circumference of the tube acted vertically, and the whole of the liquid below this level passed into the drop, the calculation would still be vitiated by the assumption that the internal pressure at the level in question is atmospheric. It would be necessary to consider the curvatures of the fluid surface at the edge of attachment. If the surface could be treated as a cylindrical prolongation of the tube (radius a), the pressure would be T/a, and the resulting force acting downwards upon the drop would amount to one-half ($\pi a T$) of the direct upward pull of the tension along the circumference. At this rate the drop would be but one-half of that above

---

* *Phil. Mag.* Vol. XXVII. p. 176 (1864).

reckoned. But the truth is that a complete solution of the statical problem
for all forms up to that at which instability sets in, would not suffice for the
present purpose. The detachment of the drop is a *dynamical* effect, and
it is influenced by collateral circumstances. For example, the bore of the
tube is no longer a matter of indifference, even though the attachment of
the drop occurs entirely at the outer edge. It will appear presently that
when the external diameter exceeds a certain value, the weight of a drop
of water is sensibly different in the two extreme cases of a very small and of
a very large bore.

But although a complete solution of the dynamical problem is impracticable, much interesting information may be obtained from the principle
of dynamical similarity. The argument has already been applied by Dupré
(*Théorie Mécanique de la Chaleur*, Paris, 1869, p. 328), but his presentation
of it is rather obscure. We will assume that when, as in most cases, viscosity
may be neglected, the mass (M) of a drop depends only upon the density (σ),
the capillary tension (T), the acceleration of gravity (g), and the linear
dimension of the tube (a). In order to justify this assumption, the formation of the drop must be sufficiently slow, and certain restrictions must be
imposed upon the shape of the tube. For example, in the case of water
delivered from a glass tube, which is cut off square and held vertically, a will
be the external radius; and it will be necessary to suppose that the ratio
of the internal radius to a is constant, the cases of a ratio infinitely small, or
infinitely near unity, being included. But if the fluid be mercury, the flat
end of the tube remains unwetted, and the formation of the drop depends
upon the internal diameter only.

The "dimensions" of the quantities on which M depends are:—

$$\sigma = (\text{Mass})^1 (\text{Length})^{-3},$$

$$T = (\text{Force})^1 (\text{Length})^{-1} = (\text{Mass})^1 (\text{Time})^{-2},$$

$$g = \text{Acceleration} = (\text{Length})^1 (\text{Time})^{-2},$$

of which M, a mass, is to be expressed as a function. If we assume

$$M \propto T^x . g^y . \sigma^z . a^u,$$

we have, considering in turn length, time, and mass,

$$y - 3z + u = 0, \quad 2x + 2y = 0, \quad x + z = 1;$$

so that $\qquad y = -x, \quad z = 1 - x, \quad u = 3 - 2x.$

Accordingly $\qquad\qquad M \propto \dfrac{Ta}{g} \left(\dfrac{T}{g\sigma a^2}\right)^{x-1}.$

Since x is undetermined, all that we can conclude is that M is of the form

$$M = \frac{Ta}{g} \cdot F\left(\frac{T}{g\sigma a^2}\right), \qquad \dots\dots\dots\dots\dots\dots\dots\dots(1)$$

where F denotes an arbitrary function.

Dynamical similarity requires that $T/g\sigma a^2$ be constant; or, if g be supposed to be so, that a^2 varies as T/σ. If this condition be satisfied, the mass (or weight) of the drop is proportional to T and to a.

If Tate's law be true, that *cæteris paribus* M varies as a, it follows from (1) that F is constant. For all fluids and for all similar tubes similarly wetted, the weight of a drop would then be proportional not only to the diameter of the tube but also to the superficial tension, and it would be independent of the density.

In order to examine how far Tate's law can be relied upon, I have thought it desirable, with the assistance of Mr Gordon, to institute fresh experiments with water, in which necessary precautions were observed, especially against the presence of grease. Attention has been given principally to the two extreme cases, (i) when the wall of the tube is thin, so that the external and internal diameters of the tube are nearly equal; (ii) when the bore is small in comparison with the external diameter. The event showed that up to an external diameter of one centimetre or more, the size of the bore is of little consequence, but that for larger diameters the weight of the drop in (ii) is sensibly less than in (i). It scarcely needs to be pointed out that in (i) the diameter can only be increased up to a certain limit, after which the tube would not remain full. In (ii) the diameter can be increased to any extent, but the drop falling from it reaches a limit. The experiments of Tate extended also to case (ii), but his results are, I believe, erroneous. For a diameter of one-half an inch (1·27 cm.) he found for the two cases drops in the ratio of 1·56 : 2·84.

In my experiments the thin-walled tubes were of glass, the ends being ground to a plane, and carefully levelled. Ten drops, following one another at intervals of about 50 seconds, were usually weighed together. As to the interval, sufficient time must be allowed for the normal formation of the drop, but the fact that evaporation is usually in progress forbids too great a prolongation. The accuracy attained was not so great as had been hoped for. Successive collections, made without disturbance, gave indeed closely accordant weights (often to one-thousandth part), but repetitions after cleaning and remounting indicated discrepancies amounting to one-half per cent., or even to one per cent. The cause of these minor variations has not been fully traced; but the results recorded, being the mean of several experiments, must be free from serious error. Attention may be called to tubes 11 and 12 of nearly the same (external) diameter. Of these 11 was plugged so as to leave only a small bore, the end being carefully ground flat. It will be seen that the difference in the weights of the drops was but small.

Again, No. 10 was of barometer-tubing, having a comparatively small bore, which accounts for the slightly diminished weight of the drop. The other tubes were thin-walled. In all cases care was taken that the cylindrical part of the tube, though clean, should remain unwetted, a condition which precluded the use of diameters much less than those recorded.

Glass			Metal		
1	·088	·0375	15	·400	·1446
2	·134	·0526	16	·450	·1662
3	·191	·0712	17	·500	·1882
4	·200	·0755	18	·530	·2023
5	·256	·0923	19	·550	·2130
6	·354	·1151	20	·559	·2167
7	·383	·1362	21	·580	·2256
8	·406	·1461	22	·597	·2295
9	·459	·1703	23	·621	·2389
10	·465	·1698	24	·640	·2454
11	·521	·1969	25	·680	·2510
12	·523	·2023	26	·730	·2531
13	·566	·2210	27	·800	·2509
14	·584	·2339			

The numbers in the second column are the external diameters measured in inches (one inch = 2·54 cm.), while the third column gives the weight in grams of a single drop, corrected for temperature to 15° C., upon the supposition (corresponding to Tate's law) that the weight is proportional to surface-tension.

The entries under the heading "Metal" relate to experiments in which the glass tubes were replaced by metal disks, bored centrally and turned true in the lathe. The water was supplied from above through a metal tube soldered to the back (upper) face of the disk (Fig. 1). At the time of use only the lower face was wetted.

Fig. 1.

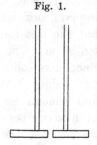

A plot of both sets of numbers is shown in the Figure (2). The two curves practically coincide up to diameters of about ·4 inch, after which that corresponding to the disks falls below. The lower curve shows some irregularities, especially in the region of diameters equal to ·6 inch. These appear to be genuine; they may originate in a sort of reflexion from the circumference of the disk of the disturbance caused by the breaking away of the drop. It is possible that at this stage the phenomenon is sensibly influenced by fluid viscosity.

Fig. 2.

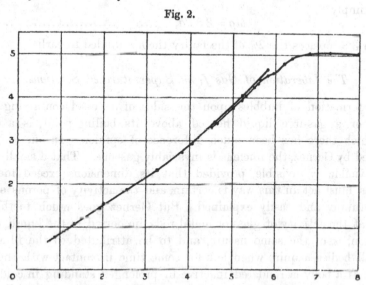

That the size of the bore should be of secondary importance is easily understood. Up to the phase of instability, the phenomenon is merely a statical one, and the element of the size of the bore does not enter. It is only the rapid motion which occurs during the separation of the drop that could be influenced. When the diameter is moderate, the most rapid motions occur at a level considerably below the tube, and the obstruction presented by the flat face of a thick-walled tube is unimportant.

The observations give materials for the determination of the function F in (1). In the following table, applicable to thin-walled tubes, the first column gives values of $T/g\sigma a^2$, and the second column those of gM/Ta, all the quantities concerned being in C.G.S. measure, or other consistent system.

$T/g\sigma a^2$	gM/Ta
2·58	4·13
1·16	3·97
·708	3·80
·441	3·73
·277	3·78
·220	3·90
·169	4·06

From this the weight of a drop of any liquid of which the density and the surface-tension are known can be calculated. For many purposes it may suffice to treat F as constant, say 3·8. The formula for the weight of a drop is then simply

$$Mg = 3·8 Ta, \qquad \dots\dots\dots\dots\dots\dots\dots\dots\dots\dots\dots\dots\dots(2)$$

in which 3·8 replaces the 2π of the faulty theory alluded to earlier.

The Liberation of Gas from Supersaturated Solutions.

The formation of bubbles upon the sides of a vessel containing "soda-water" or a gas-free liquid heated above its boiling-point, is a subject upon which there has been much difference of opinion. In one view, ably advocated by Gernez, the nucleus is invariably gaseous. That a small volume of gas, visible or invisible, provided that its dimensions exceed molecular distances, must act in this way is certain, and the activity of porous solids is thus naturally and easily explained. But Gernez goes much further, and holds that the activity of glass or metal rods, immersed in the liquid without precaution, is of the same nature, and to be attributed to the film of air which all bodies acquire when left for some time in contact with the atmosphere. If a body is rendered inactive by prolonged standing in cold water; by treatment with alcohol, ether, &c., "qui dissolvent les gaz de l'air, plus abondamment que l'eau * "; or by heating in a flame; it is because by such processes the film of air is removed. One cannot but sympathise with Tomlinson† in his repugnance to such an explanation; but the position maintained by the latter, that activity is due to contamination with grease, is also not without its difficulties.

The question whether contact with air suffices to restore the activity of a piece of glass or metal that has been rendered inactive by heat or otherwise, appears to be amenable to experiment, and should not remain an open one. In 1892 I had a number of glass tubes prepared of about 1 cm. diameter for experiments in this direction. After a thorough heating in the blowpipe-flame, the ends of the tubes were hermetically sealed. At intervals since that date some of the tubes have been opened and compared with others which had undergone no preparation. Short lengths of rubber provided with pinch-cocks are fitted to the upper ends, by means of which aerated water is easily drawn in from a shallow vessel. Three tubes remaining over from the batch above mentioned were tried a few weeks ago, and establish the conclusion that *seven years contact with air fails to restore activity*. A similar experiment may be made with iron wires. If these be heated and sealed up in glass tubes, they remain inactive, but exposure to the air of the laboratory for a day or two restores activity.

* *Annales de l'École Normale*, p. 319, 1875.

† *Phil. Mag.* Vol. XLIX. p. 305 (1875).

In opposition to the contention that grease is the primary cause of activity, Gernez brings forward a striking experiment from which it appears that a drop of olive-oil itself liberates no gas when introduced with precaution. "Quant au rôle que jouent les corps gras, il est facile de s'en rendre compte : lorsqu'on frotte un corps quelconque entre les doigts légèrement graissés, on produit à sa surface une série d'éminences linéaires séparées par les sillons qui correspondent aux lignes de l'épiderme ; les cavités forment un réseau de conduits qui contiennent de l'air, sont difficilement mouillés par l'eau et, par conséquent, constituent au sein du liquide une atmosphère éminemment favorable au dégagement des gaz*."

It seems to me that Tomlinson was substantially correct in attributing the activity of a non-porous surface to imperfect adhesion. We have to consider in detail the course of events when a surface, *e.g.* of glass, is introduced into the liquid. If the surface be clean, it is wetted by the water advancing over it, whether there be a film of air condensed upon it or not, and no gas is liberated from the liquid. But if the surface be greasy, even in a very slight degree, the behaviour is different. We know that a drop of water is reluctant to spread over a glass that is not scrupulously clean. If a large quantity of water be employed, some sort of spreading follows under the influence of gravity, but there is no proper adhesion, at least for a time, as appears at once on pouring the water off again. The precise character of the transition from glass to water when there is grease between is not well understood. It may be that there is something which can fairly be called a film of air. If so, its existence is a consequence of the presence of the grease. On the other hand, it appears at least equally probable that air is not concerned, and that the activity of the surface is directly due to the thin film of grease, whose properties, as in the case of greased water surfaces, are materially different from those of a thick layer.

On this principle, too, it is easier to understand the retention of a visible bubble when formed—a retention which often lasts for a long time. So soon as the gas is entirely surrounded by liquid of thickness exceeding the capillary limit, the bubble is bound to rise. It is difficult to see how the hypothetical film of air explains the failure of the liquid to penetrate between the bubble and the solid.

Colliding Jets.

In various papers (*Proc. Roy. Soc.* Feb. 1879, May 1879, June 1882) [Vol. I. pp. 372, 377 ; Vol. II. p. 103] I have examined the behaviour of colliding drops and jets. Experiments with drops are very simply carried out by the observation of nearly vertical fountains, rising say to two feet from nozzles $\frac{1}{20}$ inch in diameter. The scattering of the drops, when the

---

* *Loc. cit.* p. 346.

water is clean and not acted upon by electricity, shows that collision is followed by rebound. If the water is milky, or soapy with unclarified soap, or if the jet, though clean, is under the influence of feeble electricity, the apparent coherence and the heaviness of the patter made by the falling water are evidence that rebound no longer ensues, but that collision results in amalgamation. Eye observation, or photography, with the instantaneous illumination of electric sparks renders the course of events perfectly clear. [1902. The annexed illustrations are from instantaneous photographs of the same fountain with and without electrical influence.]

The form of the experiment in which are employed *jets*, issuing at moderate velocities and meeting at high obliquities, is the more instructive; but it is liable to be troublesome in consequence of the tendency of the jets to unite spontaneously. It is important to avoid dust both in the water and in the atmosphere where the collision occurs. An electromotive force of one volt suffices to determine union; but so long as the jets rebound there is complete electrical insulation between them.

As to the manner in which electricity acts, two views were suggested. It was thought probable that union was the result of actual discharge across the thin layer of intervening insulation; but it was also pointed out that the result might be due to the augmented pressure to be expected from the electrical charges upon the opposed surfaces. From observations upon the colours of thin plates exhibited at the region of contact, which he found to be undisturbed by such electrical forces as would not produce union, Mr Newall* concluded that the second of the above-mentioned explanations must be discarded.

On the other hand, as has been pointed out by Kaiser†, the progress of knowledge concerning electrical discharge has rendered the first explanation more difficult of acceptance. It would appear that some hundreds of volts are needed in order to start a spark, and that mere diminution of the interval to be crossed would not compensate for want of electromotive force.

A more attentive examination of the conditions of the experiment may perhaps remove some of the difficulties which seem to stand in the way of the second explanation. As the liquid masses approach one another, the intervening air has to be squeezed out. In the earlier stages of approximation the obstacle thus arising may not be important; but when the thickness of the layer of air is reduced to the point at which the colours of thin plates are visible, the approximation must be sensibly resisted by the viscosity of the air which still remains to be got rid of. No change in the capillary

* *Phil. Mag.* Vol. xx. p. 33 (1885).

† *Wied. Ann.* LIII. p. 667 (1894). Kaiser's own experiments were made upon the modification of the phenomenon observed by Boys, where the contact takes place between two soap-films.

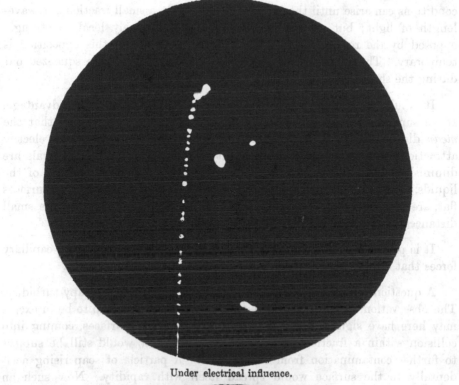

Under electrical influence.

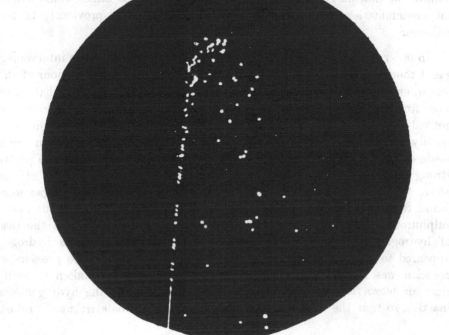

In natural condition.

conditions can arise until the interval is reduced to a small fraction of a wave-length of light; but such a reduction, unless extremely local, is strongly opposed by the remaining air. It is of course true that this opposition is temporary. The question is whether the air can be anywhere squeezed out during the short time over which the collision extends.

It would seem that the electrical forces act with peculiar advantage. If we suppose that upon the whole the air cannot be removed, so that the *mean* distance between the opposed surfaces remains constant, the electric attractions tend to produce an instability whereby the smaller intervals are diminished while the larger are increased. Extremely local contacts of the liquids, while opposed by capillary tension which tends to keep the surfaces flat, are thus favoured by the electrical forces, which moreover at the small distances in question act with exaggerated power.

It is probably by promoting local approximations in opposition to capillary forces that dust, finding its way to the surfaces, brings about union.

A question remains as to the mode of action of milk or soapy turbidity. The observation, formerly recorded, that it is possible for soap to be in excess may here have significance. It would seem that the surfaces, coming into collision within a fraction of a second of their birth, would still be subject to further contamination from the interior. A particle of soap rising acci-dentally to the surface would spread itself with rapidity. Now such an outward movement of the liquid is just what is required to hasten the removal of the intervening air. It is obvious that this effect would fail if the contamination of the surface had proceeded too far previously to the collision.

In order to illustrate the importance of the part played by the intervening gas, I thought that it would be interesting to compare the behaviour of the jets when situated in atmospheres of different gases. It seemed that gases more freely soluble in water than the atmospheric gases would be more easily got rid of in the later stages of the collision, and that thus union might more readily be brought about. This expectation has been confirmed in trials made on several different occasions. It was found sufficient to allow a pretty strong stream of the gas under examination to play upon the jets at and above the place of collision. Jets of air, of oxygen, and of coal-gas were found to be without effect. On the other hand, carbonic acid, nitrous oxide, sulphurous anhydride, and steam at once caused union. Only in the case of hydrogen was there an ambiguity. On some occasions the hydrogen appeared to be without effect, but on others (when perhaps the pressure of collision was higher) union uniformly followed. Care was taken to verify that air blown through the same tube as had supplied the hydrogen was inactive, so that the effect of the hydrogen could not be attributed to dust.

The action of hydrogen cannot be explained by its solubility. Hydrogen is, however, much less viscous than other gases, and to this we may plausibly attribute its activity in promoting union. A layer of hydrogen may be effectively squeezed out in a time that would be insufficient in the case of air and oxygen.

The Tension of Contaminated Water-Surfaces.

In my experiments upon the superficial viscosity of water (*Proc. Roy. Soc.* June 1890) [Vol. III. p. 375] I had occasion to notice that the last traces of residual contamination had very little influence upon the surface-tension, but that they became apparent when compressed in front of the vibrating needle of Plateau's apparatus. Subsequently I showed (*Phil. Mag.* Vol. XXXIII. p. 470, 1892) [Vol. III. p. 572] that according to Laplace's theory of Capillarity, in which matter is regarded as continuous, the effect of a thin surface-film in diminishing the tension of pure water should be as the *square* of the thickness of the film.

The tension of slightly contaminated surfaces was made the subject of special experiments by Miss Pockels (*Nature*, Vol. XLIII. p. 437, 1891), who concluded that a water-surface can " exist in two sharply contrasted conditions; the *normal* condition, in which the displacement of the partition [altering the density of the contamination] makes no impression upon the tension, and the *anomalous* condition, in which every increase or decrease alters the tension." It is only since I have myself made experiments upon the same lines that I have appreciated the full significance of Miss Pockels' statement. The conclusion that, judged by surface-tension, the effect of contamination comes on suddenly, seems to be of considerable importance, and I propose to illustrate it further by actual curves embodying results recently obtained.

The water is contained in a trough modelled after that of Miss Pockels. It is of tin-plate, 70 cm. long, 10 cm. broad, and 2 cm. deep, and it is filled nearly to the brim. The partitions, by which the oil is confined, are made of strips of glass resting upon the edge of the trough in such a manner that their lower surfaces are wetted while the upper surfaces remain dry. The strips may be $1\frac{1}{2}$ cm. wide, and for convenience of handling their length should exceed considerably the width of the trough. I have found advantage in cementing (with hard cement) slight webs of glass to the lower faces. The length of these is a rough fit with the width of the trough, enabling them to serve as guides preventing motion of the strips parallel to their length.

In order to observe the surface-tension Miss Pockels used a small disk (6 mm. in diameter) in contact with the surface, measuring the force necessary to detach it. In my own experiments I have employed the method of

Wilhelmy, which appears to be better adapted to the purpose. A thin
blade is mounted in a balance, its plane being vertical and its lower horizontal
edge dipping under the surface of the liquid. If absolute measures are
required, the edge of the blade should lie at the general level of the surface
when the pointer of the balance stands at zero. If m be the mass in the
other pan needed to compensate the effect of the liquid, l the length of
the blade, the surface-tension (T) may be deduced from the equation

$$2lT = mg. \quad \dots\dots\dots\dots\dots\dots\dots\dots\dots\dots\dots\dots(1)$$

When only differences of tension are concerned, the precise level of the strip
is of no consequence. As regards material, glass is to be preferred and it
should be thin in order not unduly to diminish the sensitiveness of the
balance by the displacement of water. I have used a small frame carrying
three parallel blades, the total length being 27 cm., while the thickness may
be considered nearly negligible. Before use the glass is cleaned with strong
sulphuric acid, and the angle of contact with the water when the balance is
raised appears to be zero. The total value of m for a clean surface may then
be calculated from (1), taking T at 74. We find $m = 4\cdot1$ gms. The balance
could be read without difficulty to $\cdot01$ gm., giving abundant accuracy.

The position of the barrier, giving the length of the surface to which the
grease is confined, is measured by a millimetre-scale, but is subject to a
correction needed in order to take account of the additional surface operative
when the suspended strip is raised. This amounts to about 3 cm., and is
to be added to the measured length. In a set of experiments where the
grease is successfully confined, the density is proportional to the reciprocal
of the above corrected length. It sometimes happens that continuity is
lost by the passage of grease across the barrier. This is of course most
likely to happen when the tensions on the two sides differ considerably,
and the danger may be mitigated by the use of a second barrier, so manipulated that the densities are nearly the same on the two sides of the principal
barrier.

In commencing a set of observations the first step is to secure the
cleanness of the surface. To this end the surface is *scraped*, if the expression
may be allowed, along the whole length by one of the movable partitions,
and, if thought necessary, the accumulated grease at the far end may be
removed with strips of paper. The operation should be repeated two or
three times with intermediate insertion of the balanced strips until it is
certain that no grease remains, competent to affect the tension even when
concentrated. The weights now necessary to bring the pointer to zero give
the standard with which the contaminated surfaces are to be compared.

If it be desired to begin with small contaminations, it is best to contract
the area, say to about one-half the maximum, and then to apply the grease

under examination with a previously ignited platinum wire until a small effect, such as ·02 gm., is observed at the balance. If the surface be now extended to the maximum, the attenuated grease will have lost its power, and the original reading for clean surfaces will be recovered. The barrier may now be advanced, readings being taken at intervals as the grease is concentrated. It is often more convenient to make the final adjustment by moving the barrier rather than by correcting the weights.

An example will make manifest at once the character of the results obtained. On May 15, the weight for the clean surface being 1·65 gm., the water was greased with castor-oil. With the barrier at 63 cm. this grease had no effect. The corrected length is 66, and the reciprocal of this, viz. 152, represents (for this series of observations) the density of the oil. With the barrier at 40, viz. at density 233, there was no change of the order of ·005 gm. At 36 cm., or density 256, the oil had just begun to show itself distinctly, the weight being then 1·64. At density 278 the weight became 1·62. From this point onwards increase of density tells rapidly. At 308 the weight was 1·55, and at 334 the weight was 1·40. A plot of these results is given in Fig. 3, and brings out more vividly than any description the striking character of the law discovered by Miss Pockels.

The effect of concentration beyond 571, giving ·70 gm., could not be examined in the same series. It was necessary to add more oil, and then of course the reciprocals of the corrected lengths represent the densities on a different scale from before. Corresponding to 63 cm., of which the reciprocal is 159, the weight was now 1·20 gm., falling to 1·00 at 175, ·80 at 204, ·70 at 233, ·60 at 351, ·55 at 488, and finally ·52 at 625. These values are plotted in Fig. 4, and they show that from a certain density onwards the tension falls very slowly. This curve may be continued

Figs. 3-6.

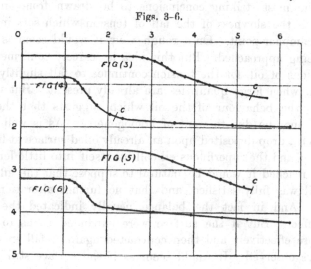

backwards by means of the results of Fig. 3, for of course the densities corresponding to any particular weight, *e.g.* 1·20 gm., are really the same in the two series.

It is of interest to inquire what point on these curves corresponds to the deadening of the movements of small particles of camphor deposited upon the surface. On a former occasion I have shown (*Phil. Mag.* Vol. XXXIII. p. 366, 1892) [Vol. III. p. 565] that whatever may be the character of the grease the cessation of the movements indicates that the tension falls short of a particular value. In the present method of experimenting there is no difficulty in determining what for brevity may be called the *camphor-point*. Two precautions should, however, be observed. It is desirable not to try the camphor until near the close of a set of experiments, and then to avoid too great a quantity. It would seem that the addition of camphor may sometimes lower the tension below the point due to the grease. The second precaution required is the raising of the balanced strip; otherwise when a weight is taken the density of the grease is altered. In several trials with castor and other oils the camphor-point was found to correspond with a drop of tension from that of clean water amounting to ·9 gm. The points thus fixed are marked in Figs. (3) and (4) with the letter *C*.

At this stage a certain discrepancy from former results should be remarked upon. Working by the method of ripples I had concluded that the camphor-point corresponded to a tension ·72 of that of pure water, *i.e.* to a drop of 28 per cent. But the ·9 gm. is only 22 per cent. of the calculated weight for pure water, *i.e.* 4·1 gms. At this rate the 72 per cent. would become 78 per cent., and the difference seems larger than can well be explained as an alteration of standard in judging when the fragments are nearly dead.

One of the most striking conclusions to be drawn from an inspection of the curves is the slowness of the fall of tension which sets in soon after passing the camphor-point. On a rough view it would seem as if a second limit were being approached. But this idea is scarcely confirmed by actual further additions of oil, for the tension continues to fall slightly after each addition, even when large quantities are already present. But there is one peculiarity in the behaviour of the oil which suggests that the failure to reach a limit may be due to want of homogeneity. As is well known, the disk into which a drop deposited upon an already oiled surface at first spreads, soon breaks up, and the superfluous oil collects itself into little lenses. After this stage is reached it would be natural to suppose that the affinity of the surface for oil was fully satisfied, and that no further alteration in tension could occur. And in fact the balance usually indicated the absence of immediate effect. But if the surface were expanded so as to spread the added oil more effectively and then contracted again, a fall in tension was almost always observed. It would seem as if the surface still retained an affinity for some minor ingredient capable of being extracted, though satiated as regards the principal ingredient.

The comparison of the present with former results throws an interesting light upon molecular magnitudes. It has been shown (*Proc. Roy. Soc.* March 1890) [Vol. III. p. 347] that the thickness of the film of olive-oil, calculated as if continuous, which corresponds to the camphor-point, is about $2 \cdot 0 \ \mu\mu$*; while from the present curves it follows that the point at which the tension *begins* to fall is about half as much, or $1 \cdot 0 \ \mu\mu$. Now this is only a moderate multiple of the supposed diameter of a gaseous molecule, and perhaps scarcely exceeds at all the diameter to be attributed to a molecule of oil. It is obvious therefore that the present phenomena lie entirely outside the scope of a theory such as Laplace's, in which matter is regarded as continuous, and that an explanation requires a direct consideration of molecules.

If we begin by supposing the number of molecules of oil upon a water surface to be small enough, not only will every molecule be able to approach the water as closely as it desires, but any repulsion between molecules will have exhausted itself. Under these conditions there is nothing to oppose the contraction of the surface—the tension is the same as that of pure water.

Castor Oil. May 15. Fig. (3)		Castor Oil. May 15. Fig. (4)		Olive Oil. May 3. Fig. (5)		Cod Liver Oil. May 11. Fig. (6)	
Density	Weight (in grams)	Density	Weight (in grams)	Density	Weight (in grams)	Density	Weight (in grams)
0	1·65	0 obs.	1·65	0	8·45	0	8·28
152	1·65	98 calc.	1·65	159	8·45	77	8·27
213	1·65	108 „	1·64	324	8·30	113	8·25
233	1·65	117 „	1·62	350	8·20	125	8·20
256	1·64	122 „	1·60	376	8·10	137	8·10
278	1·62	130 „	1·55	405	8·00	147	8·00
290	1·60	136 „	1·50	430	7·90	154	7·90
308	1·55	141 „	1·40	461	7·80	171	7·70
323	1·50	148 calc.	1·30	483	7·70	213	7·60
334	1·40	159 obs.	1·20	518	7·60	303	7·50
351	1·30	175 „	1·00			392	7·45
377	1·20	204 „	·80			465	7·40
408	1·10	233 „	·70			526	7·35
435	1·00	351 „	·60			625	7·30
472	·90	488 „	·55				
510	·80	625 obs.	·52				
571	·70						

* $\mu\mu$ = micromillimetre = 10^{-6} mm.

The next question for consideration is—at what point will an opposition to contraction arise? The answer must depend upon the forces supposed to be operative between the molecules of oil. If they behave like the smooth rigid spheres of gaseous theory, no forces will be called into play until they are closely packed. According to this view the tension would remain constant up to the point where a double layer commences to form. It would then suddenly change, to remain constant at the new value until the second layer is complete. The actual course of the curve of tension deviates somewhat widely from the above description, but perhaps not more than could be explained by heterogeneity of the oil, whereby some molecules would mount more easily than others, or by reference to the molecular motions which cannot be entirely ignored. If we accept this view as substantially true, we conclude that the first drop in tension corresponds to a complete layer one molecule thick, and that the diameter of a molecule of oil is about $1·0 \mu\mu$.

An attractive force between molecules extending to a distance of many diameters, such as is postulated in Laplace's theory, would not apparently interfere with the above reasoning. An essentially different result would seem to require a repulsive force between the molecules, resisting concentration long before the first layer is complete. In this case the tension would begin to fall as soon as the density is sufficient to bring the repulsion into play. On the whole this view appears less probable than the former, the more as it involves a molecular diameter much *exceeding* $1·0 \mu\mu$.

EXPLANATION OF FIGURES.

In the Figures (and in the tables) there is no relation between the scales of the abscissæ representing the densities in the various cases. As regards the ordinates, representing weights or tensions, the scale is the same in all the cases, but the zero point is arbitrary. It may be supposed to be situated on the line of zero densities at a point $4·1$ below the starting-point of the curve.

A Curious Observation.

[1902. The present paragraph was accidentally omitted in the original publication. In experimenting upon a shallow layer of mercury contained in a glass vessel with a flat bottom, it was noticed that a piece of iron gauze pressed under the mercury upon the bottom of the vessel unexpectedly remained down. There was no sticky substance present to which the effect could be referred, and on inspection from below it was seen that the mercury was out of contact with the bottom at places where the gauze was closest. The phenomenon was thus plainly of a capillary nature, the mercury refusing to fill up the narrowest chinks, even though the alternative was a vacuum. The experiment may be repeated in a simpler form by substituting for the gauze a piece of plate glass a few cms. square. If the bottom of the vessel be also of plate glass, the expulsion of the mercury may be observed from the whole of the contiguous areas.]

252.

THE MUTUAL INDUCTION OF COAXIAL HELICES.

[British Association Report, pp. 241, 242, 1899.]

PROFESSOR J. V. JONES* has shown that the coefficient of mutual induction (M) between a circle and a coaxial helix is the same as between the circle and a uniform circular cylindrical current-sheet of the same radial and axial dimensions as the helix, if the currents per unit length in helix and sheet be the same. This conclusion is arrived at by comparison of the integrals resulting from an application of Neumann's formula; and it may be of interest to show that it can be deduced directly from the general theory of lines of force.

In the first place, it may be well to remark that the circuit of the helix must be supposed to be completed, and that the result will depend upon the manner in which the completion is arranged. In the general case the return to the starting-point might be by a second helix lying upon the same cylinder; but for practical purposes it will suffice to treat of helices including an integral number of revolutions, so that the initial and final points lie upon the same generating line. The return will then naturally be effected along this straight line.

Let us now suppose that the helix, consisting of one revolution or of any number of complete revolutions, is situated in a field of magnetic force symmetrical with respect to the axis of the helix. In considering the number of lines of force included in the complete circuit, it is convenient to follow in imagination a radius-vector drawn perpendicularly to the axis from any point of the circuit. The number of lines cut by this radius, as the complete circuit is described, is the number required, and it is at once evident that the part of the circuit corresponding to the straight return contributes nothing to the total†. As regards any part of the helix corresponding to a rotation

---

* *Proc. Roy. Soc.* Vol. LXIII. (1897), p. 192.

† This would be true so long as the return lies anywhere in the meridianal plane. In the general case, where the number of convolutions is incomplete, the return may be made along a path composed of the extreme radii vectores and of the part of the axis intercepted between them.

of the radius through an angle $d\theta$, it is equally evident that in the limit the number of lines cut through is the same as in describing an equal angle of the circular section of the cylinder at the place in question, whence Professor Jones's result follows immediately. Every circular section is sampled, as it were, by the helix, and contributes proportionally to the result, since at every point the advance of the vector parallel to the axis is in strict proportion to the rotation. It is remarkable that the case of the helix (with straight return) is simpler than that of a system of true circles in parallel planes at intervals equal to the pitch of the helix.

The replacement of the helix by a uniform current-sheet shows that the force operative upon it in the direction of the axis (dM/dx) depends only upon the values of M appropriate to the two terminal circles.

If the field is itself due to a current flowing in a helix, the condition of symmetry about the axis is only approximately satisfied. The question whether both helices may be replaced by the corresponding current-sheets is to be answered in the negative, as may be seen from consideration of the case where there are two helices of the same pitch on cylinders of nearly equal diameters. In one relative position of the cylinders the paths are in close proximity throughout, and the value of M will be large; but this state of things may be greatly altered by a relative rotation through two right angles.

But although in strictness the helices cannot be replaced by current-sheets, the complication thence arising can be eliminated in experimental applications by a relative rotation. For instance, if the helix to which the field is supposed to be due be rotated, the *mean* field is strictly symmetrical, and accordingly the mean M is the same as if the other helix were replaced by a current-sheet. A further application of Professor Jones's theorem now proves that the first helix may also be so replaced. Under such conditions as would arise in practice, the mean of two positions distant 180°, or at any rate of four distant 90°, would suffice to eliminate any difference between the helices and the corresponding current-sheets, if indeed such difference were sensible at all.

The same process of averaging suffices to justify the neglect of spirality when the observation relates to the mutual attraction of two helices as employed in current determinations.

253.

THE LAW OF PARTITION OF KINETIC ENERGY.

[*Philosophical Magazine*, XLIX. pp. 98—118, 1900.]

THE law of equal partition, enunciated first by Waterston for the case of point molecules of varying mass, and the associated Boltzmann-Maxwell doctrine respecting steady distributions have been the subject of much difference of opinion. Indeed, it would hardly be too much to say that no two writers are fully agreed. The discussion has turned mainly upon Maxwell's paper of 1879[*], to which objections[†] have been taken by Lord Kelvin and Prof. Bryan, and in a minor degree by Prof. Boltzmann and myself. Lord Kelvin's objections are the most fundamental. He writes[‡]: "But, conceding Maxwell's fundamental assumption, I do not see in the mathematical workings of his paper any proof of his conclusion 'that the average kinetic energy corresponding to any one of the variables is the same for every one of the variables of the system.' Indeed, as a general proposition its meaning is not explained, and it seems to me inexplicable. The reduction of the kinetic energy to a sum of squares leaves the several parts of the whole with no correspondence to any defined or definable set of independent variables."

In a short note § written soon afterwards I pointed out some considerations which appeared to me to justify Maxwell's argument, and I suggested the substitution of Hamilton's *principal* function for the one employed by Maxwell‖. The views that I then expressed still commend themselves to

---

[*] *Collected Scientific Papers*, Vol. II. p. 713.

[†] I am speaking here of objections to the dynamical and statistical reasoning of the paper. Difficulties in the way of reconciling the results with a kinetic theory of matter are another question.

[‡] *Proc. Roy. Soc.* Vol. L. p. 85 (1891).

[§] *Phil. Mag.* April 1892, p. 356. [Vol. III. p. 554.]

[‖] See also Dr Watson's *Kinetic Theory of Gases*, 2nd edit. 1893.

me; and I think that it may be worth while to develop them a little further, and to illustrate Maxwell's argument by applying it to a particular case where the simplicity of the circumstances and the familiarity of the notation may help to fix our ideas.

But in the mean time it may be well to consider Lord Kelvin's "Decisive Test-case disproving the Maxwell-Boltzmann Doctrine regarding Distribution of Kinetic Energy*," which appeared shortly after the publication of my note. The following is the substance of the argument:—

"Let the system consist of three bodies, A, B, C, all movable only in one straight line, KHL:

"B being a simple vibrator controlled by a spring so stiff that when, at any time, it has very nearly the whole energy of the system, its extreme excursions on each side of its position of equilibrium are small:

"C and A, equal masses:

"C, unacted upon by force except when it strikes L, a fixed barrier, and when it strikes or is struck by B:

"A, unacted on by force except when it strikes or is struck by B, and when it is at less than a certain distance, HK, from a fixed repellent barrier, K, repelling with a force, F, varying according to any law, or constant, when A is between K and H, but becoming infinitely great when (if at any time) A reaches K, and goes infinitesimally beyond it.

"Suppose now A, B, C to be all moving to and fro. The collisions between B and the equal bodies A and C on its two sides must equalize, and keep equal, the average kinetic energy of A, immediately before and after these collisions, to the average kinetic energy of C. Hence, when the times of A being in the space between H and K are included in the average, the average of the *sum of the potential and kinetic energies of A* is equal to the average kinetic energy of C. But the potential energy of A at every point in the space HK is positive, because, according to our supposition, the velocity of A is diminished during every time of its motion from H towards K, and increased to the same value again during motion from K to H. Hence, the average kinetic energy of A is less than the average kinetic energy of C!"

The apparent disproof of the law of partition of energy in this simple problem seems to have shaken the faith even of such experts as Dr Watson and Mr Burbury†. M. Poincaré, however, considering a special case of Lord

---

* *Phil. Mag.* May 1892, p. 466.
† *Nature*, Vol. XLVI. p. 100 (1892).

Kelvin's problem*, arrives at a conclusion in harmony with Maxwell's law. Prof. Bryan† considers that the test-case " shows the impossibility of drawing general conclusions as to the distribution of energy in a *single* system from the possible law of permanent distribution in a large number of systems." It is indeed true that Maxwell's theorem relates in the first instance to a large number of systems; but, as I shall show more fully later, the extension to the time-average for a single system requires *only* the application of Maxwell's assumption that all *phases, i.e.* all states, defined both in respect to configuration and *velocity*, which are consistent with the energy condition lie on the same path, *i.e.* are attained by the system in its free motion sooner or later. This fundamental assumption, though certainly untrue in special cases, would appear to apply in Lord Kelvin's problem; and, if so, Maxwell's argument requires the equality of kinetic energies for A and C in the time-averages of a *single* system.

In view of this contradiction we may infer that there must be a weak place in one or other argument; and I think I can show that Lord Kelvin's conclusion above that the average of the sum of the potential and kinetic energies of A is equal to the average kinetic energy of C, is not generally true. In order to see this let us suppose the repulsive force F to be limited to a very thin stratum at H, so that A after penetrating this stratum is subject to no further force until it reaches the barrier K; and let us compare two cases, the whole energy being the same in both.

In case (i) F is so powerful that with whatever velocity (within the possible limits) A can approach, it is reflected at H, which then behaves like a fixed barrier. In case (ii) F is still powerful enough to produce this result, except when A approaches it with a kinetic energy nearly equal to the whole energy of the system. A then penetrates beyond H, moving slowly from H to K and back again from K to H, thus remaining for a relatively long time beyond H. Lord Kelvin's statement requires that the average total energy of A should be the same in the two cases; but this it cannot be. For during the occasional penetrations beyond H in case (ii) A has nearly the whole energy of the system; and its enjoyment of this is *prolonged* by the penetration. Hence in case (ii) A has a higher average total energy than in case (i); and a margin is provided which may allow the average *kinetic* energies to be equal. I believe that the consideration here advanced goes to the root of the matter, and shows why it is that the possession of potential energy may involve no deduction from the full share of kinetic energy.

Lord Kelvin's " decisive test-case " is entirely covered by Maxwell's reasoning—a reasoning in my view substantially correct. It would be

* *Revue générale des Sciences*, July 1894.
† "Report on Thermodynamics," Part II. § 26. *Brit. Ass. Rep.* 1894.

possible, therefore, to take this case as a typical example in illustration of the general argument; but I prefer for this purpose, as somewhat simpler, another test-case, also proposed by Lord Kelvin. This is simply that of a particle moving in two dimensions; and it may be symbolized by the motion of the ball upon a billiard-table. If there is to be potential energy, the table may be supposed to be out of level. The reconsideration of this problem may perhaps be thought superfluous, seeing that it has been ably treated already by Prof. Boltzmann*. But his method, though (I believe) quite satisfactory, is somewhat special. My object is rather to follow closely the steps of the general theory. If objections are taken to the argument of the particular case, they should be easy to specify. If, on the other hand, the argument of the particular case is admitted, the issue is much narrowed. I shall have occasion myself to make some comments relating to one point in the general theory not raised by the particular case.

In the general theory the coordinates† of the system at time t are denoted by $q_1, q_2, \ldots q_n$, and the momenta by $p_1, p_2, \ldots p_n$. At an earlier time t' the coordinates and momenta of the same motion are represented by corresponding letters accented, and the first step is the establishment of the theorem usually, if somewhat enigmatically, expressed

$$dq'_1 dq'_2 \ldots dq'_n dp'_1 dp'_2 \ldots dp'_n = dq_1 dq_2 \ldots dq_n dp_1 dp_2 \ldots dp_n. \quad \ldots\ldots(1)$$

In the present case q_1, q_2 are the ordinary Cartesian coordinates (x, y) of the particle; and if we identify the mass with unity, p_1, p_2 are simply the corresponding velocity-components (u, v); so that (1) becomes

$$dx' \, dy' \, du' \, dv' = dx \, dy \, du \, dv. \quad \ldots\ldots\ldots\ldots\ldots\ldots\ldots(2)$$

For the sake of completeness I will now establish (2) *de novo*.

In a possible motion the particle passes from the phase (x', y', u', v') at time t' to the phase (x, y, u, v) at time t. In the following discussion t' and t are absolutely fixed times, but the other quantities are regarded as susceptible of variation. These variations are of course not independent. The whole motion is determined if either the four accented, or the four unaccented, symbols be given. Either set may therefore be regarded as definite functions of the other set. Or again, the four coordinates x', y', x, y may be regarded as independent variables, of which u', v', u, v are then functions.

The relations which we require are readily obtained by means of Hamilton's principal function S, where

$$S = \int_{t'}^{t} (T - V) \, dt. \quad \ldots\ldots\ldots\ldots\ldots\ldots\ldots\ldots(3)$$

---

* *Phil. Mag.* Vol. xxxv. p. 156 (1893).

† Generalized coordinates appear to have been first applied to these problems by Boltzmann.

In this V denotes the potential energy in any position, and T is the kinetic energy, so that

$$T = \tfrac{1}{2}u^2 + \tfrac{1}{2}v^2 = \tfrac{1}{2}\dot{x}^2 + \tfrac{1}{2}\dot{y}^2. \quad\dots\dots\dots\dots\dots(4)^*$$

S may here be regarded as a function of the initial and final coordinates; and we proceed to form the expression for δS in terms of $\delta x'$, $\delta y'$, δx, δy. By (3)

$$\delta S = \int_{t'}^{t} (\delta T - \delta V)\, dt, \quad\dots\dots\dots\dots\dots(5)$$

and

$$\int \delta T\, dt = \int (\dot{x}\,\delta\dot{x} + \dot{y}\,\delta\dot{y})\, dt = \int \left(\dot{x}\frac{d\delta x}{dt} + \dot{y}\frac{d\delta y}{dt} \right) dt$$

$$= \left[\dot{x}\,\delta x + \dot{y}\,\delta y \right]_{t'}^{t} - \int (\ddot{x}\,\delta x + \ddot{y}\,\delta y)\, dt;$$

so that

$$\delta S = \left[\dot{x}\,\delta x + \dot{y}\,\delta y \right]_{t'}^{t} - \int_{t}^{t} (\ddot{x}\,\delta x \mid \ddot{y}\,\delta y + \delta V)\, dt.$$

By the general equation of dynamics the term under the integral sign vanishes throughout, and thus finally

$$\delta S = u\,\delta x + v\,\delta y - u'\,\delta x' - v'\,\delta y'. \quad\dots\dots\dots\dots(6)$$

In the general theory the corresponding equation is

$$\delta S - \Sigma p\,\delta q - \Sigma p'\,\delta q'. \quad\dots\dots\dots\dots\dots(7)$$

Equation (6) is equivalent to

$$\left. \begin{array}{ll} u' = -\,dS/dx', & u = dS/dx, \\ v' = -\,dS/dy', & v = dS/dy. \end{array} \right\} \dots\dots\dots\dots\dots(8)$$

It is important to appreciate clearly the meaning of these equations. S is in general a function of x, y, x', y'; and (*e.g.*) the second equation signifies that u is equal to the rate at which S varies with x, *when y, x', y' are kept constant*, and so in the other cases.

We have now to consider, not merely a single particle, but an immense number of similar particles, moving independently of one another under the same law (V), and distributed at time t over all possible phases (x, y, u, v). The most general expression for the law of distribution is

$$f(x,\, y,\, u,\, v)\, dx\, dy\, du\, dv, \quad\dots\dots\dots\dots\dots(9)$$

signifying that the number of particles to be found at time t within a prescribed range of phase is to be obtained by integrating (9) over the range

---

* As is not unusual in the integral calculus, we employ the same symbols x, &c. to denote the current and the final values of the variables. If desired, the final values may be temporarily distinguished as x'', &c.

in question. But such a distribution would in general be *unsteady*. If it obtained at time t, it would be departed from at time t', and *vice versâ*, owing to the natural motions of the particles. The question before us is to ascertain what distributions are steady, *i.e.* are maintained unaltered notwithstanding the motions.

It will be seen that it is the spontaneous passage of a particle from one phase to another that limits the generality of the function f. If there be no possibility of passage, say, from the phase (x', y', u', v') to the phase (x, y, u, v), or, as it may be expressed, if these phases do not lie upon the same *path*, then there is no relation imposed upon the corresponding values of f. An example, given by Prof. Bryan (*l. c.* § 17), well illustrates this point. Suppose that $V = 0$, so that every particle pursues a straight course with uniform velocity. The phases (x', y', u', v') and (x, y, u, v) can lie upon the same path only if $u' = u$, $v' = v$. Accordingly f remains arbitrary so far as regards u and v. For instance, a distribution

$$f(u, v)\, dx\, dy\, du\, dv \quad\quad\quad\quad\quad\quad\quad\quad\quad\quad (10)$$

is permanent whatever may be the form of f, understood to be independent of x and y. In this case the distribution is *uniform* in space, but uniformity is not indispensable. Suppose, for example, that *all* the particles move parallel to x, so that f vanishes unless $v = 0$. The general form (9) now reduces to

$$f(x, y, u)\, dx\, dy\, du\,; \quad\quad\quad\quad\quad\quad\quad\quad\quad\quad (11)$$

and permanency requires that the distribution be uniform along any line for which y is constant. Accordingly, f must be independent of x, so that permanent distributions are of the form

$$f(y, u)\, dx\, dy\, du, \quad\quad\quad\quad\quad\quad\quad\quad\quad\quad (12)$$

in which f is an arbitrary function of y and u. If either y or u be varied, we are dealing with a different path (in the sense here involved), and there is no connexion between the corresponding values of f. But if while y and u remain constant, x be varied, the value of f must remain unchanged, for the different values of x relate to the same path.

Before taking up the general question in two dimensions, it may be well to consider the relatively simple case of motion in one dimension, which, however, is not so simple but that it will introduce us to some of the points of difficulty. The particles are supposed to move independently upon one straight line, and the phase of any one of them is determined by the co-ordinate x and the velocity u. At time t' the phase of a particle will be denoted by (x', u'), and at time t the phase of the same particle will be (x, u), where u will in general differ from u', since we no longer suppose that V is constant, but rather that it is variable in a known manner, *i.e.* is a known function of x. The number of particles which at time t lie within the limits of phase represented by $dx\,du$ is $f(x,\,u)\,dx\,du$, and the question is whether this distribution is steady, and in particular whether it was the same at time t'. In order to find the distribution at time t', we regard x, u as known functions of x', u', and transform the multiple differential. The result of this transformation is best seen by comparison with intermediate transformations in which $dx\,du$ and $dx'\,du'$ are compared with $dx\,dx'$. We have

$$dx\,du = dx\,dx' \times \frac{du}{dx'}, \quad \dots\dots\dots\dots\dots\dots(13)$$

$$dx'\,du' = dx\,dx' \times \frac{du'}{dx}. \quad \dots\dots\dots\dots\dots\dots(14)$$

In du/dx' of (13) x is to be kept constant, and in du'/dx of (14) x' is to be kept constant. If we disregard algebraic sign, both are by (8) equal to $d^2S/dx\,dx'$, and are therefore equal to one another. Hence we may write

$$dx\,du = dx'\,du'\;; \quad \dots\dots\dots\dots\dots\dots(15)$$

and the transformation is expressed by

$$f(x,\,u)\,dx\,du = f_1(x',\,u')\,dx'\,du', \quad \dots\dots\dots\dots(16)$$

where $f_1(x',\,u')$ is the result of substituting for x, u in $f(x,\,u)$ their values in terms of x', u'. The right-hand member of (16) expresses the distribution at time t' corresponding to the distribution at time t expressed by the left-hand member, as determined by the laws of motion between the two phases. If the distribution is to be steady, $f_1(x',\,u')$ must be identical with $f(x',\,u')$; in other words $f(x,\,u)$ must be such a function of $(x,\,u)$ that it remains unchanged when $(x,\,u)$ refers to various phases of the motion of the same particle. Now, if E denote the total energy, so that

$$E = \tfrac{1}{2}u^2 + V, \quad \dots\dots\dots\dots\dots\dots(17)$$

then E remains constant during the motion; and thus, if for the moment we suppose f expressed in terms of E and x, we see that x cannot enter, or that f is a function of E only. The only permanent distributions accordingly are those included under the form

$$f(E)\,dx\,du, \quad \dots\dots\dots\dots\dots\dots(18)$$

where E is given by (17), and f is an arbitrary function.

It is especially to be noticed that the limitation to the form (18) holds only for phases lying upon the same path. If two phases have different energies, they do not lie upon the same path, but in this case the independence of the distributions in the two phases is already guaranteed by the form of (18). The question is whether all phases *of given energy* lie upon the same path. It is easy to invent cases for which the answer will be in the negative. Suppose, for example, that there are two centres of force O, O' on the line of motion which attract with a force at first proportional to distance but vanishing when the distance exceeds a certain value less than the interval OO'. A particle may then vibrate with the same (small) energy either round O or round O'; but the phases of the two motions do not lie upon the same path. Consequently f is not limited by the condition of steadiness to be the same in the two groups of phases. In all cases steadiness is *ensured* by the form (18); and if all phases of equal energy lie upon the same path, this form is *necessary* as well as sufficient.

All the essential difficulties of the theory appear to be raised by the particular case just discussed, and the reader to whom the subject is new is recommended to give it his careful attention.

In the more general problem of motion in two dimensions the discussion follows a parallel course. In order to find the distribution at time t' corresponding to (9) at time t, we have to transform the multiple differential, regarding x, y, u, v as known functions of x', y', u', v'. Here again we take the initial and final coordinates x, y, x', y' as an intermediate set of variables. Thus

$$dx'\,dy'\,du'\,dv' = dx'\,dy'\,dx\,dy \times \begin{vmatrix} \dfrac{du'}{dx}, & \dfrac{dv'}{dx} \\[2mm] \dfrac{du'}{dy}, & \dfrac{dv'}{dy} \end{vmatrix}, \quad \dots\dots\dots\dots(19)$$

$$dx\,dy\,du\,dv = dx\,dy\,dx'\,dy' \times \begin{vmatrix} \dfrac{du}{dx'}, & \dfrac{dv}{dx'} \\[2mm] \dfrac{du}{dy'}, & \dfrac{dv}{dy'} \end{vmatrix}. \quad \dots\dots\dots\dots(20)$$

In the determinants of (19), (20) the motion is regarded as a function of x, y, x', y', and the three quantities which do not appear in the denominator of any differential coefficient are to be considered constant. This was also the understanding in equations (8), from which we infer that the two determinants are equal, being each equivalent to

$$\begin{vmatrix} \dfrac{d^2S}{dx\,dx'}, & \dfrac{d^2S}{dx\,dy'} \\[2mm] \dfrac{d^2S}{dx'\,dy}, & \dfrac{d^2S}{dy\,dy'} \end{vmatrix}. \quad \dots\dots\dots\dots\dots(21)$$

Hence we may write

$$dx\,dy\,du\,dv = dx'\,dy'\,du'\,dv', \quad \dots\dots\dots\dots\dots(22)$$

an equation analogous to (15). By the same reasoning as was employed for motion in one dimension it follows that, if the distribution is to be steady, $f(x, y, u, v)$ in (9) must remain constant for all phases which lie upon the same path.   A distribution represented by

$$f(E)\,dx\,dy\,du\,dv, \dots\dots\dots\dots\dots\dots\dots(23)$$

where

$$E = \tfrac{1}{2}u^2 + \tfrac{1}{2}v^2 + V, \dots\dots\dots\dots\dots\dots(24)$$

will satisfy the conditions of steadiness whatever be the form of f; but this form is only *necessary* under the restriction known as Maxwell's assumption or postulate, viz. that all phases of equal energy lie upon the same path.

It is easy to give examples in which Maxwell's assumption is violated, and in which accordingly steady distributions are not limited to (23).   Thus, if no force act parallel to y, so that V reduces to a function of x only, the component velocity v remains constant for each particle, and no phases for which v differs lie upon the same path.   A distribution

$$f(E, v)\,dx\,dy\,du\,dv \dots\dots\dots\dots\dots\dots(25)$$

is then steady, whatever function f may be of E *and* v.

That under the distribution (23) the kinetic energy is equally divided between the component velocities u and v is evident from symmetry.   It is to be observed that the law of equal partition applies not merely upon the whole, but for every element of area $dx\,dy$, and for every value of the total energy, and at every moment of time.   When x and y are prescribed as well as E, the value of the resultant velocity itself is determined by (24).

Another feature worthy of attention is the spacial distribution; and it happens that this is peculiar in the present problem.   To investigate it we must integrate (23) with respect to u and v, x and y being constant. Since x and y are constant, V is constant; so that, if we suppose E to lie within narrow limits E and $E + dE$, the resultant velocity U will lie between limits given by

$$U\,dU = dE. \dots\dots\dots\dots\dots\dots(26)$$

If we transform from u, v to U, θ, where

$$u = U\cos\theta, \qquad v = U\sin\theta, \dots\dots\dots\dots\dots(27)$$

$du\,dv$ becomes $U\,dU\,d\theta$; so that on integration with respect to θ we have, with use of (26),

$$2\pi F(E)\,dE . dx\,dy. \dots\dots\dots\dots\dots\dots(28)$$

The spacial distribution is therefore *uniform*.

In order to show the special character of the last result, it may be well to refer briefly to the corresponding problem in three dimensions, where the coordinates of a particle are x, y, z and the component velocities are u, v, w. The steady distribution corresponding to (23) is

$$f(E)\, dx\, dy\, dz\, du\, dv\, dw, \quad \ldots\ldots\ldots\ldots\ldots\ldots(29)$$

in which

$$E = \tfrac{1}{2} U^2 + V = \tfrac{1}{2} u^2 + \tfrac{1}{2} v^2 + \tfrac{1}{2} w^2 + V. \quad \ldots\ldots\ldots\ldots(30)$$

Here equation (26) still holds good, and the transformation of $du\, dv\, dw$ is, as is well known, $4\pi U^2 dU$. Accordingly (29) becomes

$$4\pi F(E)\, dE\, .\, (2E - 2V)^{\frac{1}{2}}\, dx\, dy, \quad \ldots\ldots\ldots\ldots(31)$$

no longer uniform in space, since V is a function of x, y.

In (31) the density of distribution decreases as V increases. For the corresponding problem in *one* dimension (18) gives

$$F(E)\, dE\, .\, (2E - 2V)^{-\frac{1}{2}}\, dx, \quad \ldots\ldots\ldots\ldots(32)$$

so that in this case the density *increases* with increasing V.

The uniform distribution of the two-dimensional problem is thus peculiar. Although an immediate consequence of Maxwell's equation (41), see (41) below, I failed to remark it in the note before referred to, where I wrote as if a uniform distribution in the billiard-table example required that $V = 0$. In order to guard against a misunderstanding it may be well to say that the uniform distribution does not necessarily extend over the whole plane. Wherever $(E - V)$ falls below zero there is of course no distribution.

We have thus investigated for a particle in two dimensions the law of steady distribution, and the equal partition of energy which is its necessary consequence. And we see that "the only assumption necessary to the direct proof is that the system, if left to itself in its actual state of motion, will, sooner or later, pass through every phase which is consistent with the equation of energy" (Maxwell). It will be observed that so far nothing whatever has been said as to time-averages for a single particle. The law of equal partition, as hitherto stated, relates to a large number of particles and to a single moment of time.

The extension to time-averages, the aspect under which Lord Kelvin has always considered the problem, is important, the more that some authors appear to doubt the possibility of such extension. Thus Prof. Bryan (*Report*, § 11, 1894), speaking of Maxwell's assumption, writes :—"To discover, if possible, a general class of dynamical systems satisfying the assumption would form an interesting subject for future investigation. It is, however, doubtful how far Maxwell's law would be applicable to the *time-averages* of the energies in any such system. We shall see, in what follows, that the law of permanent distribution of a very large number of systems is in many cases not unique. Where there is more than one possible distribution it

would be difficult to draw any inference with regard to the average distribution (taken with respect to the time) for one system."

The extension to time-averages appears to me to require nothing more
than Maxwell's assumption, without which the law of distribution itself
is only an artificial arrangement, sufficient indeed but not necessary for
steadiness.   We shall still speak of the particle moving in two dimensions,
though the argument is general.   It has been shown that at any moment
the u-energy and the v-energy of the group of particles is the same ; and
it is evident that the equality subsists if we integrate over any period of
time.   But if this period be sufficiently prolonged, and if Maxwell's *assumption
be applicable*, it makes no difference whether we contemplate the whole group
of particles or limit ourselves to a single member of it.   It follows that
for a single particle the time-averages of u^2 and v^2 are equal, provided the
averages be taken over a sufficient length of time.

On the other hand, if in any case Maxwell's assumption be untrue, not
only is the special distribution unnecessary for steadiness, but even if it
be artificially arranged, the law of equal time-averages does not follow as
a consequence.

Having now considered the special problem at full—I hope it may not
be thought at undue—length, I pass on to some remarks on the general
investigation.   This proceeds upon precisely parallel lines, and the additional
difficulties are merely those entailed by the use of generalized coordinates.
Thus (1) follows from (7) by substantially the same process (given in my
former note) that (22) follows from (6).   Again, if E denote the total energy
of a system, the distribution

$$f(E)\, dq_1 \ldots dq_n dp_1 \ldots dp_n, \ldots\ldots\ldots\ldots\ldots\ldots\ldots\ldots(33)$$

where f is an arbitrary function, satisfies the condition of permanency ; and,
if Maxwell's assumption be applicable, it is the *only* form of distribution that
can be permanent.

As I hinted before, some of the difficulties that have been felt upon this
subject may be met by a fuller recognition of the invariantic character of
the expressions.   This point has been ably developed by Prof. Bryan, who
has given (*loc. cit.* § 14) a formal verification that (33) is unaltered by a change
of coordinates.   If we follow attentively the process by which (1) is established,
we see that in (3) there is no assumption that the system of coordinates is
the same at times t' and t, and that accordingly we are not tied to one system
in (33).   Indeed, so far as I can see, there would be no meaning in the
assertion that the system of generalized coordinates employed for two different
configurations was the same *.

---

* It would be like saying that two points lie upon the same curve, when the character of the
curve is not defined.

We come now to the deduction from (33) of Maxwell's law of partition of energy. On this Prof. Bryan (*loc. cit.* § 20) remarks:—"Objections have been raised to this step in Maxwell's work by myself ('Report on Thermodynamics,' Part I. § 44) on the ground that the kinetic energy cannot in general be expressed as the sum of squares of *generalized momenta* corresponding to generalized coordinates of the system, and by Lord Kelvin (*Nature*, Aug. 13, 1891) on the ground that the conclusion to which it leads has no intelligible meaning. Boltzmann (*Phil. Mag.* March 1893) has put the investigation into a slightly modified form which meets the first objection, and which imposes a certain restriction upon the generality of the result. Under this limitation the result is perfectly intelligible, and the second objection is therefore also met." At this point I find myself in disagreement with all the above quoted authorities, and in the position of maintaining the correctness of Maxwell's original deduction.

Prof. Boltzmann considers that " Maxwell committed an error in assuming that by choosing suitable coordinates the expression for the *vis viva* could always be made to contain only the squares of the momenta." This is precisely the objection which I supposed myself to have already answered in 1892. I wrote, " It seems to be overlooked that Maxwell is limiting his attention to systems *in a given configuration*, and that no dynamics is founded upon the reduced expression for T. The reduction can be effected in an infinite number of ways. We may imagine the configuration in question rendered one of stable equilibrium by the introduction of suitable forces proportional to displacements. The principal modes of isochronous vibration thus resulting will serve the required purpose."

It is possible, therefore, so to choose the coordinates that for a given configuration (and for configurations differing infinitely little therefrom) the kinetic energy T, which is always a quadratic function of the velocities, shall reduce to a sum of squares with, if we please, given coefficients. Thus *in the given configuration*

$$T = \tfrac{1}{2}\dot{q}_1{}^2 + \tfrac{1}{2}\dot{q}_2{}^2 + \ldots + \tfrac{1}{2}\dot{q}_n{}^2; \quad \ldots\ldots\ldots\ldots\ldots(34)$$

and, since in general $p = dT/d\dot{q}$,

$$p_1 = \dot{q}_1, \qquad p_2 = \dot{q}_2, \text{ &c.,}$$

so that

$$T = \tfrac{1}{2}p_1{}^2 + \tfrac{1}{2}p_2{}^2 + \ldots + \tfrac{1}{2}p_n{}^2. \quad \ldots\ldots\ldots\ldots\ldots(35)$$

Whether the coordinates required to effect a similar reduction for other configurations are the same is a question with which we are not concerned.

The mean value of $p_r{}^2$ for all the systems in the given configuration is, according to (33),

$$\frac{\int p_r{}^2 . F\{V + \tfrac{1}{2}p_1{}^2 + \ldots + \tfrac{1}{2}p_n{}^2\} \, dp_1 \ldots dp_n}{\int F\{V + \tfrac{1}{2}p_1{}^2 + \ldots + \tfrac{1}{2}p_n{}^2\} \, dp_1 \ldots dp_n} \quad \ldots\ldots\ldots\ldots(36)*$$

* *Confer* Bryan, *loc. cit.*

The limits for each variable may be supposed to be $\pm \infty$; but the large values do not really enter if we suppose $F(E)$ to be finite for moderate, perhaps for nearly definite, values of E only.

It is now evident that the mean value is the same for all the momenta p; and accordingly that for each the mean value of $\frac{1}{2}p^2$ is $1/n$ of the mean value of T. This result holds good for every moment of time, for every configuration, for every value of E, and for every system of resolution (of which there are an infinite number) which allows T to be expressed in the form (35).

In the case where the "system" consists of a single particle, (35) is justified by any system of rectangular coordinates; and although we are not bound to use the same system for different positions of the particle, it would conduce to simplicity to do so. If the system be a rigid body, we may measure the velocities of the centre of inertia parallel to three fixed rectangular axes, while the remaining momenta refer to rotations about the principal axes of the body. If Maxwell's assumption hold good, a permanent distribution is such that in one, or in any number of positions, the mean energy of each rotation and of each translation is the same. And under the same restriction a similar assertion may be made respecting the time-averages for a single rigid body.

There is much difficulty in judging of the applicability of Maxwell's assumption. As Maxwell himself showed, it is easy to find cases of exception; but in most of these the conditions strike one as rather special. It must be observed, however, that if we take it quite literally, the assumption is of a severely restrictive character; for it asserts that the system, starting from any phase, will traverse *every* other phase (consistent with the energy condition) *before* returning to the initial phase. As soon as the initial phase is recovered, a cycle is established, and no new phases can be reached, however long the motion may continue.

We return now to the question of the distribution of momenta among the systems which occupy a given configuration, still supposing the coordinates so chosen as to reduce T to a sum of squares (35). It will be convenient to fix our attention upon systems for which E lies within narrow limits, E and $E+dE$. Since E is given, there is a relation between $p_1, p_2, \ldots p_n$, and we may suppose p_n expressed in terms of E and the remaining momenta. By (35)

$$p_n dp_n = dT = dE,$$

since the configuration is given, and thus (33) becomes

$$f(E)\, dE \cdot dq_1 \ldots dq_n \cdot p_n^{-1}\, dp_1 \ldots dp_{n-1}. \quad \ldots\ldots\ldots\ldots(37)$$

For the present purpose the latter factors alone concern us, so that what we have to consider is

$$\frac{dp_1\,dp_2\ldots dp_{n-1}}{\surd\,\{2T-p_1{}^2-p_2{}^2-\ldots-p^2{}_{n-1}\}}, \qquad\ldots\ldots\ldots\ldots(38)$$

in which T, being equal to $E-V$, is given. For the moment we may suppose that $2T$ is unity.

The whole number of systems is to be found by integrating (38), the integral being so taken as to give the variables all values consistent with the condition that $p_1{}^2+p_2{}^2+\ldots+p^2{}_{n-1}$ is not greater than unity. Now

$$\int\ldots\int\frac{dp_1\,dp_2\ldots dp_{n-1}}{\surd\,\{1-p_1{}^2-\ldots-p^2{}_{n-1}\}}=\frac{\pi^{\frac{1}{2}n-\frac{1}{2}}}{\Gamma\left(\frac{1}{2}n-\frac{1}{2}\right)}\int_{-1}^{+1}(1-p_1{}^2)^{\frac{1}{2}n-\frac{3}{2}}\,dp_1,\ldots\ldots(39)$$

and

$$\int_{-1}^{+1}(1-p_1{}^2)^{\frac{1}{2}n-\frac{3}{2}}\,dp_1=\frac{\Gamma\left(\frac{1}{2}\right)\Gamma\left(\frac{1}{2}n-\frac{1}{2}\right)}{\Gamma\left(\frac{1}{2}n\right)},\ldots\ldots\ldots\ldots(40)$$

in which $\Gamma\left(\frac{1}{2}\right)=\surd\pi$. Thus the whole number of systems is

$$\frac{\{\Gamma\left(\frac{1}{2}\right)\}^n}{\Gamma\left(\frac{1}{2}n\right)},$$

or on restoration of $2T$, equal to $2E-2V$,

$$\frac{\{\Gamma\left(\frac{1}{2}\right)\}^n}{\Gamma\left(\frac{1}{2}n\right)}\{2E-2V\}^{\frac{1}{2}n-1}.\qquad\ldots\ldots\ldots\ldots\ldots(41)$$

To this we shall return later; but for the present what we require to ascertain is the distribution of one of the momenta, say p_1, irrespectively of the values of the remaining momenta. By (39), (40) the number of systems for which p_1 lies between p_1 and p_1+dp_1 in comparison with the whole number of systems is

$$\frac{\Gamma\left(\frac{1}{2}n\right)}{\Gamma\left(\frac{1}{2}\right)\Gamma\left(\frac{1}{2}n-\frac{1}{2}\right)}\left\{1-\frac{p_1{}^2}{2T}\right\}^{\frac{1}{2}n-\frac{3}{2}}\frac{dp_1}{\surd\,(2T)}.\qquad\ldots\ldots\ldots\ldots(42)$$

This is substantially Maxwell's investigation, and (42) corresponds with his equation (51). As was to be expected, the law of distribution is the same for all the momenta. From the manner of its formation, we note that the integral of (42), taken between the limits $p_1=\pm\surd\,(2T)$, is equal to unity.

Maxwell next proceeds to the consideration of the special form assumed by (42), when the number n of degrees of freedom is extremely great*. This part of the work seems to be very important; but it has been much neglected, probably because the result was not correctly stated.

---

* The particular cases where $n=2$, or $n=3$, are also worthy of notice.

Dropping the suffix as unnecessary, we have to consider the form of

$$\left\{1 - \frac{p^2}{2T}\right\}^{\frac{1}{2}n - \frac{3}{2}}$$

when n is very great, the mean value of p^2 becoming at the same time small in comparison with $2T$. If we write

$$T = nK = \tfrac{1}{2}n P^2, \quad\dots\dots\dots\dots\dots\dots(43)$$

we have

$$\text{Limit } \left\{1 - \frac{p^2}{2T}\right\}^{\frac{1}{2}n - \frac{3}{2}} = e^{-p^2/4K} = e^{-p^2/2P^2}. \quad\dots\dots\dots\dots(44)$$

The limit of the fraction containing the Γ functions may be obtained by the formula

$$\Gamma(m + 1) = e^{-m} m^m \sqrt{(2m\pi)};$$

and the limiting form of (42) becomes

$$\frac{e^{-p^2/4K}}{\sqrt{(2\pi)}} \frac{dp}{\sqrt{(2K)}}, \quad \text{or} \quad \frac{e^{-p^2/2P^2}}{\sqrt{(2\pi)}} \frac{dp}{P}. \quad\dots\dots\dots\dots(45)$$

It may be observed that the integral of (45) between the limits $\pm \infty$ is unity, and that this fact might have been used to determine the numerical factor.

Maxwell's result is given in terms of a quantity k, analogous to K, and defined by

$$\tfrac{1}{2}p^2 = k. \quad\dots\dots\dots\dots\dots\dots(46)$$

It is

$$\frac{1}{\sqrt{(2\pi)}} \frac{1}{K} e^{-\frac{k}{2K}} dk. \quad\dots\dots\dots\dots\dots(47)$$

The corresponding form from (45) is

$$\frac{1}{\sqrt{(2\pi)}} \frac{1}{2\sqrt{(kK)}} e^{-\frac{k}{2K}} dk. \quad\dots\dots\dots\dots(48)$$

In like manner if we inquire what proportion of the whole number of systems have momenta lying within the limits denoted by $dp_1 dp_2 \dots dp_r$, where r is a number very small relatively to n, we get

$$\frac{e^{-(p_1{}^2 + p_2{}^2 + \dots + p_r{}^2)/4K}}{\{\sqrt{(2\pi)}\}^r} \frac{dp_1 dp_2 \dots dp_r}{\{\sqrt{(2K)}\}^r}, \quad\dots\dots\dots\dots(49)$$

or, if we prefer it,

$$\frac{e^{-(p_1{}^2 + p_2{}^2 + \dots + p_r{}^2)/2P^2}}{\{\sqrt{(2\pi)}\}^r} \frac{dp_1 dp_2 \dots dp_r}{P^r}. \quad\dots\dots\dots\dots(50)$$

These results follow from the general expression (38), in the same way as does (45), by stopping the multiple integration at an earlier stage. The remaining variables range over values which may be considered in each case to be unlimited. If the integration between $\pm \infty$ be carried out completely, we recover the value unity.

The interest of the case where n is very great lies of course in the application to a gas supposed to consist of an immense number of similar molecules*, or of several sets of similar molecules; and the question arises whether (45) can be applied to deduce the Maxwellian law of distribution of velocities among the molecules of a single system at a given instant of time. A caution may usefully be interposed here as to the sense in which the Maxwellian distribution is to be understood. It would be absurd to attempt to prove that the distribution in a single system is necessarily such and such, for we have already assumed that every phase, including every distribution of velocities, is attainable, and indeed attained if sufficient time be allowed. The most that can be proved is that the distribution will *approximate* to a particular law for the greater part of the time, and that if sensible deviations occur they will be *transitory*.

In applying (45) to a gas it will be convenient to suppose in the first instance that all the molecules are similar. Each molecule has several degrees of freedom, but we may fix our attention upon one of them, say the x-velocity of the centre of inertia, usually denoted by u. In (45) the *whole* system is supposed to occupy a given configuration; and the expression gives us the distribution of velocity at a given time for a single molecule among all the systems. The distribution of velocity is the same for every other molecule, and thus the expression applies to the statistics of all the molecules of all the systems. Does it also apply to the statistics of all the molecules of a single system? In order to make this inference we must assume that the statistics are the same (at the same time) for all the systems, or, what comes to the same thing (if Maxwell's assumption be allowed), that they are the same for the same system at the various times when it passes through a given configuration.

Thus far the argument relates only to a single configuration. If the configuration be changed, there will be in general a change of potential energy and a corresponding change in the kinetic energy to be distributed amongst the degrees of freedom. But in the case of a gas, of which the statistics are assumed to be regular, the potential energy remains approximately constant when exclusion is made of exceptional conditions. The same law of distribution of velocity then applies to every configuration, that is, it may be asserted without reference to the question of configuration. We thus arrive at the Maxwellian law of velocities in a single gas, as well as the

---

* The terms "gas" and "molecule" are introduced for the sake of brevity. The question is still purely dynamical.

relation between the velocities in a mixture of molecules of different kinds
first laid down by Waterston.

The assumptions which we have made as to the practical regularity of
statistics are those upon which the usual theory of ideal gases is founded;
but the results are far more general. Nothing whatever has been said as
to the character of the forces with which the molecules act upon one another,
or are acted upon by external agencies. Although for distinctness a gas has
been spoken of, the results apply equally to a medium constituted as a liquid
or a solid is supposed to be. A kinetic theory of matter, as usually understood, appears to require that in equilibrium the whole kinetic energy shall
be equally shared among all the degrees of freedom, and within each degree
of freedom be distributed according to the same law. It is included in this
statement that temperature is a matter of kinetic energy only, *e.g.* that when
a vertical column of gas is in equilibrium, the mean velocity of a molecule
is the same at the top as at the bottom of the column.

Reverting to (37), (41), in order to consider the distribution of the
systems as dependent upon the coordinates independently of the velocities,
we have, omitting unnecessary factors,

$$\{E - V\}^{\frac{1}{2}n-1} \, dq_1 \, dq_2 \dots dq_n. \qquad \dots\dots\dots\dots\dots\dots(51)$$

If $n = 2$, *e.g.* in the case already considered of a single particle moving in two
dimensions, or of two particles moving in one dimension, or again whatever
n may be, provided V vanish, the first factor disappears, so that the distribution is *uniform* with respect to the coordinates $q_1 \dots q_n$. If $n > 2$ and V
be finite, the distribution is such as to favour those configurations for which
V is least.

"When the number of variables is very great, and when the potential
energy of the specified configuration is very small compared with the total
energy of the system, we may obtain a useful approximation to the value of
$\{E - V\}^{\frac{1}{2}n-1}$ in an exponential form; for if we write (as before) $E = nK$,

$$\{E - V\}^{\frac{1}{2}n-1} = E^{\frac{1}{2}n-1} e^{-V/2K} \qquad \dots\dots\dots\dots\dots\dots(52)$$

nearly, provided n is very great and V is small compared with E. The
expression is no longer approximate when V is nearly as great as E, and it
does not vanish, as it ought to do, when $V = E$." (Maxwell.)

In the case of gas composed of molecules whose mutual influence is
limited to a small distance and which are not subject to external forces,
the distribution expressed by (51) is uniform in space except near the
boundary. For if q_1 denote the x-coordinate of a particular molecule, and
if we effect the integration with respect to all the coordinates of other
molecules as well as the other coordinates of the particular molecule, we
must arrive at a result independent of x, provided x relate to a point well

in the interior. That is to say in the various systems contemplated the particular molecule is uniformly distributed with respect to x. The same is true of y and z, and thus the whole spacial distribution is uniform. If the single system constituting the gas has uniform statistics, it will follow that the distribution in it of molecules similar to the particular molecule is uniform.

The uniformity of the distribution is disturbed if an external force acts. In illustration of this we may consider the case of gravity. From (52) the distribution with respect to the coordinates of the particular molecule will be

$$e^{-gz/2K}\, dx\, dy\, dz,$$

and the same formula gives the density of molecules similar to the particular molecule in a single system.

The main purpose of this paper is now accomplished; but I will take the opportunity to make a few remarks upon some general aspects of a kinetic theory of matter. Many writers appear to commit themselves to absolute statements, but Kelvin* and Boltzmann and Maxwell fully recognize that conclusions can never be more than *probable*. The second law of thermodynamics itself is in this predicament. Indeed it might seem at first sight as if the case were even worse than this. Mr Culverwell has emphasized a difficulty, which must have been pretty generally felt, arising out of the reversibility of a dynamical system. If during one motion of a system energy is dissipated, restoration must occur when the motion is reversed. How then is one process more probable than the other? Prof. Boltzmann has replied to this objection, upon the whole I think satisfactorily, in a very interesting letter†. The available (internal) energy of a system tends to zero, or rather to a small value, only because the conditions, or phases as we have called them, corresponding to small values are more probable, *i.e.* more numerous. If there is considerable available energy at any moment, it is because the condition is then exceptional and peculiar. After a short interval of time the condition *may* become more peculiar still, and the available energy *may* increase, but this is improbable. The probability is that the available energy will, if not at once, at any rate after a short interval, decrease owing to the substitution of a more nearly normal state of things.

There is, however, another side to this question, which perhaps has been too much neglected. Small values of the available energy are indeed more

---

* Witness the following remarkable passage:—"It is a strange but nevertheless a true conception of the old well-known law of the conduction of heat to say that it is very improbable that in the course of 1000 years one-half the bar of iron shall of itself become warmer by a degree than the other half; and that the probability of this happening before 1,000,000 years pass is 1000 times as great as that it will happen in the course of 1000 years, and that it certainly will happen in the course of some very long time."—(*Nature*, Vol. IX. p. 443, 1874.)

† *Nature*, Vol. LI. p. 413 (1895).

probable than large ones, but there is a degree of smallness below which it is *improbable* that the value will lie. If at any time the value lies extremely low, it is an increase and not a decrease which is probable. Maxwell showed long ago how a being capable of dealing with individual molecules would be in a position to circumvent the second law. It is important to notice that for this end it is not necessary to deal with individual molecules. It would suffice to take advantage of local reversals of the second law, which will involve, not very rarely, a *considerable number* of neighbouring molecules. Similar considerations apply to other departures from a normal state of things, such, for example, as unequal mixing of two kinds of molecules, or such a departure from the Waterston relation (of equal mean kinetic energies) as has been investigated by Maxwell and by Tait and Burbury.

The difficulties connected with the application of the law of equal partition of energy to actual gases have long been felt. In the case of argon and helium and mercury vapour the ratio of specific heats (1·67) limits the degrees of freedom of each molecule to the three required for translatory motion. The value (1·4) applicable to the principal diatomic gases gives room for the three kinds of translation and for two kinds of rotation. Nothing is left for rotation round the line joining the atoms, nor for relative motion of the atoms in this line. Even if we regard the atoms as mere points, whose rotation means nothing, there must still exist energy of the last-mentioned kind, and its amount (according to the law) should not be inferior.

We are here brought face to face with a fundamental difficulty, relating not to the theory of gases merely, but rather to general dynamics. In most questions of dynamics a condition whose violation involves a large amount of potential energy may be treated as a *constraint*. It is on this principle that solids are regarded as rigid, strings as inextensible, and so on. And it is upon the recognition of such constraints that Lagrange's method is founded. But the law of equal partition disregards potential energy. However great may be the energy required to alter the distance of the two atoms in a diatomic molecule, practical rigidity is never secured, and the kinetic energy of the relative motion in the line of junction is the same as if the tie were of the feeblest. The two atoms, however related, remain two atoms, and the degrees of freedom remain six in number.

What would appear to be wanted is some escape from the destructive simplicity of the general conclusion relating to partition of kinetic energy, whereby the energy of motions involving larger amounts of potential energy should be allowed to be diminished in consequence. If the argument, as above set forth after Maxwell, be valid, such escape must involve a repudiation of Maxwell's fundamental postulate as practically applicable to systems with an immense number of degrees of freedom.

254.

ON THE VISCOSITY OF ARGON AS AFFECTED BY TEMPERATURE.

[*Proceedings of the Royal Society*, LXVI. pp. 68—74, 1900.]

ACCORDING to the kinetic theory, as developed by Maxwell, the viscosity of a gas is independent of its density, whatever may be the character of the encounters taking place between the molecules. In the typical case of a gas subject to a uniform shearing motion, we may suppose that of the three component velocities v and w vanish, while u is a linear function of y, independent of x and z. If μ be the viscosity, the force transmitted tangentially across unit of area perpendicular to y is measured by $\mu\, du/dy$. This represents the relative momentum, parallel to x, which in unit of time crosses the area in one direction, the area being supposed to move with the velocity of the fluid at the place in question. We may suppose, for the sake of simplicity, and without real loss of generality, that u is zero at the plane. The momentum, which may now be reckoned absolutely, does not vanish, as in the case of a gas at rest throughout, because the molecules come from a greater or less distance, where (*e.g.*) the value of u is positive. The distance from which (upon the average) the molecules may be supposed to have come depends upon circumstances. If, for example, the molecules, retaining their number and velocity, interfere less with each other's motion, the distance in question will be increased. The same effect will be produced, without a change of quality, by a simple reduction in the number of molecules, *i.e.*, in the density of the gas, and it is not difficult to recognize that the distance from which the molecules may be supposed to have come is *inversely as the density*. On this account the passage of tangential momentum *per molecule* is inversely as the density, and since the number of molecules crossing is directly as the density, the two effects compensate, and upon the whole the tangential force and therefore the viscosity remain unaltered by a change of density.

On the other hand, the manner in which this viscosity varies with temperature depends upon the nature of the encounters. If the molecules behaved like Boscovich points, which exercise no force upon one another until the distance falls to a certain value, and which then repel one another infinitely (erroneously called the theory of elastic spheres), then, as Maxwell proved, the viscosity would be proportional to the square root of the absolute temperature. Or again, if the law of repulsion were as the inverse fifth power of the distance, viscosity would be as the absolute temperature.

In the more general case where the repulsive force varies as r^{-n}, the dependence of μ upon temperature may also be given. If v be the velocity of mean square, proportional to the square root of the temperature, μ varies as $v^{\frac{n+3}{n-1}}$, a formula which includes the cases ($n = 5$, $n = \infty$) already specified. If we assume the law already discussed—that μ is independent of density—this conclusion may be arrived at very simply by the method of "dimensions."

In order to see this we note that the only quantities (besides the density) on which μ can depend are m the mass of a particle, v the velocity of mean square, and k the repulsive force at unit distance. The dimensions of these quantities are as follows :—

$$\mu = (\text{mass})^1 (\text{length})^{-1} (\text{time})^{-1},$$

$$m = (\text{mass})^1,$$

$$v = (\text{length})^1 (\text{time})^{-1},$$

$$k = (\text{mass})^1 (\text{length})^{n+1} (\text{time})^{-2}.$$

Thus, if we assume

$$\mu \propto m^x . v^y . k^z \quad \text{.............................(1)}$$

we have $\qquad 1 = x + z, \qquad -1 = y + (n+1) z, \qquad -1 = -y - 2z,$

whence $\qquad x = \dfrac{n+1}{n-1}, \qquad y = \dfrac{n+3}{n-1}, \qquad z = \dfrac{2}{n-1}.$

Accordingly $\qquad \mu = \alpha . m^{\frac{n+1}{n-1}} . v^{\frac{n+3}{n-1}} . k^{\frac{2}{n-1}}, \quad \text{.....................(2)}$

where α is a purely numerical coefficient. For a given kind of molecule, m and k are constant. Thus

$$\mu \propto v^{\frac{n+3}{n-1}} \propto \theta^{\frac{n+3}{2n-2}} . \quad \text{.............................(3)}$$

The case of sudden impacts ($n = \infty$) gives, as already remarked, $\mu \propto v \propto \theta^{\frac{1}{2}}$. Hence k disappears, and the consideration of dimensions shows that $\mu \propto d^{-2}$, where d is the diameter of the particles.

The best experiments on air show that, so far as a formula of this kind can represent the facts, $\mu \propto \theta^{0\cdot77}$. It may be observed that $n = 8$ corresponds to $\mu \propto \theta^{0\cdot79}$.

When we remember that the principal gases, such as oxygen, hydrogen, and nitrogen, are regarded as diatomic, we may be inclined to attribute the want of simplicity in the law connecting viscosity and temperature to the complication introduced by the want of symmetry in the molecules and consequent diversities of presentation in an encounter. It was with this idea that I thought it would be interesting to examine the influence of temperature upon the viscosity of argon, which in the matter of specific heat behaves as if composed of single atoms. From the fact that no appreciable part of the total energy is rotatory, we may infer that the forces called into play during one encounter are of a symmetrical character. It seemed, therefore, more likely that a simple relation between viscosity and temperature would obtain in the case of argon than in the case of the " diatomic " gases.

The best experimental arrangement for examining this question is probably that of Holman*, in which the same constant stream of gas passes in succession through two capillaries at different temperatures, the pressures being determined before the first and after the second passage, as well as between the two. But to a gas like argon, available in small quantities only, the application of this method is difficult. And it seemed unnecessary to insist upon the use of constant pressures, seeing that it was not proposed to investigate experimentally the dependence of transpiration upon pressure.

The theoretical formula for the volume of gas transpired, analogous to that first given by Stokes for an incompressible fluid, was developed by O. E. Meyer†. Although not quite rigorous, it probably suffices for the purpose in hand. If p_1, V_1 denote the pressure and volume of the gas as it enters the capillary, p_2, V_2 as it leaves the capillary, we have

$$p_1 V_1 = p_2 V_2 = \frac{\pi t R^4}{16 \mu l}(p_1{}^2 - p_2{}^2). \quad \ldots\ldots\ldots\ldots\ldots\ldots(4)$$

In this equation t denotes the time of transpiration, R the radius of the tube, l its length, and μ the viscosity measured in the usual way.

In order to understand the application of the formula for our present purpose, it will be simplest to consider first the passage of equal volumes of different gases through the capillary, the initial pressures, and the constant temperature being the same. In an apparatus, such as that about to be described, the pressures change as the gas flows, but if the pressures are definite functions of the amount of gas which at any moment has passed the

* Phil. Mag. Vol. III. p. 81 (1877).
† Pogg. Ann. Vol. CXXVII. p. 269 (1866).

capillary, this variation does not interfere with the proportionality between t and μ. For example, if the viscosity be doubled, the flow takes place precisely as before, except that the scale of time is doubled. It will take twice as long as before to pass the same quantity of gas.

Although different gases have been employed in the present experiments, there has been no attempt to compare their viscosities, and indeed such a comparison would be difficult to carry out by this method. The question has been, how is the viscosity of a given gas affected by a change of temperature? In one set of experiments the capillary is at the temperature of the room; in a closely following set the capillary is bathed in saturated steam at a temperature that can be calculated from the height of the barometer.

If the temperature were changed throughout the whole apparatus from one absolute temperature θ to another absolute temperature θ', we could make immediate application of (4); the viscosities (μ, μ') at the two temperatures would be directly as the times of transpiration (t, t'). The matter is not quite so simple when, as in these experiments, the change of temperature takes place only in the capillary. A rise of temperature in the capillary now acts in two ways. Not only does it change the viscosity, but it increases the *volume* of gas which has to pass. The ratio of volumes is θ', θ; and thus

$$\frac{\mu'}{\mu} = \frac{t'}{t} \times \frac{\theta}{\theta'}, \quad \dots\dots\dots\dots\dots\dots\dots\dots\dots\dots\dots(5)$$

subject to a small correction for the effect of temperature upon the dimensions of the capillary. It is assumed that the temperature of the reservoirs is the same in both transpirations.

The apparatus is shown in the figure. The gas flows to and fro between the bulbs A and B, the flow from A to B only being timed. It is confined by mercury, which can pass through U connexions of blown glass from A to C and from B to D. The bulbs B, C, D are supported upon their seats with a little plaster of Paris. The capillary is nearly 5 feet (150 cm.) in length and is connected with the bulbs by gas tubing of moderate diameter, all joints being blown. E represents the jacket through which steam can be passed; its length exceeds that of the capillary by a few inches.

In order to charge the apparatus, the first step is the exhaustion. This is effected through the tap, F, with the aid of a Töpler pump, and it is necessary to make a corresponding exhaustion in C and D, or the mercury would be drawn over. To this end the rubber terminal H is temporarily connected with G, while I leads to a common air-pump. When the exhaustion is complete, the gas to be tried is admitted gradually at F, the atmosphere being allowed again to exert its pressure in C and D. When the charge is sufficient, F is turned off, after which G remains open to the atmosphere, and H is connected to a manometer.

When a measurement is commenced, the first step is to read the temperatures of the bulbs and of the capillary; I is then connected to a force pump, and pressure is applied until so much of the gas is driven over that the mercury below A and in B assumes the positions shown in the diagram. I is then suddenly released so that the atmospheric pressure asserts itself in D, and the gas begins to flow back into B. The bulb J allows the flow a short time in which to establish itself before the time measurement begins as the mercury passes the connexion passage K. When the mercury reaches L, the time measurement is closed.

One of the points to be kept in view in designing the apparatus is to secure long enough time of transpiration without unduly lowering the driving pressure. At the beginning of the measured transpiration the pressure in A was about 30 cm. of mercury above atmosphere, and that in B about

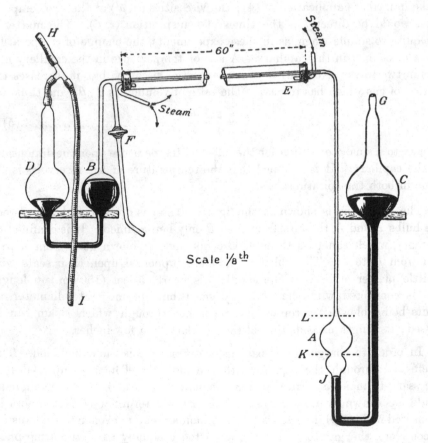

Scale ⅛ th

2 cm. below atmosphere. At the end the pressure in A was 20 cm., and in B 3 cm., both above atmosphere. Accordingly the driving pressure fell from 32 to 17 cm.

Three, or, in the case of hydrogen, five, observations of the time were usually taken, and the agreement was such as to indicate that the mean would be correct to perhaps one-tenth of a second. The time for air at the temperature of the room was about ninety seconds, and for hydrogen forty-four seconds, but these numbers are not strictly comparable.

When the low temperature observations were finished, the gas was lighted under a small boiler placed upon a shelf above the apparatus, and steam was passed through the jacket. It was necessary to see that there was enough heat to maintain a steady issue of steam, yet not so much as to risk a sensible back pressure in the jacket. The time of transpiration for air was now about 139 seconds. Care was always taken to maintain the temperature of the bulbs at the same point as in the first observations.

There are one or two matters as to which an apparatus on these lines is necessarily somewhat imperfect. In the high temperature measurements the whole of the gas in the capillary is assumed to be at the temperature of boiling water, and all that is not in the capillary to be at the temperature of the room, assumptions not strictly compatible. The compromise adopted was to enclose in the jacket the whole of the capillary and about 2 inches at each end of the approaches, and seems sufficient to exclude sensible error when we remember the rapidity with which heat is conducted in small spaces. A second weak point is the assumption that the instantaneous pressures are represented by the heights of the moving mercury columns. If the connecting U-tubes are too narrow, the resistance to the flow of mercury enters into the question in much the same way as the flow of gas in the capillary. In order to obtain a check upon this source of error the apparatus has been varied. In an earlier form the connecting U-tubes were comparatively narrow; but the result for the ratio of viscosities of hot and cold air was substantially the same as that subsequently obtained with the improved apparatus, in which these tubes were much widened. Even if there be a sensible residual error arising from this cause, it can hardly affect the comparison of temperature-coefficients of gases whose viscosity is nearly the same.

I will now give an example in detail from the observations of December 21 with purified argon. The times of transpiration at the temperature of the room (15° C.) were in seconds

$$104\tfrac{3}{4}, \qquad 104\tfrac{1}{2}, \qquad 104\tfrac{3}{4}. \qquad \text{Mean, } 104 \cdot 67.$$

When the capillaries were bathed in steam, the corresponding times were

$$167\tfrac{1}{2}, \qquad 167\tfrac{1}{2}, \qquad 167\tfrac{3}{4}. \qquad \text{Mean, } 167 \cdot 58.$$

The barometer reading (corrected) being 767·4 mm., we deduce as the temperature of the jacket 100·27° C. Thus $\theta = 287 \cdot 5$, $\theta' = 372 \cdot 8$. The reduction was effected by assuming

$$\frac{t'}{t} = \left(\frac{\theta'}{\theta}\right)^{x} . \quad \dots\dots\dots\dots\dots\dots\dots\dots(6)$$

With the above values we get

$$x = 1\cdot812.$$

As appears from (5), the integral part of x relates merely to the expansion of the gas by temperature.   If we take

$$\frac{\mu'}{\mu} = \left(\frac{\theta'}{\theta}\right)^n, \quad \dots\dots\dots\dots\dots\dots\dots\dots\dots\dots\dots(7)$$

we get $n = 0\cdot812.$

This number is, however, subject to a small correction for the expansion of the glass of the capillary.   As appears from (4), the ratio μ', μ as used above requires to be altered in the same ratio as that in which the glass expands by volume.   The value of n must accordingly be increased by $0\cdot010$, making

$$n = 0\cdot822.$$

The following table embodies the results obtained in a somewhat extended series of observations.   The numbers given are the values of n in (7), corrected for the expansion of the glass.

Air (dry)	0·754
Oxygen	0·782
Hydrogen	0·681
Argon (impure)	0·801
Argon (best)	0·815

In the last trials, the argon was probably within 1 or 2 per cent. of absolute purity.   The nitrogen lines could no longer be seen, and scarcely any further contraction could be effected on sparking with oxygen or hydrogen.

It will be seen that the temperature change of viscosity in argon does not differ very greatly from the corresponding change in air and oxygen. At any rate the simpler conditions under which we may suppose the collisions to occur, do not lead to values of n such as $0\cdot5$, or $1\cdot0$, discussed by theoretical writers.

I may recall that, on a former occasion*, I found the viscosity of argon to be $1\cdot21$ relatively to that of air, both being observed at the temperature of the room.

[1902.   See further, Vol. IV. p. 481.]

* *Roy. Soc. Proc.* January, 1896.   [Vol. IV. p. 222.]

255.

ON THE PASSAGE OF ARGON THROUGH THIN FILMS OF INDIARUBBER.

[*Philosophical Magazine*, XLIX. pp. 220, 221, 1900.]

SOON after the discovery of Argon it was thought desirable to compare the percolation of the gas through indiarubber with that of nitrogen, and Sir W. Roberts-Austen kindly gave me some advice upon the subject. The proposal was simply to allow atmospheric air to percolate through the rubber film into a vacuum, after the manner of Graham, and then to determine the proportion of argon. It will be remembered that Graham found that the percentage of *oxygen* was raised in this manner from the 21 of the atmosphere to about 40. At the time the experiment fell through, but during the last year I have carried it out with the assistance of Mr Gordon.

The rubber balloon was first charged with dry boxwood sawdust. This rather troublesome operation was facilitated by so mounting the balloon that with the aid of an air-pump the external pressure could be reduced. When sufficiently distended the balloon was connected with a large Töpler pump, into the vacuous head of which the diffused gases could collect. At intervals they were drawn off in the usual way.

The diffusion was not conducted under ideal conditions. In order to make the most of the time, the apparatus was left at work during the night, so that by the morning the internal pressure had risen to perhaps three inches of mercury. The proportion of oxygen in the gas collected was determined from time to time. It varied from 34 per cent. when the vacuum was bad to about 39 per cent. when the vacuum was good. On an average it was estimated that the proportion of oxygen would be about 37 per cent. of the whole. The total quantity of diffused gas reckoned at atmospheric pressure was about 300 c.c. per twenty-four hours.

On removal from the pump the gas was introduced into an inverted flask standing over alkali, and with addition of oxygen as required was treated with the electrical discharge from a transformer in connexion with the public supply of alternating current. In this way the nitrogen was gradually oxidized and absorbed. Towards the close of operations the gas was transferred to a smaller vessel, where it was further sparked until no further contraction occurred, and the lines of nitrogen had disappeared from the spectrum. The excess of oxygen was then removed by phosphorus.

It remains only to record the final figures. The residue, free of oxygen and nitrogen, from 3205 c.c. of diffused gas was 39 c.c. The most instructive way of stating the result is perhaps to reckon the argon as a percentage, not of the whole, but of the nitrogen and argon only. Of the 3205 c.c. total, 2020 c.c. would be nitrogen and argon, and of this the 39 c.c. argon would be 1·93 per cent. Since, according to Kellas (*Proc. Roy. Soc.* Vol. LIX. p. 67, 1895), 100 c.c. of mixed atmospheric nitrogen and argon contains 1·19 per cent. of argon, we see that in the diffused gas the proportion of argon is about half as great again as in the atmosphere. Argon then passes the indiarubber film more readily than nitrogen, but not in such a degree as to render the diffusion process a useful one for the concentration of argon from the atmosphere.

256.

ON THE WEIGHT OF HYDROGEN DESICCATED BY LIQUID AIR.

[*Proceedings of the Royal Society,* LXVI. p. 344, 1900.]

IN recent experiments by myself and by others upon the density of hydrogen, the gas has always been dried by means of phosphoric anhydride; and a doubt may remain whether on the one hand the removal of aqueous vapour is sufficiently complete, and on the other whether some new impurity may not be introduced. I thought that it would be interesting to weigh hydrogen dried in an entirely different manner, and this I have recently been able to effect with the aid of liquid air, acting as a cooling agent, supplied by the kindness of Professor Dewar from the Royal Institution. The operations of filling and weighing were carried out in the country as hitherto. I ought, perhaps, to explain that the object was not so much to make a new determination of the highest possible accuracy, as to test whether any serious error could be involved in the use of phosphoric anhydride, such as might explain the departure of the ratio of densities of oxygen and hydrogen from that of 16 : 1. I may say at once that the result was negative.

Each supply consisted of about 6 litres of the liquid, contained in two large vacuum-jacketed vessels of Professor Dewar's design, and it sufficed for two fillings with hydrogen at an interval of two days. The intermediate day was devoted to a weighing of the globe *empty*. There were four fillings in all, but one proved to be abortive owing to a discrepancy in the weights when the globe was empty, before and after the filling. The gas was exposed to the action of the liquid air during its passage in a slow stream of about half a litre per hour through a tube of thin glass.

I have said that the result was negative. In point of fact the actual weights found were $\frac{1}{10}$ to $\frac{2}{10}$ milligrams *heavier* than in the case of hydrogen dried by phosphoric anhydride. But I doubt whether the small excess is of any significance. It seems improbable that it could have been due to residual vapour, and it is perhaps not outside the error of experiment, considering that the apparatus was not in the best condition.

257.

THE MECHANICAL PRINCIPLES OF FLIGHT.

[*Manchester Memoirs*, XLIV. pp. 1—26, 1900.]

THE subject under discussion includes both natural and artificial flight. Although we are familiar with the flight of birds, there are many interesting questions which arise in connexion with natural flight, and some of them are yet very obscure.

In still air a bird, being heavier than the fluid displaced, cannot maintain his level for more than a short time without working his wings. In this matter the vicarious principle holds good. If the bird is not to fall, something must fall instead of him, and this can only be air. The maintenance of the bird thus implies the perpetual formation of a downward current of air, and involves therefore performance of work. Later we shall consider more particularly how this work is applied; but a preliminary difficulty remains to be discussed. It is well known that large birds, such as vultures and pelicans, are often observed to maintain their level for considerable periods of time, without flapping or visibly working their wings. On a smaller scale, and in more special situations, sea-gulls in these latitudes perform similar feats. This question of the soaring or sailing flight of birds has given rise to much difference of opinion. Few of the naturalists, to whom we owe the observations, are familiar with mechanical principles, and thus statements are often put forward which amount to mechanical impossibilities. The arm-chair theorist at home, on the other hand, may be too willing to discredit reports of actual observations, especially when they are made in other parts of the world. On both sides it seems to be admitted that there is no sailing flight in the absence of wind; but observers, untrained in dynamics and misled by the analogy of the kite, are apt to suppose that the existence of wind at once removes the difficulty. The doctrine of relative motion shows however that, so long as there is no connexion with the ground, a uniform horizontal wind is for this purpose the same thing as absolutely still air.

In a short paper upon this subject (*Nature*, XXVII. p. 534, 1883 [Vol. II. p. 194]) I pointed out that, " Whenever a bird pursues his course for some time without working his wings, we must conclude either (1) that the course is not horizontal, (2) that the wind is not horizontal, or (3) that the wind is not uniform. It is probable that the truth is usually represented by (1) or (2), but the question I wish to raise is whether the cause suggested by (3) may not sometimes come into operation." Case (1) is that of a rook gliding downwards from a tree in still air with motionless wings. We shall presently consider upon what conditions depend the time and distance of travel possible with a given descent. Case (2) is closely related to case (1). If the air have an upward velocity equal to that at which the rook falls through it in a vertical direction, the vertical motion is compensated, and the course of the rook relatively to the ground becomes horizontal. It is not necessary, of course, that the *whole* motion of the air be upwards; a horizontal motion of the air is simply superposed. A bird gliding into a wind having a small upward component may thus maintain relatively to the ground an absolutely fixed position, or he may advance over the ground to windward at a fixed level.

There can be no doubt that the vertical component of wind plays a large part, not merely in the flight of birds, but in general atmospheric phenomena. Living at the bottom of the atmospheric ocean, where the wind is necessarily parallel to the ground, we are liable to overlook the importance of vertical motions. This is the more remarkable when we consider that wind is due to atmospheric expansion and condensation, so that the primary movements are vertical and not horizontal. Thus the inhabitants of an oceanic island are specially interested in the so-called land and sea breezes, but the primary phenomenon is the rise and fall of air over the island as it is heated by the sun during the day and cooled by radiation at night.

A recent American observer (Huffaker, *Smithsonian Report* for 1897) has recorded many examples of vultures soaring under circumstances which suggested that they take advantage of the upward currents which rise locally from the ground when it is strongly heated by the sun. On dull days and in light winds the vultures were not seen to soar. There is no doubt that under the influence of a strong sun the layers of air near the ground approach an unstable condition, and that comparatively slight causes may determine local upward currents. Mr Huffaker suggests that in some cases the birds themselves, by flying round, may determine the upward current. Some of his observations certainly point in this direction; but it must be remembered that the immediate effect of flight will be a downward and not an upward current.

The more obvious examples of upward motion occur when an otherwise horizontal wind meets an obstruction. Some years ago I visited the north side of Madeira, where cliffs, nearly 2,000 feet high, rise perpendicularly from the sea. Being on the top of the cliff, we had difficulty in finding a sheltered spot until we noticed that *close to the edge* there was almost complete calm. Lying upon the ground and moving only one's arms, it was possible to hold a handkerchief by the corner so that a little behind the plane of the cliff it hung downwards as in still air, and a little in front of the cliff was carried upwards in the vertically rising stream. A ball of crumpled paper thrown outwards was carried up high over our heads. Of course gulls and other birds found no difficulty in rising up the face of the cliff without working their wings. During a recent visit to India, I frequently watched the effect of similar upward currents deflected by rocky fortresses which rise from the plains. Kites could be seen to maintain themselves for minutes together without a single flap of the wings. When this occurred, the birds were sailing to and fro over the *windward* side of the rock.

We now turn to the consideration of case (3).

"In a uniform wind the available energy at the disposal of the bird depends upon his velocity *relatively* to the air about him. With only a moderate waste this energy can at any moment be applied to gain elevation, the gain of elevation being proportional to the loss of relative velocity squared. It will be convenient for the moment to ignore the waste referred to, and to suppose that the whole energy available remains constant, so that however the bird may ascend or descend, the relative velocity is that due to a fall from a certain level to the actual position, the certain level being of course that to which the bird might just rise by the complete sacrifice of relative velocity."

In illustration of case (3) I instanced a wind blowing everywhere horizontally but with a velocity increasing upwards, taking for the sake of simplicity the imaginary case of a wind uniform above and below a certain plane where the velocity changes. Since a uniform motion has no effect, we may suppose without further loss of generality, that the velocities of the wind above and below the plane are $+u$ and $-u$. Let us consider how a bird, sailing somewhat above the plane of separation and endowed with an initial relative velocity v, might take advantage of the position in which he finds himself.

The first step is, if necessary, to turn round until the relative motion is down wind (in the upper stratum) and then to drop through the plane of separation. In falling down to the level of the plane there is a gain of relative velocity, but this of no significance for the present purpose, as it is purchased by the loss of elevation; but in passing through the plane there is a really effective gain. In entering the lower stratum the actual velocity is indeed unaltered, but the velocity relatively to the surrounding air (moving in the opposite direction) is *increased*.

If h denote the height above the plane of separation to which the initial relative velocity v is due, we have $v^2 = 2gh$. Here v is the velocity, relatively to the air in the upper stratum, with which the bird crosses the plane. After crossing, the velocity, now reckoned relatively to the air in the lower stratum, becomes $v + 2u$, and the new value of h is given by

$$2gh' = (v + 2u)^2,$$

so that $$2g(h' - h) = 4uv + 4u^2 = 4u(u + v).$$

Here $(h' - h)$ is the gain of potential elevation and, if u is given, it increases as v increases.

At this stage the bird is moving against the direction of the wind in the lower stratum. He next turns round—it is supposed without loss of relative velocity—until his direction is reversed so as to be with the wind of the lower stratum and contrary to the wind of the upper stratum. A passage upwards through the plane now secures another gain of relative velocity, or of potential elevation, of nearly the same value as before. The process may be repeated. At every passage through the plane (whether in the upwards or in the downwards direction) there is a gain of potential elevation, and if this gain outweighs the losses all the while in progress, the bird may maintain or improve his position without doing a stroke of work.

It may be of interest to consider a numerical example.

Suppose that

$$v = 30 \text{ miles per hour} = 1\cdot34 \times 10^3 \text{ cm. per second,}$$

and that $$h' - h = 10 \text{ feet} = 305 \text{ cm.;}$$

then in C.G.S. measure

$$(v + 2u)^2 = v^2 + 2g(h' - h) = 1\cdot80 \times 10^6 + \cdot60 \times 10^6 = 2\cdot40 \times 10^6,$$

and $$v + 2u = 1\cdot55 \times 10^3;$$

so that $$2u = \cdot21 \times 10^3 \text{ cm./sec.} = 4\cdot7 \text{ miles/hour.}$$

In this case a freshening of the wind amounting to 4·7 miles per hour is equivalent to a gain of 10 feet of potential elevation.

In order to take advantage of the gradual increase of wind with elevation usually to be met with, a bird may describe circles in an inclined plane, always descending when moving to leeward and ascending when moving to windward. Whether the differences of velocity available at considerable elevations in the atmosphere are sufficient to allow a bird to maintain his position without working his wings appears to be doubtful. Near the level of the ground or sea these differences are greater, and probably suffice to explain much of the sailing flight of albatrosses and other sea-birds.

Another way in which a bird may draw upon the internal energy of the wind has been specially discussed by Dr Langley (*Smithsonian Contributions*, 1893), who calls attention to the fact that the well-known *gustiness* of the wind, at any rate near the earth's surface, is underestimated in the usual meteorological records. The differences of horizontal velocity involved in what are commonly called gusts of wind imply in general vertical motions also, but near the ground these latter may, perhaps, be left out of account. The advantage which a bird may take of the variations in the speed of the wind is explicable upon the principles already applied, the *inertia* of the bird playing in some sort the part of the string of a kite.

If u denote the speed of the wind at any moment, and v the speed of the bird in the opposite direction, both *e.g.*, reckoned relatively to the ground, the available energy is measured by $\frac{1}{2}(v+u)^2$. Suppose now that the wind freshens, u becoming $u+du$, while v remains constant. The *increment* of available energy is

$$\tfrac{1}{2}(v+u+du)^2 - \tfrac{1}{2}(v+u)^2 = (v+u)\,du;$$

or in time t,
$$\int_0^t (v+u)\,du. \quad\dots\dots\dots\dots\dots(1)$$

The speed of the wind being supposed to be periodic, and the integration being taken over a sufficiently long period of time, we have

$$\int_0^t u\,du = 0;$$

and thus the mechanical advantage may be reckoned as

$$\int_0^t v\,du. \quad\dots\dots\dots\dots\dots(2)$$

In order that this may have a finite value, v must vary; the principle being that to get the most advantage v must be great when du is positive, that is when the wind is freshening, and smaller when the wind is failing: The higher velocity required to meet the freshening wind is to be obtained by a previous fall to a lower level.

As an example, let us suppose that u and v are periodic, so that

$$u = u_0 + u_1 \sin pt, \qquad v = v_0 + v_1 \cos(pt+e);$$

then
$$\int v\,du = p u_1 v_1 \int \cos pt \cdot \cos(pt+e)\,dt,$$

and, when t is great,
$$\int v\,du = \tfrac{1}{2} p t u_1 v_1 \cos e. \quad\dots\dots\dots\dots(3)$$

The mechanical advantage obtained in time t is greatest when e vanishes, *i.e.*, when du and v are in the same phase. This mechanical advantage is to be set against the frictional and other losses neglected in our original supposition. Were there no such losses, the value of v, or of the elevation, might continually increase.

This example shows that it is quite possible for a bird moving in a very natural manner against a strong and variable wind to maintain himself and to advance over the ground without working his wings. Observations of this kind are recorded by Mr Huffaker. It will be understood, of course, that a bird, not being interested in simplifying the calculation, will take any advantage that offers itself of the internal energy of the wind and of upward currents in order to attain his objects.

In the preceding discussions we have assumed, for the sake of simplicity, that a bird or a flying machine is able to *glide* in still air without loss of energy. It is needless to say that the truth of such an assumption can, at best, be only approximate. Apart from frictional losses, the maintenance of a given level implies the continual formation of a downward aerial current, and consequent expenditure of energy. We have next to consider the magnitude of these losses, taking the case of a plane moving at a uniform speed. And, in the first instance, we shall neglect the frictional forces, assuming that the reaction of the air upon the plane is truly normal.

Before we can advance a step in the desired direction we must know how the normal pressure upon an aeroplane is related to the size and shape of the plane, to the velocity of the motion, and above all to the angle between the plane and the direction of motion. According to an erroneous theory, to some extent sanctioned by Newton, the mean pressure would depend only upon the area of the plane and the *resolved part* of the velocity in a direction perpendicular to the plane. If V be the velocity, α the angle between V and the plane, ρ the density of the air (or other fluid concerned), the pressure p would be given by

$$p = \tfrac{1}{2}\rho V^2 \sin^2 \alpha. \quad\dots\dots\dots(4)$$

That this formula is quite erroneous, especially when α is small, has long been known*. At small angles the pressure is more nearly proportional to $\sin \alpha$ than to $\sin^2 \alpha$ and, as was strongly emphasized by Wenham in an early and important paper on aerial locomotion†, the question of shape and presentation is by no means indifferent. In the case of an elongated shape moving with given velocity V, and at a given small inclination α, the pressure is much greater when the long dimension of the plane is perpendicular than when it is (nearly) parallel to V.

* A further discussion will be found in *Phil. Mag.* Vol. ii. p. 430, 1876; *Scientific Papers*, Vol. i. p. 287; and in *Nature*, Vol. xlv. p. 108, 1892. [Vol. iii. p. 491.]
† *Report of Aeronautical Society*, 1866, p. 10.

According to a theoretical formula developed on the basis of Kirchhoff's analysis (*Phil. Mag. loc. cit.*) we should have for the mean pressure, instead of (4),

$$p = \frac{\pi \sin \alpha}{4 + \pi \sin \alpha} \rho V^2. \quad \dots\dots\dots(5)$$

This applies strictly to motion in two dimensions, or practically to the case of a very elongated blade, whose length is perpendicular to V.

At perpendicular incidence ($\alpha = 90°$) the difference between (4) and (5) is not important; but when α is small, the value of p in (5) may be enormously greater than the corresponding value from (4).

As regards numerical values, if we use C.G.S. measure, so that V is measured in centimetres per second, we have in the case of air under standard conditions $\rho = ·00128$, and p, at perpendicular incidence, measured in dynes per square centimetre, is according to (4),

$$p = ·00064 V^2. \quad \dots\dots\dots\dots(6)$$

This does not differ greatly from the data given in engineering tables. To compare with Langley's more recent experiments, we may express V in metres per second and p in grams weight per square centimetre. Thus

$$p' = ·0065 V'^2; \quad \dots\dots\dots\dots(7)$$

while the mean of Langley's numbers gives

$$p' = ·0087 V'^2, \quad \dots\dots\dots\dots(8)$$

about 30 per cent. greater. The difference is accounted for, at any rate partly, by the *suction* which experiment shows to exist at the back of the plate.

As regards the law of obliquity, the early experiments of Vince (1798) sufficed to show that the effect was more nearly as $\sin \alpha$ than as $\sin^2 \alpha$. In recent times this subject has been very thoroughly investigated by Langley, who has examined not only the influence of obliquity, but also of the shape and presentation of the plane. His results for the case to which (5) relates indicate an even greater relative effect at small angles, probably referable to the back suction. A laboratory experiment to demonstrate the reality of this suction was described in one of the papers already referred to (*Nature, loc. cit.*).

Experiments upon the law of obliquity, as executed for the case of air, by Dines* and Langley†, involve cumbrous and costly whirling machines, and if made in the open are greatly embarrassed by wind. An apparatus capable

* *Proc. Roy. Soc.* June, 1890.
† *Smithsonian Contributions to Knowledge*, 1891.

of working in the laboratory, or as a lecture illustration, has long been a desideratum. With the aid of Mr Gordon I have recently constructed one which, while very simple and inexpensive, performs sufficiently well. It may be regarded as a kind of adjustable windmill. An axis of hard steel, finely pointed at the ends, is carried by agate cups. From a central boss six spokes of round steel project symmetrically, carrying at their ends six similar vanes of tin-plate. The vanes are provided with projecting sockets of brass tubing, which fit the spokes somewhat tightly, but yet allow the vanes to be rotated when desired. The vanes are 4 inches long and $1\frac{1}{2}$ inches wide, the distance of their inner ends from the axis being about 3·7 inches. The whole apparatus is as light as may be (about 120 gm.) consistently with the necessary rigidity.

If the vanes are all inclined at the same angle, the apparatus works like an ordinary windmill, and may be set into rapid rotation by a motion through the air parallel to the axis. This motion may take place either in a horizontal or in a vertical direction. If means were provided for estimating the couple needed to prevent rotation, we should obtain the efficiency of the vanes at the given obliquity and speed. Observations at the same speed and at other obliquities would then give the means of determining the law of obliquity.

Such a procedure would be analogous to that adopted in former experiments with whirling machines. The essential feature of the present method consists in setting some of the vanes to compensate others inclined at different angles. The balance of effects is independent of the speed of the wind, so long as it is uniform over the whole section in operation. To guard against errors that might arise from a deficient fulfilment of this condition, I have preferred so to arrange that *opposite* vanes were inclined always at the same angle. For example, two pairs of opposite vanes might be set so that their planes make an angle of 6° with the axis. The remaining pair of opposite vanes would then be set at a greater angle, and this would be varied until no tendency remained to turn in either direction. The exact point of balance could be inferred either from the absence of observable effect, or by interpolation from equal slight effects in opposite directions.

As has been suggested, the motion itself may be either horizontal or vertical. Fair results may be obtained indoors at a walking speed, and my first idea was to determine balances by holding the wheel overhead while travelling in a dog-cart at 10 or 12 miles per hour. But when the axis is horizontal, much time is lost owing to the necessity of readjusting the centre of gravity after almost every shifting of the vanes. With a nearly vertical motion the position of the centre of gravity is of less consequence, and it was found that very good results could be arrived at by somewhat rapidly lowering the apparatus while held in the hands with axis vertical. It is possible that part of the delicacy obtained in this way is due to a partial annulment of gravity during the downward acceleration and consequent diminution of frictional effect at the bearings.

Some of the observations presently to be discussed were made in this way, but in most of them the arrangement was rather different. The wheel was removed from its bearings and suspended by a fine wire, whose torsion was insufficient to check the rotation seriously. The wire was pulled up vertically by a cord running over a pulley overhead. Although this arrangement offered some advantages, they were largely neutralised by disturbances due to draughts; and it is probable that equally good balances might be obtained by the simpler method.

According to an old and long discredited law, the normal pressure upon a vane moving through the air at given speed would be proportional to the square of the sine of the angle (α) between the plane of the vane and the direction of motion. The resolved part of this in the direction of rotation would be $\sin^2\alpha \cos\alpha$, which expression would represent the efficiency of the vanes of our mill as dependent upon the angle of setting. When α is small, the second factor is of little importance. A very simple experiment will now decide whether the law of $\sin^2\alpha$ is, or is not, an approximation to the truth. We find, in fact, that four vanes set at $6°$ markedly overpower two vanes set at $9°$, whereas according to the law of $\sin^2\alpha$ the reverse should happen. In order to balance the four vanes at $6°$, the two vanes need to be at about $14\frac{1}{2}°$.

By observations of this kind materials are collected for a complete plotting out of the curve of efficiency. The efficiency necessarily vanishes when $\alpha = 0$, and also on account of the resolving factor, when $\alpha = 90°$. In order to balance four vanes set at $5°$, we may set the remaining two vanes either at $10\frac{1}{2}°$ or at about $58°$. The efficiency reaches a maximum in the neighbourhood of $27°$. The results are shown in A (Fig. 1), or in the second column of the accompanying table. The scale of the ordinates is, of course, arbitrary. The efficiency for $5°$ is assumed to be 10.

α	Rotatory Efficiency	Normal Pressure	α	Rotatory Efficiency	Normal Pressure
0	0·0	0	40	27·0	88
5	10·0	25	50	23·5	91
10	19·0	48	60	19·0	94
15	24·8	64	70	13·2	96
20	28·0	75	80	6·9	99
25	29·2	80	90	0·0	100
30	29·0	84			

In order to deduce the normal pressure, the results for the rotatory efficiency must be divided by $\cos\alpha$, and accuracy is necessarily lost in the case of the larger angles. The numbers thus arrived at are plotted in curve *B*, and are given in the third column of the table, reduced so as to make the maximum (at 90°) equal to 100. As regards the relative pressures at the smaller angles, the results appear to be at least as accurate as those obtained on a larger scale with the whirling machine; but the reference to the pressure operative at 90° is probably less accurate. The principal conclusion that at small angles the pressure is proportional to sin α, and by no means to sin² α, is abundantly established.

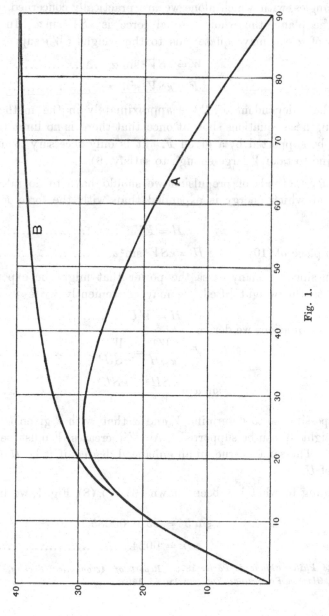

Fig. 1.

In applying these results, the first problem which suggests itself for solution is that of the gliding motion of an aeroplane. It was first successfully treated by Pénaud[*], and it may be taken under slightly different forms. We may begin by supposing the motion to be strictly horizontal, the velocity being V and the inclination of the plane to the horizon being α. Under these circumstances a propelling force F is required, which we suppose to act horizontally. The mean pressure upon the plane we will denote by $\kappa V^2 \sin \alpha$, the assumption of proportionality to $\sin \alpha$ being amply sufficient for the case of *small* angles, with which alone we are practically concerned. If S be the area of the plane, the whole normal force is $\kappa S V^2 \sin \alpha$. In view of the smallness of α, we may equate this to the weight (W) supported. Thus

$$W = \kappa S V^2 \sin \alpha, \quad \dots\dots\dots\dots\dots\dots\dots\dots(9)$$

also
$$F = \kappa S V^2 \sin^2 \alpha. \quad \dots\dots\dots\dots\dots\dots\dots(10)$$

If F be independent of V, as approximately in the method of rocket propulsion, these equations show at once that there is no limit to the weight that may be supported by a given F. It is only necessary to make α small enough, and to take V large enough to satisfy (9).

In other methods of propulsion we should have to do rather with the rate (H) at which energy is expended than with the force F itself. The relation is

$$H = FV, \quad \dots\dots\dots\dots\dots\dots\dots\dots\dots(11)$$

so that in place of (10)
$$H = \kappa S V^3 \sin^2 \alpha. \quad \dots\dots\dots\dots\dots\dots\dots(12)$$

Or, again, since in many cases the power that might be expended is proportional to the weight lifted, we may conveniently write

$$H = WU. \quad \dots\dots\dots\dots\dots\dots\dots\dots(13)$$

From these equations we derive

$$V = \frac{W^2}{\kappa SH} = \frac{W}{\kappa SU}, \quad \dots\dots\dots\dots\dots\dots\dots(14)$$

$$\sin \alpha = \frac{\kappa SH^2}{W^3} = \frac{\kappa SU^2}{W}; \quad \dots\dots\dots\dots\dots\dots(15)$$

and it is possible so to determine V and α that, with a given U and a given S, any weight W can be supported. As W increases, V must be greater and α smaller. The same is true, in an enhanced degree, if it be H that is given in place of U.

According to what has been shown (6), (7), (8), Fig. 1, we have in C.G.S. measure

$$\kappa \sin 5° = \cdot 25 \times \cdot 00085,$$

so that
$$\kappa = \cdot 0024. \quad \dots\dots\dots\dots\dots\dots\dots\dots(16)$$

---

[*] *Société Philomathique de Paris*, 1876; *Report of Aeronautical Society*, 1876. See also W. Froude, *Glasgow Proceedings*, Vol. XVIII. p. 65, 1891.

In the case of a very elongated plane the value of κ would be a little higher. We must remember that V is reckoned in centimetres per second, S in square centimetres, and the normal force in dynes.

The conclusion that a weight, however great, may be supported with a given S and a given U, or even a given H, is unpractical for more than one reason. There must be a limit below which α cannot be reduced, if only because of the high degree of instability that such an adjustment must have to contend with. Another important matter is the tangential force upon the plane, although some distinguished experimenters have expressed the opinion that it is negligible. In order to take account of it, we may add to the right-hand member of (10) a term proportional to V^2, but independent of α. Thus (12) becomes

$$H = WU = (\kappa S \sin^2 \alpha + \mu)\, V^3, \quad \dots\dots\dots\dots\dots\dots(17)$$

(9) remaining unchanged. Eliminating V, we find

$$\frac{U^3}{W} = \frac{(\kappa S \sin^2 \alpha + \mu)^2}{\kappa^3 S^3 \sin^3 \alpha}. \quad \dots\dots\dots\dots\dots\dots(18)$$

We may apply (18) to find for what value of $\sin \alpha$ the quantity U^2 attains a minimum. By the ordinary rules,

$$\sin^2 \alpha = \frac{3\mu}{\kappa S}, \quad \dots\dots\dots\dots\dots\dots\dots(19)$$

and, of course, this value of $\sin^2 \alpha$ must be *small*, if the investigation is to be applicable. If μ vanish, $\sin \alpha$ diminishes without limit. In general the minimum value of U^2 is given by

$$U^2 = \frac{16 W}{\kappa S} \left(\frac{\mu}{27 \kappa S} \right)^{\frac{1}{2}}, \quad \dots\dots\dots\dots\dots\dots(20)$$

and the corresponding value of V^2 by

$$V^2 = \frac{W}{3^{\frac{1}{2}} \kappa^{\frac{1}{2}} \mu^{\frac{1}{2}} S^{\frac{1}{2}}}. \quad \dots\dots\dots\dots\dots\dots(21)$$

These equations show that the necessary work depends entirely upon μ, and that without a knowledge of this element no numerical conclusions can be arrived at.

It might be supposed that μ, so far as it depends upon the aeroplane, would be proportional to S, but this relation is more than doubtful. In any case of a practical machine there must at any rate be a part of μ not proportional to S.

It may be well to recall that U represents the velocity at which a weight equal to W would have to be raised in order to do work equal to that done by the propelling force F. By (20), *caeteris paribus*, U varies as $S^{-\frac{1}{4}}$.

We may now pass to the case of an aeroplane gliding in still air, the path being slightly inclined downwards. If θ be the small angle between the path and the horizontal, we may regard the component of gravity in this direction, viz., $W \sin \theta$, as the propelling force F. Thus

$$H = WU = FV = WV \sin \theta, \quad \dots\dots\dots\dots\dots(22)$$

so that
$$U = V \sin \theta. \quad \dots\dots\dots\dots\dots\dots(23)$$

The same equations apply as before, with the understanding that α, being the inclination of the plane to the direction of motion through the air, is no longer identical with the inclination of the plane to the horizon. The latter angle, reckoned positive when the leading edge is downwards, will now be denoted by $(\theta - \alpha)$.

Introducing (23) into (14), (15), we get

$$V^2 = \frac{W}{\kappa S \sin \theta}, \qquad \sin \alpha = \frac{\kappa S V^2 \sin^2 \theta}{W}, \quad \dots\dots\dots(24)$$

from which it appears that whatever may be the values of W and S, θ may still be as small as we please. Thus, if frictional forces can be neglected, a high speed is all that is required in order to glide without loss of energy. This is the supposition upon which we discussed the manner in which a bird may take advantage of the internal work of the wind; and we see that the motion of the bird must be of such a character that he always retains a high velocity relatively to the surrounding air. The advantage that we showed to be obtainable must be set against losses due to friction and to imperfect fulfilment of the condition just specified.

When frictional forces are included we may use equation (18), merely substituting $V \sin \theta$ for U. The problem already considered of making U a minimum is still pertinent, since U denotes the rate of vertical descent. By (19), (20), (21)

$$\sin^2 \alpha = \frac{3\mu}{\kappa S}, \qquad \sin^2 \theta = \frac{U^2}{V^2} = \frac{16\mu}{3\kappa S}, \quad \dots\dots\dots(25)$$

so that, θ and α being small,

$$\alpha = \tfrac{3}{4}\theta, \qquad \theta - \alpha = \tfrac{1}{3}\alpha = \tfrac{1}{4}\theta. \quad \dots\dots\dots\dots(26)$$

This result, due to Pénaud, shows that when the rate of vertical descent is slowest, or when the time of falling a given height is greatest, the slope of the plane to the horizon is downwards in front, and equal to one-quarter of the slope of the line of motion. The actual minimum rate of vertical descent is given by (20). This rate is relative to still air. If there be a wind having a vertical component of the same amount, the course of the plane may be horizontal.

Another slightly different minimum problem is also treated by Pénaud, in which it is required to determine how *far* it is possible to glide while falling through a given vertical height. From (9), (17), (23), we have in general

$$\sin \theta = \frac{\sin^2 \alpha + \mu/\kappa S}{\sin \alpha}. \quad \dots\dots\dots\dots\dots(27)$$

When θ is a minimum by variation of α,

$$\sin \alpha = \tfrac{1}{2} \sin \theta = \sqrt{(\mu/\kappa S)}. \quad \dots\dots\dots\dots(28)$$

In this case the plane bisects the angle between the horizontal and the direction of motion.

In the flying machines of Pénaud, Langley, and Maxim, the propelling force is obtained by a screw, acting like the screw-propeller of a ship. A rough theory of this action is easily given and is of interest, not only in the application to the horizontal propulsion of an aeroplane, but also because a screw rotating about a vertical axis may be used for direct maintenance. The latter question may conveniently be considered first.

The screw is supposed to maintain a weight W at a fixed position in still air. This it does by creating a downward current of velocity v. If S' be the area of section of the current, equal to that swept through by the screw, the volume of air acted upon per second is $S'v$, and the momentum generated per second is $S'v.\rho v$, or $S'\rho v^2$. Hence

$$W = S'\rho v^2. \quad \dots\dots\dots\dots\dots\dots\dots(29)$$

Again, the kinetic energy generated per second is $\tfrac{1}{2}S'\rho v^3$; so that if U be the velocity at which W would have to be lifted to do a corresponding amount of work, we may, neglecting frictional losses, equate the above to UW. Thus

$$UW = \tfrac{1}{2}S'\rho v^3. \quad \dots\dots\dots\dots\dots\dots(30)$$

From (29), (30), $\tfrac{1}{2}v = U$,

and

$$S' = \frac{W}{4\rho U^2}. \quad \dots\dots\dots\dots\dots\dots(31)$$

So far as these equations are concerned, any weight can be maintained by a limited expenditure of work, but the smaller the power available the larger must be the section of the stream of air, and consequently of the screw, or other machinery, by which the air is set in motion. Again from (31)

$$WU = \frac{W^{\frac{3}{2}}}{(4\rho S')^{\frac{1}{2}}}, \quad \dots\dots\dots\dots\dots(32)$$

so that if S' be given, the whole power required varies as $W^{\frac{3}{2}}$.

To obtain numbers applicable to the case of a man supporting himself in this way by his own muscular power, we take in C.G.S. measure

$$W = 68000 \times 981, \qquad U = 15, \qquad \rho = \tfrac{1}{800},$$

thus finding

$$S' = 6{\cdot}0 \times 10^7 \text{ sq. cm.}$$

This represents the cross-section of the descending column of air. If we equate S' to $\frac{1}{4}\pi d^2$, d will be the diameter of the screw required, and we get $d = 90$ metres. It is to be observed that this assumed value of U corresponds to the power which a man may exercise when working for eight hours a day. But even if he could do ten times as much for a few minutes, d would still amount to 9 metres; and in this estimate nothing has been allowed for the weight of the mechanism, or for frictional losses. It seems safe to conclude that a man will never support himself in this manner by his own muscular power.

A screw works to better advantage when it has a forward motion through the fluid, for then a larger mass comes under its influence. Let us suppose that a screw, now rotating about a horizontal axis, is advancing through still air with horizontal velocity V. Also let v be the actual velocity with which the column of air leaves it. The volume acted on per second is $S'(V+v)$. If F be the propulsive force

$$F = S'\rho\,(V+v)\,v. \quad \dots\dots\dots\dots\dots\dots(33)$$

Again, the work per second required to generate the kinetic energy of the column is

$$\tfrac{1}{2}S'\rho\,(V+v)\,v^2. \quad \dots\dots\dots\dots\dots\dots(34)$$

The whole work expended per second (H') is accordingly

$$H' = FV + \tfrac{1}{2}S'\rho\,(V+v)\,v^2 = F(V+\tfrac{1}{2}v). \quad \dots\dots\dots\dots(35)$$

When V is great compared with v, the right-hand member of (35) reduces to its first term. We conclude that when a screw advances at a sufficiently rapid rate, the energy left behind in the fluid is negligible, so that the whole work done is available for propulsion. The distinction between H' and H, as formerly employed, then disappears.

If U denote the rate at which W would have to be lifted in order to do the work actually performed by the machine, we may now take from (15), as applicable to the *rapid* flight of an aeroplane,

$$U = \sqrt{\left(\frac{W\sin a}{\kappa S}\right)}. \quad \dots\dots\dots\dots\dots(36)$$

In the case of direct maintenance by a screw rotating about a vertical axis, (31) gives

$$U = \sqrt{\left(\frac{W}{4\rho S'}\right)}. \quad \dots\dots\dots\dots\dots(37)$$

It may be interesting to compare the powers required in the two methods, especially as some high authorities have favoured direct maintenance, without the use of an aeroplane, as the more economical. The ratio of the values of U in (36), (37) is

$$2\sqrt{\left(\frac{\rho \sin \alpha}{\kappa} \frac{S'}{S}\right)} \quad \dots\dots\dots\dots\dots(38)$$

or, in the case of air, since $\kappa = \cdot0024$, $\rho = \cdot0012$,

$$\sqrt{(2 \sin \alpha . S'/S)}. \quad \dots\dots\dots\dots\dots(39)$$

Since α may be made small, and S the area of the plane may be a large multiple of S' the area swept over by the screw, it would appear that the advantage must lie with the aeroplane, even if the object be mere maintenance, and not a rapid transit from place to place.

But although the flying machine of the future will, as it appears to me, be on the principle of the aeroplane, it cannot be denied that the method of direct maintenance by a vertically rotating screw offers certain present advantages. Among the most important of these are a much better ensured stability, and less danger in alighting owing to the absence of rapid horizontal motion. The first experiments might well be made with screws driven by electric motors, the power being supplied from the ground by means of vertical wires 30 or 40 feet long. In this way the necessary experience would be easily gained, and most of the doubtful points settled, before a completely self-contained machine was attempted.

In natural flight revolving mechanism is not, and apparently could not have been, used. As we all know, a bird flying horizontally through still air performs the necessary work by flapping his wings. The effect of a reciprocating motion in modifying the action of an aeroplane was, I believe, first considered in detail by Professor M. Fitzgerald[*]. It may be convenient to give, as naturally connected with the foregoing, an outline of this theory in a modified form, following Professor Fitzgerald in assimilating the wing to a simple aeroplane, upon which is imposed (without rotation) a vertical reciprocating motion.

We denote by u the horizontal velocity of the plane supposed uniform, by v the vertical velocity at time t, by θ the inclination of the plane to the horizon at time t, while S and W denote the area and weight as before. If we assume the same formula for the pressure as before, although the application is now to an *unsteady* motion, and further suppose that v/u and θ are always small, we get as in (9) for the whole normal pressure upon the plane at time t

$$\kappa S (u^2 + v^2)(\theta + v/u), \quad \dots\dots\dots\dots\dots(40)$$

in which however v^2 in $(u^2 + v^2)$ may be omitted.

---

[*] *Proc. Roy. Soc.* Vol. LXIV. p. 420, 1899.

We now assume that θ and v are periodic, for example that

$$\theta = \theta_0 + \theta_1 \cos pt, \quad \dots\dots\dots\dots(41)$$

$$v/u = \beta \cos (pt + \epsilon), \quad \dots\dots\dots\dots(42)$$

where the periodic time τ is related to p according to

$$\tau = 2\pi/p.$$

At this stage the criticism may present itself that the assumed motion involves a reaction for which we have made no provision. In practice the reaction is supplied by the inertia of the body of the bird to which the wings are attached. The difficulty would be got over by supposing that there are several planes executing similar movements, but in different phases regularly disposed. It seems hardly worth while to complicate the present investigation by introducing a vertical movement of the weight.

By (40) the whole pressure at time t, perpendicular to the plane, is

$$\kappa S u^2 \{\theta_0 + \theta_1 \cos pt + \beta \cos (pt + \epsilon)\}. \quad \dots\dots\dots(43)$$

Of this the mean value is to be equated to the weight W supported, so that

$$W = \kappa S u^2 \theta_0. \quad \dots\dots\dots\dots(44)$$

The horizontal component of the whole pressure at time t is

$$S . \kappa u^2 . \{\theta + v/u\} \theta, \quad \dots\dots\dots\dots(45)$$

and of this the mean value is to be supposed to be zero, in order that the plane may move with uniform horizontal velocity. Thus

$$\theta_0^2 + \tfrac{1}{2}\theta_1^2 + \tfrac{1}{2}\beta\theta_1 \cos \epsilon = 0. \quad \dots\dots\dots(46)$$

Again, if WU be the (mean) rate of expenditure of work,

$$WU = S . \kappa u^2 . \int_0^\tau (\theta + v/u) v \, d(t/\tau) = S . \tfrac{1}{2}\kappa u^3 (\beta\theta_1 \cos \epsilon + \beta^2). \quad \dots(47)$$

If we eliminate β between (46), (47), we get

$$WU = S . \tfrac{1}{2}\kappa u^3 \frac{(2\theta_0^2 + \theta_1^2)(2\theta_0^2 + \theta_1^2 \sin^2 \epsilon)}{\theta_1^2 \cos^2 \epsilon}, \quad \dots\dots\dots(48)$$

from which we see that if θ_1 be given (as well as S, W, u), U is least when $\epsilon = 0$, viz., when the phase of maximum vertical velocity coincides with the phase of greatest inclination. In this case by use of (44) we have

$$U = \frac{W}{\kappa S u}\left(1 + \frac{2\theta_0^2}{\theta_1^2}\right). \quad \dots\dots\dots(49)$$

If we regard W, S, u as given, the smallest value of U corresponds to θ_1 being large in comparison with θ_0 which is given by (44)*.

* It must not be forgotten that θ_1 itself has been assumed to be small.

The smallest value is

$$U = \frac{W}{\kappa S u}. \qquad \dots\dots\dots\dots\dots\dots\dots\dots\dots(50)$$

The work required to be done is here the same function of S, W, and the horizontal velocity as was found in (14), where V has the meaning here assigned to u.

We see from (46) that, under the circumstances supposed, $\theta_1 + \beta$ is numerically small in comparison with θ_0, and à *fortiori* in comparison with θ_1. Accordingly the forward edge of the plane is inclined downwards when the motion of the plane is downwards.

As regards the pressure, it is by (43) proportional to

$$\theta_0 + (\theta_1 + \beta) \cos pt,$$

in which the second term is relatively small. The pressure acts always upon the under side of the plane, and the weight is approximately supported in all phases.

258.

ON THE LAW OF RECIPROCITY IN DIFFUSE REFLEXION.

[*Philosophical Magazine*, XLIX. pp. 324, 325, 1900.]

IN the current number of the *Philosophical Magazine* (Vol. XLIX. p. 199) Dr Wright discusses the question of the amount of light diffusely reflected from a given area of a matt surface as dependent upon the angle of incidence (i) and the angle of emission (ϵ). According to Lambert's law the function of i and ϵ is

$$\cos i \cos \epsilon; \dots\dots\dots\dots\dots\dots\dots\dots\dots\dots\dots\dots(1)$$

and this law, though in the present case without theoretical foundation, appears approximately to represent the facts. The question may indeed be raised whether it is possible so to define an ideally matt surface that Lambert's law may become strictly applicable.

The conclusion drawn by Dr Wright from his experiments with compressed powders upon which I desire to comment is that numbered (4) in his *résumé* of results, viz. "A law for the intensity of reflected scattered light cannot be symmetric in reference to i and ϵ." It appears to me that this statement is in contradiction to a fundamental principle of reciprocity, of such generality that escape from it is difficult. This principle is discussed at length in my book on the *Theory of Sound*, § 109. Its application to the present question may be thus stated :—Suppose that in any direction (i) and at any distance r from a small surface (S) reflecting in *any manner* there be situated a radiant point (A) of given intensity, and consider the intensity of the reflected vibrations at any point B situated in direction ϵ and at distance r' from S. The theorem is to the effect that the intensity is the same as it would be at A if the radiant point were transferred to B*. The conclusion follows that whatever may be its character in other respects, the function of i and ϵ which represents the intensity of the reflected scattered light *must* be symmetrical with respect to these quantities.

The actual departures from the reciprocal relation found by Dr Wright were not very large, and they may possibly be of the nature of experimental errors. In any case it seems desirable that the theoretical difficulty in accepting Dr Wright's conclusion should be pointed out.

---

* I have not thought it necessary to enter into questions connected with polarization, but a more particular statement could easily be made.

259.

ON THE VISCOSITY OF GASES AS AFFECTED BY TEMPERATURE.

[*Proceedings of the Royal Society*, LXVII. pp. 137—139, 1900.]

A FORMER paper* describes the apparatus by which I examined the influence of temperature upon the viscosity of argon and other gases. I have recently had the opportunity of testing, in the same way, an interesting sample of gas prepared by Professor Dewar, being the residue, uncondensed by *liquid hydrogen*, from a large quantity collected at the Bath springs. As was to be expected†, it consists mainly of helium, as is evidenced by its spectrum when rendered luminous in a vacuum tube. A line, not visible from another helium tube, approximately in the position of D_5 (Neon) is also apparent‡.

The result of the comparison of viscosities at about 100° C. and at the temperature of the room was to show that the temperature effect was the same as for *hydrogen*.

* *Roy. Soc. Proc.* Vol. LXVI. (1900), p. 68. [Vol. IV. p. 452.]

† *Roy. Soc. Proc.* Vol. LIX. (1896), p. 207; Vol. LX. (1896), p. 56. [Vol. IV. p. 225.]

‡ I speak doubtfully, because to my eye the interval from D_1 to D_3 (helium) appeared about equal to that between D_3 and the line in question, whereas, according to the measurements of Ramsay and Travers (*Roy. Soc. Proc.* Vol. LXIII. (1898), p. 438), the wave-lengths are—

$$D_1 \quad . \quad . \quad . \quad . \quad . \quad . \quad . \quad 5895 \cdot 0$$
$$D_2 \quad . \quad . \quad . \quad . \quad . \quad . \quad . \quad 5889 \cdot 0$$
$$D_3 \quad . \quad . \quad . \quad . \quad . \quad . \quad . \quad 5875 \cdot 9$$
$$D_5 \quad . \quad . \quad . \quad . \quad . \quad . \quad . \quad 5849 \cdot 6,$$

so that the above-mentioned intervals would be as $19 \cdot 1 : 26 \cdot 3$. [*June* 23.—Subsequent observations with the aid of a scale showed that the intervals above spoken of were as $20 : 21$. According to this the wave-length of the line seen, and supposed to correspond to D_5, would be about 5855 on Rowland's scale, where $D_1 = 5896 \cdot 2$, $D_2 = 5890 \cdot 2$, $D_3 = 5876 \cdot 0$.] I may record that the refractivity of the gas now under discussion is $0 \cdot 132$ relatively to air.

In the former paper the results were reduced so as to show to what power (n) of the absolute temperature the viscosity was proportional.

	n	c
Air	0·754	111·3
Oxygen	0·782	128·2
Hydrogen⎱	0·681	72·2
Helium ⎰		
Argon	0·815	150·2

Since practically only two points on the temperature curve were examined, the numbers obtained were of course of no avail to determine whether or no any power of the temperature was adequate to represent the complete curve. The question of the dependence of viscosity upon temperature has been studied by Sutherland*, on the basis of a theoretical argument which, if not absolutely rigorous, is still entitled to considerable weight. He deduces from a special form of the kinetic theory as the function of temperature to which the viscosity is proportional

$$\frac{\theta^{\frac{3}{2}}}{1+c/\theta}, \dots\dots\dots\dots\dots\dots\dots\dots(1)$$

c being some constant proper to the particular gas. The simple law $\theta^{\frac{3}{2}}$, appropriate to "hard spheres," here appears as the limiting form when θ is very great. In this case, the collisions are sensibly uninfluenced by the molecular forces which may act at distances exceeding that of impact. When, on the other hand, the temperature and the molecular velocities are lower, the mutual attraction of molecules which pass near one another increases the number of collisions, much as if the diameter of the spheres was increased. Sutherland finds a very good agreement between his formula (1) and the observations of Holman and others upon various gases.

If the law be assumed, my observations suffice to determine the values of c. They are shown in the table, and they agree well with the numbers for air and oxygen calculated by Sutherland from observations of Obermayer.

* *Phil. Mag.* Vol. xxxvi. (1893), p. 507.

260.

REMARKS UPON THE LAW OF COMPLETE RADIATION.

[*Philosophical Magazine*, XLIX. pp. 539, 540, 1900.]

BY complete radiation I mean the radiation from an ideally black body, which according to Stewart* and Kirchhoff is a definite function of the absolute temperature θ and the wave-length λ. Arguments of (in my opinion†) considerable weight have been brought forward by Boltzmann and W. Wien leading to the conclusion that the function is of the form

$$\theta^5 \phi\,(\theta\lambda)\,d\lambda, \quad \dots\dots\dots\dots\dots\dots\dots\dots\dots(1)$$

expressive of the energy in that part of the spectrum which lies between λ and $\lambda + d\lambda$. A further specialization by determining the form of the function ϕ was attempted later‡. Wien concludes that the actual law is

$$c_1\lambda^{-5}e^{-c_2/\lambda\theta}\,d\lambda, \quad \dots\dots\dots\dots\dots\dots\dots\dots(2)$$

in which c_1 and c_2 are constants, but viewed from the theoretical side the result appears to me to be little more than a conjecture. It is, however, supported upon general thermodynamic grounds by Planck§.

Upon the experimental side, Wien's law (2) has met with important confirmation. Paschen finds that his observations are well represented, if he takes

$$c_2 = 14{,}455,$$

---

* Stewart's work appears to be insufficiently recognized upon the Continent.   [See *Phil. Mag.* I. p. 98, 1901; p. 494 below.]

† *Phil. Mag.* Vol. XLV. p. 522 (1898).

‡ *Wied. Ann.* Vol. LVIII. p. 662 (1896).

§ *Wied. Ann.* Vol. I. p. 74 (1900).

θ being measured in centigrade degrees and λ in thousandths of a millimetre (μ). Nevertheless, the law seems rather difficult of acceptance, especially the implication that as the temperature is raised, the radiation of given wavelength approaches a limit. It is true that for visible rays the limit is out of range. But if we take $\lambda = 60\,\mu$, as (according to the remarkable researches of Rubens) for the rays selected by reflexion at surfaces of Sylvin, we see that for temperatures over 1000° (absolute) there would be but little further increase of radiation.

The question is one to be settled by experiment; but in the meantime I venture to suggest a modification of (2), which appears to me more probable *à priori*. Speculation upon this subject is hampered by the difficulties which attend the Boltzmann-Maxwell doctrine of the partition of energy. According to this doctrine every mode of vibration should be alike favoured; and although for some reason not yet explained the doctrine fails in general, it seems possible that it may apply to the graver modes. Let us consider in illustration the case of a stretched string vibrating transversely. According to the Boltzmann-Maxwell law the energy should be equally divided among all the modes, whose frequencies are as 1, 2, 3, ... Hence if k be the reciprocal of λ, representing the frequency, the energy between the limits k and $k + dk$ is (when k is large enough) represented by dk simply.

When we pass from one dimension to three dimensions, and consider for example the vibrations of a cubical mass of air, we have (*Theory of Sound*, § 267) as the equation for k^2,

$$k^2 = p^2 + q^2 + r^2,$$

where p, q, r are integers representing the number of subdivisions in the three directions. If we regard p, q, r as the coordinates of points forming a cubic array, k is the distance of any point from the origin. Accordingly the number of points for which k lies between k and $k + dk$, proportional to the volume of the corresponding spherical shell, may be represented by $k^2 dk$, and this expresses the distribution of energy according to the Boltzmann-Maxwell law, so far as regards the wave-length or frequency. If we apply this result to radiation, we shall have, since the energy in each mode is proportional to θ,

$$\theta k^2 dk, \quad \dots\dots\dots\dots\dots\dots\dots\dots\dots\dots(3)$$

or, if we prefer it,

$$\theta \lambda^{-4} d\lambda. \quad \dots\dots\dots\dots\dots\dots\dots\dots(4)$$

It may be regarded as some confirmation of the suitability of (4) that it is of the prescribed form (1).

The suggestion is that (4) rather than, as according to (2),

$$\lambda^{-5} d\lambda \quad \dots\dots\dots\dots\dots\dots\dots\dots\dots(5)$$

may be the proper form when $\lambda\theta$ is great*.   If we introduce the exponential factor, the complete expression will be

$$c_1\theta\lambda^{-4}e^{-c_2/\lambda\theta}\,d\lambda. \quad\dots\dots\dots\dots\dots\dots\dots\dots(6)$$

If, as is probably to be preferred, we make k the independent variable, (6) becomes

$$c_1\theta k^2 e^{-c_2 k/\theta}\,dk. \quad\dots\dots\dots\dots\dots\dots\dots\dots(7)$$

Whether (6) represents the facts of observation as well as (2) I am not in a position to say.   It is to be hoped that the question may soon receive an answer at the hands of the distinguished experimenters who have been occupied with this subject.

* [1902. This is what I intended to emphasize. Very shortly afterwards the anticipation above expressed was confirmed by the important researches of Rubens and Kurlbaum (*Drude Ann.* IV. p. 649, 1901), who operated with exceptionally long waves. The formula of Planck, given about the same time, seems best to meet the observations. According to this modification of Wien's formula, $e^{-c_2/\lambda\theta}$ in (2) is replaced by $1\div(e^{c_2/\lambda\theta}-1)$. When $\lambda\theta$ is great, this becomes $\lambda\theta/c_2$, and the complete expression reduces to (4).]

261.

ON APPROXIMATELY SIMPLE WAVES.

[*Philosophical Magazine*, L. pp. 135—139, 1900.]

THE first question that arises is as to the character of *absolutely* simple waves; and here "it may be well to emphasize that a simple vibration implies *infinite* continuance, and does not admit of variations of phase or amplitude. To suppose, as is sometimes done in optical speculations, that a train of simple waves may begin at a given epoch, continue for a certain time involving it may be a large number of periods, and ultimately cease, is a contradiction in terms*." A like contradiction is involved if we speak of unpolarized light as homogeneous, really homogeneous light being necessarily polarized.

This much being understood, approximately simple waves might be defined as waves which *for a considerable succession* deviate but little from a simple train. Under this definition large changes of amplitude and frequency would not be excluded, provided only that they entered slowly enough. More frequently further limitation would be imposed, and approximately simple waves would be understood to mean waves which for a considerable succession can be approximately identified with a simple train of given frequency, if not of given amplitude. But the *phase†* of the simple train approximately representing the given waves would vary from place to place, slowly indeed but to any extent.

Thus if we take, as analytically expressing the dependence of the displacement upon time,

$$H \cos pt + K \sin pt, \quad \dots\dots\dots\dots\dots\dots(1)$$

---

* *Theory of Sound*, 2nd ed. § 65 *a*, 1894.

† What is here called for brevity the phase is more properly the *deviation* of phase from that of an absolutely simple train of waves.

where H and K are slowly varying functions of t, the frequency may be regarded as constant, while the amplitude $\sqrt{(H^2 + K^2)}$ and the phase $\tan^{-1}(K/H)$ vary slowly but without limit. It scarcely needs to be pointed out that a *slow* uniform progression of phase is equivalent to a *small* change of frequency.

In one important class of cases the phase remains constant and then, since a constant addition to t need not be regarded, (1) is sufficiently represented by

$$H \cos pt \dots\dots\dots\dots\dots\dots\dots\dots\dots(2)$$

simply. If the changes of amplitude are periodic, we may write

$$H = H_0 + H_1 \cos qt + H_1' \sin qt + H_2 \cos 2qt + H_2' \sin 2qt + \dots, \dots(3)$$

in which q is supposed to be small. The vibration (2) is then always equivalent to a combination of simple vibrations of frequencies represented by

$$p, \quad p+q, \quad p-q, \quad p+2q, \quad p-2q, \text{ \&c.}$$

Under this head may be mentioned the case of ordinary beats, so familiar in Acoustics. Here

$$H = H_1 \cos qt, \quad \dots\dots\dots\dots\dots\dots(4)$$

and

$$H \cos pt = \tfrac{1}{2} H_1 \cos (p+q) t + \tfrac{1}{2} H_1 \cos (p-q) t. \dots\dots\dots(5)$$

It may be observed that although the phase is regarded as constant, the change of sign in the amplitude has the same effect as an alteration of phase of 180°.

Another important example is that of intermittent vibrations. If we put

$$H = 2(1 + \cos qt), \quad \dots\dots\dots\dots\dots\dots(6)$$

the amplitude is always of one sign, and

$$H \cos pt = 2 \cos pt + \cos (p+q) t + \cos (p-q) t. \quad \dots\dots\dots(7)$$

Three simple vibrations are here required to represent the effect.

Again (*Theory of Sound*, § 65 a), if

$$H = 4 \cos^4 qt, \quad \dots\dots\dots\dots\dots\dots\dots(8)$$

we have

$$H \cos pt = \tfrac{3}{2} \cos pt + \cos (p+q) t + \cos (p-q) t$$
$$+ \tfrac{1}{4} \cos (p+2q) t + \tfrac{1}{4} \cos (p-2q) t. \quad \dots\dots(9)$$

If K also be variable and periodic in the same period as H, so that

$$K = K_0 + K_1 \cos qt + K_1' \sin qt + K_2 \cos 2qt + K_2' \sin 2qt + \dots, \dots(10)$$

we have the most general periodicity expressed when we substitute these val.ies in (1); and the general conclusion as to the periods of the simple vibrations required to represent the effect remains undisturbed.

If K and H vary together in such a manner that the amplitude $\sqrt{(H^2 + K^2)}$ remains constant, the sole variation is one of phase. My object at present is to call attention to this class of cases, so far as I know hitherto neglected, unless an example (*Phil. Mag.* xxxiv. p. 409, 1892 [Vol. iv. p. 16]) in which an otherwise constant amplitude is periodically and suddenly reversed be considered an exception.

If we take

$$H = \cos\,(\alpha \sin qt), \qquad K = \sin\,(\alpha \sin qt), \dots\dots\dots\dots(11)$$

H and K are of the required periodicity, and the condition of a constant amplitude is satisfied. In fact (1) becomes

$$\cos\,(pt - \alpha \sin qt). \quad\dots\dots\dots\dots\dots\dots\dots(12)$$

Now, since

$$e^{ia\cos\theta} = J_0\,(\alpha) + 2iJ_1\,(\alpha)\cos\theta + 2i^2 J_2\,(\alpha)\cos 2\theta + \dots$$
$$+ 2i^n J_n\,(\alpha)\cos n\theta + \dots,$$

we have

$$e^{ia\sin qt} = J_0 + 2iJ_1 \sin qt + 2J_2 \cos 2qt + 2iJ_3 \sin 3qt$$
$$+ 2J_4 \cos 4qt + \dots, \quad\dots\dots\dots\dots(13)$$

and thus

$$\cos\,(\alpha \sin qt) = J_0\,(\alpha) + 2J_2\,(\alpha)\cos 2qt + 2J_4\,(\alpha)\cos 4qt + \dots, \quad\dots(14)$$

$$\sin\,(\alpha \sin qt) = 2J_1\,(\alpha)\sin qt + 2J_3\,(\alpha)\sin 3qt + \dots, \quad\dots\dots\dots(15)$$

where J_0, J_1, &c. denote (as usual) the Bessel's functions of the various orders. In the notation of (3) and (10)

$$H_1 = H_3 = \dots = 0, \qquad H_1' = H_2' = H_3' = \dots = 0,$$

$$H_0 = J_0\,(\alpha), \qquad H_2 = 2J_2\,(\alpha), \qquad H_4 = 2J_4\,(\alpha), \quad \&c.,$$

$$K_0 = K_1 = K_2 = \dots = 0, \qquad K_2' = K_4' = \dots = 0,$$

$$K_1' = 2J_1\,(\alpha), \qquad K_3' = 2J_3\,(\alpha), \qquad K_5' = 2J_5\,(\alpha), \quad \&c.$$

Accordingly (12), expressed as a combination of simple waves, is

$$J_0\,(\alpha)\cos pt + J_2\,(\alpha)\,\{\cos\,(p - 2q)\,t + \cos\,(p + 2q)\,t\}$$
$$+ J_4\,(\alpha)\,\{\cos\,(p + 4q)\,t + \cos\,(p - 4q)\,t\} + \dots$$
$$+ J_1\,(\alpha)\,\{\cos\,(p - q)\,t - \cos\,(p + q)\,t\}$$
$$+ J_3\,(\alpha)\,\{\cos\,(p - 3q)\,t - \cos\,(p + 3q)\,t\} + \dots. \quad\dots\dots\dots(16)$$

n	$J_n(3)$	$J_n(6)$	$J_n(12)$	$J_n(18)$	$J_n(24)$
0	$-\cdot26005$	$+\cdot15065$	$+\cdot04769$	$-\cdot01336$	$-\cdot05623$
1	$+\cdot33906$	$-\cdot27668$	$-\cdot22345$	$-\cdot18799$	$-\cdot15404$
2	$+\cdot48609$	$-\cdot24287$	$-\cdot08493$	$-\cdot00753$	$+\cdot04339$
3	$+\cdot30906$	$+\cdot11477$	$+\cdot19514$	$+\cdot18632$	$+\cdot16127$
4	$+\cdot13203$	$+\cdot35764$	$+\cdot18250$	$+\cdot06964$	$-\cdot00308$
5	$+\cdot04303$	$+\cdot36209$	$-\cdot07347$	$-\cdot15537$	$-\cdot16230$
6	$+\cdot01139$	$+\cdot24584$	$-\cdot24372$	$-\cdot15596$	$-\cdot06455$
7	$+\cdot00255$	$+\cdot12959$	$-\cdot17025$	$+\cdot05140$	$+\cdot13002$
8	$+\cdot00049$	$+\cdot05653$	$+\cdot04510$	$+\cdot19593$	$+\cdot14039$
9	$+\cdot00008$	$+\cdot02117$	$+\cdot23038$	$+\cdot12276$	$-\cdot03643$
10	$+\cdot00001$	$+\cdot00696$	$+\cdot30048$	$-\cdot07317$	$-\cdot16771$
11	....	$+\cdot00205$	$+\cdot27041$	$-\cdot20406$	$-\cdot10333$
12	....	$+\cdot00055$	$+\cdot19528$	$-\cdot17624$	$+\cdot07299$
13	....	$+\cdot00013$	$+\cdot12015$	$-\cdot03092$	$+\cdot17632$
14	....	$+\cdot00003$	$+\cdot06504$	$+\cdot13157$	$+\cdot11803$
15	....	$+\cdot00001$	$+\cdot03161$	$+\cdot23559$	$-\cdot03863$
16	....	....	$+\cdot01399$	$+\cdot26108$	$-\cdot16631$
17	....	....	$+\cdot00570$	$+\cdot22855$	$-\cdot18312$
18	....	....	$+\cdot00215$	$+\cdot17063$	$-\cdot09311$
19	....	....	$+\cdot00076$	$+\cdot11271$	$+\cdot04345$
20	....	....	$+\cdot00025$	$+\cdot06731$	$+\cdot16191$
21	....	....	$+\cdot00008$	$+\cdot03686$	$+\cdot22640$
22	....	....	$+\cdot00002$	$+\cdot01871$	$+\cdot23429$
23	....	....	$+\cdot00001$	$+\cdot00886$	$+\cdot20313$
24	....	....	....	$+\cdot00395$	$+\cdot15504$
25	....	....	....	$+\cdot00166$	$+\cdot10695$
26	....	....	....	$+\cdot00066$	$+\cdot06778$
27	....	....	....	$+\cdot00025$	$+\cdot03990$
28	....	....	....	$+\cdot00009$	$+\cdot02200$
29	....	....	....	$+\cdot00003$	$+\cdot01143$
30	....	....	....	$+\cdot00001$	$+\cdot00563$
31	....	....	....	....	$+\cdot00263$
32	....	....	....	....	$+\cdot00118$
33	....	....	....	....	$+\cdot00050$
34	....	....	....	....	$+\cdot00021$
35	....	....	....	....	$+\cdot00008$
36	....	....	....	....	$+\cdot00003$
37	....	....	....	....	$+\cdot00001$

If α, representing the maximum disturbance of phase, be small, we may write approximately

$$J_0 = 1 - \tfrac{1}{4}\alpha^2, \qquad J_1 = \tfrac{1}{2}\alpha, \qquad J_2 = \tfrac{1}{8}\alpha^2,$$

while J_3 &c. are of higher powers in α than α^2. Thus if we stop at the first power of α, we are concerned only with the multiples of t represented by

$$p, \qquad p - q, \qquad p + q;$$

while if we include α^2 we have

$$p, \qquad p - q, \qquad p + q, \qquad p - 2q, \qquad p + 2q.$$

But when α is not small, the convergence is slow, and a large number of terms will be required even for a moderately close approximation. The preceding table, due to Meissel, is condensed from Gray and Mathew's *Bessel's Functions*. So far as π can be identified with 3, the values of α equal to 3, 6, 12, 18, 24 correspond to maximum deviations of phase (in both directions) equal to $\tfrac{1}{2}$, 1, 2, 3, 4 periods respectively. It appears that the largest value of $J_n(\alpha)$ occurs for a value of n somewhat less than α. Indeed, it is at once evident from (12) that frequencies in the neighbourhood of $p \pm q\alpha$ will be important elements.

262.

ON A THEOREM ANALOGOUS TO THE VIRIAL THEOREM.

[*Philosophical Magazine*, L. pp. 210—213, 1900.]

As an example of the generality of the theorem of Clausius, Maxwell*
mentions that "in any framed structure consisting of struts and ties, the sum
of the products of the pressure in each strut into its length, exceeds the
sum of the products of the tension of each tie into its length, by the product
of the weight of the whole structure into the height of its centre of gravity
above the foundations." It will be convenient to sketch first the proof of
the purely statical theorem of which the above is an example, and afterwards
of the corresponding statical applications of the analogue. The proof of the
general dynamical theorem will then easily follow.

If X, Y, Z denote the components, parallel to the axes, of the various
forces which act upon a particle at the point x, y, z, then since the system
is in equilibrium,

$$\Sigma X = 0, \qquad \Sigma Y = 0, \qquad \Sigma Z = 0.$$

If we multiply these equations by x, y, z respectively, and afterwards effect
a summation over all the particles of the system, we obtain a result which
may be written

$$\Sigma [x \cdot \Sigma X + y \cdot \Sigma Y + z \cdot \Sigma Z] = 0. \quad \dots\dots\dots\dots\dots(1)$$

The utility of the equation depends upon an alteration in the manner
of summation, and in particular upon a separation of the forces R (considered
positive when repellent) which act mutually between two particles along their

---

* *Nature*, Vol. x. p. 477, 1874; [Maxwell's] *Scientific Papers*, Vol. II. p. 410.

line of junction ρ. If x, y, z and x', y', z' be the coordinates of the particles, we have so far as regards the above-mentioned forces,

$$X(x' - x) + Y(y' - y) + Z(z' - z) = R\rho;$$

or with summation over every pair of particles $\Sigma R\rho$. The complete equation may now be written

$$\Sigma(Xx + Yy + Zz) + \Sigma R\rho = 0, \quad \dots\dots\dots\dots(2)$$

where in the first summation X, Y, Z represent the components of the *external* forces operative at the point x, y, z. In Maxwell's example the only external forces are the weights of the various parts of the system (supposed to be concentrated at the junctions of the struts and ties), and the reactions at the foundations.

The analogous theorem, to which attention is now called, is derived in a similar manner from the equally evident equation

$$\Sigma[x \cdot \Sigma Y + y \cdot \Sigma X] = 0. \quad \dots\dots\dots\dots(3)$$

We have to extract from the summation on the left the force R mutually operative between the particles at x, y, z and at x', y', z'; and we shall limit ourselves to the case of two dimensions. If X, Y be the components of force acting upon the latter particle, ρ the distance between the particles, and ϕ the inclination of ρ to the axis of x, we have

$$Y(x' - x) + X(y' - y) = R\rho \sin 2\phi;$$

so that if now X, Y represent the total *external* force acting at x, y, (3) becomes

$$\Sigma[xY + yX] + \Sigma R\rho \sin 2\phi = 0, \quad \dots\dots\dots\dots(4)$$

where the first summation extends to every particle and the second to every *pair* of particles.

If the external force at x, y be P and be inclined at an angle α, we have

$$X = P\cos\alpha, \qquad Y = P\sin\alpha;$$

so that, if $x = r\cos\theta$, $y = r\sin\theta$ as usual, (4) may be written

$$\Sigma Pr\sin(\theta + \alpha) + \Sigma R\rho \sin 2\phi = 0. \quad \dots\dots\dots\dots(5)$$

As simple examples of these equations, consider the square framework with one diagonal represented in Figs. 1 and 2, and take the coordinate axes parallel to the sides of the square. Since $\sin 2\phi = 0$ for all four sides of the square, the only R that occurs is that which acts along the diagonal where $\sin 2\phi = -1$. In Fig. 1 opposed forces P act at the middle points of the sides, but since in each case $\theta + \alpha = 0$, the terms containing P disappear. Hence $R = 0$.

In Fig. 2, where external forces P act diagonally at the unconnected corners, $\sin(\theta + \alpha) = -1$, and since $\rho = 2r$, $R = -P$, signifying that the diagonal piece acts as a tie under tension P. In neither case would the *weight* of the members disturb the conclusion.

Fig. 1.   Fig. 2.

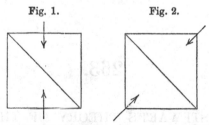

The forces exercised by the containing vessel upon a liquid confined under hydrostatic pressure p contribute nothing to the left-hand member of (4). The normal force acting inwards upon the element of boundary ds is $p\,ds$, so that

$$X = -p\,dy, \qquad Y = p\,dx,$$

and accordingly

$$\Sigma\,[xY + yX] = \tfrac{1}{2}p \int d\,(x^2 - y^2),$$

vanishing when the integration extends over the whole boundary.

Abandoning now the supposition that the particle at x, y is at rest, we have

$$\frac{d^2\,(xy)}{dt^2} = 2\,\frac{dx}{dt}\frac{dy}{dt} + x\,\frac{d^2 y}{dt^2} + y\,\frac{d^2 x}{dt^2},$$

so that if m be the mass of the particle, X, Y the components of force acting upon it,

$$2m\,\frac{dx}{dt}\frac{dy}{dt} = m\,\frac{d^2\,(xy)}{dt^2} + xY + yX; \quad \dots\dots\dots\dots(6)$$

or with summation over all the particles of the system,

$$2\Sigma\,m\,\frac{dx}{dt}\frac{dy}{dt} = \frac{d^2}{dt^2}\,\Sigma\,(mxy) + \Sigma\,(xY + yX). \quad \dots\dots\dots(7)$$

We now take the mean values with respect to time of the various terms in (7). If the system be such that

$$\frac{d}{dt}\,\Sigma\,(mxy)$$

does not continually increase, we obtain, as in the case of the virial theorem,

$$2\Sigma\,m\,\frac{dx}{dt}\frac{dy}{dt} = \Sigma\,(xY + yX). \quad \dots\dots\dots\dots(8)$$

It would seem that this equation has application to the molecular theory of the viscosity of gases, analogous to that of the virial as applied to hydrostatic pressure.

263.

ON BALFOUR STEWART'S THEORY OF THE CONNEXION BETWEEN RADIATION AND ABSORPTION.

[*Philosophical Magazine*, I. pp. 98—100, 1901.]

On a recent occasion* I remarked that Stewart's work appeared to me to be insufficiently recognized upon the Continent. One reason for this is probably the comparative inaccessibility of the *Edinburgh Transactions* in which his first paper appeared†. Another may be found in the fact that the paper itself is not well arranged, and that the principal conclusion is put forward in the first instance as if it were the result of Stewart's special experiments. The experiments were indeed of great value; but this course gave an opening to Kirchhoff's objection that " this proof [of the law that *the absorption of a plate equals its radiation and that for every description of heat‡*] cannot be a *strict* one, because experiments which have only taught us concerning *more* and *less*, cannot strictly teach us concerning *equality*§." I am inclined to think that Stewart would have received more recognition if he had never experimented at all!

While yielding to no one in admiration for Kirchhoff, I can hardly regard him as in this matter an impartial critic. In a paper‖ which should be studied by the historical inquirer, Stewart himself protests against some of Kirchhoff's remarks, and to my judgment makes out his case. In his excellent *Handbuch der Spectroscopie*, recently published, Prof. Kayser, with evident desire to be impartial, gives Stewart much, but not all, of the credit that I would claim for him. But, so far as I have seen, neither Stewart himself nor any of his critics favourable or unfavourable have cited the paragraph upon which he mainly relies. It may be of service to readers who are unlikely to see the original, if I reproduce it here, exactly as it stood :—

---

* *Phil. Mag.* S. 5, Vol. XLIX. p. 539 (1900). [Vol. IV. p. 483.]

† *Edin. Trans.* Vol. XXII. p. 1, March 1858.

‡ The italics are Stewart's.

§ Kirchhoff, " On the History of Spectrum Analysis," &c., *Phil. Mag.* Vol. XXV. p. 258 (1863).

‖ *Phil. Mag.* Vol. XXV. p. 354 (1863).

20.   A more rigid demonstration may be given thus:—Let AB, BC be two contiguous, equal, and similar plates in the interior of a substance of indefinite extent, kept at a uniform temperature.
The accumulated radiation from the interior impinges on the upper surface of the upper plate; let us take that portion of it which falls upon the particles A, in the direction DA. This ray, in passing from A to B will have been partly absorbed by the substance between A and B; but the radiation of the upper plate being equal to its absorption (since its temperature remains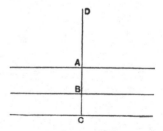
the same), the ray will have been just as much recruited by the united radiation of the particles between A and B, as it was diminished in intensity by their absorption. It will therefore reach B with the same *intensity* as it had at A. But the *quality* of the ray at B will also be the same as its *quality* at A. For, if it were different, then either a greater or less proportion would be absorbed in its passage from B to C, than was absorbed of the equally intense ray at A, in its passage between A and B. The amount of heat absorbed by the particles between B and C would therefore be different from that absorbed by the particles between A and B. But this cannot be; for, on the hypothesis of an equal and independent radiation of each particle, the radiation of the particles between B and C is equal to that of the particles between A and B, and their absorption equals their radiation. Hence the radiation impinging on B, in the direction of DB, must be equal in quality as well as quantity to that impinging upon A; and, consequently, the radiation of the particles between A and B must be equal to their absorption, as regards quality as well as quantity; that is, this equality between the radiation and absorption must hold for every individual description of heat[*]."

Surely this goes to the root of the matter, and it presents the argument in its most natural form. Kirchhoff's independent investigation of a year and a half later[†] is more formal and elaborate, but scarcely more convincing.

No one in England or elsewhere disputes the great obligations under which Spectrum Analysis lies to Kirchhoff. In a passage quoted by Dr Kayser (*loc. cit.* p. 92) from Lord Kelvin—"To Kirchhoff belongs, I believe, solely the great credit of having first actually sought for and found other metals than sodium in the sun"—the force of "solely" seems to have been misunderstood. I have Lord Kelvin's authority for interpreting this to mean that the *entire* credit of the discovery mentioned belongs to Kirchhoff, not that he is entitled to no credit in other directions.

[*] *Edin. Trans.* Vol. XXII. p. 13, March 1858.
[†] *Monatsbericht der Akad. d. Wiss. zu Berlin*, Dec. 1859.

264.

SPECTROSCOPIC NOTES CONCERNING THE GASES OF THE ATMOSPHERE.

[*Philosophical Magazine*, I. pp. 100—105, 1901.]

On the Visibility of Hydrogen in Air.

MY first experiments upon this question were made in July 1897. The sparks were taken between platinum points in a small chamber through which dried air at atmospheric pressure could be led, and the spectrum was examined with a spectroscope of two prisms. The C-line could be very nearly obliterated by careful drying. But if $\frac{1}{2000}$ part by volume of hydrogen were added and the mixture passed afterwards through the phosphoric anhydride, the increased visibility of C was very marked. At that time I was occupied with the density of carbonic oxide and interested in the question as to whether it contained appreciable quantities of hydrogen*. When carbonic oxide, prepared from prussiate of potash and dried as for weighing, was passed through the apparatus, the C-line became nearly invisible; but the test with carbonic oxide was thought to be less delicate than with air in consequence of the proximity of another bright line in the former case.

I have lately resumed these experiments, induced thereto principally by the remarkable results of M. Gautier. This observer, working by chemical methods, finds that air normally contains about $\frac{2}{10,000}$ of hydrogen in addition to variable amounts of hydrocarbons. It appeared to me that a spectroscopic confirmation would be interesting.

---

* *Proc. Roy. Soc.* Vol. LXII. p. 205 (1897). [Vol. IV. p. 348.]

Using the old apparatus, in which the tubes conveying the gas and the electrodes were fitted into a rubber cork, I could not succeed in getting quit of C from the spectrum of somewhat powerful sparks, however carefully the air were dried. The coil was excited with five Grove cells and a large leyden-jar was connected with the secondary in the usual way. This observation was of course consistent with the presence of hydrogen in the atmosphere; but it was suspicious that the best approach to evanescence was obtained with a somewhat brisk rather than with a slow current of air, indicating that the source of the hydrogen was in the apparatus rather than in the atmosphere. As it seemed desirable to apply heat, I discarded the old apparatus, substituting for it a simpler one consisting merely of a small bulbous enlargement of the gas-leading tube. Into this the platinum elec-trodes were sealed. The gases under examination were stored under slight pressure and on leaving the reservoirs were partially dried with sulphuric acid. A three-way tap allowed the easy substitution of one gas for another. After passing this tap the gas was further dried by phosphoric anhydride on its way to the sparking-bulb.

The application of heat to the bulb and to the short length of tubing between the bulb and the phosphoric anhydride led, as was expected, to a recrudescence of C. Subsequently there seemed to be an improvement. Not only was C less conspicuous, but its visibility remained about the same although the rate of flow were varied. It is difficult to describe in words the effect upon the eye, but I may say that with the actual spectroscopic arrangements including a somewhat wide slit the line could be certainly and steadily seen.

The above was the appearance with a stream of (country) air. When air to which $\frac{1}{5000}$ part of hydrogen (by volume) had been added was substituted, the visibility of C was markedly increased; and the difference was such that one could easily believe that the proportion of hydrogen actually operative had been doubled. This conclusion would be in precise agreement with M. Gautier, could we assume that the smaller quantity of hydrogen really accompanied the air. But the facts now to be recorded render this assump-tion extremely doubtful.

In the first place the visibility of C with ordinary air was not perceptibly diminished by passage of the air over red-hot cupric oxide included between the sulphuric acid and the three-way tap. It may be argued that cupric oxide is not competent in moderate length to remove the last traces of hydrogen from air, even though the air be passed over it in a slow stream. I found, however, on a former occasion* that hydrogen purposely introduced

---

* On an Anomaly encountered in Determinations of the Density of Nitrogen Gas, *Proc. Roy. Soc.* Vol. LV. p. 343 (1894). [Vol. IV. p. 107.]

into nitrogen could be so far removed in this way that the weight remained sensibly unaffected, although $\frac{1}{10,000}$ of residual hydrogen might be expected to manifest itself.

Moreover, when air purposely contaminated as above with $\frac{1}{5000}$ of hydrogen was passed over the copper oxide, the additional hydrogen appeared to be removed, the visibility of C reducing itself to that corresponding to untreated air.

Being desirous of testing the matter as far as possible, I have experimented also with nitrous oxide and with oxygen. In the former case the general appearance of the spectrum is much the same as with air. The C-line was thought to be, if anything, *more* visible than in the case of air, but the difference could not be depended upon. Two samples of gas were tried, one from an iron bottle as supplied commercially, the other prepared in the laboratory from ammonium nitrate. Oxygen from permanganate of potash also showed the C-line more distinctly than air, but this may probably be attributed to the elimination of a neighbouring nitrogen line. It is possible, of course, that these gases may have contained traces of hydrogen, but in that case it is strange that the proportion should be so nearly the same as in air.

These observations certainly seem to leave a minimum of room for the hydrogen found by M. Gautier, but I should be unwilling to call his conclusion in question on the strength of what are after all but eye estimates. I have not been able to find a detailed account of M. Gautier's experiments or of what precautions he took to assure himself that the water collected could not have had its origin in the glass or copper oxide of his hot tubes*. The most satisfactory test would be comparison experiments in which oxygen or nitrous oxide is substituted for air, or, perhaps better still, in which air is used over and over again.

If, as I should suppose were I to judge from my own experiments alone, the residual C-line was not wholly or even principally due to hydrogen in the air, it would have to be explained by hydrogen evolved from the glass of the sparking-chamber or from the platinum electrodes. In view of what is known respecting the behaviour of vacuum-tubes, such an explanation does not appear improbable.

Experiments upon the visibility of C in vacuum-tubes have shown a much smaller degree of sensibility. The tube was in connexion with a Töpler pump and was traversed by a stream of air. The passage from high (atmospheric) to low pressure took place at a glass capillary which allowed about 30 c.c. per hour (reckoned at atmospheric pressure) to leak past. When

* [1902. See *Annales de Chimie*, xxii. Jan. 1901. Further experiments of my own are detailed in *Phil. Mag.* iii. p. 416, 1902.]

moist air from the room on a damp day (15° C.) was admitted, the hydrogen C-line was very bright, nearly obliterating one of the dark bands of nitrogen. On drying the air with phosphoric anhydride, the C-line disappeared. Air mixed with 1 per cent. of hydrogen showed it doubtfully, $1\frac{1}{2}$ per cent. plainly, 2 per cent. perhaps equally with the moist air. The much smaller sensibility (about 50 times) in these experiments may be partly due to the less favourable character of the ground upon which the hydrogen line has to show itself.

Demonstration at Atmospheric Pressure of Argon from very small quantities of Air.

Success in reducing the necessary amount of air depends a good deal upon the form of tube employed. That sketched (Fig. 1) allows a minimum residuo to be sparked and examined. In some experiments the tube, standing over weak alkali, was charged with 5 c.c. only of air. The first part of the sparking is with electrodes ending in platinum points and brought up in U-shaped tubes of which the bends are filled with mercury. A Ruhmkorff actuated by two or three Grove cells is employed at this stage, and the sparks pass just under the shoulder of the containing tube, oxygen being supplied as required from a small electrolytic generator. When the volume is sufficiently reduced and most of the nitrogen has disappeared, the electrodes above spoken of are removed.

Fig. 1.

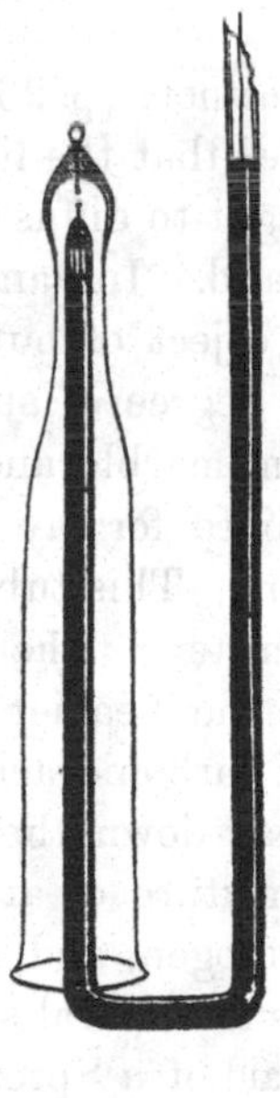

Scale $\frac{2}{3}$.

In the next stage the sparks are taken between a sealed-in electrode at the top of the containing tube, and another sealed into the top of a single U-tube brought round through the alkali, and rising (as shown) through the narrow part of the containing tube. In order to avoid splashing and consequent risk of fracture from sudden cooling of the heated glass, it was thought an advantage that the tubes through part of their length should make a tolerably close fit. But the most important precaution appears to be the use of very short sparks and a reduction of the battery to two cells. When it is desired to observe the spectrum, a small jar must be connected in the usual way.

The spectroscope employed had two prisms, and the sparks were focused upon a somewhat wide slit by a 2-inch lens. A low-power eyepiece was favourable.

The group of lines in the argon spectrum first observed by Schuster* [for figure, see Vol. IV. p. 199] is easily seen. Owing to the warmth F is very diffuse, sometimes nearly filling up the interval between 4879 and 4847. On one occasion when the original air taken was only 5 c.c., the group was almost as distinct as with pure argon. The residual gas, measured cold, was probably no more than ·1 c.c. This was rather an extreme case, and it would not have been possible to renew the sparking without an addition of oxygen, to be afterwards removed by careful additions of hydrogen. But the argon spectrum shows fairly well even when the gas is diluted with two or three times its volume of oxygen.

I have described this experiment at some length because I think that it would make a good exercise for students, requiring no special apparatus but what they should be able to construct for themselves. Although, as I have said, 5 c.c. of air is ample, a novice would naturally begin with 10 or 15 c.c.

Concentration of Helium from the Atmosphere.

In a footnote [p. 266] to a paper on the Separation of Gases by Diffusion†, I suggested that the lighter constituent of a mixture might be concentrated by causing it to diffuse against a stream of an easily absorbable gas, such as carbonic acid. In Jan. 1899 a good many trials upon these lines were made with the object of putting in evidence the helium of the atmosphere, and a certain degree of success was attained. A stream of carbonic acid (prepared from marble and hydrochloric acid and reckoned at 3 litres per hour) is maintained for say 14 hours through a diffusion-tube open above to the atmosphere. This tube, placed vertically, is about 40 cm. long and of about 5 cm. diameter. The gases of the atmosphere diffuse downwards into the tube, but the heavier constituents are held almost entirely at bay by the stream of carbonic acid. If we draw off continuously a supply from a point say halfway down the diffusion-tube, we shall obtain carbonic acid with a small admixture of atmospheric gases in which the lighter ingredients, e.g. water, hydrogen, and helium, are relatively much concentrated. In my experiments the lateral stream was about 250 c.c. per hour, and was manipulated with the aid of a Sprengel pump. Between the pump and the diffusion-tube was interposed a short length of tobacco-pipe through the walls of which the gas had to pass and which presented the right degree of obstruction. After passage to the low-pressure side, the bulk of the CO_2 was absorbed with alkali, and the residual gases collected over alkali at the foot of the Sprengel in the usual way.

* Rayleigh and Ramsay, *Phil. Trans.* 186, p. 224 (1895). [Vol. IV. p. 169.] See also *Nature*, Vol. LII. p. 163 (1895). [Vol. IV. p. 199.]

† *Phil. Mag.* Vol. XLII. p. 493 (1896). [Vol. IV. p. 266.]

The subsequent treatment for removal of nitrogen by the electric discharge was conducted as usual, towards the close in the tube described and figured above. The final residue on the occasion when D_3 was best seen (under the jar discharge) was about ·25 c.c. Argon was also plainly visible and probably constituted the greater part of the bulk. When the volume was doubled by addition of oxygen, D_3 was seen less well.

Success depended a good deal upon precautions to avoid the presence of gases, and especially of argon, which had not undergone diffusion. It was necessary to eliminate the dissolved gases of the dilute hydrochloric acid with which the CO_2 was prepared, and to keep an atmosphere of CO_2 in the supply-vessel. Until these precautions were taken, D_3, though frequently suspected, was not clearly and steadily seen. Even at the best, good measurements could hardly have been taken; but the line appeared to be in the right place for helium, as distinguished for example from neon.

265.

ON THE STRESSES IN SOLID BODIES DUE TO UNEQUAL HEATING, AND ON THE DOUBLE REFRACTION RESULTING THEREFROM*.

[*Philosophical Magazine*, I. pp. 169—178, 1901.]

THE phenomena of light and colour exhibited in the polariscope when strained glass is interposed between crossed nicols are well known to every student of optics. The strain may be of a permanent character, as in glass imperfectly annealed or specially unannealed, or it may be temporary, due to variations of temperature or to mechanical force applied from without. One of the best examples under the last head is that of a rectangular bar subjected to flexure, the plane of the flexure being perpendicular to the course of the light. The full effect is obtained when the length of the bar is at 45° to the direction of polarization. The revival of light is a maximum at the edges, where the material traversed is most stretched or compressed, while down the middle a dark bar is seen representing the "neutral axis." It is especially to be noted that the effect is due to the glass being *unequally* stretched in the two directions perpendicular to the line of vision. Thus in the case under discussion no force is operative perpendicular to the length of the bar. Under a purely hydrostatic pressure the singly refracting character of the material would not be disturbed.

When a piece of glass, previously in a state of ease, is unequally heated, double refraction usually ensues. This is due, not directly to the heat, but to the stresses, different in different directions and at different places, caused by the unequal expansions of the various parts. The investigation of these stresses is a problem in Elasticity first attacked, I believe, by

---

* From the Lorentz Collection of Memoirs.

J. Hopkinson*. It will be convenient to repeat in a somewhat different notation his formulation of the general theory, and afterwards to apply it to some special problems to which the optical method of examination is applicable.

In the usual notation† if P, Q, R, S, T, U be tho components of stress; u, v, w the displacements at the point x, y, z; λ, μ the elastic constants; we have such equations as

$$P = \lambda\left(\frac{du}{dx} + \frac{dv}{dy} + \frac{dw}{dz}\right) + 2\mu\frac{du}{dx}, \quad \dots\dots\dots\dots(1)$$

$$S = \mu\left(\frac{dw}{dy} + \frac{dv}{dz}\right). \quad \dots\dots\dots\dots\dots\dots(2)$$

These hold when the material is at the standard temperature. If we suppose that the temperature is raised by θ and that no stresses are applied,

$$\frac{du}{dx} = \frac{dv}{dy} = \frac{dw}{dz} = \kappa\theta,$$

while dw/dy &c. vanish. The stresses that would be needed to produce the same displacements without change of temperature are

$$P = Q = R = (3\lambda + 2\mu)\,\kappa\theta, \qquad S = T = U = 0.$$

Hence, so far as the principle of superposition holds good, we may write in general

$$P = \lambda\left(\frac{du}{dx} + \frac{dv}{dy} + \frac{dw}{dz}\right) + 2\mu\frac{du}{dx} - (3\lambda + 2\mu)\,\kappa\theta, \quad \dots\dots(3)$$

$$S = \mu\left(\frac{dw}{dy} + \frac{dv}{dz}\right), \quad \dots\dots\dots\dots\dots\dots(4)$$

with similar equations for Q, R, T, U.

If there be no bodily forces, the equation of equilibrium is

$$\frac{dP}{dx} + \frac{dU}{dy} + \frac{dT}{dz} = 0, \quad \dots\dots\dots\dots\dots(5)$$

with two similar equations; or with use of (3) and (4)

$$(\lambda + \mu)\frac{d}{dx}\left(\frac{du}{dx} + \frac{dv}{dy} + \frac{dw}{dz}\right) + \mu\nabla^2 u - \gamma\frac{d\theta}{dx} = 0, \quad \dots\dots(6)$$

if

$$\gamma = (3\lambda + 2\mu)\,\kappa. \quad \dots\dots\dots\dots\dots(7)$$

One of the simplest cases that can be considered is that of a plate, bounded by infinite planes parallel to xy, and so heated that θ is a function of z only.

* *Mess. of Math.* Vol. VIII. p. 168 (1879). [1902. From a notice by W. König (*Beiblätter*, 1901) I gather that some of the problems here dealt with had already been treated by Neumann in 1841.]
† See, for example, Love's *Theory of Elasticity*, Cambridge University Press, 1892.

If, further, θ be symmetrical with respect to the middle surface, the plate will remain unbent; and if the mean value of θ be zero, the various plane sections will remain unextended. Assuming, therefore, that u, v vanish while w is variable, we get from (3) and (4)

$$R = (\lambda + 2\mu)\frac{dw}{dz} - \gamma\theta = 0, \quad \dots\dots\dots\dots\dots\dots(8)$$

$$P = Q = \lambda\frac{dw}{dz} - \gamma\theta, \quad \dots\dots\dots\dots\dots\dots\dots(9)$$

$$S = T = U = 0. \quad \dots\dots\dots\dots\dots\dots\dots\dots(10)$$

In (8) R is assumed to vanish, since no force is supposed to act upon the faces. From (8), (9)

$$P = Q = -\frac{2\mu\gamma\theta}{\lambda + 2\mu}. \quad \dots\dots\dots\dots\dots\dots(11)$$

If the plate be examined in the polariscope by light traversing it in the direction of y, the double refraction, depending upon the difference between R and P, of which the former is zero, is represented simply by (11). Dark bars will be seen at places where $\theta = 0$. If the direction of the light be across the plate, *i.e.* parallel to z, there is no tendency to double refraction, since everywhere $P = Q$.

In the above example where every layer parallel to xy remains unextended, the local alteration of temperature produces its full effect. But in general the circumstances are such that the plate is able to relieve itself to a considerable extent. A uniform elevation of temperature, for instance, would entail no stress. And again, a uniform temperature gradient, such as would finally establish itself if the two surfaces of the plate were kept at fixed temperatures, is compensated by *bending* and entails no stress. In such cases before calculating the stress by (11) we must throw out the mean value of θ so as to make $\int P dz = 0$, and also such a term proportional to the distance from the middle surface as shall ensure that $\int Pz dz = 0$. Otherwise the edges of the plate could not be regarded as free from imposed stress in the form of a force or couple.

The assumption in (1), (2) that $u = v = 0$ is now replaced by

$$u = (\alpha + \beta z)x, \qquad v = (\alpha + \beta z)y, \quad \dots\dots\dots\dots(12)$$

and
$$w = w' - \tfrac{1}{2}\beta(x^2 + y^2), \quad \dots\dots\dots\dots\dots(12')$$

where w' is a function of z only. We find

$$R = (\lambda + 2\mu)\frac{dw'}{dz} + 2\lambda(\alpha + \beta z) - \gamma\theta, \quad \dots\dots\dots\dots(13)$$

$$P = Q = \lambda\frac{dw'}{dz} + (2\lambda + 2\mu)(\alpha + \beta z) - \gamma\theta, \quad \dots\dots\dots(14)$$

$$S = T = U = 0. \quad \dots\dots\dots\dots\dots\dots\dots\dots(15)$$

Since R is supposed to vanish, we get

$$P = Q = \frac{2\mu\gamma}{\lambda + 2\mu}\left[\frac{\alpha + \beta z}{\kappa} - \theta\right]. \quad \dots\dots\dots\dots\dots(16)$$

In (16) α and β are to be determined by the conditions

$$\int P\,dz = 0, \qquad \int Pz\,dz = 0;$$

or, which comes to the same, we are to reject from θ such linear terms as will leave

$$\int \theta\,dz = 0, \qquad \int \theta z\,dz = 0. \quad \dots\dots\dots\dots\dots\dots(17)$$

Since w' and θ are independent of x and y, the equations of equilibrium (5) are satisfied.

It is of interest to trace the influence of time upon the double refraction of the heated plate when light passes through it edgeways, e.g. parallel to y. Initially θ may be supposed to be an arbitrary function of z, while the faces of the plate, say at 0 and c, are maintained at given temperatures. Ultimately the distribution of temperature is expressed by a linear function of z, say $H' + Kz$; and, as is known from Fourier's theory, the distribution at time t may be expressed by

$$\theta = H' + Kz + \Sigma A_n e^{-p_n t}\sin(n\pi z/c), \quad \dots\dots\dots\dots(18)$$

where n is an integer and p_n, depending also upon the conductivity, is proportional to n^2. After a moderate interval, the terms corresponding to the higher values of n become unimportant.

In the subsequent calculation it is convenient to take the origin of z in the middle surface, instead of as in (18) at one of the faces. Thus

$$\theta = H + Kz + A_1 e^{-p_1 t}\cos\frac{\pi z}{c} - A_3 e^{-p_3 t}\cos\frac{3\pi z}{c} + \dots$$

$$- A_2 e^{-p_2 t}\sin\frac{2\pi z}{c} + A_4 e^{-p_4 t}\sin\frac{4\pi z}{c} - \dots. \quad \dots\dots\dots(19)$$

If θ' represent the value of θ when reduced by the subtraction of the proper linear terms as already explained, we find

$$\theta' = A_1 e^{-p_1 t}\left(\cos\frac{\pi z}{c} - \frac{2}{\pi}\right) - A_3 e^{-p_3 t}\left(\cos\frac{3\pi z}{c} + \frac{2}{3\pi}\right) + \dots$$

$$- A_2 e^{-p_2 t}\left(\sin\frac{2\pi z}{c} - \frac{6z}{\pi c}\right) + A_4 e^{-p_4 t}\left(\sin\frac{4\pi z}{c} + \frac{6z}{2\pi c}\right) - \dots. \quad \dots(20)$$

After a moderate time the term in A_1 usually acquires the preponderance, and then $\theta' = 0$ when $\cos(\pi z/c) = 2/\pi$. When the plate is looked at edgeways in the polariscope, dark bars are seen where $z = \pm\,·280c$, c being the whole thickness of the plate.

As a particular case of (19), (20) let us suppose that the distribution of temperature is symmetrical, or that K vanishes as well as the coefficients of even suffix A_2, A_4, &c. H then represents the temperature at which the two faces are maintained, and (19) reduces to

$$\theta = H + A_1 e^{-p_1 t} \cos \frac{\pi z}{c} - A_3 e^{-p_3 t} \cos \frac{3\pi z}{c} + \ldots \qquad \ldots \ldots (21)$$

If we suppose further that the initial temperature is uniform and equal to Θ, we find by Fourier's methods

$$A_1 = \frac{4}{\pi}(\Theta - H), \qquad A_3 = \frac{4}{3\pi}(\Theta - H), \qquad A_5 = \frac{4}{5\pi}(\Theta - H), \quad \&c. \quad \ldots (22)$$

and

$$\frac{\pi}{4} \frac{\theta'}{\Theta - H} = e^{-p_1 t}\left(\cos \frac{\pi z}{c} - \frac{2}{\pi}\right) - \tfrac{1}{3} e^{-p_3 t}\left(\cos \frac{3\pi z}{c} + \frac{2}{3\pi}\right)$$

$$+ \tfrac{1}{5} e^{-p_5 t}\left(\cos \frac{5\pi z}{c} - \frac{2}{5\pi}\right) - \ldots, \qquad \ldots \ldots (23)$$

where also

$$p_3 = 9 p_1, \qquad p_5 = 25 p_1, \qquad \&c. \quad \ldots \ldots \ldots (24)$$

At the middle surface, where $z = 0$, the right-hand member of (23) becomes

$$e^{-p_1 t}\left(1 - \frac{2}{\pi}\right) - \tfrac{1}{3} e^{-9 p_1 t}\left(1 + \frac{2}{3\pi}\right) + \ldots \qquad \ldots \ldots \ldots (25)$$

Initially

$$(25) = 1 - \frac{1}{3} + \frac{1}{5} - \ldots \ldots - \frac{2}{\pi}\left(1 + \frac{1}{3^2} + \frac{1}{5^2} + \ldots\right) = \frac{\pi}{4} - \frac{2}{\pi}\cdot\frac{\pi^2}{8} = 0,$$

as was required. If we put $e^{-p_1 t} = T$, (25) may be written

$$T\left(1 - \frac{2}{\pi}\right) - \tfrac{1}{3}T^9\left(1 + \frac{2}{3\pi}\right) + \tfrac{1}{5}T^{25}\left(1 - \frac{2}{5\pi}\right) - \ldots ; \qquad \ldots \ldots (26)$$

and (26) may be tabulated as a function of T, and thence of t. It vanishes when $T = 1$ and when $T = 0$. The maximum value occurs when $T = \cdot 747$. When T is less than this, which corresponds to an increased value of t, only the first two or three terms in (26) need be regarded. The above value of T gives

$$p_1 t = \cdot 292 ;$$

and if, as for glass, the diffusivity for heat in C.G.S. measure be $\cdot 004$, we get

$$T = \frac{\cdot 292\, c^2}{\cdot 004\, \pi^2}. \qquad \ldots \ldots \ldots \ldots \ldots \ldots \ldots (27)$$

Thus if a plate of glass be one centimetre thick, so that $c = 1$, the light seen in the polariscope at the centre of the thickness is a maximum about $7\frac{1}{2}$ seconds after heat is applied to the faces.

The following small table will give an idea of the relation between (26) and T.

T	(26)	T	(26)
0·0	0·0000	0·6	0·2139
0·1	0·0363	0·7	0·2381
0·2	0·0727	0·8	0·2371
0·3	0·1090	0·9	0·1823
0·4	0·1453	1·0	0·0000
0·5	0·1809		

In his paper above referred to, Hopkinson considered the strains produced by unequal heating in a spherical mass, under the supposition that the temperature was everywhere the same at the same distance from the centre. A similar analysis applies in the two-dimensional problem, which is of greater interest from the present point of view. We suppose that everything is symmetrical with respect to an axis, taken as axis of z, and that θ is a function of r, equal to $\sqrt{(x^2 + y^2)}$, only. The displacements in the directions of z and r will be denoted by w and u; in the third direction, perpendicular to z and r, there is supposed to be no displacement.

We may commence with the strictly two-dimensional case where $w = 0$ throughout. This implies a stress R whose magnitude is given by

$$R = \lambda \left(\frac{du}{dr} + \frac{u}{r} \right) - \gamma\theta, \quad\quad\quad\quad\quad (28)$$

in which

$$\frac{du}{dr} + \frac{u}{r} \quad\quad\quad\quad\quad\quad (29)$$

represents the dilatation.

The other principal stresses operative radially and tangentially are

$$P = (\lambda + 2\mu) \frac{du}{dr} + \lambda \frac{u}{r} - \gamma\theta, \quad\quad\quad (30)$$

$$Q = \lambda \frac{du}{dr} + (\lambda + 2\mu) \frac{u}{r} - \gamma\theta. \quad\quad\quad (31)$$

The equation of equilibrium, analogous to (5), is obtained by considering the stresses operative upon the polar element of area. It is

$$\frac{d(rP)}{dr} = Q. \quad\quad\quad\quad\quad\quad (32)$$

Substituting from (30), (31), we get

$$\frac{d^2u}{dr^2} + \frac{1}{r}\frac{du}{dr} - \frac{u}{r^2} = \frac{\gamma}{\lambda + 2\mu}\frac{d\theta}{dr},$$

so that

$$\frac{du}{dr} + \frac{u}{r} = \frac{\gamma\theta}{\lambda + 2\mu} + \alpha, \quad\dots\dots\dots\dots\dots\dots(33)$$

where α is an arbitrary constant. Integrating a second time we find

$$ru = \frac{\gamma}{\lambda + 2\mu}\int_0^r \theta r\,dr + \tfrac{1}{2}\alpha r^2 + \beta, \quad\dots\dots\dots\dots(34)$$

in which, however, β must vanish, if the cylinder is complete through $r = 0$.
From (34)

$$P = (\lambda + \mu)\alpha - \frac{2\mu\gamma}{\lambda + 2\mu}\frac{1}{r^2}\int_0^r \theta r\,dr, \quad\dots\dots\dots\dots\dots(35)$$

$$Q = (\lambda + \mu)\alpha + \frac{2\mu\gamma}{\lambda + 2\mu}\frac{1}{r^2}\int_0^r \theta r\,dr - \frac{2\mu\gamma\theta}{\lambda + 2\mu}, \quad\dots\dots\dots(36)$$

and

$$P - Q = \frac{2\mu\gamma}{\lambda + 2\mu}\left\{\theta - \frac{2}{r^2}\int_0^r \theta r\,dr\right\}. \quad\dots\dots\dots\dots(37)$$

It is on $(P - Q)$ that the double refraction depends when light traverses the cylinder in a direction parallel to its axis.

In (35), (36), (37)

$$\frac{2}{r^2}\int_0^r \theta r\,dr$$

represents the mean temperature (above the standard) of the solid cylinder of radius r. It is to be remarked that the double refraction of the ray at r is independent of the values of θ beyond r, and also of any boundary-pressure. If θ increases (or decreases) continuously from the centre outwards, the double refraction never vanishes, and no dark circle is seen in the polariscope.

In the above solution if the cylinder is terminated by flat faces, we must imagine suitable forces R, given by (28), to be operative over the faces. The integral of these forces may be reduced to zero by allowing a suitable expansion parallel to the axis. Regarding dw/dz as a constant (not necessarily zero), independent of r and z, we have in place of (28)

$$R = \lambda\left(\frac{du}{dr} + \frac{u}{r}\right) + (\lambda + 2\mu)\frac{dw}{dz} - \gamma\theta. \quad\dots\dots\dots\dots(38)$$

The additions to P and Q are $\lambda dw/dz$, while $(P - Q)$ remains unchanged.

If the cylinder is long relatively to its diameter, the last state of things may be supposed to remain approximately unchanged, even though the terminal faces be free from applied force. In the neighbourhood of the ends there will be local disturbances, requiring a more elaborate analysis for their calculation, but the simple solution will apply to the greater part of the length.

The case of a thin plate whose faces are everywhere free from applied force is more difficult to treat in a rigorous manner, but the following is probably a sufficient account of the matter. By supposing $R = 0$ in (38) we get

$$(\lambda + 2\mu)\frac{dw}{dz} = \gamma\theta - \lambda\left(\frac{du}{dr} + \frac{u}{r}\right); \qquad \dots\dots\dots\dots(39)$$

and using this value of dw/dz,

$$P = \frac{2\lambda\mu}{\lambda + 2\mu}\left(\frac{du}{dr} + \frac{u}{r}\right) + 2\mu\frac{du}{dr} - \frac{2\mu\gamma\theta}{\lambda + 2\mu}, \qquad \dots\dots\dots(40)$$

$$Q = \frac{2\lambda\mu}{\lambda + 2\mu}\left(\frac{du}{dr} + \frac{u}{r}\right) + 2\mu\frac{u}{r} - \frac{2\mu\gamma\theta}{\lambda + 2\mu}. \qquad \dots\dots\dots(41)$$

Comparing these with (30), (31), we see that the only difference is that λ and γ of those equations are now replaced by

$$\frac{2\lambda\mu}{\lambda + 2\mu} \quad \text{and} \quad \frac{2\mu\gamma}{\lambda + 2\mu}.$$

Hence, instead of (37), we should have

$$P - Q = \frac{\mu\gamma}{\lambda + \mu}\left\{\theta - \frac{2}{r^2}\int_0^r \theta r\,dr\right\}, \qquad \dots\dots\dots\dots(42)$$

and the same general conclusions follow.

In the preceding calculations we have supposed that the solid is free from stress at a uniform standard temperature when u, v, w vanish. In the case of unannealed glass, it would require a variable temperature to relieve the material from stress. To meet this, θ in the above equations would have to be reckoned from the variable temperature corresponding to the state of ease, rather than from a uniform standard temperature.

Some of the questions above considered are easily illustrated experimentally. A slab of glass about 8 cm. square and 1 cm. thick, polished upon opposite edges, when placed in the polariscope shows but little revival of light so long as the temperature is uniform. The contact of the hands with the two faces suffices to cause an almost instantaneous illumination, rising to a maximum at the middle of the thickness after a few seconds. Dark bands situated about halfway between the middle and the faces are a conspicuous feature. After about 30 or 40 seconds the light fades greatly, a result more rapidly attained if the hands be removed after 10 or 20 seconds' contact. In the earlier stages of the heating the outside layers are the warmer, and being prevented from expanding fully are in a condition of *compression*. The inner layers at the same time are in tension, a conclusion that may be verified by interposition of another piece of glass, of which the mechanical condition is known, and of which the effect may be either an augmentation or a diminution of the light.

An examination in the polariscope of the so-called *toughened* glass, introduced a few years ago, is interesting. It was understood to be prepared by a sudden cooling in oil while still plastic with heat. When it is examined through the thickness of the sheet, a great want of uniformity is manifested. In spite of the shortness of the distance traversed, there is in places considerable revival of light with intermediate irregularly disposed dark bands. The course of these bands is altered when by fracture any part is relieved from the constraining influence of neighbouring parts. To make an examination by light transmitted edgewise, it was necessary to immerse the glass in a liquid of nearly equal refractivity (benzole with a little bisulphide of carbon) contained in a small tank. The width, traversed by the light, was about 1 cm. In this way, and with the aid of a magnifier, the condition of the various layers could be well made out. The dark bands of no double refraction seemed to be nearer to the faces than according to the calculation made above, but the whole thickness is so small that this observation is scarcely to be relied upon. The interior was in a state of tension, and the double refraction was nearly sufficient at the middle to give the yellow or brown of the first order. By the action of hydrofluoric acid on the lower end of one of the strips the outermost layers were dissolved away. This caused a drawing together of the dark bands towards the middle, and though a good deal remained the light was much reduced.

The cause of the *toughening* has been sought in a special crystalline condition due to the sudden cooling. There may be something of this nature ; but it would seem that most of the peculiarities manifested may be explained by reference to the known condition of stress. The fracture of glass is usually due to bending, and the failure occurs at the surface which is under tension. If, initially, the superficial layers are under strong compression, a degree of bending may be harmless which otherwise would cause fatal results. It seems possible also that the superficial compression may be the explanation of the special hardness observed.

A short length of glass rod in its natural imperfectly annealed condition may be used to illustrate symmetrical stress. The ends may be ground, and [then] either polished or provided with cover-glasses cemented with Canada balsam. In the specimen examined by me the colours varied from the black of the first order on the axis to the red of the second order near the surface. The length of the cylinder was 1·6 cm. and the diameter 1·8 cm.

266.

ON A NEW MANOMETER, AND ON THE LAW OF THE PRESSURE OF GASES BETWEEN 1·5 AND 0·01 MILLIMETRES OF MERCURY.

[*Phil. Trans.* CXCVI A. pp. 205—223, 1901.]

Received January 15,—Read February 21, 1901.

Introduction.

THE behaviour of air and other gases at low densities is a subject which presents peculiar difficulties to the experimenter, and highly discrepant results have been arrived at as to the relations between density and pressure. While Mendeleef and Siljerström have announced considerable deviations from Boyle's law, Amagat* finds that law verified in the case of air to the full degree of accuracy that the observations admit of. In principle Amagat's method is very simple. The reservoir consists mainly of two nearly equal bulbs, situated one above the other and connected by a comparatively narrow passage. By the rise of mercury from a mark below the lower bulb to another on the connecting passage, the volume is altered in a known ratio which is nearly that of 2 : 1. The corresponding pressures are read with a specially constructed differential manometer. Of this the lower part which penetrates the mercury of the cistern is single. Near the top it divides into a **U**, widening at the level of the surface of the mercury into tubes of 2 centims. diameter. Higher up again these tubes re-unite and by means of a three-way tap can be connected either with an air-pump or with the upper bulb. Suitable taps are provided by which the two branches can be isolated from one another. During the observations one branch is vacuous and the other communicates with the enclosed gas, so that the difference of levels represents the pressure. This difference is measured by a cathetometer.

It is evident that when the pressure is very low the principal difficulty relates to the measurement of this quantity, and that the errors to be feared in respect to volume and temperature are of little importance. Amagat, fully alive to this aspect of the matter, took extraordinary pains with the manometer and with the cathetometer by which it was read. An insidious

* *Ann. de Chimie*, Vol. XXVIII. p. 480 (1883).

error may enter from the refraction of the walls of the tubes through which the mercury surfaces are seen. But after all his precautions Amagat found that he could not count upon anything less than $\frac{1}{100}$ millim., even in the means of several readings. It may be well to give his exact words (p. 494):—

"Dans les expériences dont je donnerai plus loin les résultats numériques, les déterminations sont faites en général en alternant cinq fois les lectures sur chaque menisque; les lectures étaient faites au demi-centième, et les divergences dans les séries régulières oscillent ordinairement entre un centième et un centième et demi; en prenant la moyenne, il ne faut pas compter sur plus d'un centième; et cela, bien entendu, sans tenir compte des causes d'erreur indépendantes de la lecture cathétométrique. . . . . . . Les résultats numériques consignés aux Tableaux que je vais donner maintenant sont eux-mêmes la moyenne de plusieurs expériences; car, outre que les lectures ont été faites en général cinq fois en alternant, on est toujours, après avoir réduit le volume à moitié, revenu au volume primitif, puis au volume moitié: chaque expérience a donc été faite aux moins deux fois, et souvent trois et quatre."

The following are the final results for air:—

Pression initiale en millims.	$\frac{pv}{p'v'}$	Pression initiale en millims.	$\frac{pv}{p'v'}$	Pression initiale en millims.	$\frac{pv}{p'v'}$
millims.		millims.		millims.	
12·297	0·9986	3·770	1·0019	1·377	1·0042
12·260	1·0020	3·663	0·9999	1·316	1·0137
10·727	0·9992	3·165	1·0015	1·182	1·0030
7·462	1·0013	2·531	1·0013	1·140	1·0075
7·013	1·0015	2·180	1·0015	1·100	0·9999
6·210	1·0021	1·898	1·0050	0·978	1·0160
6·160	1·0025	1·852	0·9986	0·958	1·0100
4·946	1·0010	1·751	[1]·0030	0·860	1·0045
4·275	1·0048	1·457	1·0150	0·295	0·9680
3·841	1·0027	1·414	1·0143		

Since, as it would appear, the "initial" pressure is the smaller of a pair, the lowest pressure concerned is about ·3 millim. of mercury, and the error at this stage is about 3 per cent. It is not quite clear which is which of pv and $p'v'$. For while it is expressly stated that p is smaller than p', the value of v'/v is given at 2·076. I think that this is really the value of v/v'. But any lingering doubt that may be felt upon this point is of no consequence here, inasmuch as Amagat's comment upon the tabular numbers is "On ne saurait donc se prononcer, ni sur les sens ni même sur l'existence de ces écarts."

After such elaborate treatment by the greatest authority in these matters, the question would probably have long remained where Amagat left it, had not C. Bohr found reason to suspect the behaviour of *oxygen* at low pressures. This led to a prolonged and apparently very careful investigation, of which the conclusion was that at a pressure of ·7 millim. of mercury the law connecting pressure and volume is subject to a *discontinuity*.

" 1.   Bei einer Temperatur zwischen 11° und 14° C. weicht der Sauerstoff innerhalb der beobachteten Druckgrenzen von dem Boyle-Mariotte'schen Gesetze ab. Die Abhängigkeit zwischen Volumen und Druck für einen Werth des letztgenannten, grösser als 0·70 mm., kann man annähernd durch die Formel

$$(p + 0·109)\, v = k$$

ausdrücken, während die Formel für Werthe der Drucke, welche kleiner als 0·70 mm. sind:

$$(p + 0·070)\, v = k$$

ist.

2.   Sinkt der Druck unterhalb 0·70 mm., so erleidet der Sauerstoff eine Zustandsveränderung; er kann wieder durch ein Erhöhen des Druckes bis über 0·70 mm. die ursprüngliche Zustandsform übergeführt werden*."

Fig. 1.

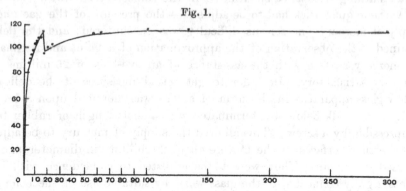

Fig. 1 is a reproduction of one of Bohr's curves, in which the ordinate represents pv and the abscissa represents p on such a scale that 1 millim. of mercury corresponds to the number 20.   It will be seen that at the place of discontinuity a change of pv to no less than $\frac{1}{10}$ of its amount occurs with no perceptible concomitant change in the value of p.   In the neighbourhood of the discontinuity the pressure is uncertain.   Thus (p. 475) " Wenn man bei einer gewissen Sauerstoffmenge im Rohre α das Quecksilber erst in der Art einstellt, dass der Druck einen etwas geringeren Werth als 0·70 millim. hat, und dann durch Verringern des Volumens den Druck über 0·70 millim. steigert (z.B. bis 0·8 millim.), so zeigt sich, dass dieser Druck nicht constant bleibt, sondern im Verlaufe von 3—5 Stunden bis zu einem Werthe sinkt, der ungefähr 10 Proc. kleiner ist, als der ursprüngliche."

* *Wied. Ann.* Vol. xxvii. p. 479 (1886).

So far as I am aware, no attempt to repeat Bohr's difficult and remarkable experiments has been recorded, but some confirmation of anomalous behaviour of oxygen in this region of pressure is afforded by the observations of Ramsay and Baly*. Sutherland† interprets the results as a " Spontaneous Change of Oxygen into Ozone and a Remarkable Type of Dissociation," and connects therewith some observations of Crookes relating to radiometer effects in oxygen gas. On the other hand, chemical tests applied by Professor Threlfall and Miss Martin‡ failed to indicate the presence of ozone in suitably expanded oxygen.

Improved Apparatus for Measuring very small Pressures.

In spite of the interest attaching to the anomaly encountered by Bohr, I should hardly have ventured to attack the question experimentally myself, had I not seen my way to what promised to be an improved method of dealing with very small pressures. In operations connected with the weighing of gases, extending over a series of years, I have had much experience of a specially constructed manometric gauge in which an iron rod, provided above and below with suitable points, is actually applied to the two mercury surfaces arranged so as to be situated in the same vertical line§. Although *two* variable quantities had to be adjusted—the pressure of the gas *and* the supply of mercury—no serious difficulty was encountered; and the delicacy obtained in the observation of the approximation of a point and its image in the mercury surface with the assistance of an eye-lens of 25 millims. focus was very satisfactory. In order to get actual measures of the delicacy, a hollow glass apparatus in the form of a fork was mounted upon a levelling table. The stalk below was terminated with a short length of rubber tubing compressible by a screw. This allowed the supply of mercury to be adjusted. The mercury surfaces in the U were about 20 millims. in diameter, and were exposed to the air. They were to be adjusted to coincidence with needle points, rigidly connected to the glass-work, by suitable use of the compressor and of the screw of the levelling table. Readings of the latter in successive and independent settings showed that a degree of accuracy was attainable much superior to the limit fixed by Amagat for the best work with the cathetometer. It is unnecessary to record the numbers obtained at this stage of the work, inasmuch as the final results to be given below prove that the errors of setting are considerably less than $\frac{1}{1000}$ millim.

It will now be possible to form a preliminary idea of the proposed mano-meter. The readings of the levelling screw, obtained as above, may be

---

* *Phil. Mag.* Vol. xxxviii. p. 301 (1894).

† *Phil. Mag.* Vol. xliii. p. 201 (1897).

‡ *Proc. Roy. Soc. of New South Wales*, 1897.

§ " On the Densities of the Principal Gases," *Proc. Roy. Soc.* Vol. liii. p. 134, 1893. [Vol. iv. p. 39.]

regarded as corresponding to the zero of pressure, or rather of pressure difference. If the pressures operative upon the mercury surfaces be slightly different, the setting is disturbed; and the change of reading at the screw required to re-establish the adjustment represents the difference of pressures. In order to interpret the result absolutely it is only necessary to know further the pitch of the levelling screw, the leverage with which it acts, and the distance between the points to which the mercury surfaces are set. If the space over one mercury surface be vacuous, the change of reading at the levelling screw represents the absolute pressure in the space over the other mercury surface.

The difficulty, which will at once present itself to the mind of the reader, in the use of a manometer on this plan, is the necessity for a flexible connexion between the instrument and the rest of the apparatus, such as the air-pump and the vessel in which the pressure is required to be known. With the aid of short lengths of rubber tubing this requirement could be easily met, but the class of work for which such a manometer is wanted would usually preclude the use of rubber. In my apparatus the requisite flexibility is obtained by the insertion of considerable lengths (3 metres) of glass tubing between the manometer and the parts which cannot turn with it. Although the adjustment was made by the screw of a levelling table as described, the actual readings were taken by the mirror method, the supports of the mirror being connected as directly as possible with the points whose angular motion is to be registered. In this way we become independent of the rigidity of the glass-work, and are permitted to use wood freely in the levelling table and in its supports. It frequently happened that an adjustment left correct was found to be out after an interval. The screw had not been

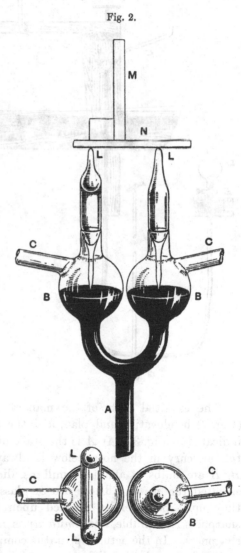

Fig. 2.

moved, but the mirror-reading was altered. On resetting by use of the screw, the original mirror-reading was recovered within the limits of error.

Fig. 3.

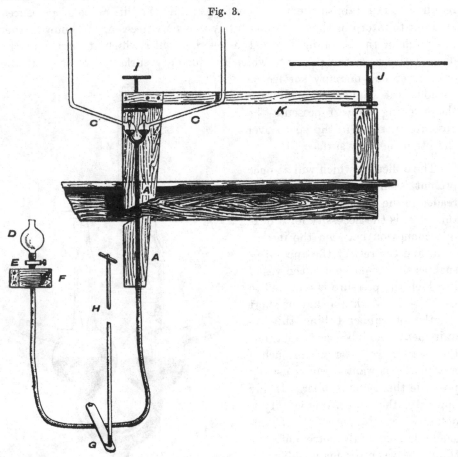

The essential parts of the manometer, as finally employed, are shown (Fig. 2) in elevation and plan, and the general scheme of the mounting is indicated in Fig. 3. At *A* is the stalk of the glass fork, of such length that the mercury in the hose below is always at a pressure above atmosphere; *B, B* are bulbs of about 25 millims. diameter, at the centres of which are situated the *points*. These are of glass*, which need not be opaque; and they must be carefully finished upon a stone. A considerable degree of sharpness is desirable, but *similarity* is more important than the extreme of sharpness. In the actual apparatus complete similarity was not attained, and in the first trials the difference was rather embarrassing. However, after a little practice the eye becomes educated to set the mercury to each point in

---

* At first iron needle points were tried.

a constant manner, and this is all that is really required. The same consideration shows that minute outstanding capillary differences should not lead to error. It may be remarked that the mercury is always on the rise at the time of adjustment, and in fact it was found best to make it a rule not to allow the points to be drowned at any time when it could be avoided. After such a drowning it was usually (perhaps always) found that the mercury surface was disturbed by the *proximity* of the points without actual contact, an effect attributed to electrification.

The presentation of the point to the mercury, or rather of the point to its image as seen by reflection in the mercury, was examined with the aid of two similar eye-lenses (not shown) of 22 millims. focus. The illumination, from a small gas flame suitably reflected by mirrors, was from behind, and it and the lenses were so arranged that both points could be seen without a motion of the head. Precautions were required to prevent the radiation from the gas flame and from the observer from producing disturbance, especially by unequal heating of the two limbs of the U. The U itself was well bandaged up, and between it and the observer were interposed sheets of copper and of insulating material so as to ensure that at all events there should be no want of symmetry in any heating that might take place.

The adjustment itself is a *double* one, requiring both the use of the levelling screw J and an accurate feed of mercury. The hose terminates as usual in a small mercury reservoir D. This facilitates the preliminary arrangements, but in use the reservoir is cut off by a screw clamp E just below it. The rough adjustment of the supply of mercury is effected by a large wooden compressor F. The fine adjustment required for the actual setting is a more delicate matter. The first attempts were by fine screw compressors acting upon the pendent part of the hose, but the tremors thence arising were found very disturbing. A remedy was eventually applied by operating upon the part of the hose which lies flat upon the floor or rather on the bottom of a mercury tray. The compressor is shown at G, Fig. 3; the screw being provided with a long handle H to bring it within convenient reach. The advantage accruing from this small device would scarcely be credited.

The glass-work is attached by cement to a board, which hangs downwards in face of the observer and is itself fixed rigidly to the levelling stand K. This is supported at two points I, which define the axis of rotation, and by a finely adjustable screw J, within reach of the observer. The whole stands in a very steady position upon the floor of an underground cellar in my country house.

The arrangements for the connexion of the mirror must now be described. The glass stems, whose lower extremities form the "points," are prolonged upwards by substantial tubing, and terminate above in three

slightly rounded ends, L, L, suitable for the support of the mirror platform N. The two supports necessary on the left are obtained by a
symmetrical branching of the tube on that side. The platform is of worked
glass, so that a slight displacement of the contacts has no effect on the
slope of the mirror. The latter is of worked glass silvered in front. Suitable
stops are provided to guide the mirror platform into the right position and to
prevent accidents, but these exercise no constraint.

The axis II about which the apparatus rotates is horizontal and parallel
to the face of the mirror, so that the sine of the angle θ of rotation from the
zero position represents the difference of levels of the mercury surfaces.
The axis II lies approximately in the mirror surface and at about the middle
of the height of the operative part. The rotation of the mirror is observed
in the usual way by means of a telescope and vertical millimetre scale. The
aperture of the object-glass is 30 millims., and the distance from the mirror
3150 millims. The readings can be taken to about ·1 millim.

In many kinds of observation the zero can only be verified at intervals,
as it requires the pressures over the mercury to be equalised. On the whole
the zero was tolerably constant to within two or three-tenths of a millimetre
of the scale. A delicate level was attached to the telescope to give warning
of any displacement of the stand (all of metal) or of the ground.

The differences of pressure to be evaluated are not quite in simple proportion to the scale reading from zero. The latter varies as $\tan 2\theta$, while the
former varies as $\sin \theta$. The correcting factor is therefore

$$\frac{\sin \theta}{\frac{1}{2} \tan 2\theta} = 1 - \tfrac{3}{2}\theta^2 \text{ approximately.}$$

If the zero reading (in millimetres) be a, and the current reading x, D the
distance between telescope and mirror,

$$\theta = \frac{x - a}{2D} \text{ approximately;}$$

so that the correcting factor is

$$1 - \tfrac{3}{2}\frac{(x - a)^2}{4D^2}.$$

The actual correction to be applied to $(x - a)$ is thus

$$-\tfrac{3}{2}\frac{(x - a)^3}{4D^2}.$$

In practice $(x - a)$ rarely exceeded 350, for which the correction would
be -1·6. When $(x - a)$ falls below 120, the correction is insensible.

The next question is the reduction to absolute measure. What (corrected) scale-reading corresponds to 1 millim. actual difference of mercury
levels? The distance between the points is 27·3 millims., so that 1 millim.

mercury corresponds to 231 millims. of the telescope scale. The highest pressure that could be dealt with is about 1½ millims. of mercury.

The above reckoning proceeds upon the supposition that the distance between the points can be regarded as invariable. Certain small discrepancies manifested at the higher slopes of the apparatus induced me to examine the question more particularly, for it seemed not impossible that owing to the bending of the glass-work some displacement might occur. But a rather troublesome measurement of the actual distance in various positions by means of microscopes negatived this idea. I would however recommend that this point be kept specially in view in the design of any subsequent apparatus of this kind.

Experiments to determine the Relation of Pressure and Volume at given Temperature.

In order to test Boyle's law one of the lateral branches C is connected to the air-pump and the other to the chamber in which the gas is contained. The pump is of the Töpler form, and is provided with a bulb containing phosphoric anhydride. No tap or contracted passage intervenes between the pump-head and B. A lateral channel communicates with a three-way tap, by which this side of the apparatus can be connected with the gas-generating vessel. The third way leads to a blow-off under mercury more than a barometer-height below.

The two sides of the apparatus are connected by a cross-tube which can be closed or opened by means of a tap. The plug of this tap is provided with a wide bore. When it is intended to read the zero, the tap is open. If desired, the mercury may be raised in the Töpler so as to prevent the penetration of gas into the pump-head. When pressures are to be observed, the tap of the cross-tube is closed, and a good vacuum is made on the pump side. No particular difficulty was experienced with the vacuum. In the use of the Töpler the mercury was allowed to flow out below, and was transferred at intervals to the movable reservoir. The latter was protected from atmospheric moisture by a chloride of calcium tube. When, after standing five or ten minutes, the mercury was put over, and, on impact, gave a hard metallic sound with inclusion of no more than a small speck of gas, the vacuum was nearly sufficient, and no further change could be detected at the manometer. The capacity of the pump-head was two or three times that of the remaining space to be exhausted.

In the earlier experiments the gas-containing tube, placed vertically, was graduated to 50 cub. centims. at intervals of 10 cub. centims. Prolonged below by more than a barometer-height of smaller tubing, it terminated in a hose and mercury reservoir, the latter protected by chloride of calcium. In order to get rid of most of the adherent moisture and carbonic anhydride, the tubes on both sides of the apparatus were heated pretty strongly in a vacuous condition. The first trial was with oxygen, in the hope of at once obtaining a confirmation of Bohr's anomaly; but not succeeding in this, I fell back upon nitrogen and hydrogen. With a vacuum on the pump side, readings of pressure were taken with the mercury in the chamber at 0 and at 50 cub. centims., and the ratio of pressures (about 2 : 1) was deduced. When this had been repeated, some of the gas was allowed to escape by opening the cross-tap, the zero was again observed, and the vacuum re-established on the pump side. Another ratio of pressures could now be obtained, corresponding to the same (unknown) volumes as before, but to a different total pressure.

In utilising the ratios of pressure thus obtained, it was of course necessary to consider how far the temperature could be assumed to be unchanged within each pair of pressures brought into comparison. The general temperature of the cellar was extremely uniform, and no difference could be read upon a thermometer worth taking into account. Passing over this question for the present, we may consider how far the results conformed to Boyle's law. The agreement of the ratios, except, perhaps, at the highest pressures of about $1\frac{1}{2}$ millims. of mercury, was sufficiently good, and of itself goes a long way to confirm Boyle's law. In strictness, all that the constancy of the ratio can prove is that the relation between pressure (p) and density (ρ) is of the form

$$p = \kappa \rho^{n}, \quad\dots\dots\dots\dots\dots\dots\dots\dots\dots\dots\dots\dots(1)$$

where n is some numerical quantity. To limit n to the value unity, the constancy of the ratios might be followed up into the region of pressure for which Boyle's law is known to hold, but this can scarcely be said to have been done here. Otherwise, we need to know what the ratio of densities in the two positions of the mercury really is, and not merely that it remains constant.

In the case of the original volume chamber the first was the method employed. The smaller volume, defined by the upper mark in the volume tube and by the "point" in the manometer, was filled with dry air at a known atmospheric pressure. The included air was then isolated and expanded until it occupied the larger (approximately double) volume, and the new pressure determined by observation of the difference of levels in the tube and in a mercury reservoir similarly fashioned. The operation was rather a difficult one, and the result was only barely accurate enough. The ratio of volumes thus determined by use of Boyle's law, as applied to air at atmospheric and half atmospheric pressures, agreed sufficiently well with the ratio of pressures found by the manometer for rare hydrogen and nitrogen;

and thus Boyle's law may be considered to be extended to these rare gases. The rarefaction was carried down to a total pressure of only ·02 millim. At this stage discrepancies of the order of 5 per cent. are to be expected.

Having obtained fairly satisfactory results with hydrogen and nitrogen, I returned to oxygen, fully expecting to verify the anomalous behaviour described by Bohr. In this I have totally failed. The gas was prepared by heating permanganate of potash, dried by phosphoric anhydride, and may be regarded as fairly pure. The region of pressure round ·7 millim. was carefully examined, use being made of the intermediate divisions of the 50 cub. centims. range of volume. No unsteadiness of the kind indicated by Bohr, or appreciable departure from Boyle's law, was detected. And when the pressures were diminished down to a few hundredths of a millimetre, there was no falling off in the product of pressure and volume. The observations were repeated a second time with a fresh supply of oxygen.

The experience gained up to this date (August, 1900) showed that the manometer worked well, and that there was no difficulty about the vacuum, but I was not altogether satisfied with the way in which the volumes had been determined. There was some want of elegance, to say the least, in using Boyle's law for this purpose, and barely adequate accuracy in the application itself. The latter objection might have been overcome by the use of a suitable cathetometer, but such was not to hand. The most direct method by actually gauging with mercury the spaces concerned being scarcely feasible, I devised another method which has the advantage of easy execution and is practically independent of the assumption of Boyle's law. The opportunity was taken to increase the range over which the volume could be varied.

The new chamber, composed mainly of tubing of 18 millims. diameter, is graduated at intervals of 10 cub. centims. over a total range of 200 cub. centims. It is prolonged above and below by narrow tubing in order to connect it with the sloping manometer bulb and with the hose and mercury reservoir as before. The zero mark is situated on the upper tube a few centimetres above its junction with the wider one. It is scarcely necessary to say that no rubber was employed except for the hoses, and that these were always occupied by mercury under a pressure above atmosphere. The mercury reservoirs themselves were protected against damp by chloride of calcium.

If we call the ungauged volume (from the zero mark to the bulb of the sloping manometer with "point" set) V, and the gauged volume v, the total volume occupied by the gas is $V + v$; and the problem is how to determine V. If we may assume the correctness of Boyle's law for rare gases and may rely upon the sloping manometer, the process is simple enough. We have only to find the pressures exerted by the included gas at volumes V and $V + v$, whence by Boyle's law the ratio of these volumes is known and thus V determined in terms of v. In order to avoid the use of Boyle's law, further observations are necessary.

The requisite data can be obtained by changing the quantity of gas. Suppose that with the original quantity of gas certain pressures, P, P', correspond to total volumes, $V + v_1$, $V + v_2$, and that with a reduced amount of gas the *same* pressures are recorded with volumes $V + v_3$, $V + v_4$. Since the pressure is a function of the density, whether Boyle's law be applicable or not, it must follow that

$$\frac{V + v_1}{V + v_2} = \frac{V + v_3}{V + v_4}, \quad \dots\dots\dots\dots\dots\dots(2)$$

whence V is determined in terms of the known volumes v_1, v_2, v_3, v_4. It may be remarked that this argument does not assume even the correctness of the scale of pressures.

In carrying out the method practically it was necessary to work to the fixed marks of the volume chamber, and thus the same pressures could not be recovered *exactly*. But the use of Boyle's law in order to make what is equivalent to small corrections is unobjectionable.

With this explanation it may suffice to give the details of an actual determination executed with nitrogen. With the original quantity of gas, volumes $V + 70$, $V + 170$ gave pressures proportional to 345·4, 184·9. Sufficient gas was now removed to allow the remainder to give nearly the same higher pressure as before with $v = 0$. Thus, corresponding to volumes $V + 0$, $V + 40$ the pressures were 344·9, 183·3. We have now only to calculate V from the equation

$$\frac{V + 40}{V} = \frac{344 \cdot 9}{183 \cdot 3} \frac{184 \cdot 9}{345 \cdot 4} \frac{V + 170}{V + 70},$$

or
$$V^2 + 110V + 2800 = 1 \cdot 0072 \, (V^2 + 170V);$$

whence
$$V = 45 \cdot 5 \text{ cub. centims.}$$

The adopted value, derived from observations upon nitrogen and hydrogen, is

$$V = 45 \cdot 6 \text{ cub. centims.}$$

In charging the apparatus, the first step is to make a good vacuum throughout, the cross-tap being open. The gas supply being started, the first portions are allowed to blow off from under mercury, and then, by use of the three-way tap, a sufficiency is introduced into the apparatus to an absolute pressure of, perhaps, 10 centims. of mercury. The gas-leading tube would then be sealed off. Ultimately the remainder of the supply tube and the blow-off tube were exhausted to diminish the risk of leakage.

The " nitrogen" was prepared from air by passage over red-hot copper and desiccation with phosphoric anhydride. Accordingly it contained argon to the amount of about 1 per cent.

In taking a set of observations the procedure would be as follows. Assurance having been obtained that the vacuum was good, the next step would be to set the mercury in the volume chamber so that $v = 190$ cub. centims., then after a few minutes to adjust the sloping manometer and to read the telescope scale. It was of course necessary to ensure that sufficient time was allowed for uniformity of pressure to establish itself, and observations were frequently renewed after a quarter of an hour or longer. In the case of oxygen, to be considered later, several hours were sometimes allowed. If operations were leisurely conducted, with first a rough setting of the volume and then a rough setting of the manometer followed by accurate settings in the same order, little or no change could afterwards be detected. Indeed I was rather surprised to find how rapidly equilibrium seemed to be established. The next smaller volume, e.g., $v = 150$, would then be observed, and so on until $v = 40$. In observations to be used for the examination of Boyle's law v was not further reduced, as too much stress might thereby be thrown upon the accuracy of V. The same observations were then repeated in reverse order and the mean taken. The numbers recorded are thus the mean of two settings only of the manometer.

The next step was to allow about half the gas to escape. The mercury at the pump was allowed to rise so as to cut off the pump-head and $V + v$ was so adjusted as to be equal to the volume remaining upon the other side, about 130 cub. centims. The cross-tap was then opened, and after a sufficient interval of time the zero, corresponding to no pressure, was read. In the course of the observations upon nitrogen, extending over ten days, the zero varied from 43·5 to 43·8. Whenever possible the zero used for a set was the mean of values found before and after.

The annexed tables give the results for nitrogen in detail. In Table I., dealing with the highest quantity of gas, the first column gives the volume ($V = 45·6$ cub. centims.); the second represents the pressure, being the mean of the two actually read numbers (expressing millimetres of telescope scale) less the zero reading 43·7 and corrected to infinitely small arcs as already explained. The third column is the logarithm of the product of the first two, and should be constant if Boyle's law holds. The fourth column gives the approximate value of the pressure in millimetres of mercury; the fifth the deviation of pv from the mean taken as unity. In the sixth column is shown the amount by which the observed value of p exceeds that requisite in order to make pv constant, expressed in millimetres of mercury.

TABLE I.—Nitrogen.

November 9–11,   Zero = 43·7.

Volume in cub. centims.	Pressure in scale divisions	Log. product	Pressure in millims. Hg	Deviation of pv	Error of p in millims.
$V+$ 70	345·4	·6013	1·49	+ ·0002	+ ·0003
$V+$ 80	318·3	·6018	1·38	+ ·0014	+ ·0019
$V+$ 90	294·1	·6007	1·27	− ·0012	− ·0015
$V+$110	256·8	·6016	1·11	+ ·0009	+ ·0010
$V+$130	227·4	·6013	·98	+ ·0002	+ ·0002
$V+$150	203·7	·6004	·88	− ·0018	− ·0016
$V+$170	184·9	·6005	·80	− ·0014	− ·0011
$V+$190	169·8	·6021	·73	+ ·0021	+ ·0015
		·6012			

TABLE II.—Nitrogen.

November 11–12,   Zero = 43·7.

Volume in cub. centims.	Pressure in scale divisions	Log. product	Pressure in millims. Hg	Deviation of pv	Error of p in millims.
$V+$ 0	344·9	·1966	1·49	+ ·0007	+ ·0010
$V+$ 10	282·3	·1958	1·22	− ·0012	− ·0015
$V+$ 20	239·5	·1962	1·04	− ·0002	− ·0002
$V+$ 40	183·3	·1956	·79	− ·0016	− ·0013
$V+$ 60	148·8	·1963	·64	·0000	·0000
$V+$ 80	125·2	·1966	·54	+ ·0007	+ ·0004
$V+$110	101·1	·1968	·44	+ ·0012	+ ·0005
$V+$150	80·2	·1955	·35	− ·0018	− ·0006
$V+$190	66·9	·1976	·29	+ ·0030	+ ·0009
		·1963			

TABLE III.—Nitrogen.

November 13, Zero = 43·6.

Volume in cub. centims.	Pressure in scale divisions	Log. product	Pressure in millims. Hg	Deviation of pv	Error of p in millims.
$V+$ 40	91·1	·892	·394	·000	·0000
$V+$ 60	73·9	·892	·320	·000	·0000
$V+$ 80	62·3	·893	·269	+·002	+·0005
$V+110$	50·2	·893	·217	+·002	+·0004
$V+150$	39·6	·889	·171	−·007	−·0012
$V+190$	33·1	·892	·143	·000	·0000
		·892			

TABLE IV.—Nitrogen.

November 14, Zero = 43·5.

Volume in cub. centims.	Pressure in scale divisions	Log. product	Pressure in millims. Hg	Deviation of pv	Error of p in millims.
$V+$ 40	46·0	·595	·199	+·005	+·0010
$V+$ 60	37·1	·593	·160	·000	·0000
$V+$ 80	31·1	·592	·135	−·002	−·0003
$V+110$	25·1	·592	·109	−·002	−·0002
$V+150$	20·1	·595	·087	+·005	+·0004
$V+190$	16·5	·590	·071	−·007	−·0005
		·593			

TABLE V.—Nitrogen.

November 16, Zero = 43·5.

Volume in cub. centims.	Pressure in scale divisions	Log. product	Pressure in millims. Hg	Deviation of pv	Error of p in millims.
$V+$ 40	22·8	·290	·099	·000	·0000
$V+$ 60	18·6	·293	·081	+·007	+·0006
$V+$ 80	15·6	·292	·067	+·005	+·0003
$V+110$	12·7	·296	·055	+·014	+·0008
$V+150$	9·9	·287	·043	−·007	−·0003
$V+190$	8·15	·283	·035	−·016	−·0006
		·290			

TABLE VI.—Nitrogen.

November 17–18,   Zero = 43·7.

Volume in cub. centims.	Pressure in scale divisions	Log. product	Pressure in millims. Hg	Deviation of pv	Error of p in millims.
$V+$ 40	11·40	·989	·049	+ ·005	+ ·0002
$V+$ 60	9·10	·983	·039	− ·009	− ·0004
$V+$ 80	7·65	·983	·033	− ·009	− ·0003
$V+$110	6·25	·988	·027	+ ·002	+ ·0001
$V+$150	5·10	·999	·022	+ ·028	+ ·0006
$V+$190	4·05	·980	·017	− ·016	− ·0003
		·987			

TABLE VII.—Nitrogen.

November 18–19,   Zero = 43·8.

Volume in cub. centims.	Pressure in scale divisions	Log. product	Pressure in millims. Hg	Deviation of pv	Error of p in millims.
$V+$ 40	5·90	·703	·026	+ ·014	+ ·0004
$V+$ 60	4·60	·686	·020	− ·026	− ·0005
$V+$ 80	4·15	·717	·018	+ ·047	+ ·0008
$V+$110	3·10	·683	·013	− ·033	− ·0004
$V+$150	2·55	·698	·011	+ ·002	·0000
		·697			

In the second set the quantity of gas had been adjusted to give a suitable pressure with $v = 0$. It is from it and from Table I. that the data were obtained for the calculation of V already given.

These tables give a fairly complete account of the behaviour of nitrogen from a pressure of about 1·5 millims. down to ·01 millim. of mercury. In each set the range of pressure is nearly in the ratio of 3 : 1, and overlaps the range of the preceding and following sets. An examination of the fifth column shows no indication of departure from Boyle's law. The sixth column allows a judgment to be formed of the degree of accuracy to which the law is verified. It gives the amount by which p exceeds the value necessary in order that pv should be absolutely constant, expressed in millimetres of mercury. The errors thus exhibited include not only those arising in the setting of the manometer and the reading of the telescope, but also those entailed in the measurements of volume, and in consequence of fluctuations of temperature. The latter source of error is of course more important at the higher pressures. It will be seen that the accuracy attained is very remarkable. Even at the higher pressures the mean error is only about ·001 millim., while at the lower pressures of Tables III.—VII. the mean error is less than ·0004 millim. And it must be remembered that the numbers to which these errors relate are the means of *two* observations only.

As a means of dealing with very small pressures, the sloping manometer has proved itself in a high degree satisfactory, the performance being some twenty-five times better than Amagat's standard. It could hardly have been expected that the mean error would prove to be less than one wave-length of yellow light*. Considered as a pressure, the mean error corresponds to the change of barometric pressure accompanying an elevation of 4 millims.

On hydrogen more than one series of observations have been carried out. The specimen that will be given is not in some respects the most satisfactory, but it is chosen as having been pursued to the greatest rarefactions. The gas was dried carefully with phosphoric anhydride and was introduced into the apparatus as already described. It is thought sufficient to record only numbers corresponding to the three last columns of Tables I.—VII., the first column giving the pressure in millims. of mercury, the second the deviation of pv from the mean value of the set taken as unity, the third the error in p from what would be required to make pv absolutely constant.

In several of the sets of observations recorded in Table VIII., there would seem to be a tendency for the positive errors to concentrate towards the beginning, *i.e.*, for pv to diminish slightly with p. It was at this stage that a suspicion arose that the distance between the glass points of the manometer might not be quite constant, but, as has been related, the suspicion was not verified. It is just possible that at the higher pressures and smaller

* I had at one time contemplated an apparatus from which a further ten-fold increase in accuracy might be expected. Two beams of light, reflected nearly perpendicularly from the mercury surfaces, would be brought to interference by an arrangement similar to that used in investigating the refractivity of gases (*Proc. Roy. Soc.* Vol. LIX. p. 200, 1896 [Vol. IV. p. 218]; Vol. LXIV. p. 97, 1898 [Vol. IV. p. 364]). Preliminary trials proved that the method is feasible; but the delicacy is excessive in view of the fact that according to Hertz the pressure of mercury vapour at common temperatures itself amounts to ·001 millim.

TABLE VIII.—Hydrogen.

October—November, 1900.

Pressure in millims. Hg	Deviation of pv	Error of p in millims.	Pressure in millims. Hg	Deviation of pv	Error of p in millims.
1·43	+ ·0025	+ ·0036	1·44	+ ·0018	+ ·0026
1·31	+ ·0030	+ ·0039	1·18	− ·0005	− ·0006
1·20	+ ·0002	+ ·0002	1·00	+ ·0009	+ ·0009
1·11	− ·0012	− ·0013	·87	+ ·0007	+ ·0006
·97	− ·0005	− ·0005	·77	+ ·0005	+ ·0004
·86	− ·0002	− ·0002	·62	·0000	·0000
·77	− ·0016	− ·0012	·57	− ·0028	− ·0016
·70	− ·0025	− ·0017	·52	− ·0009	− ·0005
·64	+ ·0007	+ ·0004	·48	− ·0018	− ·0009
—	—	—	·42	+ ·0018	+ ·0008
·769	+ ·0021	+ ·0016	·386	·0000	·0000
·624	+ ·0028	+ ·0017	·315	+ ·0044	+ ·0014
·524	·0000	·0000	·264	+ ·0023	+ ·0006
·423	+ ·0002	+ ·0001	·213	+ ·0014	+ ·0003
·335	− ·0037	− ·0012	·168	− ·0072	− ·0012
·279	− ·0018	− ·0005	·140	− ·0014	− ·0002
·196	+ ·0079	+ ·0015	·098	− ·009	− ·0009
·158	+ ·0046	+ ·0007	·080	·000	·0000
·133	+ ·0053	+ ·0007	·068	+ ·005	+ ·0003
·106	− ·0053	− ·0006	·055	+ ·007	+ ·0004
·085	− ·0037	− ·0003	·044	+ ·007	+ ·0003
·070	− ·0083	− ·0006	·036	− ·005	− ·0002
·051	+ ·007	+ ·0004	·027	− ·047	− ·0013
·041	+ ·002	+ ·0001	·023	+ ·016	+ ·0004
·034	− ·009	− ·0003	·018	− ·054	− ·0010
·027	− ·023	− ·0006	·016	+ ·021	+ ·0003
·023	+ ·036	+ ·0009	·013	+ ·040	+ ·0005
·018	− ·014	− ·0003	·010	+ ·019	+ ·0002

volumes the temperature changes were not insensible. It is probable that they would operate in the direction mentioned, inasmuch as at the smaller volumes a larger proportion of the gas would be in the connecting tubes at a higher level in the room, and therefore warmer. Considerable precaution was taken, and I was not able to satisfy myself that disturbance due to temperature really existed. In another series of observations on hydrogen the tendency was scarcely apparent, and it remains doubtful whether there is any real indication of departure from Boyle's law. It may be noted that interest was concentrated rather upon the lower pressures, and that perhaps less pains were taken over the readings of the higher pressures, where in any case the error would be a smaller proportion of the whole. Also some of the observations were not repeated. Another point that may be noted is that the means are chosen with respect to the values of pv, and that a different choice would in many cases materially reduce the mean error in the last column.

Having thoroughly tested the apparatus and the method of experimenting with hydrogen and nitrogen, I returned with curiosity to the case of oxygen Special pains were taken to ensure that the gas should be pure and above all dry. To this end glass tubes were prepared containing permanganate of potash and phosphoric anhydride, and these were connected by sealing to one of the branches of the three-way tap. A high vacuum having been made throughout, heat was gradually applied, and some of the oxygen allowed to blow off. The phosphoric tube (of considerable capacity) was then allowed to stand full of gas for some little time, after which the necessary gas to a pressure of about 10 centims. was allowed to enter the apparatus by means of the three-way tap. With regard to the maintenance of the purity of the gas under rarefaction, it may be remarked that the method of experimenting was favourable, inasmuch as the last stages were not reached until the apparatus had been exposed to the gas under trial for a week or two. Any contamination that might be communicated from the glass during the first few days would for the most part be removed before the final stages were reached.

Before the regular series was commenced, special observations extending over several days were made in the region of pressure (from 1 millim. to ·5 millim.) where Bohr found anomalies. No unsteadiness could be detected. Whatever reading was obtained within a few minutes of a change of pressure was confirmed after an interval of an hour or more. For example, on November 29, at $12^h 25^m$ the pressure which had stood for some time at ·80 millim. was lowered to ·65 millim. At $8^h 0^m$ the pressure was unaltered. In no case was the behaviour in any way different to that which had been observed with the other gases. It is true that when the observations were reduced one preliminary set showed an excess of pressure at the smaller volumes similar to that recorded in the case of hydrogen, but the tendency is scarcely visible in the regular series now to be given, which extended from November 27 to December 9.

An examination of the numbers in the Table IX. shows that Boyle's law was observed, practically up to the limits of the accuracy of the measurements, and in particular that there was no such falling off in the value of pv at low pressures as was encountered by Bohr. What can be the cause of the difference of our experiences I am at a loss to conjecture. I can only suppose that it must be connected somehow with the quality of the gas, complicated perhaps by interaction with the glass or with the mercury.

TABLE IX.—Oxygen.

Pressure in millims. Hg	Deviation of pv	Error of p in millims.	Pressure in millims. Hg	Deviation of pv	Error of p in millims.
1·53	+ ·0016	+ ·0024	·580	− ·0035	− ·0020
1·17	− ·0012	− ·0014	·472	+ ·0005	+ ·0002
·95	+ ·0005	+ ·0005	·396	− ·0007	− ·0003
·80	+ ·0007	+ ·0006	·321	+ ·0016	+ ·0005
·65	+ ·0012	+ ·0008	·255	+ ·0012	+ ·0003
·57	− ·0009	− ·0005	·212	+ ·0016	+ ·0003
·51	− ·0014	− ·0007	—	—	—
·47	− ·0014	− ·0007	—	—	—
·43	+ ·0009	+ ·0004	—	—	—
·288	+ ·002	+ ·0007	·142	+ ·005	+ ·0007
·233	·000	·0000	·115	+ ·009	+ ·0011
·196	·000	·0000	·094	− ·019	− ·0018
·159	+ ·005	+ ·0008	·077	·000	·0000
·125	− ·002	− ·0003	·062	+ ·012	+ ·0007
·103	− ·009	− ·0010	·051	− ·012	− ·0006
·068	− ·002	− ·0002	·034	·000	·0000
·056	+ ·005	+ ·0003	·029	+ ·059	+ ·0017
·048	+ ·019	+ ·0009	·022	− ·042	− ·0009
·038	+ ·009	+ ·0004	·019	+ ·023	+ ·0004
·029	− ·019	− ·0005	·014	− ·035	− ·0005
·025	− ·009	− ·0002	—	—	—

The final result of the observations on the three gases may be said to be the full confirmation of Boyle's law between pressures of 1·5 millims. and ·01 millim. of mercury. If there is any doubt, it relates to the case of hydrogen, which appears to press somewhat in excess at the highest pressures. But when we consider the smallness of the amount and the various complications to which it may be due, as well as à *priori* probabilities, we may well hesitate to accept the departure from Boyle's law as having a real existence.

So far as the present results can settle the question, they justify to the full the ordinary use of McLeod's gauge within the limits of pressure mentioned and for nitrogen and hydrogen gases. The same might be said for oxygen; but until the discrepancy with the conclusions of Bohr can be explained, the necessity for some reserves must be admitted.

In any case the new manometer has done its work successfully, and is proved to be capable of measuring small pressures to about $\frac{1}{2000}$ of a millimetre of mercury. It was constructed under my direction by Mr Gordon.

267.

ON A PROBLEM RELATING TO THE PROPAGATION OF SOUND BETWEEN PARALLEL WALLS.

[*Philosophical Magazine*, I. pp. 301—311, 1901.]

THE influence of viscosity and heat conduction in modifying the propagation of sound in circular tubes of moderate dimensions has been treated by Kirchhoff * in his usual masterly style, but he passes over the case when the diameter is very large. In my book on the *Theory of Sound*, 2nd edition, § 348, I have given a full statement of Kirchhoff's theory, and have indicated the alterations required when the boundary is supposed to take the form of two parallel planes instead of a cylindrical surface. In any case the action of the wall is supposed to be such as to annihilate variation of temperature, and tangential as well as normal motion. In connexion with the problem of the propagation of sound over water I recently had occasion to extend the analysis to the case of a layer of very great thickness; and though, as the result showed, the solution fails to answer the question which I had then in view, it is of some interest in itself. In this case the practical question differs somewhat from that proposed by Kirchhoff, who assumes not only complete periodicity with respect to time, but also a quasi-periodicity with respect to x, the direction of propagation, all the functions being supposed proportional to e^{mx}, where m is a complex constant, and not otherwise to depend upon x. This assumption is retained in the present paper. It seems advisable to give a brief recapitulation of Kirchhoff's theory, referring for more detailed explanation to the original paper or to the account of it in *Theory of Sound*.

* *Pogg. Ann.* Vol. CXXXIV. 1868; *Collected Memoirs*, p. 540.

The condition of the gas at any point x, y, z being defined by the component velocities u, v, w, and θ', where θ' is proportional to the excess of temperature, the equation for θ' is found to be

$$h^2\theta' - \{a^2 + h(\mu' + \mu'' + \nu)\}\nabla^2\theta' + \frac{\nu}{h}\{b^2 + h(\mu' + \mu'')\}\nabla^4\theta' = 0. \quad ...(1)$$

In this equation ∇^2 stands for $d^2/dx^2 + d^2/dy^2 + d^2/dz^2$; h is such that all the variables (u, v, w, θ') are proportional to e^{ht}; a is the velocity of sound as reckoned on Laplacean principles, b the corresponding Newtonian value; μ', μ'', ν are coefficients of viscosity and of heat conduction.

A solution of (1) may be obtained in the form

$$\theta' = A_1Q_1 + A_2Q_2, \quad ....(2)$$

where Q_1, Q_2 are functions of x, y, z satisfying respectively

$$\nabla^2 Q_1 = \lambda_1 Q_1, \qquad \nabla^2 Q_2 = \lambda_2 Q_2, \quad ....(3)$$

λ_1, λ_2 being the roots of

$$h^2 - \{a^2 + h(\mu' + \mu'' + \nu)\}\lambda + \frac{\nu}{h}\{b^2 + h(\mu' + \mu'')\}\lambda^2 = 0; \quad ......(4)$$

while A_1, A_2 denote arbitrary constants.

In correspondence with this value of θ', particular solutions are obtained by equating u, v, w to the differential coefficients of

$$B_1Q_1 + B_2Q_2,$$

taken with respect to x, y, z. The relation of the constants B_1, B_2 to A_1, A_2 is

$$B_1 = A_1\left(\nu - \frac{h}{\lambda_1}\right), \qquad B_2 = A_2\left(\nu - \frac{h}{\lambda_2}\right). \quad ....(5)$$

More general solutions may be obtained by addition to u, v, w respectively of u', v', w', where u', v', w' satisfy

$$\nabla^2 u' = \frac{h}{\mu}u', \qquad \nabla^2 v' = \frac{h}{\mu}v', \qquad \nabla^2 w' = \frac{h}{\mu}w'. \quad ....(6)$$

Thus

$$\begin{aligned}
u &= u' + B_1dQ_1/dx + B_2dQ_2/dx, \\
v &= v' + B_1dQ_1/dy + B_2dQ_2/dy, \\
w &= w' + B_1dQ_1/dz + B_2dQ_2/dz,
\end{aligned} \right\} \quad ....(7)$$

where B_1, B_2 have the values given above.

It appears that

$$\frac{du'}{dx} + \frac{dv'}{dy} + \frac{dw'}{dz} = 0. \quad ....(8)$$

These results were applied by Kirchhoff to the case of plane waves, supposed to be propagated in infinite space in the direction of $+x$, and it may conduce to clearness to deal first with this case. Here v' and w' vanish, while u', Q_1, Q_2 are independent of y and z. It follows from (8) that u' also vanishes. The equations for Q_1 and Q_2 are

$$d^2 Q_1/dx^2 = \lambda_1 Q_1, \qquad d^2 Q_2/dx^2 = \lambda_2 Q_2; \quad \dots\dots\dots\dots(9)$$

so that we may take

$$Q_1 = e^{-x\sqrt{\lambda_1}}, \qquad Q_2 = e^{-x\sqrt{\lambda_2}}, \quad \dots\dots\dots\dots(10)$$

where the signs of the square roots are to be so chosen that the real parts are positive. Accordingly

$$u = A_1 \lambda_1^{\frac{1}{2}} \left(\frac{h}{\lambda_1} - v \right) e^{-x\sqrt{\lambda_1}} + A_2 \lambda_2^{\frac{1}{2}} \left(\frac{h}{\lambda_2} - v \right) e^{-x\sqrt{\lambda_2}}, \quad \dots\dots(11)$$

$$\theta' = A_1 e^{-x\sqrt{\lambda_1}} + A_2 e^{-x\sqrt{\lambda_2}}, \quad \dots\dots\dots\dots\dots\dots(12)$$

in which the constants A_1, A_2 may be regarded as determined by the values of u and θ' when $x = 0$.

The solution, as expressed by (11), (12), is too general for our purpose, providing as it does for arbitrary communication of heat at $x = 0$. From the quadratic (4) in λ we see that if μ', μ'', v be regarded as small quantities, one of the values of λ, say λ_1, is approximately equal to h^2/a^2, while the other (λ_2) is very great. The solution which we now require is that corresponding to λ_1 simply. The second approximation to λ_1 is by (4)

$$\lambda_1 = \frac{h^2}{a^2} \left\{ 1 - \frac{h(\mu' + \mu'' + v)}{a^2} \right\} + \frac{v b^2 h^3}{a^6},$$

so that

$$\pm \sqrt{\lambda_1} = \frac{h}{a} - \frac{h^2}{2a^3} \{ \mu' + \mu'' + v(1 - b^2/a^2) \}. \quad \dots\dots\dots(13)$$

If we now write in for h, we see that the typical solution is

$$u = e^{-m'x} e^{in(t - x/a)}, \quad \dots\dots\dots\dots\dots\dots(14)$$

where

$$m' = \frac{n^2}{2a^3} \left\{ \mu' + \mu'' + v \left(1 - \frac{b^2}{a^2} \right) \right\}. \quad \dots\dots\dots\dots(15)$$

In (14) an arbitrary multiplier and an arbitrary addition to t may, as usual, be introduced; and, if desired, the solution may be realized by omitting the imaginary part.

In passing on to consider the influence of walls, by which gas is confined, upon the propagation of sound, it is here proposed to take the case of two dimensions, rather than the tube of circular section treated by Kirchhoff.

The analysis, however, is nearly the same. We suppose that sound is propagated in the layer of gas bounded by fixed walls at $y = \pm y_1$, so that $w = 0$, while u, v, θ' are functions of x and y only. The like may be assumed respecting u', v', Q_1, Q_2. We suppose further that as functions of x these quantities are proportional to e^{mx}, where m is a complex constant to be determined. The equations (3) for Q_1, Q_2 become

$$d^2 Q_1/dy^2 = (\lambda_1 - m^2) Q_1, \qquad d^2 Q_2/dy^2 = (\lambda_2 - m^2) Q_2. \quad \dots (16, 17)$$

For u', v' equations (6), (8) give

$$\frac{d^2 u'}{dy^2} = \left(\frac{h}{\mu} - m^2\right) u', \qquad \frac{d^2 v'}{dy^2} = \left(\frac{h}{\mu} - m^2\right) v', \qquad mu' + \frac{dv'}{dy} = 0. \quad \dots (18, 19, 20)$$

These three equations are satisfied if u' be determined by means of the first, and v' is chosen so that

$$v' = -\frac{m}{h/\mu' - m^2} \frac{du'}{dy}, \qquad \dots (21)$$

a relation obtained by subtracting from (19) the result of differentiating (20) with respect to y. The solution of (18) may be written $u' = AQ$, in which A is a constant, and Q a function of y satisfying

$$\frac{d^2 Q}{dy^2} = \left(\frac{h}{\mu} - m^2\right) Q. \qquad \dots (22)$$

Thus, by (5), (7),

$$u = AQ - A_1 m \left(\frac{h}{\lambda_1} - \nu\right) Q_1 - A_2 m \left(\frac{h}{\lambda_2} - \nu\right) Q_2, \qquad \dots (23)$$

$$v = -A \frac{m}{h/\mu' - m^2} \frac{dQ}{dy} - A_1 \left(\frac{h}{\lambda_1} - \nu\right) \frac{dQ_1}{dy} - A_2 \left(\frac{h}{\lambda_2} - \nu\right) \frac{dQ_2}{dy}, \quad \dots (24)$$

$$\theta' = A_1 Q_1 + A_2 Q_2. \qquad \dots (25)$$

On the walls at $y = \pm y_1$, u, v, θ' must satisfy certain conditions. It will here be supposed that there is neither motion of the gas nor change of temperature; so that when $y = \pm y_1$, u, v, θ' vanish. The condition of which we are in search is thus expressed by the evanescence of the determinant of (23), (24), (25), viz. :

$$\frac{m^2 h}{h/\mu' - m^2} \left(\frac{1}{\lambda_1} - \frac{1}{\lambda_2}\right) \frac{d\log Q}{dy} + \left(\frac{h}{\lambda_1} - \nu\right) \frac{d\log Q_1}{dy} - \left(\frac{h}{\lambda_2} - \nu\right) \frac{d\log Q_2}{dy} = 0, \dots (26)$$

which is to be satisfied when $y = \pm y_1$.

Since u is an even function of y, we have from (3), (22),

$$\left. \begin{aligned} Q &= \cos\{y\sqrt{(m^2 - h/\mu')}\} \\ Q_1 &= \cos\{y\sqrt{(m^2 - \lambda_1)}\} \\ Q_2 &= \cos\{y\sqrt{(m^2 - \lambda_2)}\} \end{aligned} \right\}. \qquad \dots (27)$$

From (23), (25), and from the fact that $u=0$ when $y=y_1$, we get as the general value of u, without regard to the constant multiplier,

$$u = \frac{Q(y)}{Q(y_1)} + \frac{v-h/\lambda_1}{h/\lambda_1 - h/\lambda_2}\frac{Q_1(y)}{Q_1(y_1)} - \frac{v-h/\lambda_2}{h/\lambda_1 - h/\lambda_2}\frac{Q_2(y)}{Q_2(y_1)}. \quad \ldots(28)$$

In equation (26) the values of λ_1, λ_2 are independent of y_1, being determined by (4). In the application to air under normal conditions μ', μ'', v may be regarded as small, and we have approximately

$$\lambda_1 = h^2/a^2, \qquad \lambda_2 = ha^2/vb^2. \quad \ldots(29)$$

A second approximation to the value of λ_1 is given in (13). It is here assumed that the velocity of propagation of viscous effects of the pitch in question, viz. $\sqrt{(2\mu'n)}$, is small in comparison with that of sound, so that $in\mu'/a^2$, or $h\mu'/a^2$, is a small quantity—a condition abundantly satisfied in practice.

In interpreting the solution we limit ourselves here to the case which arises when μ', μ'', v are treated as very small—so small that the layer of gas immediately affected by the walls is but an insignificant fraction of the whole. When μ' &c. vanish, we have

$$\lambda_1 = h^2/a^2, \qquad m^2 = h^2/a^2,$$

so that unless y be great $y\sqrt{(m^2-\lambda_1)}$ is small. On the other hand, $y_1\sqrt{(m^2-h/\mu')}$, $y_1\sqrt{(m^2-\lambda_2)}$ are large. For the moment we leave the value of $y\sqrt{(m^2-\lambda_1)}$ open, and merely introduce the simplifications arising out of the largeness of the arguments in Q and Q_2.

If z be a complex quantity of the form $\xi+i\eta$, we have in general

$$\cos z = \cos\xi\cosh\eta - i\sin\xi\sinh\eta, \quad \ldots(30)$$

$$\tan z = \frac{\sin 2\xi + i\sinh 2\eta}{\cos 2\xi + \cosh 2\eta}. \quad \ldots(31)$$

Thus, when η is large,

$$\frac{d\log\cos z}{dz} = -\tan z = -i;$$

so that when $y=y_1$, since h is a pure imaginary,

$$\frac{d\log Q}{dy} = \sqrt{\left(\frac{h}{\mu'}\right)}, \qquad \frac{d\log Q_2}{dy} = \sqrt{\left(\frac{ha^2}{vb^2}\right)}. \quad \ldots(32)$$

The introduction into (26) of these values and those of λ_1 and λ_2 from (29), gives

$$\frac{d\log Q_1}{dy} = -\frac{\gamma'h^{\frac{3}{2}}}{a^2},$$

where

$$\gamma' = \sqrt{\mu'} + (a/b - b/a)\sqrt{v}; \quad \ldots(33)$$

or, if $z = y_1 \sqrt{(m^2 - \lambda_1)}$,

$$z \tan z = \gamma' h^{\frac{3}{2}} y_1 / a^2. \quad\ldots\ldots\ldots\ldots\ldots\ldots\ldots(34)$$

This is the equation by which z, and thence m^2, is to be determined.

In the case corresponding to that treated by Kirchhoff, y_1 is not so large but that the right-hand member of (34) is a small quantity. The solution of (34) is then

$$z^2 = \gamma' h^{\frac{3}{2}} y_1 / a^2; \quad\ldots\ldots\ldots\ldots\ldots\ldots\ldots(35)$$

whence

$$m^2 = \lambda_1 + \frac{\gamma' h^{\frac{3}{2}}}{a^2 y_1} = \frac{h^2}{a^2}\left(1 + \frac{\gamma'}{y_1 \sqrt{h}}\right). \quad\ldots\ldots\ldots\ldots(36)$$

We now write $h = ni$, so that the frequency is $n/2\pi$. Thus

$$\sqrt{h} = \sqrt{(\tfrac{1}{2}n)}.(1 + i) \quad\ldots\ldots\ldots\ldots\ldots\ldots(37)$$

and

$$m = \pm (m' + im''), \quad\ldots\ldots\ldots\ldots\ldots\ldots(38)$$

where by (36)

$$m' = \frac{\sqrt{n}.\gamma'}{\sqrt{2}.2ay_1}, \qquad m'' = \frac{n}{a} + \frac{\sqrt{n}.\gamma'}{\sqrt{2}.2ay_1}. \quad\ldots\ldots\ldots(39)$$

The solution differs from that found by Kirchhoff for a circular tube of radius r merely by the substitution of $2y_1$ for r*.

So far, then, as it depends on t and x, the typical solution is

$$e^{int} e^{-m'x - im''x},$$

or when realized,

$$e^{-m'x} \cos (nt - m''x), \quad\ldots\ldots\ldots\ldots\ldots(40)$$

where m', m'' have the values given in (39). This is for waves travelling in the positive direction.

As a function of y, u is given by (28); but this may now be simplified in virtue of the supposition that the layer directly influenced by the viscosity is but a small fraction of y_1. By (27)

$$Q(y_1) = \cos (y_1 \sqrt{n/\mu'} . \sqrt{-i}) = \cos \{y_1 \sqrt{n/2\mu'}.(1 - i)\}$$

$$= \tfrac{1}{2} e^{y_1 \sqrt{(n/2\mu')}} \left\{\cos \left(y_1 \sqrt{\frac{n}{2\mu'}}\right) + i \sin \left(y_1 \sqrt{\frac{n}{2\mu'}}\right)\right\}, \quad\ldots\ldots(41)$$

use being made of (30), in which η is large. In consequence of (41), $Q(y) \div Q(y_1)$ vanishes unless y be nearly equal to y_1, viz. unless the point considered be within the frictional layer. In like manner, and under the same restriction, $Q_2(y) \div Q_2(y_1)$ may be neglected. Except in the immediate neighbourhood of the walls, (28) now reduces to

$$u = -\frac{Q_1(y)}{Q_1(y_1)}. \quad\ldots\ldots\ldots\ldots\ldots(42)$$

* *Theory of Sound*, 2nd edit. § 350.

In the case considered by Kirchhoff, where the argument of Q_1 is small, we have from (27) approximately

$$Q_1(y) = Q_1(y_1) = 1,$$

and accordingly $u = -1$. To this approximation the velocity is uniform across the whole section until it begins to fall off as the walls are closely approached.

As a first step towards the consideration of what occurs when y_1 is great, we may proceed to a second approximation. Thus, from (27),

$$Q(y) = 1 - \tfrac{1}{2}y^2(m^2 - \lambda_1) = 1 - \tfrac{1}{2}y^2 \frac{\gamma' h^{\frac{3}{2}}}{a^2 y_1}; \quad \ldots\ldots\ldots\ldots(43)$$

so that

$$-u = 1 + \tfrac{1}{2}(y_1^2 - y^2)\frac{\gamma' h^{\frac{3}{2}}}{a^2 y_1}$$

$$= 1 - \frac{(y_1^2 - y^2)\gamma' n^{\frac{3}{2}}}{2a^2 y_1 \sqrt{2}} + i\frac{(y_1^2 - y^2)\gamma' n^{\frac{3}{2}}}{2a^2 y_1 \sqrt{2}}. \quad \ldots\ldots\ldots\ldots(44)$$

This equation expresses the dependence of u upon y. The dependence on t and x is given by the factors already considered, viz.

$$e^{int} \cdot e^{-m'x - im''x}.$$

That (44) is complex indicates that the *phase* varies with y. The realized expression will be

$$u = -\left\{1 - \frac{(y_1^2 - y^2)\gamma' n^{\frac{3}{2}}}{a^2 y_1 \sqrt{2}}\right\} \cdot e^{-m'x} \times \cos\left\{nt - m''x + \frac{(y_1^2 - y^2)\gamma' n^{\frac{3}{2}}}{2a^2 y_1 \sqrt{2}}\right\}, \quad \ldots(45)$$

from which we infer (i) that the intensity is *least* in the middle where $y = 0$, and increases towards the walls until the frictional layer is approached; (ii) that, as y^2 increases from the centre, constancy of phase demands a diminishing x, or, in other words, that the wave-surface is convex towards $+x$ and the wave *divergent*.

We have now to trace the solution of (34) when the right-hand member becomes large. Writing it in the form

$$\frac{z \tan z}{i^{\frac{3}{2}}} = \frac{\gamma' n^{\frac{3}{2}} y_1}{a^2}, \quad \ldots\ldots\ldots\ldots\ldots\ldots(46)$$

we have to find such a complex value of z, say $\xi + i\eta$, that the function on the left is *real*. Initially, when y_1 is small,

$$\xi + i\eta = \rho(\cos\theta + i\sin\theta) = \rho(\cos 67\tfrac{1}{2}° + i\sin 67\tfrac{1}{2}°),$$

and

$$\rho^2 = \gamma' n^{\frac{3}{2}} y_1 / a^2.$$

If we retain the angle $67\tfrac{1}{2}°$ and increase ρ, we find, calculating by means of (31), that $i^{-\frac{3}{2}}z \cdot \tan z$ becomes complex with imaginary part positive. Thus if $\rho = 1$, we get

$$i^{-\frac{3}{2}}z \cdot \tan z = \cdot 80 \,(\cos 9°\,54' + i \sin 9°\,54').$$

This is a sign that θ must be reduced. If we take $\rho = 1$, $\theta = 60°$, we find

$$i^{-\frac{3}{2}}z \cdot \tan z = \cdot 83 \,(\cos 2°\,7' - i \sin 2°\,7').$$

If $\rho = 1\cdot5$, while $\theta = 60°$, we get in detail

$$z = \xi + i\eta = \cdot75 + i \times 1\cdot299\,;$$

$$\sin 2\xi = \sin 85°\,57' = \cdot998, \qquad \sinh 2\eta = 6\cdot695,$$

$$\cos 2\xi = \cos 85°\,57' = \cdot071, \qquad \cosh 2\eta = 6\cdot769,$$

whence

$$\tan z = \frac{\cdot998 + i \times 6\cdot695}{\cdot071 + 6\cdot769} = 1\cdot0 \,(\cos 81°\,31' + i \sin 81°\,31'),$$

so that

$$i^{-\frac{3}{2}}z \cdot \tan z = 1\cdot5 \,(\cos 6°\,31' + i \sin 6°\,31').$$

The course of the calculation makes it clear that as z increases, $\tan z$ approaches the limit i, so that ultimately $\theta = \tfrac{1}{4}\pi$, or the angle reduces from $67\tfrac{1}{2}°$ to $45°$. Hence

$$z^2 = i\gamma'^2 n^3 y_1{}^2/a^4, \quad\dots\dots\dots\dots\dots\dots\dots(47)$$

and

$$m^2 = h^2/a^2 - in^3\gamma'^2/a^4, \quad\dots\dots\dots\dots\dots(48)$$

independent of y_1.

In order to obtain u as a function of y, we have now to interpret (42) for the case where y_1 is great. It may be written

$$u = -\frac{(\cos zy/y_1)}{\cos z}, \quad\dots\dots\dots\dots\dots\dots\dots(49)$$

where z, given by (47), is

$$z = \frac{\gamma' n^3 y_1}{a^2\sqrt{2}}\,(1 + i). \quad\dots\dots\dots\dots\dots\dots(50)$$

By (30), since η is large, we have approximately

$$\cos z = \tfrac{1}{2}e^\eta \,(\cos \xi + i \sin \xi), \quad\dots\dots\dots\dots\dots(51)$$

where

$$\xi = \eta = \frac{\gamma' n^3 y_1}{a^2\sqrt{2}}. \quad\dots\dots\dots\dots\dots\dots(52)$$

Accordingly, $\cos z$ is large. If, as near the middle of the layer of gas, y be not large, $\cos (2y/y_1) = 1$, and

$$u = -\,1/\cos z, \quad\dots\dots\dots\dots\dots\dots\dots(53)$$

a small quantity. When y is so large that $2y/y_1$ is large, as well as z, we may write

$$u = -\,e^{\eta'-\eta}\cdot e^{i(\xi'-\xi)}, \quad\dots\dots\dots\dots\dots\dots(54)$$

where

$$\xi' = \eta' = \frac{\gamma' n^3 y}{a^2\sqrt{2}}. \quad\dots\dots\dots\dots\dots\dots(55)$$

As the walls are approached u rapidly increases, and at last $e^{\eta'-\eta}$ becomes nearly equal to unity.   We must bear in mind, however, that (42) and therefore (54) must not be applied within the frictional layer lying quite close to the walls, so that we are not at liberty to suppose y actually equal to y_1.

Under normal conditions the thickness of the frictional layer is very small. If in C.G.S. measure we take $\mu' = {\cdot}16$, $n = 2\pi \times 256$, we find $\sqrt{(n/2\mu')} = 67$. Thus if we suppose the thickness of the frictional layer in (41) to be defined by

$$(y_1 - y)\sqrt{(n/2\mu')} = 1,$$

we get

$$y_1 - y = {\cdot}15 \text{ millim.}$$

If the point under consideration be a few multiples of this (say 1 millim.) from the walls, the ratio $Q(y) \div Q(y_1)$ may be neglected.

The thickness of the layer through which $Q_2(y) \div Q_2(y_1)$ is sensible is of the same order of magnitude.

Let us next consider what value of $(y_1 - y)$ makes $(\eta' - \eta)$ in (54) equal to unity.   By (52), (55)

$$y_1 - y = \frac{\sqrt{2}\,.\,a^2}{\gamma' n^{\frac{3}{2}}}.$$

If we take $\mu' = {\cdot}16$, $\nu = {\cdot}256$, we find from (33) $\gamma' = {\cdot}6$; and $a = 33200$; so that for a frequency of 256 we get

$$y_1 - y = \frac{\sqrt{2}\,.\,33200^2}{{\cdot}6 \times (2\pi \times 256)^{\frac{3}{2}}} = 40000.$$

For air and for a sound of this pitch the falling off becomes important at a distance of about 400 metres from the walls.

As has already been suggested, this solution fails to answer the practical question for the sake of which it was originally attempted.   It was desired to know whether in the propagation of sound for long distances over smooth water, there was any important *shadow* formed near the surface under the influence of viscosity and heat conduction.   It would apparently be a matter of some difficulty to formulate and solve a definite problem in which this question is involved.   But, as Lord Kelvin has pointed out to me, a sufficient answer to the practical question may be arrived at by very simple reasoning on the basis of a solution originally given by Stokes (see *Theory of Sound*, § 347).   If U be the tangential velocity of a plane vibrating rigidly in an atmosphere of viscous fluid with a frequency $n/2\pi$, the work required to maintain the motion is, for unit of area,

$$\int \sqrt{({\textstyle\frac{1}{2}}\rho n\mu)}\,.\,U^2 dt,$$

or

$$\tfrac{1}{2}U_m{}^2\,.\,t\,.\,\sqrt{({\textstyle\frac{1}{2}}\rho n\mu)},$$

where U_m denotes the maximum value of U during the period. The same expression may be applied to find the work lost by the presence of a fixed plane in air vibrating with velocity U. The energy of this motion is, per unit of volume, $\frac{1}{2}\rho U_m^2$, or for a stratum of height y resting upon unit of area,

$$\tfrac{1}{2}\rho y\, U_m^2.$$

If we equate the two expressions we get a superior limit to the thickness of the stratum whose energy could be absorbed in time t. We find thus

$$y = \sqrt{\left(\frac{n\mu}{2\rho}\right)} . t \,;$$

or, if we take $n = 2\pi \times 256$, and as for air $\mu/\rho = \mu' = \cdot 16$, $y = 11t$.

Thus in 9 seconds the thickness of a stratum of shadow could not reach 1 metre, and must, in fact, be very much less. It would appear therefore that this effect may be neglected in practice, unless it be in the case of an observer extremely close to the water.

POLISH.

[*Proceedings of the Royal Institution*, XVI. pp. 563—570, 1901;
Nature, LXIV. pp. 385—388, 1901.]

THE lecture commenced with a description of a home-made spectroscope of considerable power. The lens, a plano-convex of 6 inches aperture and 22 feet focus, received the rays from the slit, and finally returned them to a pure spectrum formed in the neighbourhood. The skeleton of the prism was of lead; the faces, inclined at 70°, were of thick plate-glass cemented with glue and treacle. It was charged with bisulphide of carbon, of which the free surface (of small area) was raised above the operative part of the fluid. The prism was traversed twice, and the effective thickness was $5\frac{1}{2}$ inches, so that the resolving power corresponded to 11 inches, or 28 cm., of CS_2. The liquid was stirred by a perforated triangular plate, nearly fitting the prism, which could be actuated by means of a thread within reach of the observer. The reflector was a *flat*, chemically silvered in front.

So far as eye observations were concerned, the performance was satisfactory, falling but little short of theoretical perfection. The stirrer needed to be in almost constant operation, the definition usually beginning to fail within about 20 seconds after stopping the stirrer. But although the stirrer was quite successful in maintaining uniformity of temperature as regards *space, i.e.* throughout the dispersing fluid, the temperature was usually somewhat rapidly variable with *time*, so that photographs, requiring more than a few seconds of exposure, showed inferiority. In this respect a grating is more manageable.

The lens and the faces of the prism were ground and polished (in 1893) upon a machine kindly presented by Dr Common. The flat surfaces were tested with a spherometer, in which a movement of the central screw through $\frac{1}{100000}$ inch could usually be detected by the touch. The external surfaces of the prism faces were the only ones requiring accurate flatness. In polishing, the operation was not carried as far as would be expected of a professional optician. A few residual pittings, although they spoil the appearance of a surface, do not interfere with its performance, at least for many purposes.

In the process of grinding together two glass surfaces, the particles of emery, even the finest, appear to act by *pitting* the glasses, *i.e.* by breaking out small fragments. In order to save time and loss of accuracy in the polishing, it is desirable to carry the grinding process as far as possible, using towards the close only the finest emery. The limit in this direction appears to depend upon the tendency of the glasses (6 inches diameter) to *seize*, when they approach too closely, but with a little care it is easy to attain such a fineness that a candle is seen reflected at an angle of incidence not exceeding 60°, measured as usual from the perpendicular.

The fineness necessary, in order that a surface may reflect and refract regularly without diffusion, viz. in order that it may appear *polished*, depends upon the wave-length of the light and upon the angle of incidence. At a grazing incidence all surfaces behave as if polished, and a surface which reflects red light pretty well may fail signally when tested with blue light at the same angle. If we consider incidences not too far removed from the perpendicular, the theory of gratings teaches that a regularly corrugated surface behaves as if absolutely plane, provided that the *wave-length* of the corrugations is less than the wave-length of the light, and this without regard to the *depth* of the corrugations. Experimental illustrations, drawn from the sister science of Acoustics, were given. The source was a bird-call from which issued vibrations having a wave-length of about 1·5 cm., and the percipient was a high-pressure sensitive flame. When the bird-call was turned away, the flame was silent, but it roared vigorously when the vibrations were reflected back upon it from a plate of glass. A second plate, upon which small pebbles had been glued so as to constitute an ideally rough surface, acted nearly as well, and so did a piece of tin plate suitably corrugated. In all these cases the reflection was *regular*, the flame becoming quiet when the plates were turned out of adjustment through a very small angle. In another method of experimenting the incidence was absolutely perpendicular, the flame being exposed to both the incident and the reflected waves. It is known that under these circumstances the flame remains quiescent at the *nodes* and flares most vigorously at the *loops*. As the reflector is drawn slowly back, the flame passes alternately through the nodes and loops, thus executing a cycle of changes as the reflector moves through *half* a wavelength. The effects observed were just the same whether the reflector were smooth or covered with pebbles, or whether the corrugated tin plate were substituted. All surfaces were smooth *enough* in relation to the wave-length of the vibration to give substantially a specular reflection.

Finely-ground surfaces are still too coarse for perpendicular specular re-flection of the longest visible waves of light. Here the material may be metal, or glass silvered chemically on the face subsequently to the grinding. But experiment is not limited by the capabilities of the eye; and it seems certain that a finely ground surface would be smooth enough to reflect with-out sensible diffusion the longest waves, such as those found by Rubens to be nearly 100 times longer than the waves of red light. An experiment may be tried with radiation from a Leslie cube containing hot water, or from a Welsbach mantle (without a chimney). In the lecture the latter was em-ployed, and it fell first at an angle of about 45° upon a finely ground flat glass silvered in front. By this preliminary reflection, the radiation was purified from waves other than those of considerable wave-length. The second reflection (also at 45°) was alternately from polished and finely ground silvered surfaces of the same size, so mounted as to permit the accurate sub-stitution of the one for the other. The heating-power of the radiation thus twice reflected was tested with a thermopile in the usual manner. Repeated comparisons proved that the reflection from the ground surface was about ·76 of that from the polished surface, showing that the ground surface re-flected the waves falling upon it with comparatively little diffusion. A slight rotation of any of the surfaces from their proper positions at once cut off the effect. It is probable that the device of submitting radiation to preliminary reflections from one or more merely ground surfaces might be found useful in experiments upon the longest waves.

In view of these phenomena we recognise that it is something of an accident that polishing processes, as distinct from grinding, are needed at all; and we may be tempted to infer that there is no essential difference between the operations. This appears to have been the opinion of Herschel[*], whom we may regard as one of the first authorities on such a subject. But, although, perhaps, no sure conclusion can be demonstrated, the balance of evidence appears to point in the opposite direction. It is true that the same powders may be employed in both cases. In one experiment a glass surface was polished with the same emery as had been used effectively a little earlier in the grinding. The difference is in the character of the backing. In

_____

[*] _Enc. Met._, Art. Light, p. 447, 1830: "The intensity and regularity of reflection at the external surface of a medium is found to depend not merely on the nature of the medium, but very essentially on the degree of smoothness and polish of its surface. But it may reasonably be asked, how any regular reflection can take place on a surface polished by art, when we recollect that the process of polishing is, in fact, nothing more than grinding down large asperities into smaller ones by the use of hard gritty powders, which, whatever degree of mechanical comminu-tion we may give them, are yet vast masses, in comparison with the ultimate molecules of matter, and their action can only be considered as an irregular tearing up by the roots of every projection that may occur in the surface. So that, in fact, a surface artificially polished must bear some-what of the same kind of relation to the surface of a liquid, or a crystal, that a ploughed field does to the most delicately polished mirror, the work of human hands."

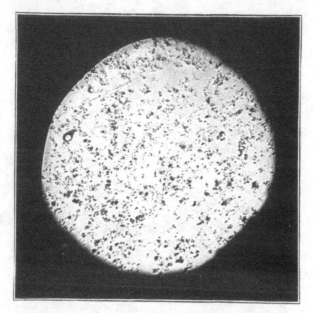

Fig. 1.

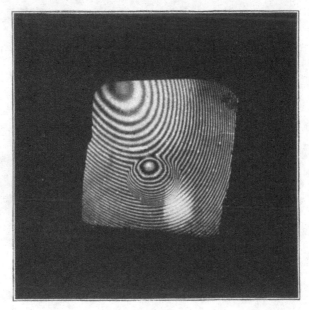

Fig. 2.

grinding, the emery is backed by a hard surface, *e.g.* of glass, while during the polishing the powder (mostly rouge in these experiments) is imbedded in a comparatively yielding substance, such as pitch. Under these conditions, which preclude more than a moderate pressure, it seems probable that no pits are formed by the breaking out of fragments, but that the material is worn away (at first, of course, on the eminences) almost molecularly.

The progress of the operation is easily watched with a microscope, provided, say, with a $\frac{1}{4}$-inch object-glass. The first few minutes suffice to effect a very visible change. Under the microscope it is seen that little facets, parallel to the general plane of the surface, have been formed on all the more prominent eminences*. The facets, although at this stage but a very small fraction of the whole area, are adequate to give a sensible specular reflection, even at perpendicular incidence. On one occasion five minutes' polishing of a rather finely ground glass surface was enough to qualify it for the formation of interference bands, when brought into juxtaposition with another polished surface, the light being either white or from a soda flame; so that in this way an optical test can be applied almost before the polishing has begun†.

As the polishing proceeds, the facets are seen under the microscope to increase both in number and in size, until they occupy much the larger part of the area. Somewhat later the parts as yet untouched by the polisher appear as pits, or spots, upon a surface otherwise invisible. Fig. 1 represents a photograph of a surface at this stage taken with the microscope. The completion of the process consists in rubbing away the whole surface down to the level of the deepest pits. The last part of the operation, while it occupies a great deal of time, and entails further risk of losing the "truth" of the surface, adds very little to the effective area, or to the intensity of the light regularly reflected or refracted.

Perhaps the most important fact taught by the microscope is that the polish of individual parts of the surface does not improve during the process. As soon as they can be observed at all, the facets appear absolutely structureless. In its subsequent action the polishing tool, bearing only upon the parts already polished, extends the boundary of these parts, but does not enhance their quality. Of course, the mere fact that no structure can be perceived does not of itself prove that pittings may not be taking place of a character too fine to be shown by a particular microscope or by any possible microscope. But so much discontinuity, as compared with the grinding action, has to be admitted in any case, that one is inevitably led to the conclusion that in all probability the operation is a molecular one, and that no coherent fragments

---

* The interpretation is facilitated by a thin coating of aniline dye which attaches itself mainly to the hollows.

† With oblique incidence, as in Talbot's experiments (see *Phil. Mag.* xxviii. p. 191, 1889 [Vol. iii. p. 308]), achromatic bands may be observed from a surface absolutely unpolished, but this disposition would not be favourable for testing purposes.

containing a large number of molecules are broken out. If this be so, there would be much less difference than Herschel thought between the surfaces of a polished solid and of a liquid.

Several trials have been made to determine how much material is actually removed during the polishing of glass. In one experiment a piece 6 inches in diameter, very finely ground, was carefully weighed at intervals during the process. Losses of ·070, ·032, ·045, ·026, ·032 gms. were successively registered, amounting in all to ·205 gms. Taking the specific gravity of the glass as 3, this corresponds to a thickness of $3·6 \times 10^{-4}$ cm., or to about 6 wave-lengths of mean light, and it expresses the distance between the original *mean* surface and the final plane. But the polish of this glass, though sufficient for most practical purposes, was by no means perfect. Probably the 6 wave-lengths would have needed to be raised to 10 in order to satisfy a critical eye. It may be interesting to note for comparison that, in the grinding, one charge of emery, such as had remained suspended in water for seven or eight minutes, removed a thickness of glass corresponding to 2 wave-lengths.

In other experiments the thickness removed in polishing was determined optically. A very finely ground disc was mounted in the lathe and polished locally in rings. Much care was needed to obtain the desired effect of a ring showing a continuously increasing polish from the edges inwards. To this end it was necessary to keep the polisher (a piece of wood covered with resin and rouge) in constant motion, otherwise a number of narrow grooves developed themselves.

The best ring was about half an inch wide. When brought into contact with a polished flat and examined at perpendicular incidence with light from a soda flame, the depression at its deepest part gave a displacement of three bands, corresponding to a depth of $1\frac{1}{2}\lambda$. On a casual inspection this central part appeared well polished, but examination under the microscope revealed a fair number of small pits. Further working increased the maximum depth to $2\frac{1}{2}\lambda$, when but very few pits remained. In this case, then, polish was effected during a lowering of the mean surface through 2 or 3 wave-lengths, but the grinding had been exceptionally fine.

It may be well to emphasize that the observations here recorded relate to a *hard* substance. In the polishing of a soft substance, such as copper, it is possible that material may be loosened from its original position without becoming detached. In such a case pits may be actually filled in, by which the operation would be much quickened. Nothing suggestive of this effect has been observed in experiments upon glass.

Another method of operating upon glass is by means of hydrofluoric acid. Contrary to what is generally supposed, this action is extremely regular, if proper precautions are taken. The acid should be weak, say one part of commercial acid to two hundred of water, and it should be kept in constant

motion by a suitable rocking arrangement. The parts of the glass not intended to be eaten into are, as usual, protected with wax. The effect upon
a polished flat surface is observed by the formation of Newton's rings with
soda-light. After perhaps three-quarters of an hour, the depression corresponds to half a band, *i.e.* amounts to $\frac{1}{4}\lambda$, and it appears to be uniform over
the whole surface exposed. Two pieces of plate glass, 3 inches square, and
flat enough to come into fair contact all over, were painted with wax in
parallel stripes, and submitted to the acid for such a time, previously ascertained, as would ensure an action upon the exposed parts of $\frac{1}{4}\lambda$. After
removal of the wax, the two plates, crossed and pressed into contact so as to
develop the colours, say of the second order, exhibited a chess-board pattern.
Where two uncorroded, or where two corroded parts, are in contact, the
colours are nearly the same, but where a corroded and an uncorroded surface
overlap, a strongly contrasted colour is developed. The combination lends
itself to lantern projection, and the pattern upon the screen [shown] is very
beautiful, if proper precautions are taken to eliminate the white light reflected
from the first and fourth surfaces of the plates.

In illustration of the action of hydrofluoric acid, photographs* were
shown of interference bands as formed by soda-light between glass surfaces,
one optically flat and the other ordinary plate, upon which a drop of dilute
acid had been allowed to stand (Fig. 2). Truly plane surfaces would give
bands straight, parallel, and equidistant.

Hydrofluoric acid has been employed with some success to correct ascertained errors in optical surfaces. But while improvements in actual optical
performance have been effected, the general appearance of a surface so treated
is unprepossessing. The development of latent scratches has been described
on a former occasion†.

A second obvious application of hydrofluoric acid has hitherto been less
successful. If a suitable stopping could be found by which the deeper pits
could be protected from the action, corrosion by acid could be used in substitution for a large part of the usual process of polishing.

In connexion with experiments of this sort, trial was made of the action
of the acid upon finely ground glass, such for example as is used as a backing
for stereoscopic transparencies, and very curious results were observed. For
this purpose the acid may conveniently be used much stronger, say one part
of commercial acid to 10 parts of water, and the action may be prolonged
for hours or days. The general appearance of the glass after treatment is
smoother and more translucent, but it is only under the microscope that the
remarkable changes which the surface has undergone become intelligible.
Fig. 3 is from a photograph taken in the microscope, the focus being upon
the originally ground surface itself. The whole area is seen to be divided

* The plates were sensitised in the laboratory with cyanine.
† *Proc. Roy. Inst.* March 1893. [Vol. IV. p. 59.]

into cells. These cells increase as the action progresses, the smaller ones being, as it were, eaten up by the bigger. The division lines between the cells are *ridges*, raised above the general level, and when seen in good focus appear absolutely sharp. The general surface within the cells shows no structure, being as invisible as if highly polished.

That each cell is in fact a concave lens, forming a separate image of the source of light, is shown by slightly screwing out the object-glass. Fig. 4 was taken in this way from the same surface, the source of light being the flame of a paraffin lamp, in front of which was placed a cross cut from sheet-metal.

The movement required to pass from the ridge to the image of the source, equal to the focal length (f) of the lens, may be utilised to determine the depth (t) of a cell. In one experiment the necessary movement was ·005 inch. The semi-aperture (y) of the "lens" was ·0015 inch, whence by the formula $y^2 = ft$, we find $t = ·00045$ inch. This represents the depth of the cell, and it amounts to about 8 wave-lengths of yellow light.

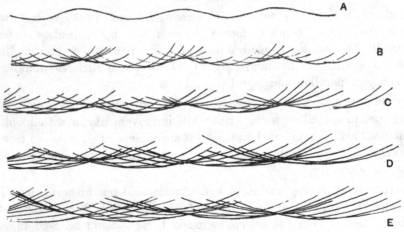

Fig. 5.

The action of the acid seems to be readily explained if we make the very natural supposition that it eats in everywhere, at a fixed rate, normally to the actual surface. If the amount of the normal corrosion after a proposed time be known, the new surface can be constructed as the "envelope" of spheres having the radius in question and centres distributed over the old surface. Ultimately, the new surface becomes identified with a series of spherical segments having their centres at the deeper pits of the original surface. The construction is easily illustrated in the case of two dimensions. In the figure A is supposed to be the original surface; B, C, D, E surfaces formed by corrosion, being constructed by circles having their centres on A. In B the ridges are still somewhat rounded, but they become sharp in D and E. The general tendency is to sharpen elevations and to smooth off depressions.

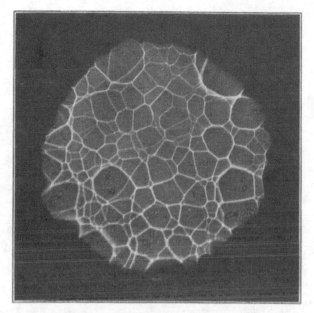

Fig. 3.

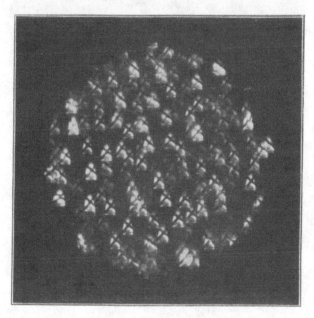

Fig. 4.

269.

DOES CHEMICAL TRANSFORMATION INFLUENCE WEIGHT?

[*Nature*, LXIV. p. 181, June, 1901.]

CAREFUL experiments by Heydweiller, published in the last number of *Drude's Annalen* (Vol. v. p. 394), lead their author to the conclusion that in certain cases chemical action is accompanied by a minute, but real, alteration of weight. The chemical actions here involved must be regarded as very mild ones, *e.g.* the mere dissolution of cupric sulphate in water, or the substitution of iron for copper in that salt.

The evidence for the reality of these changes, which amount to 0·2 or 0·3 mg., and are accordingly well within the powers of a good balance to demonstrate, will need careful scrutiny; but it may not be premature to consider what is involved in the acceptance of it. The first question which arises is—does the *mass* change as well as the *weight*? The affirmative answer, although perhaps not absolutely inconsistent with any well ascertained fact, will certainly be admitted with reluctance. The alternative—that mass and weight are not always in proportion—involves the conclusion, in contradiction to Newton, that the length of the seconds' pendulum at a given place depends upon the material of which the bob is composed. Newton's experiment was repeated by Bessel, who tried a number of metals, including gold, silver, lead, iron, zinc, as well as marble and quartz, and whose conclusion was that the length of the seconds' pendulum formed of these materials did not vary by one part in 60,000. At the present day it might be possible to improve even upon Bessel, or at any rate to include more diverse substances in the comparisons; but in any case the accuracy obtainable would fall much short of that realized in weighings.

As regards Heydweiller's experiments themselves, there is one suggestion which I may make as to a possible source of error. Is the chemical action sufficiently in abeyance at the time of the first weighing? If there is copper sulphate in one branch of an inverted U and water in the other, the equilibrium can hardly be complete. The water all the time tends to distil over into the salt, and any such distillation must be attended by thermal effects which would interfere with the accuracy of the weighing.

[See further *Nature*, May 15, 1902.]

270.

ACOUSTICAL NOTES.—VI.

[*Philosophical Magazine*, II. pp. 280—285, 1901.]

Forced Vibrations.

IF free vibrations be represented by $\cos nt$, and if the forced vibration due to a force acting in a very long period be $\cos pt$, then the actual forced vibration will be

$$\frac{n^2 \cos pt}{n^2 - p^2}.$$

It is here implied :—

(1) That in all cases the forced vibration takes its period from the force, whatever may be the natural period.

(2) That if the forced vibration be the slower, viz. if $p < n$, the phase is the same as if the vibration were infinitely slow, in which case the vibrator would be situated at any instant of time in the position where the momentary force would permanently maintain it.

(3) That if the forced vibration be the quicker ($p > n$), the phase of the actual vibration is the opposite of that defined in (2).

(4) That if the force have nearly the period of the free vibrations, the effect is much enhanced. Indeed, according to the formula it would become infinite, which means that forces of a viscous character, never really absent, must now be brought into the reckoning.

So far as I am aware, illustrations of this important theory* have usually been wanting in lecture demonstrations, except as regards (4). I have found that if we employ as vibrator a magnet with attached mirror, as used for example in Thomson galvanometers, the whole may readily be brought before a large audience.

---

* Young's *Lectures on Natural Philosophy*, p. 578 (1807).

With the aid of an external magnet, whose distance could be varied, the frequency of (complete) vibration was adjusted to 10 per minute, the vibrations being manifested by the motion of a spot of light reflected from the mirror on to a scale in the usual manner. The force brought to bear upon the vibrator had its origin in the revolution of a rather long permanent magnet, situated at some little distance, and so mounted as to be capable of rotation. No particular situation is necessary, but the action of the magnet is simplest in certain special cases, as when its centre is at the level of the suspended magnet and in the direction of the screen. The plane of revolution being horizontal, the deflecting action is then greatest when the revolving magnet points towards the suspended magnet. In one of these positions, say when the spot is deflected to the right, a bell rings automatically. Uniform rotation at any desired speed is maintained by hand with the aid of gearing, diminishing the speed in the ratio of 5 : 1, and of a metronome set as required.

To illustrate propositions (1) and (2) the long magnet is caused to rotate with a frequency of 8 per minute, i.e. with a frequency somewhat less than that natural to the suspended system. At first the phenomenon is complicated by the interaction of natural and forced vibrations; but the former soon die away. It is then recognised that the vibrations observed upon the screen are isochronous with the revolution of the magnet, and that the bell rings at the moment when the spot of light attains its greatest elongation towards the right.

In the next experiment the speed of revolution is altered to 12 per minute, so as to bring about the condition of things contemplated in (3). After a little interval of settling down the bell rings always at the moment when the spot is most deflected to the *left*, showing that the phase has been altered by half a period.

To illustrate (4) the speed of revolution may now be adjusted to 8 per minute. The arc of vibration is seen gradually to increase until it reaches a large value, the bell now ringing, not at either extreme elongation, but as the spot passes from left to right through its position of equilibrium.

Vibrations of Strings.

At the Royal Institution it is usual to illustrate this subject by experiments after the method of Melde and Tyndall. The string is connected with a large tuning-fork, whose prongs stand vertically, and the vibrations are maintained electrically in the well-known manner. The electric contact is between solids (of platinum), one attached to the prong, the other forming the point of an adjustable screw carried by the framework.

The string, 10 feet long, is stretched horizontally and the tension is adjusted until a vigorous vibration ensues, which happens when one of the modes of vibration has a period in simple relation to that of the fork. There is here an important distinction according as the length of the string is parallel or perpendicular to the motion of the point of attachment. In the latter case the vibrations are of the character commonly classified as *forced*, and the period is the same as that of the fork. But if the fork be so situated that the motion of the point of attachment is along the length of the string, the vibrations are of an entirely different character, and are executed in a period the *double* of that of the fork. The theory of vibrations of this class was discussed in a paper on Maintained Vibrations* published many years ago, reference to which must here suffice.

A convenient device for demonstrating the relationship of periods is to illuminate the string by sparks synchronous with the vibrations of the fork itself. For this purpose an induction-coil is included in the circuit by which the fork is driven, so that every break at the fork causes a spark between the secondary terminals, to which a small jar is connected in the usual manner. If then the vibrations of the string be isochronous with the fork, and therefore with the sparks, the intermittent illumination exhibits what is ordinarily seen as a gauzy spindle resolved into the appearance corresponding to a single phase of the vibration; that is, the string is seen apparently fixed (in a displaced position) and *single*. But if, as when the point of attachment moves parallel to the length of the string, the vibrations are only half as fast as those of the fork, the string is found in two (opposite) phases at the moments of illumination, and is consequently seen *double*. The effect is improved by a piece of ground glass, which may be held either between the sparks and the string, or between the string and the eye. In the latter case it is a *shadow* that is seen. It is desirable to retain enough continuous light to allow the form of the gauzy spindle to remain visible. In this way the difference between the two kinds of vibration may be exhibited to many persons at once. [1902. The stroboscopic method of observation had already been very similarly applied to this experiment by Oosting, *Onder houden trillungen van gespannen draden*, Helder, 1889.]

A detail of some importance relates to the use of the *condenser*, associated as usual with the primary circuit of the coil. If its poles be connected simply with the outer terminals of the fork-apparatus regarded as an interrupter, the secondary sparks will be inferior or may fail altogether. The explanation is to be sought in the self-induction of the magnet associated with the fork, which apparently interferes with the suddenness of the break. The poles of the condenser should be connected as directly as possible with the two pieces of metal between which the break takes place. In the

---

* *Phil. Mag.* Vol. xv. p. 229 (1883); *Scientific Papers*, Vol. ii. p. 188.

apparatus at the Royal Institution it makes all the difference on which side of the small electromagnet the pole of the condenser is attached.

Beats of Sounds led to the Two Ears separately.

When two approximately pure tones, of equal intensity and of approximately equal frequency, are conveyed to one ear, beats are perceived according to a well-known elementary theory, the frequency of the beats being the difference of the frequencies of the tones. When the beats are somewhat slow, the phase of *silence* is distinctly recognisable, and indeed the moment of the occurrence of this phase is capable of being fixed with great accuracy.

The question whether the beats are still audible when one sound is led to one ear alone, and the second sound to the second ear alone, is of great importance. A careful experiment of this sort is described by Prof. S. P. Thompson *, in which the sounds were conveyed to the ears by rubber tubes; and the conclusion was that in spite of all precautions the beats were most distinctly heard, although there was no phase of "silence," such as is perceived when both sounds are conveyed to the same ear.

I have lately tried a somewhat similar experiment, using telephones and electrical conveyance, by which perhaps the risk of the sounds reaching the wrong ears is reduced to a minimum. Two entirely independent, electrically driven, forks of about 128 vibrations per second were the sources of sound. Near the electromagnet of each fork was placed a small coil of wire in connexion with a telephone. The higher harmonics were greatly moderated by the interposition of thick sheets of copper; but the sounds were doubtless no more than rough approximations to pure tones. Both forks were placed at a great distance from the observer; and in one case the double connecting wire was passed through a hole in a thick wall specially arranged many years ago for this sort of experimenting. When the telephones were pressed closely to the ears, the utmost possible was done to secure that each sound should have access only to its proper ear.

The results depended somewhat upon the frequency of the beats. When this exceeded one per second, the beats were very easily audible. When, on the other hand, the frequency was reduced to $\frac{1}{2}$ or $\frac{1}{4}$ beat per second, the beats were not easily perceived at first. After a little while the attention seemed to concentrate itself upon the variable element in the aggregate effect, and the cycle became clear. But even after some practice neither Mr Gordon nor I could hear slow beats during the first 10 or 15 seconds of observation.

The general results of the experiments do not appear to me to exclude the view that the comparatively feeble beats heard under these conditions

* *Phil. Mag.* Vol. IV. p. 274 (1877).

may be due to the passage of sound from one ear to the other through the bones of the head or perhaps through the Eustachian tube.

Loudness of Double Sounds.

Observations upon the double syrens (with separate horns) used by the Trinity House have given the impression that as heard from a distance the two syrens are no better than one, even though the horns are parallel, and the observer situated in the direction of the axis. Dr Tyndall's experience was similar. In his Report of 1874 he remarks (June 2), "There was no sensible difference of intensity between the single horn and the two horns"; and again (June 10), "Subsequent comparative experiments even proved the sound of the two horns to be more effective than that of the three."

These conclusions are rather startling, suggesting the query as to what then can be the use of multiplying pipes in an organ or voices in a chorus. In order to clear the ground a little, I have recently tried some small-scale experiments with organ-pipes.

Two stopped pipes of pitch about 256 were mounted near the window of a room on the ground-floor. When the window was open the sounds could be heard (over grass) to about 200 metres; but when the window was closed the range was much less. Some difficulty was experienced in getting equal effects from the two pipes. According to the instructions of the observer, one or other supply-pipe was more or less throttled with wax.

With approximate equality of intensities and with such tuning that the beats were at the rate of about two per second, the results were very distinct. The beats were much more easily audible than either of the component sounds. Doubtless part of the advantage was due to the contrast provided by the silences; but it was thought that, apart from this, the swell of the beat was distinctly louder than either sound alone.

The result of the experiment is, of course, just what was to be expected from a mechanical point of view. According to theory the intensity (reckoned according to energy propagated) at the loudest part of the beat should be *four* times that of the (equal) component sounds heard separately.

In another set of experiments the pipes were mistuned until the interval was about a minor third, no distinct beats being audible. In this case the intensity of the compound sound might be expected to be double of that of the (equal) component sounds. The impression upon the observer hardly corresponded to this anticipation. It was difficult to say that the compound sound was decidedly the louder; although the accession of the second sound as an addition to the first could always be distinguished, and this whether the higher or the lower sound were the one added. It may be remarked that the question involved in this experiment is partly physiological, and not merely mechanical as in the case of sounds nearly in unison.

ON THE MAGNETIC ROTATION OF LIGHT AND THE SECOND
LAW OF THERMODYNAMICS.

[*Nature*, LXIV. pp. 577, 578, 1901.]

IN a paper published sixteen years ago I drew attention to a peculiarity of the magnetic rotation of the plane of polarisation arising from the circumstance that the rotation is in the same absolute direction whichever way the light may be travelling. "A consequence remarkable from the theoretical point of view is the possibility of an arrangement by which the otherwise general optical law of reciprocity shall be violated. Consider, for example, a column of diamagnetic medium exposed to such a force that the rotation is 45°, and situated between two Nicols whose principal planes are inclined to one another at 45°. Under these circumstances light passing one way is completely stopped by the second Nicol, but light passing the other way is completely transmitted. A source of light at one point A would thus be visible at a second point B, when a source at B would be invisible at A ; a state of things *at first sight*[*] inconsistent with the second law of thermodynamics." (*Phil. Trans.* CLXXVI. p. 343, 1885; *Scientific Papers*, Vol. II. p. 360.) It is here implied that the inconsistency is apparent only, but I did not discuss it further.

In his excellent report ("Les Lois théoriques du Rayonnement, Rapports présentés au Congrès International de Physique," Paris, 1900, Vol. II. p. 29), W. Wien, considering the same experimental combination of Nicols and magnetised dielectric, arrives at a contrary conclusion. It may be well to quote his statement of the case. "La rotation magnétique du plan de polarisation constitue un cas exceptionnel digne de remarque, et l'on pourrait ici imaginer un dispositif qui mettrait en échec le principe de Carnot s'il n'existait pas une compensation inconnue.

---

[*] The italics are in the original. That magnetic rotation may interfere with the law of reciprocity had already been suggested by Helmholtz.

" Faisons, en effet, les suppositions suivantes: Deux corps de température égale sont entourés d'une enveloppe adiabatique. Les rayons qu'ils s'envoient réciproquement traversent deux prismes de nicol. Entre ces prismes se trouve une substance non absorbante sur laquelle agissent des forces magnétiques qui font tourner le plan de polarisation d'un angle déterminé. La radiation émanant du corps 1 pénètre dans le nicol 1. Nous supposerons que le rayon subissant la réflexion totale n'est pas absorbé, mais renvoyé dans sa propre direction par des miroirs convenablement disposés. Admettons que le plan de polarisation soit tourné de 45° par les forces magnétiques. La section principale du deuxième nicol étant orientée dans la direction parallèle au plan de polarisation du rayon émergent, toute la lumière transmise par la substance absorbante (sic) traversera le nicol. Par conséquent, la moitié des rayons émis par le corps 1 frappera le corps 2.

" Les rayons émis par le corps 2 se divisent en deux parties égales, dans le nicol 2. Une moitié est, comme précédemment, renvoyée par réflexion. L'autre moitié, après que son plan de polarisation a subi une rotation de 45° dans le même sens que les rayons émis par le corps 1, vient frapper le premier nicol. La section principale de ce nicol étant perpendiculaire au plan de polarisation, aucune radiation ne le traverse, et nous pouvons renvoyer toute la lumière au corps 2.

" Le corps 2 reçoit ainsi trois fois plus d'énergie que le corps 1. [That is, 2 receives the whole of its own radiation and the half of that of 1, while 1 receives only the half of its own radiation.] L'un de ces corps s'échauffera par conséquent de plus en plus aux dépens de l'autre."

Wien then suggests certain ways of escape from this conclusion, but it appears to me that the difficulty itself depends upon an oversight. It is *not* possible to send back to 2 the whole of its radiation in the manner proposed. The second half, which after passage of Nicol 2 is totally reflected at Nicol 1 and then returned upon its course, on its arrival at Nicol 2 is not transmitted (as Wien seems to suppose) but is totally reflected. When again returned upon its course by a perpendicular reflector, and again reflected through 45° by the magnetised medium, it is in a condition to be completely transmitted by Nicol 1, and thus finds its way to body 1, and not to body 2 as the argument requires. The two bodies receive altogether the *same* amount of radiation, and there is therefore no tendency to a change of temperature.

Although I have not been able to find any note of it, I feel assured that the above reasoning was present to my mind when I wrote the passage already cited.

272.

ON THE INDUCTION-COIL*.

[*Philosophical Magazine*, ii. pp. 581—594, 1901.]

ALTHOUGH several valuable papers relating to this subject have recently been published by Oberbeck†, Walter‡, Mizuno§, Beattie‖, and Klingelfuss¶, it can hardly be said that the action of the instrument is well understood. Perhaps the best proof of this assertion is to be found in the fact that, so far as I am aware, there is no *à priori* calculation, determining from the data of construction and the value of the primary current, even the order of magnitude of the length of the secondary spark. I need hardly explain that I am speaking here (and throughout this paper) of an induction-coil working by a *break* of the primary circuit, not of a transformer in which the primary circuit, remaining unbroken, is supplied with a continuously varying alternating current.

The complications presented by an actual coil depend, or may depend, upon several causes. Among these we may enumerate the departure of the *iron* from theoretical behaviour, whether due to circumferential eddy-currents or to a failure of proportionality between magnetism and magnetizing force. A second, and a very important, complication has its origin in the manner of break, which usually occupies too long a time, or at least departs too much from the ideal of an instantaneous abolition of the primary current. A third complication arises from the *capacity* of the secondary coil, in virtue of which the currents need not be equal at all parts of the length, even at the same

---

* From the Jubilee volume presented to Prof. Bosscha.

† *Wied. Ann.* LXII. p. 109 (1897); LXIV. p. 193 (1898).

‡ *Wied. Ann.* LXII. p. 300 (1897) ; LXVI. p. 623 (1898).

§ *Phil. Mag.* XLV. p. 447 (1898).

‖ *Phil. Mag.* L. p. 139 (1900).

¶ *Wied. Ann.* v. p. 837 (1901).

moment of time. If we ignore these complications, treating the break as instantaneous, the iron as ideal, and the secondary as closed and without capacity, the theory, as formulated by Maxwell*, is very simple. In his notation, if x, y denote the primary and secondary currents, L, M, N the coefficients of self and mutual induction, the energy of the field is

$$\tfrac{1}{2}Lx^2 + Mxy + \tfrac{1}{2}Ny^2. \quad\dots\dots\dots\dots\dots\dots(1)$$

If c be the primary current before the break, the secondary current at time t after the break has the expression

$$y = c\,\frac{M}{N}\,e^{-S/N\,.\,t}, \quad\dots\dots\dots\dots\dots\dots\dots(2)$$

S being the resistance of the secondary circuit. The current begins with a value $c\,.\,M/N$, and gradually disappears.

The formation of the above initial current is best understood in the light of Kelvin's theorem, as explained by me in an early paper†. For this purpose it is more convenient to consider the reversed phenomenon, viz., the instantaneous *establishment* of a primary current c. The theorem teaches that subject to the condition $x = c$ the kinetic energy (1) is to be made a minimum; so that

$$Mc + Ny = 0$$

gives the initial secondary current. In the case of the *break* we have merely to reverse the sign of y.

Immediately *after* the break, when $x = 0$ and y has the above value, the kinetic energy is

$$\tfrac{1}{2}Ny^2, \quad \text{or} \quad \tfrac{1}{2}\cdot\frac{M^2c^2}{N}.$$

Immediately *before* the break the kinetic energy is $\tfrac{1}{2}Lc^2$, so that the loss of energy at break—the energy of the primary spark—is

$$\tfrac{1}{2}c^2\frac{LN - M^2}{N}, \quad\dots\dots\dots\dots\dots\dots(3)$$

vanishing when the primary and secondary circuits are closely intertwined— the case of no " magnetic leakage."

If we maintain the suppositions as to the behaviour of the iron and the suddenness of the break, the above calculated secondary current may be supposed to be instantaneously formed, even although the secondary circuit be not closed. This is most easily seen when a condenser, such as a leyden-

---

* " Electromagnetic Field," *Phil. Trans.* 1864; Maxwell's *Scientific Papers*, I. p. 546.

† " On some Electromagnetic Phenomena considered in connexion with the Dynamical Theory," *Phil. Mag.* XXXVIII. p. 1 (1869); *Scientific Papers*, I. p. 6.

jar, is associated with the ends of the secondary. Even when no jar is applied, the capacity of the secondary itself acts in the same direction and allows the formation of the current. Whether partly due to a jar or not, it will be convenient for the present to regard the capacity as associated with the terminals only of the secondary wire. Under these circumstances the secondary current follows the laws laid down by Kelvin in 1853, the same in fact as govern all vibrations in which there is but one degree of freedom. If the resistance is not too high, the current is oscillatory. After the lapse of one quarter of a complete period of these oscillations, the current vanishes, and the whole remaining energy is the potential energy of electric charge. If the resistance of the secondary wire can be neglected (so far as its influence during this short time is concerned), the potential energy of charge is the equivalent of the original energy of the secondary current at the moment after the break. In the case of no magnetic leakage, this is again the same as the energy of the primary current before break.

On these principles it is easy to calculate a limit for the maximum potential-difference at the terminals of the secondary, or for the spark-length, so far as this is determined by the potential-difference. For if q be the capacity at the secondary terminals, V the maximum potential-difference, the energy of the charge is $\frac{1}{2}qV^2$, and this can never exceed the energy of the primary current before break, viz., $\frac{1}{2}Lc^2$. The limit to the value of V is accordingly

$$V = c \cdot \sqrt{(L/q)}, \quad \dots\dots\dots\dots\dots\dots\dots(4)$$

and it is proportional to the primary current.

So long as the iron can be treated as ideal, the above formula holds good, and upon the supposition of a sufficiently sudden break there seems to be no reason why it should not afford a tolerable approximation to the actual maximum value of V. The proportionality between spark-length and primary current was found to hold good in Walter's experiments over a considerable range.

When the core is very long in proportion to its diameter, or when it approximates to a closed circuit, the behaviour of the iron may deviate widely from that described as ideal, and the quantity denoted by L has no existence. But the principle remains that the energy of charge at the moment preceding the secondary spark cannot exceed, though it may somewhat closely approach, the energy of the primary current before break.

We have next to consider how the energy of the primary current is to be reckoned, and here we encounter questions as to which opinion is not yet undivided. The general opinion would, I suppose, be that the bodily magnetization of the iron represents a large store of available energy. If this be correct, the inference would be irresistible in favour of a very long,

or a completely closed, iron core.   Some years ago*, reasoning on the basis of
the theory of Warburg and Hopkinson, I endeavoured to show that highly
magnetized iron could not be regarded as a store of energy—that the energy
expended in producing the magnetization was recoverable but to a small
extent, or not at all.   Although this conclusion does not appear to have been
accepted, perhaps in consequence of an erroneous application to alternating
current transformers, I still see no means of escape from it.   The available
energy of a highly magnetized closed circuit of iron is insignificant.   If the
length be limited, there is available energy, in virtue of the free polarity at
the ends.

The theory is best illustrated by the case of an ellipsoid of revolution
exposed to uniform *external* magnetizing force $\mathfrak{H}'$ acting parallel to the axis.
"If $\mathfrak{I}$ be the magnetization parallel to the axis of symmetry ($2c$), the de-
magnetizing effect of $\mathfrak{I}$ is $N\mathfrak{I}$, where N is a numerical constant, a function
of the eccentricity (e).   When the ellipsoid is of the ovary or elongated form,

$$a = b = c \sqrt{(1 - e^2)},$$

$$N = 4\pi \left(\frac{1}{e^2} - 1 \right) \left(\frac{1}{2e} \log \frac{1 + e}{1 - e} - 1 \right),$$

becoming in the limiting case of the sphere ($e = 0$),

$$N = \tfrac{4}{3}\pi ;$$

and at the other extreme of elongation assuming the form

$$N = 4\pi \frac{a^2}{c^2} \left(\log \frac{2c}{a} - 1 \right). \quad \dots\dots\dots\dots\dots\dots(5)$$

"The force actually operative upon the iron is found by subtracting $N\mathfrak{I}$
from that externally imposed, so that

$$\mathfrak{H} = \mathfrak{H}' - N\mathfrak{I} ;$$

and if from experiments on very elongated ellipsoids ($N = 0$) we know the
relation between $\mathfrak{H}$ and $\mathfrak{I}$, then the above equation gives us the relation
between $\mathfrak{H}'$ and $\mathfrak{I}$ for any proposed ellipsoid of moderate elongation.   If we
suppose that $\mathfrak{H}$ is plotted as a function of $\mathfrak{I}$, we have only to add in the
ordinates $N\mathfrak{I}$, proper to a straight line, in order to obtain the appropriate
curve for $\mathfrak{H}'$."

The work expended in magnetizing the iron is per unit of volume

$$\int \mathfrak{H} d\mathfrak{I} + \tfrac{1}{2} N \mathfrak{I}^2,$$

* "On the Energy of Magnetized Iron," *Phil. Mag.* xxII. p. 175 (1886) ; *Scientific Papers*,
II. p. 543.

if we reckon from the condition of zero magnetization. The first part is practically wasted; the second, which in most cases of open magnetic circuits is much the larger, is completely recovered when the iron is demagnetized.

If it appear paradoxical that the large integral electromotive force which would accompany the disappearance of high magnetization in a closed iron circuit should be so inefficient, we must remember that the mechanical value of electromotive force depends upon the magnitude of the current which it drives, and that in the present case the existence of more than a very small current is inconsistent with that drop of magnetization upon which the electromotive force depends.

The considerations above explained are of interest in the present question as affording a limit depending only upon the iron core and the secondary capacity. For $\mathfrak{I}$ cannot exceed a value estimated at about 1700 c.g.s., whatever may be the magnetizing force of the primary current. Thus if v be the volume of the core, the maximum energy* is

$$\tfrac{1}{2} N \times v \times 1700^2;$$

and the limit to V is found by equating this to $\tfrac{1}{2} q V^2$, so that

$$V = 1700 \sqrt{(Nv/q)}. \quad\dots\dots\dots\dots\dots\dots\dots\dots\dots (6)$$

I have made a rough application of this formula to a coil in my possession, with results that may be here recorded. The core had a diameter of 3 cm. and a length of 27 cm., so that $v = 180$ c.c. From (5), properly applicable only to an ellipsoid, we get by setting $2a = 3$, $2c = 27$, $N = \cdot 30$.

The capacity of the secondary is more difficult to deal with. In modern coils the greater part would appear to arise from the positive and negative potentials at the ends of the coil as opposed to the zero potential of the *primary* wire. The capacity between the primary and secondary wires, considered as poles of a condenser, can be calculated and in many cases determined experimentally. The axial dimension of the secondary of the coil above referred to is about 18 cm., and the external diameter of the primary wire is about 5 cm., making the area of each of the opposed surfaces 270 sq. cm. The interval between the primary and secondary wires is ·25 cm.; so that, taking the specific inductive capacity of the intervening layer at 3, we get for the capacity in electrostatic measure of the condenser so constituted

$$\frac{1}{4\pi} \times 3 \times \frac{270}{\cdot 25} = 258 \text{ cm.}\dagger$$

---

* The energy of the primary current without a core is here neglected.

† Another coil by Apps, in which the insulation was sufficiently good to allow the application of electrostatic methods, was tested experimentally. The capacity between primary and secondary wires was thus found to be 120 cm., less than the half of that calculated for the first coil. But in this case an ebonite tube separated the two wires.

Only a fraction of this, however, is operative in the present case. On the supposition of a coil constructed in numerous sections, the potential in the middle will be zero, the same as that of the primary wire, and will increase numerically towards either end. The factor of reduction on this account will be $\int_{-\frac{1}{2}}^{+\frac{1}{2}} x^2\, dx$, or $\frac{1}{12}$, so that we may take as the value of q in (6) about 23 cm.—probably rather an underestimate. Calculating from these data, we get in (6)

$$V = 2600.$$

This is in electrostatic measure. The corresponding volts are 7.9×10^5. If we reckon 33,000 volts to the cm., the spark-length will stand at 24 cm. The coil in question is supposed to be capable of an 8 or 10 cm. spark, and doubtless was capable when new. It is remarkable that the limit, fixed by the iron and secondary capacity alone, should exceed so moderately the actual capability of the coil.

The limiting formula (6), in which neither the value of the primary current nor the number of secondary windings appears, is arrived at by supposing the iron to be magnetically saturated. It illustrates, no doubt with much exaggeration, the disadvantage of too great a length. If a be given, while c varies, v and q are both proportional to c, so that $V \propto \sqrt{N}$. And $\sqrt{N} \propto c^{-1}$ nearly. In somewhat the same way the increase of effective capacity explains the comparative failure of attempts to increase spark-length by combining similar coils in series, in spite of the augmented energy at the moment of break*.

If the object be a rough estimate rather than a limit, a more practical formula will be obtained by substituting for $\mathfrak{I}$ in (6) its approximate value $\mathfrak{H}'/N$; so that

$$V = \mathfrak{H}' \sqrt{\left(\frac{v}{Nq}\right)}, \quad \dots\dots\dots\dots\dots\dots\dots(7)$$

$\mathfrak{H}'$ denoting the external magnetizing force, due to the primary current. The actual magnetizing force, required to magnetize the soft iron, is here regarded as relatively negligible. According to (7) the spark-length is proportional, *cæteris paribus*, to the primary current; and it increases with the length of the coil, since N now occurs in the denominator. The application must not be pushed into the region where the iron becomes approximately saturated.

In the above discussion the capacity q of the secondary will probably be thought to play an unexpectedly important part, and the question may be raised whether it is really this capacity which limits the spark-length in

---

* I am indebted to Mr Swinton for the details of some experiments in this direction made for Lord Armstrong.

actual coils. It is not difficult to prove by experiment that capacities of the order above estimated, applied to the secondary terminals, do in fact reduce the spark-length, though not, so far as I have seen, to the extent demanded by the law of $q^{-\frac{1}{2}}$. But we must remember that this law has been obtained on the assumptions, not to be fulfilled in practice, of absolute suddenness of break and of entire absence of eddy-currents in the iron. If under these conditions secondary capacity were also absent, it would seem that there could be no limit to the maximum potential developed. The experiments of Prof. J. J. Thomson* may be considered to show that even in extreme cases, such as the present, the iron, as a magnetic body, would not fail to respond.

As regards the eddy-currents, it may be well to consider a little further upon what their importance depends. If there were no secondary circuit, the magnetism of each wire of the iron core would be continued at the moment after break, supposed infinitely sudden, by a superficial eddy-current. A secondary circuit, closely intertwined with the primary, would transfer these eddy-currents to itself, and so continue for the first moment the magnetism of the core. But a little later, as the magnetism diminished, eddy-currents would tend to be formed, and their importance for our purpose depends upon their duration. If this be short, compared with the time-constants of the secondary circuit, their influence may be neglected. Otherwise the electromotive force of the falling magnetism lags, and acts to less advantage. The time-constant, viz. the time in which the current falls in the ratio $e : 1$, for the principal eddy-current in a cylinder of radius R is given by

$$\tau = \frac{4\pi\mu C R^2}{(2\cdot 404)^2}, \quad \dots\dots\dots\dots\dots\dots\dots\dots\dots(8)$$

where C represents the conductivity and μ the permeability†. If d be the thickness of a thin sheet having the same time-constant as the wire of radius R, it is easily shown in the same way that

$$d : R = \pi : 2\cdot 404.$$

If we take for iron in C.G.S. measure

$$C = 1/9611, \quad \mu = 500,$$

we get approximately

$$\tau = \tfrac{1}{10} R^2; \quad \dots\dots\dots\dots\dots\dots\dots\dots\dots\dots(9)$$

so that for a wire of 1 mm. diameter $\tau = \tfrac{1}{4000}$ second. It may be doubted whether this would be small enough to prevent the eddy-currents reacting injuriously upon the secondary circuit.

* *Recent Researches*, p. 323.

† *Brit. Assoc. Rep.* p. 446 (1882); *Scientific Papers*, II. p. 128.

We will now consider the third of the causes which impose a limit upon the secondary spark, viz. want of suddenness in the break, supposed for the present to be unprovided with a condenser. After the cessation of metallic contact the primary current is prolonged by the formation of a sort of arc, the duration of which depends among other things, such as the character of the metals, upon the magnitude of the current itself. If we again suppose the behaviour of the iron to be ideal, we may treat the secondary circuit as a simple vibrator, upon which acts a force (U) proportional to the rate of fall of the primary current. The equation of such a vibrator is, as usual,

$$\frac{d^2u}{dt^2} + \kappa \frac{du}{dt} + n^2u = U; \quad \dots\dots\dots\dots\dots(10)$$

and the solution corresponding to $u = 0$ (no charge), $du/dt = 0$ (no current), when $t = 0$, is*

$$u = \frac{1}{n'} \int_0^t e^{-\frac{1}{2}\kappa(t-t')} \sin n' (t - t') . U dt', \quad \dots\dots\dots(11)$$

where

$$n' = \sqrt{(n^2 - \tfrac{1}{4}\kappa^2)}. \quad \dots\dots\dots\dots\dots(12)$$

The various elements of (11) represent in fact the effects at time t of the velocities $U dt'$ communicated $(t - t')$ earlier. In the present case we are to suppose that U is positive throughout, and that $\int U dt'$ is given.

The integral simplifies in the case of $\kappa = 0$, that is of evanescent secondary resistance. We have then $n' = n$, and

$$u = \frac{1}{n} \int_0^t \sin n (t - t') . U dt'. \quad \dots\dots\dots\dots(13)$$

It is easy to see that the integral, representing the potential at the secondary terminals, is a maximum when U is concentrated at some one time t', and t is such that

$$\sin n (t - t') = 1,$$

that is, when the break is absolutely sudden and the time considered is one quarter period later. If the break be not sudden, $\sin n (t - t')$ will depart from its maximum value during part of the range of integration, and the highest possible value of u will not be attained.

The theory is substantially the same if κ be finite. There is some value of $(t - t')$ for which

$$e^{-\frac{1}{2}\kappa(t-t')} \sin n' (t - t')$$

is a maximum; and the greatest value of u will be arrived at by concentrating U at some time t', and by so choosing t that $(t - t')$ has the value above defined. The conclusion is that if the primary current fall to zero

---

* *Theory of Sound*, Vol. I. § 66.

from its maximum value without oscillation, the potential at the secondary terminals will be greatest when this fall is absolutely sudden, and that this greatest value begins to be sensibly departed from when the break occupies a time comparable with one of the time-constants of the secondary circuit. In the case of no resistance we have to deal merely with the time of secondary oscillation; but if the resistance is high, the other time-constant, N/S, may be the smaller {see equation (2)}.

It is here that the character of the secondary coil, especially as regards the number of its windings, enters into the question. On the supposition of an absolutely sudden break, we arrived at the rather paradoxical conclusion that the limit of spark-length depended only upon the capacity of the secondary without regard to the number of windings—a number which could be changed in a high ratio without sensibly influencing the capacity. We see now, at any rate, that a reduction in the number of windings, and the accompanying diminution in the time of oscillation, would necessitate a greater and greater suddenness of break, if the full effect is to be retained.

We will now consider the action of the primary condenser—a question, the reader may be inclined to think, already too long postponed. For it is well known that in most actual coils the condenser is an auxiliary of the utmost importance, increasing the spark-length 5 or 10 times, even when the break is made at pieces of platinum. And, although it has been customary to say, no doubt correctly, that the condenser acts by absorbing into itself the primary spark, and so increasing the suddenness of break, it is usual to attribute to it a further virtue, and not unnaturally when it is remembered that the effect may be not merely to stop, but actually to reverse, the primary current. If, however, the theory of the foregoing pages is correct, we shall be constrained to take a different view.

The action of the condenser, and especially the most advantageous capacity, has been studied experimentally by Walter and by Mizuno. That there must be a most advantageous capacity is evident beforehand, inasmuch as a very small capacity is continuous with no condenser at all, and a very large capacity is continuous with an uninterrupted flow of the primary current. It is more instructive that the former observer found the most advantageous capacity to vary with the manner of break (whether in air or under oil), and that the latter found a dependence upon the strength of the primary current, a larger current demanding a larger condenser.

When a condenser is employed, it is important that it be connected as directly as possible with the points between which the break is made to occur. A comparatively small electromagnet, included between one of the break-points and the associated condenser-terminal, suffices to diminish, or even to annul, the advantage which the use of the condenser otherwise presents*. The explanation is, of course, that the current in an electro-magnet so situated tends to flow on across the break-gap, and so to establish an arc, with a force which the condenser is powerless to relieve.

Returning to the theoretical aspect of the question, and inquiring whether there is any reason for expecting a condenser to give an advantage as compared with an absolutely sudden cessation of the primary current, it is difficult to see ground for other than a negative answer. In the case of no magnetic leakage, somewhat closely approached, one would suppose, in practice, an instantaneous abolition of the primary current throws the *whole* available energy into the secondary circuit, and thus, doing all that is possible, allows no room for an improvement. Under such conditions a condenser can only do harm.

In the opposite extreme case of but a relatively small mutual induction between primary and secondary, it is indeed conceivable that the action of a condenser may be advantageous. The two currents would then be com-paratively independent and, if the resistances were low, they might execute numerous oscillations. If the primary current were simply stopped, the effect in the secondary would be small; whereas, especially if there were synchronism, the vibrations of the primary current rendered possible by the condenser might cause an accumulation of effect in the secondary. The case would be that of "intermittent vibrations†," such as may occur when a large tuning-fork is clamped in a vice. A vibration, started by a blow, in one prong gradually transfers itself to the other. But it is difficult to believe that any-thing of this sort occurs in an induction-coil as actually used.

I do not know how far the theoretical arguments here advanced will convince the reader that the use of a condenser in the primary circuit should offer no advantage as compared with a sufficiently sudden simple break; but I may confess that I should have hesitated to put them forward had I not obtained experimental confirmation of them. My earlier attempts in this direction were unsuccessful. A quick break was constructed in which a spring, bearing upwards against a stop, could be knocked away by a blow with a staff, or by a falling weight. Although the contacts were of platinum, but little advantage was gained in comparison with the ordinary platinum break of the coil. Thus in one set of experiments, where the coil was excited by a single Grove cell, a break made quickly by hand gave a spark about 8 mm. long. The use of a weight, hung by a cotton thread, and falling through about 12 feet when the thread was burned, increased the length only to 8½ mm. This was without a condenser. When the condenser was applied, the spark-length was 14 mm., and it made no perceptible difference whether or not the falling weight was employed. Considering that the velocity of the

---

* *Phil. Mag.* Vol. ii. p. 282 (1901). [Vol. iv. p. 552.]
† *Theory of Sound*, Vol. i. § 114.

weight at impact must have been about 30 feet per second and that its mass was large compared with that of the spring, these results were far from promising.   With a stronger primary current the advantage gained from the condenser was much greater, and the utility of the quicker break, with or without condenser, seemed to be *nil*.

But, in spite of the failure of the quick break, one or two observations presented themselves which seemed worthy of being followed up.   It was noticed that, with one Grove cell in the primary, the spark, although very inferior when no condenser at all was employed, was improved when the usual condenser (of large capacity) was replaced by a single sheet of coated glass (Franklin's pane).   And, what was perhaps more instructive still, when the already weak primary current was further reduced by the insertion of one or two ohms extra resistance, the spark-length (now very small) was *less* with than without the usual coil condenser.   This observation was repeated, with like result, upon another coil (by Apps) and its associated condenser.   At any rate in the case of very weak primary currents, the usual condenser did harm rather than good.

The view, suggested by the foregoing results, that while the ordinary break was quick enough in the case of weak currents to allow a condenser to be dispensed with, the superior arcing power of strong currents demanded a much more rapid break, encouraged further efforts.   An attempt to secure suddenness by forcibly breaking with a jerk a length of rather thin copper wire, forming part of the primary circuit, failed entirely, as did also, perhaps for want of sufficiently powerful appliances, an attempt to *blow up* a portion of the primary circuit by electric discharge.   Another method, however, at once allowed an advance to be secured.   This consisted in cutting the primary circuit by a pistol-bullet; and it was found that this form of break without condenser was about as efficient as the usual platinum break with condenser, although the primary current was increased to that supplied by three or four Grove cells and the spark-length to 40 mm., that is, under about the ordinary conditions of working.

A further improvement was effected by cutting away about half of the bullet with the intention of raising its velocity.   The following results were recorded with an Apps' coil excited by three Grove cells.   The spark-gap being 50 mm., the usual platinum break and condenser were not able to send a spark across.   Even with the somewhat more efficient break provided by a pot of mercury well drowned in oil and condenser, only about one break in fifteen succeeded.   On the other hand, of three bullets fired so as to cut the primary wire (no condenser) two succeeded; while for the failure of the third there was some explanation.   The bullet without condenser was now distinctly *superior* to the best ordinary break with condenser.

The next step was the substitution of a *rifle*-bullet, fired from a service rifle.   Here again the bullets were reduced to about one-half, and after cutting the wire were received in a long box packed with wet sawdust. At 60 mm., while the mercury-under-oil break with condenser gave only feeble brush-discharges, good sparks were nearly uniformly obtained from the bullet working without a condenser. At 70 mm. the bullet without condenser was about upon a level with the mercury-under-oil break with condenser at 60 mm. As regards the strength of the primary current, if there was any difference, the advantage was upon the side of the ordinary break with condenser, inasmuch as in the case of the bullet the leads were longer and included about 8 cm. of finer copper wire where the bullet passed.

In the next set of experiments upon the same Apps' coil excited by three Groves, the bullet was used each time, and the comparison was between the effect with and without the usual coil condenser. At 55 mm. the bullet without condenser gave each time a fair or a good spark, while with the condenser there was nothing more than a feeble brush scarcely visible in a good light.

The single pane of coated glass was next substituted for the usual condenser of the coil, with the idea that possibly this might be useful although the larger capacity was deleterious. But no distinct difference was detected when the bullet was fired with this or without any condenser.

In the last set of experiments now recorded the primary current was raised, six Grove cells being employed partly in parallel, and the wire was cut each time by a rifle-bullet. At 90 mm. no spark could be got when the coil condenser was in connexion; when it was disconnected, a spark, good or fair, was observed nearly every shot.

Altogether these experiments strongly support the view that the only use of a condenser, in conjunction with an ordinary break, is to quicken it by impeding the development of an arc, so that when a sufficient rapidity of break can be obtained by other means, the condenser is deleterious, operating in fact in the reverse direction, and prolonging the period of decay of the primary current. It is hoped that the establishment of this fact will inspire confidence in the theory, and perhaps suggest improvements in the design of coils. The first requirement is evidently the existence of sufficient energy at break, and this implies a considerable mass of iron, well magnetized, and not forming a circuit too nearly closed. The full utilization of this energy is impeded by want of suddenness in the break, by eddy-currents in the iron, and (in respect of spark-length) by capacity in the secondary. It is to be presumed that in a well-designed coil these impediments should operate somewhat equally. It would be useless to subdivide the iron, or to reduce the secondary capacity, below certain limits, unless at the same time the break could be made more sudden. It would not be surprising if it were found that the tentative efforts of skilful instrument-makers have already led to a suitable compromise, at least in the case of coils of moderate size. The design of larger instruments may leave more to be accomplished.

CONTENTS OF VOLUMES I.–IV.

CLASSIFIED ACCORDING TO SUBJECT.

MATHEMATICS.

GENERAL MECHANICS.

ELASTIC SOLIDS.

HYDRODYNAMICS.

THERMODYNAMICS.

DYNAMICAL THEORY OF GASES.

PROPERTIES OF GASES.

ELECTRICITY AND MAGNETISM.

OPTICS.

MISCELLANEOUS.

INDEX OF NAMES.

CAMBRIDGE : PRINTED BY J. AND C. F. CLAY, AT THE UNIVERSITY PRESS.

Printed in the United States
By Bookmasters